OXFORD MASTER SERIES IN PHYSICS

OXFORD MASTER SERIES IN PHYSICS

The Oxford Master Series is designed for final year undergraduate and beginning graduate students in physics and related disciplines. It has been driven by a perceived gap in the literature today. While basic undergraduate physics texts often show little or no connection with the huge explosion of research over the last two decades, more advanced and specialized texts tend to be rather daunting for students. In this series, all topics and their consequences are treated at a simple level, while pointers to recent developments are provided at various stages. The emphasis is on clear physical principles like symmetry, quantum mechanics, and electromagnetism which underlie the whole of physics. At the same time, the subjects are related to real measurements and to the experimental techniques and devices currently used by physicists in academe and industry. Books in this series are written as course books, and include ample tutorial material, examples, illustrations, revision points, and problem sets. They can likewise be used as preparation for students starting a doctorate in physics and related fields, or for recent graduates starting research in one of these fields in industry.

CONDENSED MATTER PHYSICS

1. M. T. Dove: *Structure and dynamics: an atomic view of materials*
2. J. Singleton: *Baud theory and electronic properties of solids*
3. A. M. Fox: *Optical properties of solids*
4. S. J. Blundell: *Magnetism in condensed matter*
5. J. F. Annett: *Superconductivity*
6. R. A. L. Jones: *Soft condensed matter*

ATOMIC, OPTICAL, AND LASER PHYSICS

7. C. J. Foot: *Atomic Physics*
8. G. A. Brooker: *Modern classical optics*
9. S. M. Hooker, C. E. Webb: *Laser physics*
15. A. M. Fox: *Quantum optics: an introduction*

PARTICLE PHYSICS, ASTROPHYSICS, AND COSMOLOGY

10. D. H. Perkins: *Particle astrophysics*
11. Ta-Pei Cheng: *Relativity, gravitation, and cosmology*

STATISTICAL, COMPUTATIONAL, AND THEORETICAL PHYSICS

12. M. Maggiore: *A modern introduction to quantum field theory*
13. W. Krauth: *Statistical mechanics: algorithms and computations*
14. J. P. Sethna: *Entropy, order parameters, and complexity*

Quantum Optics

An Introduction

MARK FOX

Department of Physics and Astronomy
University of Sheffield

OXFORD

UNIVERSITY PRESS

OXFORD
UNIVERSITY PRESS

Great Clarendon Street, Oxford, OX2 6DP,
United Kingdom

Oxford University Press is a department of the University of Oxford.
It furthers the University's objective of excellence in research, scholarship,
and education by publishing worldwide. Oxford is a registered trade mark of
Oxford University Press in the UK and in certain other countries

First Edition published in 2006
Reprinted 2007 (twice, once with corrections), 2009, 2010, 2011, 2012, 2013,
2014 twice

Published in the United States of America by Oxford University Press
198 Madison Avenue, New York, NY 10016, United States of America

British Library Cataloguing in Publication Data

Data available

Library of Congress Cataloging in Publication Data

Data available

ISBN 978–0–19–856673–1

Preface

Quantum optics is a subject that has come to the fore over the last 10–20 years. Formerly, it was regarded as a highly specialized discipline, accessible only to a small number of advanced students at selected universities. Nowadays, however, the demand for the subject is much broader, with the interest strongly fuelled by the prospect of using quantum optics in quantum information processing applications.

My own interest in quantum optics goes back to 1987, when I attended the Conference on Lasers and Electro-Optics (CLEO) for the first time. The ground-breaking experiments on squeezed light had recently been completed, and I was able to hear invited talks from the leading researchers working in the field. At the end of the conference, I found myself sufficiently interested in the subject that I bought a copy of Loudon's *Quantum theory of light* and started to work through it in a fairly systematic way. Nearly 20 years on, I still consider Loudon's book as my favourite on the subject, although there are now many more available to choose from. So why write another?

The answer to this question became clearer to me when I tried to develop a course on quantum optics as a submodule of a larger unit entitled 'Aspects of Modern Physics'. This course is taken by undergraduate students in their final semester, and aims to introduce them to a number of current research topics. I set about designing a course to cover a few basic ideas about photon statistics, quantum cryptography, and Bose–Einstein condensation, hoping that I would find a suitable text to recommend. However, a quick inspection of the quantum optics texts that were available led me to conclude that they were generally pitched at a higher level than my target audience. Furthermore, the majority were rather mathematical in their presentation. I therefore reluctantly concluded that I would have to write the book I was seeking myself. The end result is what you see before you. My hope is that it will serve both as a useful basic introduction to the subject, and also as a tasty *hors d'oeuvre* for the more advanced texts like Loudon's.

In developing my course notes into a full-length book, the first problem that I encountered was the selection of topics. Traditional quantum optics books like Loudon's assume that the subject refers primarily to the properties of light itself. At the same time, it is apparent that the subject has broadened considerably in its scope, at least to many people working in the field. I have therefore included a broad range of topics that probably would not have found their way into a quantum optics text 20 years ago. It is probable that someone else writing a similar text would make a different selection of topics. My selection has been based mainly on my perception of the key subject areas, but it also reflects my own research interests to some extent. For this reason, there are probably more examples of quantum optical effects in solid state systems than might normally have been expected.

Some of the subjects that I have selected for inclusion are still developing very rapidly at the time of writing. This is especially true of the topics in quantum information technology covered in Part IV. Any attempt to give a detailed overview of the present status of the experiments in these fields would be relatively pointless, as it would date very quickly. I have therefore adopted the strategy of trying to explain the basic principles and then illustrating them with a few recent results. It is my hope that the chapters I have written will be sufficient to allow students who are new to the subjects to understand the fundamental concepts, thereby allowing them to go to the research literature should they wish to pursue any topics in more detail.

At one stage I thought about including references to a good number of internet sites within the 'Further Reading' sections, but as the links to these sites frequently change, I have actually only included a few. I am sure that the modern computer-literate student will be able to find these sites far more easily than I can, and I leave this part of the task to the student's initiative. It is a fortunate coincidence that the book is going to press in 2005, the centenary of Einstein's work on the photoelectric effect, when there are many articles available to arouse the interest of students on this subject. Furthermore, the award of the 2005 Nobel Prize for Physics to Roy Glauber "for his contribution to the quantum theory of optical coherence" has generated many more widely-accessible information resources.

An issue that arose after receiving reviews of my original book plan was the difficulty in making the subject accessible without gross over-simplification of the essential physics. As a consequence of these reviews, I suspect that some sections of the book are pitched at a slightly higher level than my original target of a final-year undergraduate, and would in fact be more suitable for use in the first year of a Master's course. Despite this, I have still tried to keep the mathematics to a minimum as far as possible, and concentrated on explanations based on the physical understanding of the experiments that have been performed.

I would like to thank a number of people who have helped in the various stages of the preparation of this book. First, I would like to thank all of the anonymous reviewers who made many helpful suggestions and pointed out numerous errors in the early versions of the manuscript. Second, I would like to thank several people for critical reading of parts of the manuscript, especially Dr Brendon Lovett for Chapter 13, and Dr Gerald Buller and Robert Collins for Chapter 12. I would like to thank Dr Ed Daw for clarifying my understanding of gravity wave interferometers. A special word of thanks goes to Dr Geoff Brooker for critical reading of the whole manuscript. Third, I would like to thank Sonke Adlung at Oxford University Press for his support and patience

throughout the project and Anita Petrie for overseeing the production
of the book. I am also grateful to Dr Mark Hopkinson for the TEM picture in Fig. D.3, and to Dr Robert Taylor for Fig. 4.7. Finally, I would
like to thank my doctoral supervisor, Prof. John Ryan, for originally
pointing me towards quantum optics, and my numerous colleagues who
have helped me to carry out a number of quantum optics experiments
during my career.

Sheffield
June 2005

Contents

List of symbols   xv

List of abbreviations   xviii

I   Introduction and background   1

1   Introduction   3
  1.1   What is quantum optics?   3
  1.2   A brief history of quantum optics   4
  1.3   How to use this book   6

2   Classical optics   8
  2.1   Maxwell's equations and electromagnetic waves   8
    2.1.1   Electromagnetic fields   8
    2.1.2   Maxwell's equations   10
    2.1.3   Electromagnetic waves   10
    2.1.4   Polarization   12
  2.2   Diffraction and interference   13
    2.2.1   Diffraction   13
    2.2.2   Interference   15
  2.3   Coherence   16
  2.4   Nonlinear optics   19
    2.4.1   The nonlinear susceptibility   19
    2.4.2   Second-order nonlinear phenomena   20
    2.4.3   Phase matching   23

3   Quantum mechanics   26
  3.1   Formalism of quantum mechanics   26
    3.1.1   The Schrödinger equation   26
    3.1.2   Properties of wave functions   28
    3.1.3   Measurements and expectation values   30
    3.1.4   Commutators and the uncertainty principle   31
    3.1.5   Angular momentum   32
    3.1.6   Dirac notation   34
  3.2   Quantized states in atoms   35
    3.2.1   The gross structure   35
    3.2.2   Fine and hyperfine structure   39
    3.2.3   The Zeeman effect   41
  3.3   The harmonic oscillator   41
  3.4   The Stern–Gerlach experiment   43
  3.5   The band theory of solids   45

4 Radiative transitions in atoms   48

4.1 Einstein coefficients   48

4.2 Radiative transition rates   51

4.3 Selection rules   54

4.4 The width and shape of spectral lines   56

    4.4.1 The spectral lineshape function   56

    4.4.2 Lifetime broadening   56

    4.4.3 Collisional (pressure) broadening   57

    4.4.4 Doppler broadening   58

4.5 Line broadening in solids   58

4.6 Optical properties of semiconductors   59

4.7 Lasers   61

    4.7.1 Laser oscillation   61

    4.7.2 Laser modes   64

    4.7.3 Laser properties   67

II Photons   73

5 Photon statistics   75

5.1 Introduction   75

5.2 Photon-counting statistics   76

5.3 Coherent light: Poissonian photon statistics   78

5.4 Classification of light by photon statistics   82

5.5 Super-Poissonian light   83

    5.5.1 Thermal light   83

    5.5.2 Chaotic (partially coherent) light   86

5.6 Sub-Poissonian light   87

5.7 Degradation of photon statistics by losses   88

5.8 Theory of photodetection   89

    5.8.1 Semi-classical theory of photodetection   90

    5.8.2 Quantum theory of photodetection   93

5.9 Shot noise in photodiodes   94

5.10 Observation of sub-Poissonian photon statistics   99

    5.10.1 Sub-Poissonian counting statistics   99

    5.10.2 Sub-shot-noise photocurrent   101

6 Photon antibunching   105

6.1 Introduction: the intensity interferometer   105

6.2 Hanbury Brown–Twiss experiments and classical intensity fluctuations   108

6.3 The second-order correlation function $g^{(2)}(\tau)$ 111

6.4 Hanbury Brown–Twiss experiments with photons   113

6.5 Photon bunching and antibunching   115

    6.5.1 Coherent light   116

    6.5.2 Bunched light   116

    6.5.3 Antibunched light   117

6.6 Experimental demonstrations of photon antibunching   117

6.7 Single-photon sources   120

7   Coherent states and squeezed light   126
   7.1   Light waves as classical harmonic oscillators   126
   7.2   Phasor diagrams and field quadratures   129
   7.3   Light as a quantum harmonic oscillator   131
   7.4   The vacuum field   132
   7.5   Coherent states   134
   7.6   Shot noise and number–phase uncertainty   135
   7.7   Squeezed states   138
   7.8   Detection of squeezed light   139
      7.8.1   Detection of quadrature-squeezed
             vacuum states   139
      7.8.2   Detection of amplitude-squeezed light   142
   7.9   Generation of squeezed states   142
      7.9.1   Squeezed vacuum states   142
      7.9.2   Amplitude-squeezed light   144
   7.10 Quantum noise in amplifiers   146

8   Photon number states   151
   8.1   Operator solution of the harmonic oscillator   151
   8.2   The number state representation   154
   8.3   Photon number states   156
   8.4   Coherent states   157
   8.5   Quantum theory of Hanbury Brown–Twiss
         experiments   160

III Atom–photon interactions   165

9   Resonant light–atom interactions   167
   9.1   Introduction   167
   9.2   Preliminary concepts   168
      9.2.1   The two-level atom approximation   168
      9.2.2   Coherent superposition states   169
      9.2.3   The density matrix   171
   9.3   The time-dependent Schrödinger equation   172
   9.4   The weak-field limit: Einstein's B coefficient   174
   9.5   The strong-field limit: Rabi oscillations   177
      9.5.1   Basic concepts   177
      9.5.2   Damping   180
      9.5.3   Experimental observations of
             Rabi oscillations   182
   9.6   The Bloch sphere   187

10   Atoms in cavities   194
   10.1   Optical cavities   194
   10.2   Atom–cavity coupling   197
   10.3   Weak coupling   200
      10.3.1   Preliminary considerations   200
      10.3.2   Free-space spontaneous emission   201

	10.3.3	Spontaneous emission in a single-mode cavity: the Purcell effect	202
	10.3.4	Experimental demonstrations of the Purcell effect	204
10.4	Strong coupling		206
	10.4.1	Cavity quantum electrodynamics	206
	10.4.2	Experimental observations of strong coupling	209
10.5	Applications of cavity effects		211

11 Cold atoms — 216
11.1	Introduction		216
11.2	Laser cooling		218
	11.2.1	Basic principles of Doppler cooling	218
	11.2.2	Optical molasses	221
	11.2.3	Sub-Doppler cooling	224
	11.2.4	Magneto-optic atom traps	226
	11.2.5	Experimental techniques for laser cooling	227
	11.2.6	Cooling and trapping of ions	229
11.3	Bose–Einstein condensation		230
	11.3.1	Bose–Einstein condensation as a phase transition	230
	11.3.2	Microscopic description of Bose–Einstein condensation	232
	11.3.3	Experimental techniques for Bose–Einstein condensation	233
11.4	Atom lasers		236

IV Quantum information processing — 241

12 Quantum cryptography — 243
12.1	Classical cryptography		243
12.2	Basic principles of quantum cryptography		245
12.3	Quantum key distribution according to the BB84 protocol		249
12.4	System errors and identity verification		253
	12.4.1	Error correction	253
	12.4.2	Identity verification	254
12.5	Single-photon sources		255
12.6	Practical demonstrations of quantum cryptography		256
	12.6.1	Free-space quantum cryptography	257
	12.6.2	Quantum cryptography in optical fibres	258

13 Quantum computing — 264
13.1	Introduction		264
13.2	Quantum bits (qubits)		267
	13.2.1	The concept of qubits	267
	13.2.2	Bloch vector representation of single qubits	269
	13.2.3	Column vector representation of qubits	270

13.3	Quantum logic gates and circuits	270
	13.3.1 Preliminary concepts	270
	13.3.2 Single-qubit gates	272
	13.3.3 Two-qubit gates	274
	13.3.4 Practical implementations of qubit operations	275
13.4	Decoherence and error correction	279
13.5	Applications of quantum computers	281
	13.5.1 Deutsch's algorithm	281
	13.5.2 Grover's algorithm	283
	13.5.3 Shor's algorithm	286
	13.5.4 Simulation of quantum systems	287
	13.5.5 Quantum repeaters	287
13.6	Experimental implementations of quantum computation	288
13.7	Outlook	292

14 Entangled states and quantum teleportation — 296
14.1	Entangled states	296
14.2	Generation of entangled photon pairs	298
14.3	Single-photon interference experiments	301
14.4	Bell's theorem	304
	14.4.1 Introduction	304
	14.4.2 Bell's inequality	305
	14.4.3 Experimental confirmation of Bell's theorem	308
14.5	Principles of teleportation	310
14.6	Experimental demonstration of teleportation	313
14.7	Discussion	316

Appendices

A	**Poisson statistics**	321
B	**Parametric amplification**	324
	B.1 Wave propagation in a nonlinear medium	324
	B.2 Degenerate parametric amplification	326
C	**The density of states**	330
D	**Low-dimensional semiconductor structures**	333
	D.1 Quantum confinement	333
	D.2 Quantum wells	335
	D.3 Quantum dots	337
E	**Nuclear magnetic resonance**	339
	E.1 Basic principles	339
	E.2 The rotating frame transformation	341
	E.3 The Bloch equations	344

F   Bose–Einstein condensation   346
    F.1   Classical and quantum statistics   346
    F.2   Statistical mechanics of Bose–Einstein condensation 348
    F.3   Bose–Einstein condensed systems   350

Solutions and hints to the exercises   352

Bibliography   360

Index   369

List of symbols

The alphabet only contains 26 letters, and the use of the same symbol to represent different quantities is unavoidable in a book of this length. Whenever this occurs, it should be obvious from the context which meaning is intended.

$\hat{a}$	annihilation operator	$g_\omega(\omega)$	spectral lineshape function
$\hat{a}^\dagger$	creation operator	g_F	hyperfine g-factor
a	length parameter	g_J	Landé g-factor
$\vec{a}$	unit vector	g_N	nuclear g-factor
a_0	Bohr radius	g_s	electron spin g-factor
A	area	g_0	atom–cavity coupling constant
A_{ij}	Einstein A coefficient	$g^{(1)}(\tau)$	first-order correlation function
$\vec{b}$	unit vector	$g^{(2)}(\tau)$	second-order correlation function
$\boldsymbol{B}$	magnetic field (flux density)	G	gain; Grover operator
B_{ij}	Einstein B coefficient	h	strain
B'	magnetic field gradient	$\boldsymbol{H}$	magnetic field
c_i	amplitude coefficient	$\hat{H}$	Hamiltonian
C	capacitance	H	Hadamard operator
C_V	heat capacity at constant volume	H'	perturbation
d	distance; slit width	$H_n(x)$	Hermite polynomial
d_{ij}	nonlinear optical coefficient tensor	$\hat{\mathbf{i}}$	unit vector along the x-axis
D	diameter	i	electrical current
$\boldsymbol{D}$	electric displacement	I	optical intensity; nuclear spin
D_p	momentum diffusion coefficient	I_rot	moment of inertia
E	energy	I_s	saturation intensity
E_g	band-gap energy	$\boldsymbol{I}$	nuclear angular momentum
E_X	exciton binding energy	$\mathbf{I}$	identity matrix
$\boldsymbol{\mathcal{E}}$	electric field	I_z	z-component of nuclear angular momentum
$\mathcal{E}_0$	electric field amplitude	j	current density; angular momentum (single electron)
f	frequency		
$f(T)$	fraction of condensed particles	$\hat{\mathbf{j}}$	unit vector along the y-axis
f_{ij}	oscillator strength	J	angular momentum
$\boldsymbol{F}$	force; total angular momentum	k	wave vector
$\mathcal{F}$	finesse	k	modulus of wave vector; spring constant
F_Fano	Fano factor	$\hat{\mathbf{k}}$	unit vector along the z-axis
F_P	Purcell factor	l	orbital angular momentum (single electron)
g	degeneracy; nonlinear coupling	l_z	z-component of orbital angular momentum (single electron)
$g(E)$	density of states at energy E		
$g(k)$	state density in k-space	L	length; mean free path
$g(\omega)$	density of states at angular frequency ω	$\boldsymbol{L}$	orbital angular momentum
$g_\nu(\nu)$	spectral lineshape function	L_c	coherence length

L_{w}	quantum well thickness
m	mass
m_0	electron rest mass
m^*	effective mass
m_{e}^*	electron effective mass
m_{H}	mass of hydrogen atom
$\mathbf{M}$	matrix
$\boldsymbol{M}$	magnetization
M_x	x-component of the magnetization
M_y	y-component of the magnetization
M_z	z-component of the magnetization
n	refractive index; photon number; number of events
n_2	nonlinear refractive index
n_{o}	refractive index for ordinary ray
n_{e}	refractive index for extraordinary ray
$\overline{n}$	mean photon number
$n(E)$	thermal occupancy of level at energy E
$n_{\mathrm{BE}}(E)$	Bose–Einstein distribution function
$n_{\mathrm{FD}}(E)$	Fermi–Dirac distribution function
N	number of atoms, particles, photons, counts, time intervals, data bits
$\mathsf{N}_{\mathrm{stop}}$	stopping number of absorption–emission cycles
$\hat{O}$	operator
p	momentum; probability
$\boldsymbol{p}$	electric dipole moment
$\hat{\boldsymbol{p}}$	momentum operator
P	pressure; power
$\mathcal{P}$	probability
$\mathcal{P}_{ij}$	probability for $i \to j$ transition
$\boldsymbol{P}$	electric polarization
q	charge; generalized position coordinate; qubit
Q	quality factor
r	radius; amplitude reflection coefficient
$\boldsymbol{r}$	position vector
$\hat{\boldsymbol{r}}$	position operator
R	reflectivity; net absorption rate; electrical resistance
$\mathcal{R}$	pumping rate; count rate
$R_i(\theta)$	rotation operator about Cartesian axis i
s	squeeze parameter; saturation parameter
$\boldsymbol{s}$	spin angular momentum (single electron)
s_z	z-component of spin angular momentum (single electron)
S	Clauser, Horne, Shimony, and Holt parameter
$\boldsymbol{S}$	spin angular momentum

t	time; amplitude transmission coefficient
t_{e}	expansion time
T	temperature; time interval
$\hat{T}$	kinetic energy operator
$\mathcal{T}$	time interval; transmission
T_{c}	critical temperature
T_{op}	gate operation time
T_{osc}	oscillation period
T_{p}	pulse duration
T_1	longitudinal (spin–lattice) relaxation time
T_2	transverse (spin–spin) relaxation time; dephasing time
u	initial velocity
$u(\nu)$	spectral energy density at frequency ν
$u(\omega)$	spectral energy density at angular frequency ω
U	energy density
$\hat{U}$	unitary operator
$\boldsymbol{v}$	velocity
V	volume; potential energy
$\hat{V}$	perturbation; potential energy operator
V_{ij}	perturbation matrix element
w	Gaussian beam radius
W	count rate in time interval T
W_{ij}	transition rate
x	position coordinate
$\hat{\mathbf{x}}$	unit vector along the x-axis
$\hat{x}$	position coordinate operator
X	X operator
$X_{1,2}$	quadrature field
y	position coordinate
$\hat{\mathbf{y}}$	unit vector along the y-axis
Y_{l,m_l}	spherical harmonic function
z	position coordinate
$\hat{\mathbf{z}}$	unit vector along the z-axis
Z	atomic number; Z operator; partition function; impedance
α	coherent state complex amplitude; damping coefficient
β	spontaneous emission coupling factor
γ	gyromagnetic ratio; damping rate; decay rate; linewidth; gain coefficient
$\boldsymbol{\Gamma}$	torque
δ	frequency detuning
$\delta(x)$	Dirac delta function
δ_{ij}	Kronecker delta function
Δ	detuning in angular frequency units
ε	error probability

ϵ_{r}	relative permittivity	τ	lifetime
θ	angle; polar angle	τ_{c}	coherence time
Θ	rotation angle; pulse area	$\tau_{\mathrm{collision}}$	time between collisions
η	quantum efficiency	τ_{D}	detector response time
κ	photon decay rate	τ_{G}	gravity wave period
λ	wavelength	τ_{R}	radiative lifetime
λ_{deB}	de Broglie wavelength	τ_{NR}	non-radiative lifetime
μ	reduced mass; chemical potential; mean value	ϕ	optical phase
$\boldsymbol{\mu}$	magnetic dipole moment	φ	wave function; optical phase; azimuthal angle
μ_{ij}	dipole moment for $i \rightarrow j$ transition	χ	electric susceptibility; spin wave function
μ_{R}	relative magnetic permeability		
ν	frequency	$\chi^{(n)}$	nth-order nonlinear susceptibility
ν_{L}	laser frequency	$\chi^{(2)}_{ijk}$	second-order nonlinear susceptibility tensor
ν_{vib}	vibrational frequency		
ξ	dipole orientation factor; optical loss; emission probability per unit time per unit intensity	χ_{M}	magnetic susceptibility
		Φ	photon flux; wave function
		Ψ	wave function
$\boldsymbol{\rho}$	density matrix	ψ	wave function
ρ_{ij}	element of density matrix	ω	angular frequency
ρ	energy density of black-body radiation	ω_{L}	Larmor precession angular frequency
ϱ	charge density	Ω	solid angle; angular frequency
σ	standard deviation; electrical conductivity	$\boldsymbol{\Omega}$	angular velocity vector
σ_{s}	scattering cross-section	Ω_{R}	Rabi angular frequency

List of quantum numbers

In atomic physics, lower and upper case letters refer to individual electrons or whole atoms respectively.

F	total angular momentum (with nuclear spin included)	m_j, M_J	magnetic (z-component of total angular momentum)
I	nuclear spin	m_l, M_L	magnetic (z-component of orbital angular momentum)
j, J	total electron angular momentum		
l, L	orbital angular momentum	m_s, M_S	magnetic (z-component of spin angular momentum)
M_F	magnetic (z-component of total angular momentum including hyperfine interactions)		
		n	principal
M_I	magnetic (z-component of nuclear spin)	s, S	spin

List of abbreviations

AC	alternating current
AOS	acousto-optic switch
APD	avalanche photodiode
B92	Bennett 1992
BB84	Bennett–Brassard 1984
BBO	β-barium borate
BS	beam splitter
BSM	Bell-state measurement
CHSH	Clauser–Horne–Shimony–Holt
CW	continuous wave
DBR	distributed Bragg reflector
DC	direct current
EPR	Einstein–Podolsky–Rosen
EPRB	Einstein–Podolsky–Rosen–Bohm
FWHM	full width at half maximum
HBT	Hanbury Brown–Twiss
LD	laser diode
LED	light-emitting diode
LHV	local hidden variables
LIGO	light interferometer gravitational wave observatory
LISA	laser interferometer space antenna
LO	local oscillator
MBE	molecular beam epitaxy
MOCVD	metalorganic chemical vapour epitaxy
NMR	nuclear magnetic resonance
PBS	polarizing beam splitter
PC	Pockels cell
PD	photodiode
PMT	photomultiplier tube
QED	quantum electrodynamics
RF	radio frequency
rms	root mean square
SNL	shot-noise level
SNR	signal-to-noise ratio
SPAD	single-photon avalanche photodiode
STP	standard temperature and pressure
TEM	transmission electron microscope
VCSEL	vertical-cavity surface-emitting laser

Part I

Introduction and background

Introduction

1

1.1 What is quantum optics ?

Quantum optics is the subject that deals with optical phenomena that can only be explained by treating light as a stream of photons rather than as electromagnetic waves. In principle, the subject is as old as quantum theory itself, but in practice, it is a relatively new one, and has really only come to the fore during the last quarter of the twentieth century.

In the progressive development of the theory to light, three general approaches can be clearly identified, namely the **classical**, **semi-classical**, and **quantum** theories, as summarized in Table 1.1. It goes without saying that only the fully quantum optical approach is totally consistent both with itself and with the full body of experimental data. Nevertheless, it is also the case that semi-classical theories are quite adequate for most purposes. For example, when the theory of absorption of light by atoms is first considered, it is usual to apply quantum mechanics to the atoms, but treat the light as a classical electromagnetic wave.

The question that we really have to ask to define the subject of quantum optics is whether there are any effects that cannot be explained in the semi-classical approach. It may come as a surprise to the reader that there are relatively few such phenomena. Indeed, until about 30 years ago, there were only a handful of effects—mainly those related to the vacuum field such as spontaneous emission and the Lamb shift—that really required a quantum model of light.

Let us consider just one example that seems to require a photon picture of light, namely the **photoelectric effect**. This describes the ejection of electrons from a metal under the influence of light. The explanation of the phenomenon was first given by Einstein in 1905, when he realized that the atoms must be absorbing energy from the light beam in quantized packets. However, careful analysis has subsequently shown that the results can in fact be understood by treating only the atoms as quantized objects, and the light as a classical electromagnetic wave. Arguments along the same line can explain how the individual pulses emitted by 'single-photon counting' detectors do not necessarily imply that light consists of photons. In most cases, the output pulses can in fact be explained in terms of the probabilistic ejection of an individual electron from one of the quantized states in an atom under the influence of a classical light wave. Thus although these experiments point us towards the photon picture of light, they do not give conclusive evidence.

1.1 **What is quantum optics ?**   3

1.2 **A brief history of quantum optics**   4

1.3 **How to use this book**   6

Table 1.1 The three different approaches used to model the interaction between light and matter. In classical physics, the light is conceived as electromagnetic waves, but in quantum optics, the quantum nature of the light is included by treating the light as photons.

Model	Atoms	Light
Classical	Hertzian dipoles	Waves
Semi-classical	Quantized	Waves
Quantum	Quantized	Photons

Table 1.2 Subtopics of recent European Quantum Optics Conferences

Year	Topic
1998	Atom cooling and guiding, laser spectroscopy and squeezing
1999	Quantum optics in semiconductor materials, quantum structures
2000	Experimental technologies of quantum manipulation
2002	Quantum atom optics: from quantum science to technology
2003	Cavity QED and quantum fluctuations: from fundamental concepts to nanotechnology

Source: European Science Foundation, http://www.esf.org.

It was not until the late 1970s that the subject of quantum optics as we now know it started to develop. At that time, the first observations of effects that give direct evidence of the photon nature of light, such as photon antibunching, were convincingly demonstrated in the laboratory. Since then, the scope of the subject has expanded enormously, and it now encompasses many new topics that go far beyond the strict study of light itself. This is apparent from Table 1.2, which lists the range of specialist topics selected for recent European Quantum Optics Conferences. It is in this widened sense, rather than the strict one, that the subject of quantum optics is understood throughout this book.

1.2   A brief history of quantum optics

We can obtain insight into the way the subject of quantum optics fits into the wider picture of quantum theory by running through a brief history of its development. Table 1.3 summarizes some of the most important landmarks in this development, together with a few recent highlights.

In the early development of optics, there were two rival theories, namely the corpuscular theory proposed by Newton, and the wave theory expounded by his contemporary, Huygens. The wave theory was convincingly vindicated by the double-slit experiment of Young in 1801 and by the wave interpretation of diffraction by Fresnel in 1815. It was then given a firm theoretical footing with Maxwell's derivation of the electromagnetic wave equation in 1873. Thus by the end of the nineteenth century, the corpuscular theory was relegated to mere historical interest.

The situation changed radically in 1901 with Planck's hypothesis that black-body radiation is emitted in discrete energy packets called *quanta*. With this supposition, he was able to solve the ultraviolet catastrophe problem that had been puzzling physicists for many years. Four years later in 1905, Einstein applied Planck's quantum theory to explain the photoelectric effect. These pioneering ideas laid the foundations for the quantum theories of light and atoms, but in themselves did not give direct experimental evidence of the quantum nature of the light. As mentioned above, what they actually prove is that *something* is quantized, without definitively establishing that it is the *light* that is quantized.

Table 1.3 Selected landmarks in the development of quantum optics, including a few recent highlights. The final column points to the appropriate chapter of the book where the topic is developed

Year	Authors	Development	Chapter
1901	Planck	Theory of black-body radiation	5
1905	Einstein	Explanation of the photoelectric effect	5
1909	Taylor	Interference of single quanta	14
1909	Einstein	Radiation fluctuations	5
1927	Dirac	Quantum theory of radiation	8
1956	Hanbury Brown and Twiss	Intensity interferometer	6
1963	Glauber	Quantum states of light	8
1972	Gibbs	Optical Rabi oscillations	9
1977	Kimble, Dagenais, and Mandel	Photon antibunching	6
1981	Aspect, Grangier, and Roger	Violations of Bell's inequality	14
1985	Slusher *et al.*	Squeezed light	7
1987	Hong, Ou, and Mandel	Single-photon interference experiments	14
1992	Bennett, Brassard *et al.*	Experimental quantum cryptography	12
1995	Turchette, Kimble *et al.*	Quantum phase gate	10, 13
1995	Anderson, Wieman, Cornell *et al.*	Bose–Einstein condensation of atoms	11
1997	Mewes, Ketterle *et al.*	Atom laser	11
1997	Bouwmeester *et al.*, Boschi *et al.*	Quantum teleportation of photons	14
2002	Yuan *et al.*	Single-photon light-emitting diode	6

The first serious attempt at a real quantum optics experiment was performed by Taylor in 1909. He set up a Young's slit experiment, and gradually reduced the intensity of the light beam to such an extent that there would only be one quantum of energy in the apparatus at a given instant. The resulting interference pattern was recorded using a photographic plate with a very long exposure time. To his disappointment, he found no noticeable change in the pattern, even at the lowest intensities.

In the same year as Taylor's experiment, Einstein considered the energy fluctuations of black-body radiation. In doing so, he showed that the discrete nature of the radiation energy gave an extra term proportional to the average number of quanta, thereby anticipating the modern theory of photon statistics.

The formal theory of the quantization of light came in the 1920s after the birth of quantum mechanics. The word 'photon' was coined by Gilbert Lewis in 1926, and Dirac published his seminal paper on the quantum theory of radiation a year later. In the following years, however, the main emphasis was on calculating the optical spectra of atoms, and little effort was invested in looking for quantum effects directly associated with the light itself.

The modern subject of quantum optics was effectively born in 1956 with the work of Hanbury Brown and Twiss. Their experiments on correlations between the starlight intensities recorded on two separated detectors provoked a storm of controversy. It was subsequently shown that their results could be explained by treating the light classically and only applying quantum theory to the photodetection process. However, their experiments are still considered a landmark in the field because they were the first serious attempt to measure the fluctuations in the light intensity on short time-scales. This opened the door to more sophisticated experiments on photon statistics that would eventually lead to the observation of optical phenomena with no classical explanation.

The invention of the laser in 1960 led to new interest in the subject. It was hoped that the properties of the laser light would be substantially different from those of conventional sources, but these attempts again proved negative. The first clues of where to look for unambiguous quantum optical effects were given by Glauber in 1963, when he described new states of light which have different statistical properties to those of classical light. The experimental confirmation of these non-classical properties was given by Kimble, Dagenais, and Mandel in 1977 when they demonstrated photon antibunching for the first time. Eight years later, Slusher *et al.* completed the picture by successfully generating squeezed light in the laboratory.

In recent years, the subject has expanded to include the associated disciplines of quantum information processing and controlled light–matter interactions. The work of Aspect and co-workers starting from 1981 onwards may perhaps be conceived as a landmark in this respect. They used the entangled photons from an atomic cascade to demonstrate violations of Bell's inequality, thereby emphatically showing how quantum optics can be applied to other branches of physics. Since then, there has been a growing number of examples of the use of quantum optics in ever widening applications. Some of the recent highlights are listed in Table 1.3.

This brief and incomplete survey of the development of quantum optics makes it apparent that the subject has 'come of age' in recent years. It is no longer a specialized, highly academic discipline, with few applications in the real world, but a thriving field with ever broadening horizons.

1.3 How to use this book

The structure of the book is shown schematically in Fig. 1.1. The book has been divided into four parts:

Part I Introduction and background material.
Part II Photons.
Part III Atom–photon interactions.
Part IV Quantum information processing.

Part I contains the introduction and the background information that forms a starting point for the rest of the book, while Parts II–IV contain the new material that is being developed.

The background material in Part I has been included both for revision purposes and to fill in any small gaps in the prior knowledge that has

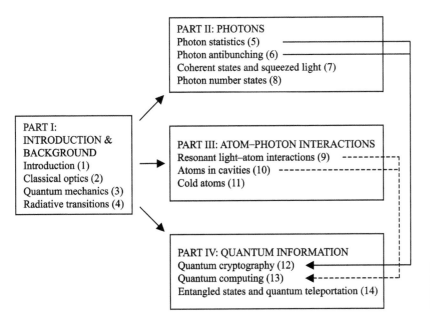

Fig. 1.1 Schematic representation of the development of the themes within the book. The figures in brackets refer to the chapter numbers.

been assumed. A few exercises are provided at the end of each chapter to help with the revision process. There are, however, two sections in Chapter 2 that might need more careful reading. The first is the discussion of the first-order correlation function in Section 2.3, and the second is the overview of nonlinear optics in Section 2.4. These topics are not routinely covered in introductory optics courses, and it is recommended that readers who are unfamiliar with them should study the relevant sections before moving on to Parts II–IV.

The new material developed in the book has been written in such a way that Parts II–IV are more or less independent of each other, and can be studied separately. At the same time, there are inevitably a few cross-references between the different parts, and the main ones have been indicated by the arrows in Fig. 1.1. All of the chapters in Parts II–IV contain worked examples and a number of exercises. Outline solutions to some of these exercises are given at the back of the book, together with the numerical answers for all of them. A solutions manual is available for instructors on request. The book concludes with six appendices, which expand on selected topics, and also present a brief summary of several related subjects that are connected to the main themes developed in Parts II–IV.

2

Classical optics

2.1	**Maxwell's equations and electromagnetic waves**	**8**
2.2	**Diffraction and interference**	**13**
2.3	**Coherence**	**16**
2.4	**Nonlinear optics**	**19**
Further reading		**24**
Exercises		**24**

It is appropriate to start a book on quantum optics with a brief review of the classical description of light. This description, which is based on the theory of electromagnetic waves governed by Maxwell's equations, is adequate to explain the majority of optical phenomena and forms a very persuasive body of evidence in its favour. It is for this reason that most optics texts are developed in terms of wave and ray theory, with only a brief mention of quantum optics. The strategy adopted in this book will therefore be that quantum theory will be invoked only when the classical explanations are inadequate.

In this chapter we give an overview of the results of electromagnetism and classical optics that are relevant to the later chapters of the book. It is assumed that the reader is already familiar with these subjects, and the material is only presented in summary form. The chapter also includes a short overview of the subject of classical nonlinear optics. This may be less familiar to some readers, and is therefore developed at slightly greater length. A short bibliography is provided in the Further Reading section for those readers who are unfamiliar with any of the topics that are described here.

2.1 Maxwell's equations and electromagnetic waves

The theory of light as electromagnetic waves was developed by Maxwell in the second half of the nineteenth century and is considered as one of the great triumphs of classical physics. In this section we give a summary of Maxwell's theory and the results that follow from it.

2.1.1 Electromagnetic fields

Maxwell's equations are formulated around the two fundamental electromagnetic fields:

- the **electric field** $\mathcal{E}$;
- the **magnetic field** B.

Two other variables related to these fields are also defined, namely the **electric displacement** D, and the equivalent magnetic quantity H. Since both include the effects of the medium, we must briefly review

Older electromagnetism texts tend to call H the magnetic field and B either the **magnetic flux density** or the **magnetic induction**. However, it is now common practice to specify magnetic fields in units of flux density, namely Tesla. Moreover, it can be argued that B is the more fundamental quantity, since the force experienced by a charge with velocity v in a magnetic field depends on B through $F = qv \times B$. A more detailed explanation of the difference between B and H and a justification for the use of B for the magnetic field may be found in Brooker (2003, §1.2). The distinction is of little practical importance in optics, because the two quantities are usually linearly related to each other through eqn 2.8.

how we quantify the way the medium responds to the fields before formulating the equations that have to be solved.

The dielectric response of a medium is determined by the **electric polarization** P, which is defined as the electric dipole moment per unit volume. The electric displacement D is related to the electric field $\mathcal{E}$ and the electric polarization P through:

$$D = \epsilon_0 \mathcal{E} + P. \tag{2.1}$$

In an isotropic medium, the microscopic dipoles align along the direction of the applied electric field, so that we can write:

$$P = \epsilon_0 \chi \mathcal{E}, \tag{2.2}$$

where ϵ_0 is the **electric permittivity** of free space (8.854×10^{-12} F m^{-1} in SI units) and χ is the **electric susceptibility** of the medium. By combining eqns 2.1 and 2.2, we then find:

$$D = \epsilon_0 \epsilon_{\mathrm{r}} \mathcal{E}, \tag{2.3}$$

where

$$\epsilon_{\mathrm{r}} = 1 + \chi. \tag{2.4}$$

ϵ_{r} is the **relative permittivity** of the medium.

The equivalent of eqn 2.1 for magnetic fields is

$$H = \frac{1}{\mu_0} B - M, \tag{2.5}$$

where μ_0 is the magnetic permeability of the vacuum ($4\pi \times 10^{-7}$ H m^{-1} in SI units) and M is the **magnetization** of the medium, which is defined as the magnetic moment per unit volume. In an isotropic material, the **magnetic susceptibility** χ_{M} is defined according to:

$$M = \chi_{\mathrm{M}} H, \tag{2.6}$$

so that eqn 2.5 can be rearranged to give:

$$\begin{aligned} B &= \mu_0(H + M) \\ &= \mu_0(1 + \chi_{\mathrm{M}})H \\ &= \mu_0 \mu_{\mathrm{r}} H, \end{aligned} \tag{2.7}$$

where $\mu_{\mathrm{r}} = 1 + \chi_{\mathrm{M}}$ is the **relative magnetic permeability** of the medium. In free space, where $\chi_{\mathrm{M}} = 0$, this reduces to:

$$B = \mu_0 H. \tag{2.8}$$

In optics, it is usually assumed that the magnetic dipoles that contribute to χ_{M} are too slow to respond, so that $\mu_{\mathrm{r}} = 1$. It is therefore normal to relate B to H through eqn 2.8, and to use them interchangeably.

In anisotropic materials, the value of χ depends on the direction of the field relative to the axes of the medium. It is therefore necessary to use a tensor to represent the electric susceptibility. In nonlinear materials, the polarization depends on higher powers of the electric field. See Section 2.4.

Magnetic materials are too slow to respond at optical frequencies because the magnetic response time T_1 (see eqn E.21 in Appendix E) is much longer than the period of an optical wave ($\sim 10^{-15}$ s). By contrast, the electric susceptibility is non-zero at optical frequencies because it includes the contributions of the dipoles produced by oscillating electrons, which can easily respond on these time-scales.

2.1.2   Maxwell's equations

The laws that describe the combined electric and magnetic response of a medium are summarized in **Maxwell's equations** of electromagnetism:

$$\boldsymbol{\nabla} \cdot \boldsymbol{D} = \varrho, \tag{2.9}$$

$$\boldsymbol{\nabla} \cdot \boldsymbol{B} = 0, \tag{2.10}$$

$$\boldsymbol{\nabla} \times \boldsymbol{\mathcal{E}} = -\frac{\partial \boldsymbol{B}}{\partial t}, \tag{2.11}$$

$$\boldsymbol{\nabla} \times \boldsymbol{H} = \boldsymbol{j} + \frac{\partial \boldsymbol{D}}{\partial t}, \tag{2.12}$$

where ϱ is the free charge density, and $\boldsymbol{j}$ is the free current density. The first of these four equations is Gauss's law of electrostatics. The second is the equivalent of Gauss's law for magnetostatics with the assumption that free magnetic monopoles do not exist. The third equation combines the Faraday and Lenz laws of electromagnetic induction. The fourth is a statement of Ampere's law, with the second term on the right-hand side to account for the displacement current.

2.1.3   Electromagnetic waves

Wave-like solutions to Maxwell's equations are possible with no free charges ($\varrho = 0$) or currents ($\boldsymbol{j} = 0$). To see this, we substitute for $\boldsymbol{D}$ and $\boldsymbol{H}$ in eqn 2.12 using eqns 2.3 and 2.8 respectively, giving:

$$\frac{1}{\mu_0} \boldsymbol{\nabla} \times \boldsymbol{B} = \epsilon_0 \epsilon_\mathrm{r} \frac{\partial \boldsymbol{\mathcal{E}}}{\partial t}. \tag{2.13}$$

We then take the curl of eqn 2.11 and eliminate $\boldsymbol{\nabla} \times \boldsymbol{B}$ using eqn 2.13:

$$\boldsymbol{\nabla} \times (\boldsymbol{\nabla} \times \boldsymbol{\mathcal{E}}) = -\mu_0 \epsilon_0 \epsilon_\mathrm{r} \frac{\partial^2 \boldsymbol{\mathcal{E}}}{\partial t^2}. \tag{2.14}$$

Finally, by using the vector identity

$$\boldsymbol{\nabla} \times (\boldsymbol{\nabla} \times \boldsymbol{\mathcal{E}}) = \boldsymbol{\nabla}(\boldsymbol{\nabla} \cdot \boldsymbol{\mathcal{E}}) - \boldsymbol{\nabla}^2 \boldsymbol{\mathcal{E}}, \tag{2.15}$$

The equation for the displacement of a wave with velocity v propagating in the x-direction is:

$$\frac{\partial^2 y}{\partial x^2} = \frac{1}{v^2} \frac{\partial^2 y}{\partial t^2}.$$

Equation 2.16 represents a generalization of this to a wave that propagates in three dimensions.

and the fact that $\boldsymbol{\nabla} \cdot \boldsymbol{\mathcal{E}} = 0$ (see eqn 2.9 with $\varrho = 0$ and $\boldsymbol{D}$ given by eqn 2.3) we obtain the final result:

$$\boldsymbol{\nabla}^2 \boldsymbol{\mathcal{E}} = \mu_0 \epsilon_0 \epsilon_\mathrm{r} \frac{\partial^2 \boldsymbol{\mathcal{E}}}{\partial t^2}. \tag{2.16}$$

Equation 2.16 describes electromagnetic waves with a speed v given by

$$\frac{1}{v^2} = \mu_0 \epsilon_0 \epsilon_\mathrm{r}. \tag{2.17}$$

In free space $\epsilon_\mathrm{r} = 1$ and the speed c is given by:

$$c = \frac{1}{\sqrt{\mu_0 \epsilon_0}} = 2.998 \times 10^8 \ \mathrm{m\,s}^{-1}. \tag{2.18}$$

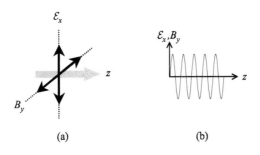

In a dielectric medium, the speed is given instead by:

$$v = \frac{1}{\sqrt{\epsilon_r}} c \equiv \frac{c}{n},\qquad (2.19)$$

where n is the **refractive index**. It is apparent from eqn 2.19 that

$$n = \sqrt{\epsilon_r},\qquad (2.20)$$

which allows us to relate the optical properties of a medium to its dielectric properties.

The usual solutions to Maxwell's equations are transverse waves with the electric and magnetic fields at right angles to each other. Consider a wave of angular frequency ω propagating in the z-direction with the electric field along the x-axis, as shown in Fig. 2.1. With $\mathcal{E}_y = \mathcal{E}_z = 0$ and $B_x = B_z = 0$, the Maxwell equations 2.11 and 2.13 reduce to:

$$\frac{\partial \mathcal{E}_x}{\partial z} = -\frac{\partial B_y}{\partial t}$$

$$-\frac{\partial B_y}{\partial z} = \mu_0 \epsilon_0 \epsilon_r \frac{\partial \mathcal{E}_x}{\partial t}.\qquad (2.21)$$

These have solutions of the form:

$$\mathcal{E}_x(z,t) = \mathcal{E}_{x0} \cos(kz - \omega t + \phi)$$

$$B_y(z,t) = B_{y0} \cos(kz - \omega t + \phi).\qquad (2.22)$$

where $\mathcal{E}_{x0}$ is the **amplitude**, ϕ is the **optical phase**, and k is the **wave vector** given by:

$$k = \frac{2\pi}{\lambda_m} = \frac{\omega}{v} = \frac{n\omega}{c},\qquad (2.23)$$

λ_m being the wavelength inside the medium. On substitution of eqn 2.22 into eqn 2.21, we find:

$$B_{y0} = \frac{k}{\omega} \mathcal{E}_{x0} = \frac{n}{c} \mathcal{E}_{x0}.\qquad (2.24)$$

The equivalent relationship for H_{y0} is

$$H_{y0} = \mathcal{E}_{x0}/Z,\qquad (2.25)$$

where Z is the **wave impedance**:

$$Z = \sqrt{\frac{\mu_0}{\epsilon_0 \epsilon_r}},\qquad (2.26)$$

which takes the value of 377 Ω in free space.

It is possible to find solutions to Maxwell's equations that are not transverse in some special situations. One of these is the case of a metal waveguide. Another is that of a material with $\epsilon_r = 0$ at some particular frequency. (See Exercise 2.1.)

The electric and magnetic fields can also be described by complex fields with

$$\mathcal{E}_x(z,t) = \mathcal{E}_{x0}\, e^{i(kz-\omega t+\phi)},$$

and

$$B_y(z,t) = B_{y0}\, e^{i(kz-\omega t+\phi)}.$$

The use of complex solutions simplifies the mathematics and is used extensively throughout this book. Physically measurable quantities are obtained by taking the real part of the complex wave. The optical phase ϕ is determined by the starting conditions of the source that produces the light.

The energy flow in an electromagnetic wave can be calculated from the **Poynting vector**:

$$I = \mathcal{E} \times H .\qquad(2.27)$$

The Poynting vector gives the **intensity** (i.e. energy flow (power) per unit area in $\mathrm{W\,m^{-2}}$) of the light wave. On substituting eqns 2.22–2.26 into eqn 2.27, and taking the time average over the cycle, we obtain:

$$\langle I \rangle = \frac{1}{Z}\langle \mathcal{E}(t)^2\rangle_{\mathrm{rms}} = \frac{1}{2}c\epsilon_0 n\mathcal{E}_{x0}^2 ,\qquad(2.28)$$

where $\langle \mathcal{E}(t)^2\rangle_{\mathrm{rms}}$ represents the root-mean-square of the electric field. This shows that the intensity of a light wave is proportional to the square of the amplitude of the electric field.

2.1.4 Polarization

The word 'polarization' is used both for the dielectric polarization P and for the direction of the electric field in an electromagnetic wave. It is usually obvious from the context which meaning is appropriate.

Circularly polarized light is also called 'positive' or 'negative' depending on whether it rotates clockwise or anticlockwise as seen from the source. This makes positive circular polarization equivalent to left circular polarization, and vice versa.

The direction of the electric field of an electromagnetic wave is called the **polarization**. Several different types of polarization are possible.

- **Linear**: the electric field vector points along a constant direction.
- **Circular**: the electric field vector rotates as the wave propagates, mapping out a circle for each cycle of the wave. The light is called **right circularly polarized** if the electric field vector rotates to the right (clockwise) in a fixed plane as the observer looks towards the light source, and **left circularly polarized** if it rotates in the opposite sense. Circularly polarized light can be decomposed into two orthogonal linearly polarized waves of equal amplitude with a 90° phase difference between them.
- **Elliptical**: this is similar to circular polarization, except that the amplitudes of the two orthogonal linearly polarized waves are different, or the phase between them is neither 0° nor 90°, so that the electric field maps out an ellipse as it propagates.
- **Unpolarized**: the light is randomly polarized.

Figure 2.1 thus depicts a linearly polarized wave with the polarization along the x-axis.

In free space the polarization of a wave is constant as it propagates. However, in certain anisotropic materials, the polarization can change as the wave propagates. A common manifestation of optical anisotropy found in non-absorbing materials is the phenomenon of **birefringence**. Birefringent crystals separate arbitrarily polarized beams into two orthogonally polarized beams called the **ordinary ray** and the **extra-ordinary ray**. These two rays experience different refractive indices of n_{o} and n_{e}, respectively.

The **polarizing beam splitter** (PBS) is an important component in a number of quantum optical experiments. A PBS is commonly made by cementing together two birefringent materials like calcite or quartz, and has the property of splitting a light beam into its orthogonal linear polarizations as shown in Fig. 2.2. The figure shows the effect on a linearly polarized light beam propagating along the z-axis when the axes of the crystals are oriented so that the output polarizations are horizontal (h) and vertical (v). The beam splitter resolves the electric field into its two components along the crystal axes, so that the output fields are given by:

$$\mathcal{E}_{v} = \mathcal{E}_{0} \cos \theta,$$
$$\mathcal{E}_{h} = \mathcal{E}_{0} \sin \theta, \tag{2.29}$$

where $\mathcal{E}_{0}$ is the amplitude of the incoming wave, and θ is the angle of the input polarization with respect to the vertical axis. Since the intensity is proportional to the square of the amplitude (cf. eqn 2.28), the intensities of the two orthogonally polarized output beams are given by:

$$I_{v} = I_{0} \cos^{2} \theta,$$
$$I_{h} = I_{0} \sin^{2} \theta, \tag{2.30}$$

where I_{0} is the intensity of the incoming beam. The intensity splitting ratio is $50:50$ when θ is set at $45°$. A similar splitting ratio is also obtained when the incoming light is unpolarized, where we have to take the average values of $\cos^{2} \theta$ and $\sin^{2} \theta$ for all possible angles, namely $1/2$ in both cases.

2.2 Diffraction and interference

The wave nature of light is most clearly demonstrated by the phenomena of **diffraction** and **interference**. We shall not discuss these phenomena at any length here, since they are included in all classical optics texts, but merely quote a few important results that will be needed later in the book.

2.2.1 Diffraction

Let us consider the diffraction of plane parallel light of wavelength λ from a single slit of width d as illustrated in Fig. 2.3. Two general regimes can be distinguished, namely those for **Fresnel diffraction** and **Fraunhofer diffraction**. The distinction between the two is determined by the distance L between the screen and the slit. When L is much larger

With the Cartesian axes set up as in Fig. 2.1 and the beam travelling parallel to a horizontal optical bench, the waves polarized along the x-axis are called **vertically polarized** and those in the y-z plane **horizontally polarized**, respectively, for obvious reasons.

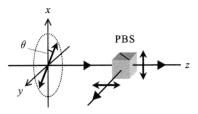

Fig. 2.2 A polarizing beam splitter (PBS) splits an incoming wave into two orthogonally polarized beams. The figure shows the case where the orientation of the beam splitter is set to give vertically and horizontally polarized output beams.

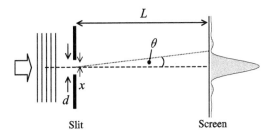

Fig. 2.3 Plane parallel waves incident at a slit of width d are diffracted and produce an intensity pattern on a screen. The diffraction pattern illustrated here corresponds to the Fraunhofer limit, which occurs when the distance L between the slit and the screen is large.

The Fraunhofer condition is often produced experimentally by inserting a lens between the slit and screen and observing in the lens's focal plane.

In describing diffraction and interference phenomena, and hence also the effects of coherence, the mathematics is more compact when the complex-exponential representation of the electric field is used. It is implicitly assumed throughout that measurable quantities are obtained by taking the real part of the complex quantities that are calculated, wherever appropriate.

than the **Rayleigh distance** (d^2/λ), the diffraction pattern is said to be in the **far-field** (Fraunhofer) limit. On the other hand, when $L \lesssim d^2/\lambda$, we are in the **near-field** (Fresnel) regime. In what follows, we consider only Fraunhofer diffraction.

In the Fraunhofer limit, the pattern on the screen observed at angle θ is obtained by summing the field contributions over the slit:

$$\mathcal{E}(\theta) \propto \int_{-d/2}^{+d/2} \exp(-\mathrm{i}kx\sin\theta)\,\mathrm{d}x\,, \tag{2.31}$$

where $kx\sin\theta$ is the relative phase shift at a position x across the slit, k being the wave vector defined in eqn 2.23. On performing the integral and taking the modulus squared to obtain the intensity, we find:

$$I(\theta) \propto \left(\frac{\sin\beta}{\beta}\right)^2\,, \tag{2.32}$$

where

$$\beta = \frac{1}{2}kd\sin\theta\,. \tag{2.33}$$

This diffraction pattern is illustrated in Fig. 2.3. The principal maximum occurs at $\theta = 0$, and there are minima whenever $\beta = m\pi$, m being an integer. Subsidiary maxima occur just below $\beta = (2m+1)\pi/2$, for $m \geq 1$. The intensity at the first subsidiary maximum is less than 5% of that of the principal maximum, and the intensity decreases steadily for all higher-order maxima. The angle at which the first minimum occurs is given by

$$\sin\theta_{\min} = \pm\frac{\lambda}{d}\,. \tag{2.34}$$

If the small-angle approximation is appropriate, this reduces to:

$$\theta_{\min} = \pm\frac{\lambda}{d}\,. \tag{2.35}$$

It is therefore normal to consider the diffraction from a slit as causing an angular spread of $\sim\lambda/d$.

The diffraction patterns obtained from apertures of other shapes can be calculated by similar methods. One important example is that of a circular hole of diameter D. The intensity pattern has circular symmetry about the axis, with a principal maximum at $\theta = 0$ and the first minimum at $\theta_{\min}$, where:

$$\sin\theta_{\min} = 1.22\frac{\lambda}{D}\,. \tag{2.36}$$

This result is commonly used to calculate the resolving power of optical instruments like telescopes and microscopes.

2.2.2 Interference

Interference patterns generally occur when a light wave is divided and then recombined with a phase difference between the two paths. There are many different examples of interference, the most stereotypical probably being the Young's double-slit experiment. The basic principles can, however, be conveniently understood by reference to the **Michelson interferometer** illustrated in Fig. 2.4. This will also serve as a useful framework for discussing the concept of coherence in the following section.

The simplest version of the Michelson interferometer consists of a 50:50 beam splitter (BS) and two mirrors M1 and M2, with air paths throughout. Light is incident on the input port of the beam splitter, where it is divided and directed towards the mirrors. The light reflected off M1 and M2 recombines at BS, producing an interference pattern at the output port. The path length of one of the arms can be varied by translating one of the mirrors (say M2) in the direction parallel to the beam.

Let us assume that the input beam consists of parallel rays from a linearly polarized monochromatic source of wavelength λ and amplitude $\mathcal{E}_0$. The output field is obtained by summing the two contributions from the waves reflected back from M1 and M2 with their phases determined by the path lengths:

$$\mathcal{E}^{\text{out}} = \mathcal{E}_1 + \mathcal{E}_2$$

$$= \frac{1}{2}\mathcal{E}_0 e^{i2kL_1} + \frac{1}{2}\mathcal{E}_0 e^{i2kL_2} e^{i\Delta\phi}$$

$$= \frac{1}{2}\mathcal{E}_0 e^{i2kL_1}\left(1 + e^{i2k\Delta L} e^{i\Delta\phi}\right), \qquad (2.37)$$

where $\Delta L = L_2 - L_1$ and $k = 2\pi/\lambda$ as usual. $\Delta\phi$ is a factor that accounts for the possibility that there are phase shifts between the two paths even

Both beams exiting at the output port have been transmitted once and reflected once by the beam splitter. Let us assume that the beam splitter is a 'half-silvered mirror' consisting of a plate of glass with a semi-reflective coating on one side and an anti-reflection coating on the other. One of the reflections will take place with the light beam incident from the air, and the other with the light incident from within the glass. The phase shifts introduced by these two reflections are not the same. In particular, the requirement to conserve energy at the beam splitter will usually be satisfied if $\Delta\phi = \pi$. (See Exercise 7.14.)

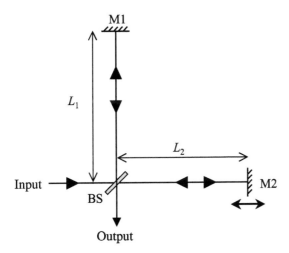

Fig. 2.4 The Michelson interferometer. The apparatus consists of a 50:50 beam splitter (BS) and two mirrors M1 and M2. Interference fringes are observed at the output port as the length of one of the arms (arm 2 in this case) is varied.

when $L_1 = L_2$. Field maxima occur whenever

$$\frac{4\pi}{\lambda}\Delta L + \Delta\phi = 2m\pi\,, \qquad (2.38)$$

and minima when

$$\frac{4\pi}{\lambda}\Delta L + \Delta\phi = (2m+1)\pi\,, \qquad (2.39)$$

where m is again an integer. Thus as L_2 is scanned, bright and dark fringes appear at the output port with a period equal to $\lambda/2$. The interferometer thus forms a very sensitive device to measure differences in the optical path lengths of the two arms.

A typical application of a Michelson interferometer is the measurement of the refractive indices of dilute media such as gases. The interferometer is configured with $L_1 \approx L_2$, and an evacuated cell of length L in one of the arms is then slowly filled with a gas of refractive index n. By recording the shifting of the fringes at the output port as the gas is introduced, the change of the relative path length between the two arms, namely $2(n-1)L$, can be determined, and hence n.

2.3   Coherence

The discussion of the interference pattern produced by a Michelson interferometer in the previous section assumed that the phase shift between the two interfering fields was determined only by the path difference $2\Delta L$ between the arms. However, this is an idealized scenario that takes no account of the frequency stability of the light. In realistic sources, the output contains a range $\Delta\omega$ of angular frequencies, which leads to the possibility that bright fringes for one frequency occur at the same position as the dark fringes for another. Since this washes out the interference pattern, it is apparent that the frequency spread of the source imposes practical limits on the maximum path difference that will give observable fringes.

The property that describes the stability of the light is called the **coherence**. Two types of coherence are generally distinguished:

- temporal coherence,
- spatial coherence.

The discussion below is restricted to temporal coherence. The concept of spatial coherence is discussed briefly in Section 6.1 in the context of the Michelson stellar interferometer.

The temporal coherence of a light beam is quantified by its **coherence time** τ_c. An analogous quantity called the **coherence length** L_c can be obtained from:

$$L_\mathrm{c} = c\tau_\mathrm{c}\,. \qquad (2.40)$$

The coherence time gives the time duration over which the phase of the wave train remains stable. If we know the phase of the wave at some

See Section 4.4 for a discussion of spectral line broadening mechanisms.

Some authors use an alternative nomenclature in which temporal coherence is called **longitudinal coherence** and spatial coherence is called **transverse coherence**. A clear discussion of spatial coherence may be found in Brooker (2003) or Hecht (2002).

position z at time t_1, then the phase at the same position but at a different time t_2 will be known with a high degree of certainty when $|t_2 - t_1| \ll \tau_c$, and with a very low degree when $|t_2 - t_1| \gg \tau_c$. An equivalent way to state this is to say that if, at some time t we know the phase of the wave at z_1, then the phase at the same time at position z_2 will be known with a high degree of certainty when $|z_2 - z_1| \ll L_c$, and with a very low degree when $|z_2 - z_1| \gg L_c$. This means, for example, that fringes will only be observed in a Michelson interferometer when the path difference satisfies $2\Delta L \lesssim L_c$.

Insight into the factors that determine the coherence time can be obtained by considering the filtered light from a single spectral line of a discharge lamp. Let us suppose that the spectral line is pressure-broadened, so that its spectral width $\Delta\omega$ is determined by the average time $\tau_{\text{collision}}$ between the atomic collisions. (See Section 4.4.3.) We model the light as generated by an ensemble of atoms randomly excited by the electrical discharge and then emitting a burst of radiation with constant phase until randomly interrupted by a collision. It is obvious that in this case the coherence time will be limited by $\tau_{\text{collision}}$. Furthermore, since $\tau_{\text{collision}}$ also determines the width of the spectral line, it will also be true that:

> This type of radiation is an example of **chaotic light**. The name refers to the randomness of the excitation and phase interruption processes.

$$\tau_c \approx \frac{1}{\Delta\omega}. \qquad (2.41)$$

The result in eqn 2.41 is in fact a general one and shows that the coherence time is determined by the spectral width of the light. This clarifies that a perfectly monochromatic source with $\Delta\omega = 0$ has an infinite coherence time (perfect coherence), whereas the white light emitted by a thermal source has a very short coherence time. A filtered spectral line from a discharge lamp is an intermediate case, and is described as **partially coherent**.

> The derivation of eqn 2.41 for a general case may be found, for example, in Brooker (2003, §9.11).

The temporal coherence of light can be quantified more accurately by the **first-order correlation function** $g^{(1)}(\tau)$ defined by:

$$g^{(1)}(\tau) = \frac{\langle \mathcal{E}^*(t)\mathcal{E}(t + \tau) \rangle}{\langle |\mathcal{E}(t)|^2 \rangle}. \qquad (2.42)$$

The symbol $\langle \cdots \rangle$ used here indicates that we take the average over a long time interval T:

> In Chapter 6 we shall study the properties of the *second-order* correlation function $g^{(2)}(\tau)$. This correlation function is so-called because it characterizes the properties of the optical intensity, which is proportional to the *second* power of the electric field. (cf. eqn 2.28.)

$$\langle \mathcal{E}^*(t)\mathcal{E}(t + \tau) \rangle = \frac{1}{T}\int_T \mathcal{E}^*(t)\mathcal{E}(t + \tau)\, dt. \qquad (2.43)$$

$g^{(1)}(\tau)$ is called the *first-order* correlation function because it is based on the properties of the *first* power of the electric field. It is also called the **degree of first-order coherence**.

Let us assume that the input field $\mathcal{E}(t)$ is quasi-monochromatic with a centre frequency of ω_0 so that it varies with time according to:

$$\mathcal{E}(t) = \mathcal{E}_0\, e^{-i\omega_0 t} e^{i\phi(t)}. \qquad (2.44)$$

On substituting into eqn 2.42 we then find that $g^{(1)}(\tau)$ is given by:

$$g^{(1)}(\tau) = e^{-i\omega_0\tau} \left\langle e^{i[\phi(t+\tau)-\phi(t)]} \right\rangle . \qquad (2.45)$$

This means that the real part of $g^{(1)}(\tau)$ is an oscillatory function of τ with a period of $2\pi/\omega_0$. This rapid oscillatory variation produces the fringe pattern in an interference experiment, and it is the variation of the modulus of $g^{(1)}(\tau)$ due to the second factor in eqn 2.45 that contains the information about the coherence of the light.

Light that has $|g^{(1)}(\tau) = 1|$ for all values of τ is said to be perfectly coherent. Such idealized light has an infinite coherence time and length. The highly monochromatic light from a single longitudinal mode laser is a fairly good approximation to perfectly coherent light for most practical purposes.

It is clear from eqn 2.42 that $|g^{(1)}(0)| = 1$ for all cases. For $0 < \tau \ll \tau_c$, we expect $\phi(t + \tau) \approx \phi(t)$, and the value of $|g^{(1)}(\tau)|$ will remain close to unity. As τ increases, $|g^{(1)}(\tau)|$ decreases due to the increased probability of phase randomness. For $\tau \gg \tau_c$, $\phi(t + \tau)$ will be totally uncorrelated with $\phi(t)$, and $\exp i[\phi(t + \tau) - \phi(t)]$ will average to zero, implying $|g^{(1)}(\tau)| = 0$. Hence $|g^{(1)}(\tau)|$ drops from 1 to 0 over a time-scale of order τ_c.

The detailed form of $g^{(1)}(\tau)$ for partially coherent light depends on the type of spectral broadening that applies. For light with a Lorentzian lineshape of half width $\Delta\omega$ in angular frequency units, $g^{(1)}(\tau)$ is given by:

See Section 4.4 for a discussion of spectral lineshapes. The derivation of eqns 2.46–2.49 may be found, for example, in Loudon (2000, §3.4).

$$g^{(1)}(\tau) = e^{-i\omega_0\tau} \exp\left(-|\tau|/\tau_c\right) , \qquad (2.46)$$

where

$$\tau_c = 1/\Delta\omega . \qquad (2.47)$$

The equivalent formulae for a Gaussian lineshape are:

$$g^{(1)}(\tau) = e^{-i\omega_0\tau} \exp\left[-\frac{\pi}{2}\left(\frac{\tau}{\tau_c}\right)^2\right] , \qquad (2.48)$$

where

$$\tau_c = (8\pi \ln 2)^{1/2}/\Delta\omega . \qquad (2.49)$$

A typical variation of the real part of $g^{(1)}(\tau)$ with τ for Gaussian light is shown in Fig. 2.5. The coherence time in this example has been set at the artificially short value of 20 times the optical period.

The **visibility** of the fringes observed in an interference experiment is defined as:

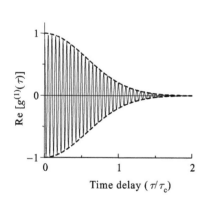

$$\text{visibility} = \frac{I_{\max} - I_{\min}}{I_{\max} + I_{\min}} , \qquad (2.50)$$

Fig. 2.5 Typical variation of the real part of the first-order correlation function $g^{(1)}(\tau)$ as a function of time delay τ for Gaussian light with a coherence time of τ_c. The coherence time in this example has been chosen to be 20 times longer than the optical period.

where $I_{\max}$ and $I_{\min}$ are the intensities recorded at the fringe maxima and minima, respectively. It is qualitatively obvious that the visibility is determined by the coherence of the light, and this point can be quantified by deriving an explicit relationship between the visibility and the first-order correlation function.

We consider again a Michelson interferometer and assume that we have a light source with a constant average intensity, so that the fringe pattern only depends on the time difference τ between the fields that interfere rather than the absolute time. We can therefore write the output field as:

$$\mathcal{E}^{\text{out}}(t) = \frac{1}{\sqrt{2}}(\mathcal{E}(t) - \mathcal{E}(t+\tau)) \,. \tag{2.51}$$

The time-averaged intensity observed at the output is proportional to the average of the modulus squared of the field:

$$
\begin{aligned}
I(\tau) &\propto \langle \mathcal{E}^{\text{out}*}(t)\mathcal{E}^{\text{out}}(t)\rangle \\
&\propto (\langle \mathcal{E}^*(t)\mathcal{E}(t)\rangle + \langle \mathcal{E}^*(t+\tau)\mathcal{E}(t+\tau)\rangle \\
&\quad - \langle \mathcal{E}^*(t)\mathcal{E}(t+\tau)\rangle - \langle \mathcal{E}^*(t+\tau)\mathcal{E}(t)\rangle)/2 \,.
\end{aligned}
\tag{2.52}
$$

The constant nature of the source implies that the first and second terms are identical. Furthermore, the third and fourth are complex conjugates of each other. We therefore find:

$$I(\tau) \propto \langle \mathcal{E}^*(t)\mathcal{E}(t)\rangle - \text{Re}[\langle \mathcal{E}^*(t)\mathcal{E}(t+\tau)\rangle] \,. \tag{2.53}$$

We can then substitute from eqn 2.42 to find:

$$
\begin{aligned}
I(\tau) &\propto \langle \mathcal{E}^*(t)\mathcal{E}(t)\rangle \left(1 - \text{Re}[g^{(1)}(\tau)]\right) \\
&= I_0 \left(1 - \text{Re}[g^{(1)}(\tau)]\right) \,,
\end{aligned}
\tag{2.54}
$$

where I_0 is the input intensity. Substitution into eqn 2.50 with $I_{\text{max/min}} = I_0(1 \pm |g^{(1)}(\tau)|)$ readily leads to the final result that:

$$\text{visibility} = |g^{(1)}(\tau)| \,. \tag{2.55}$$

Hence the intensity observed at the output of a Michelson interferometer as ΔL is scanned would, in fact, look like Fig. 2.5, with $\tau = 2\Delta L/c$.

   A summary of the main points of this section may be found in Table 2.1.

We have assumed a $50 : 50$ power splitting ratio in eqn 2.51, which gives a $1/\sqrt{2}$ amplitude combining ratio. We have also assumed that the phase shift $\Delta\phi$ introduced in eqn 2.37 is equal to π. The path difference ΔL is related to τ through $\tau = 2\Delta L/c$.

2.4   Nonlinear optics

2.4.1   The nonlinear susceptibility

The linear relationship between the electric polarization of a dielectric medium and the electric field of a light wave implied by eqn 2.2 is an

Table 2.1 Coherence properties of light as quantified by the coherence time τ_{c} and the first-order correlation function $g^{(1)}(\tau)$. In the final column we assume $|\tau| > 0$.

Description of light	Spectral width	Coherence	Coherence time	$|g^{(1)}(\tau)|$
Perfectly monochromatic	0	Perfect	Infinite	1
Chaotic	$\Delta\omega$	Partial	$\sim 1/\Delta\omega$	$1 > |g^{(1)}(\tau)| > 0$
Incoherent	Effectively infinite	None	Effectively zero	0

approximation that is valid only when the electric field amplitude is small. With the widespread use of large-amplitude beams from powerful lasers, it is necessary to consider a more general form of eqn 2.2 in which the relationship between the polarization and electric field is *nonlinear*:

$$P = \epsilon_0 \chi^{(1)} \mathcal{E} + \epsilon_0 \chi^{(2)} \mathcal{E}^2 + \epsilon_0 \chi^{(3)} \mathcal{E}^3 + \cdots . \qquad (2.56)$$

The first term in eqn 2.56 is the same as in eqn 2.2 and describes the linear response of the medium. $\chi^{(1)}$ can thus be identified with the linear electric susceptibility χ in eqn 2.2. The other terms describe the *nonlinear* response of the medium. The term in $\mathcal{E}^2$ is called the second-order nonlinear response and $\chi^{(2)}$ is called the **second-order nonlinear susceptibility**. Similarly, the term in $\mathcal{E}^3$ is called the third-order nonlinear response and $\chi^{(3)}$ is called the **third-order nonlinear susceptibility**. In general, we can write

$$P^{(1)} = \epsilon_0 \chi^{(1)} \mathcal{E} \qquad (2.57)$$

$$P^{(2)} = \epsilon_0 \chi^{(2)} \mathcal{E}^2 \qquad (2.58)$$

$$P^{(3)} = \epsilon_0 \chi^{(3)} \mathcal{E}^3 \qquad (2.59)$$

$$\vdots$$

$$P^{(n)} = \epsilon_0 \chi^{(n)} \mathcal{E}^n, \qquad (2.60)$$

where, for $n \geq 2$, $P^{(n)}$ is the nth-order **nonlinear polarization** and $\chi^{(n)}$ is the nth-order **nonlinear susceptibility**.

It is usually the case that the nonlinear susceptibilities have a rather small magnitude. This means that when the electric field amplitude is small, the nonlinear terms are negligible and we revert to the linear relationship between $\boldsymbol{P}$ and $\boldsymbol{\mathcal{E}}$ that is assumed in **linear optics**. On the other hand, when the electric field is large, the nonlinear terms in eqn 2.56 cannot be ignored and we enter the realm of **nonlinear optics**, in which many new phenomena occur.

In the subsections that follow, we briefly describe some of the more common second-order nonlinear phenomena, and also introduce the concept of phase matching. Length considerations preclude a discussion of the phenomena that are caused by the third-order nonlinear susceptibility, such as frequency tripling, self-phase modulation, two-photon absorption, the Raman effect, and the intensity-dependence of the refractive index.

2.4.2   Second-order nonlinear phenomena

The second-order nonlinear polarization is given by eqn 2.58. If the medium is excited by cosinusoidal waves at angular frequencies ω_1 and ω_2 with amplitudes $\mathcal{E}_1$ and $\mathcal{E}_2$, respectively, then the nonlinear polarization will be equal to:

$$P^{(2)}(t) = \epsilon_0 \chi^{(2)} \times \mathcal{E}_1 \cos \omega_1 t \times \mathcal{E}_2 \cos \omega_2 t$$

$$= \epsilon_0 \chi^{(2)} \mathcal{E}_1 \mathcal{E}_2 \frac{1}{2} [\cos (\omega_1 + \omega_2)t + \cos (\omega_1 - \omega_2)t] . \qquad (2.61)$$

The intensities produced by conventional sources such as thermal or discharge lamps are usually too small to produce nonlinear effects, and it is valid to assume that the optical phenomena are well described by the laws of linear optics.

The nonlinear refractive index is considered briefly in Exercise 2.11.

We have switched back to using sine and cosine functions to represent the fields here to ensure that we keep track of all the frequencies correctly. If we were to use the complex exponential representation, we would have to be careful to write:

$$\mathcal{E}(t) = \mathrm{Re}[\mathrm{e}^{-\mathrm{i}\omega t}] = \tfrac{1}{2}(\mathrm{e}^{-\mathrm{i}\omega t} + \mathrm{e}^{\mathrm{i}\omega t}).$$

This shows that the second-order nonlinear response generates an oscillating polarization at the sum and difference frequencies of the input fields according to:

$$\omega_{\text{sum}} = \omega_1 + \omega_2, \tag{2.62}$$

$$\omega_{\text{diff}} = |\omega_1 - \omega_2|. \tag{2.63}$$

The medium then reradiates at ω_{sum} and ω_{diff}, thereby emitting light at frequencies $(\omega_1 + \omega_2)$ and $|\omega_1 - \omega_2|$. The generation of these new frequencies by nonlinear processes is called **sum frequency mixing** and **difference frequency mixing**, respectively. If $\omega_1 = \omega_2$, the sum frequency is at twice the input frequency, and the effect is called **frequency doubling** or **second harmonic generation**. The nonlinear process can also work in reverse, splitting a beam of frequency ω into two beams with frequencies of ω_1 and ω_2, where $\omega = \omega_1 + \omega_2$. Table 2.2 lists some of the more important second-order nonlinear phenomena.

Second-order nonlinear processes can be represented by Feynman diagrams involving three photons as indicated in Fig. 2.6. Conservation of energy applies at each vertex. In sum-frequency mixing, two input photons at frequencies ω_1 and ω_2 are annihilated and a third one at frequency $\omega_1 + \omega_2$ is created, as shown in Fig. 2.6(a). In frequency doubling, the two input photons are at the same frequency, and the output photon is at double the input frequency, as shown in Fig. 2.6(b). Figure 2.6(c) shows the Feynman diagram for down conversion in which an input photon at the **pump** frequency ω_p is annihilated and two new photons at the **signal** and **idler** frequencies ω_s and ω_i, respectively, are created. Conservation of energy requires that

$$\omega_p = \omega_s + \omega_i. \tag{2.64}$$

Down-conversion processes are very important in quantum optics.

Figure 2.7 illustrates schematically two common applications of second-order nonlinear optics, namely second-harmonic generation and **parametric amplification**. In the former, a powerful pump beam at frequency ω generates a new beam at frequency 2ω by frequency doubling, as shown in Fig. 2.7(a). In the latter, a weak signal field at

(a)
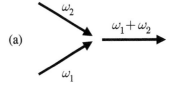

(b)
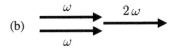

(c)
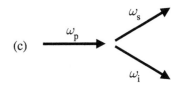

Fig. 2.6 Feynman diagrams for second-order nonlinear processes. (a) Sum frequency mixing. (b) Frequency doubling. (c) Down conversion.

Table 2.2 Second-order nonlinear effects. The second column lists the frequencies of the light beams incident on the nonlinear crystal, while the third gives the frequency of the output beam(s). For down conversion, the output frequencies must satisfy eqn 2.64. In the case of degenerate parametric amplification, the beam at frequency ω is amplified or de-amplified depending on its phase relative to the pump beam at frequency 2ω.

Effect	Input	Output
Frequency doubling	ω	2ω
Sum frequency mixing	ω_1, ω_2	$(\omega_1 + \omega_2)$
Difference frequency mixing	ω_1, ω_2	$|\omega_1 - \omega_2|$
Down conversion	ω_p	ω_s, ω_i
Degenerate parametric amplification	$2\omega, \omega$	ω

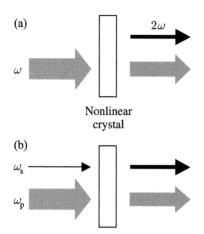

(a)

2ω

ω

Nonlinear
crystal

(b)

ω_s

ω_p

Fig. 2.7 (a) Second-harmonic generation. (b) Parametric amplification. In
(b), a weak signal beam at frequency ω_s
is amplified by nonlinear mixing with
a strong pump beam at frequency ω_p.
In both (a) and (b), the energy of the
beam generated by the nonlinear process is derived from the pump beam.
This is illustrated schematically by the
fact that the transmitted pump beam
is shown with a smaller arrow.

It is not necessary that $\mathcal{E}_j$ and $\mathcal{E}_k$
in eqn 2.66 should be derived from
different light beams. For example, in
frequency doubling there is only a single light beam incident on the nonlinear
crystal, and $\mathcal{E}_k$ and $\mathcal{E}_l$ are taken from
this one beam.

frequency ω_s experiences amplification when a powerful pump beam of
frequency ω_p is present, as shown in Fig. 2.7(b). In both processes,
the energy to generate the new beams is taken from the pump. The
parametric amplification process works by repeated difference-frequency
mixing. The nonlinear medium first generates idler photons at frequency
$\omega_i = (\omega_p - \omega_s)$, and these idler photons then generate photons at frequency $(\omega_p - \omega_i) = \omega_s$ by further mixing with the pump field. If the
phase-matching conditions discussed in Section 2.4.3 are satisfied, it is
possible to transfer power from the pump to the signal beam, generating
amplification at the signal frequency.

The nonlinear medium acts as a **degenerate parametric amplifier**
in the special case when

$$\omega_s = \omega_i = \omega_p/2. \tag{2.65}$$

In this case, the amplification experienced by the signal depends on its
phase relative to the pump. When the signal is in phase with the pump,
it is amplified, but deamplification occurs when the phase of the signal
is shifted by $\pm 90°$. The degenerate parametric amplifier therefore acts
as a **phase-sensitive amplifier** in which the 'parameter' is the phase
of the signal beam relative to the pump. (See Appendix B.) This effect is
important for the generation of quadrature squeezed states as discussed
in Section 7.9.

The well-defined axes of crystalline materials make it necessary to consider the directions in which the fields are applied. This type of behaviour
can be described by generalizing eqn 2.58 and writing the components
of the second-order nonlinear polarization $\boldsymbol{P}^{(2)}$ in the following form:

$$P_i^{(2)} = \epsilon_0 \sum_{j,k} \chi_{ijk}^{(2)} \mathcal{E}_j \mathcal{E}_k. \tag{2.66}$$

The quantity $\chi_{ijk}^{(2)}$ that appears here is the **second-order nonlinear
susceptibility tensor**, and the subscripts i, j, and k correspond to the
Cartesian coordinate axes x, y, and z. It will usually be convenient to
define these axes so that they coincide with the principal axes of the
crystal whenever this is possible.

The second-order nonlinear response defined in eqn 2.66 can be written in a contracted form involving the **nonlinear optical coefficient
tensor** d_{ij} by making use of the fact that $\chi_{xyz}^{(2)} \mathcal{E}_y \mathcal{E}_z$ must be the same as
$\chi_{xzy}^{(2)} \mathcal{E}_z \mathcal{E}_y$, etc. Written out explicitly, the components of the nonlinear
polarizations are given by:

$$\begin{pmatrix} P_x^{(2)} \\ P_y^{(2)} \\ P_z^{(2)} \end{pmatrix} = \begin{pmatrix} d_{11} & d_{12} & d_{13} & d_{14} & d_{15} & d_{16} \\ d_{21} & d_{22} & d_{23} & d_{24} & d_{25} & d_{26} \\ d_{31} & d_{32} & d_{33} & d_{34} & d_{35} & d_{36} \end{pmatrix} \begin{pmatrix} \mathcal{E}_x \mathcal{E}_x \\ \mathcal{E}_y \mathcal{E}_y \\ \mathcal{E}_z \mathcal{E}_z \\ 2\mathcal{E}_y \mathcal{E}_z \\ 2\mathcal{E}_z \mathcal{E}_x \\ 2\mathcal{E}_x \mathcal{E}_y \end{pmatrix}.$$

$$\tag{2.67}$$

By comparing this with eqn 2.66, we see that $d_{11} = \epsilon_0 \chi_{xxx}^{(2)}$, $d_{14} = \epsilon_0 \chi_{xyz}^{(2)}$, etc. Tables of optical properties of crystals usually quote the values of d_{ij} rather than $\chi_{ijk}^{(2)}$.

In many crystals the nonlinear optical coefficient tensor can be simplified considerably because the crystal symmetry requires that many of the terms are zero, and some others are the same. For example, the uniaxial nonlinear crystal KDP (potassium dihydrogen phosphate: KH_2PO_4) has tetragonal ($\overline{4}$2m) crystal symmetry. This means that the only nonzero components of d_{ij} are d_{14}, d_{25}, and d_{36}, with d_{14} equal to d_{25}. Similarly, in BBO (beta-barium borate: β-BaB_2O_4), which has rhombohedral (3m) symmetry, we have four different nonlinear coefficients, namely $d_{22} = -d_{21} = -d_{16}$, $d_{31} = d_{32}$, $d_{24} = d_{15}$, and d_{33}. All the other tensor elements are zero.

2.4.3 Phase matching

Nonlinear effects are usually small, and a long length of the nonlinear medium is therefore needed in order to obtain a useful nonlinear conversion efficiency. This only works effectively if the newly generated waves have the same phase relations between them throughout the whole crystal, so that the fields add together constructively. When this is achieved, we are in a regime called **phase matching**. Phase matching usually only occurs for very specific orientations of the nonlinear crystal.

The reason why phase matching is such an important issue in nonlinear optics is that the refractive index of the crystal invariably changes with the frequency. This means that the waves generated by the nonlinear interaction travel at different velocities from that of the pump beam. In frequency doubling, for example, the second harmonic waves at angular frequency 2ω will normally have a slower phase velocity than the fundamental waves at ω. This means that the waves generated at the front of the crystal will arrive at the back at a different time from the fundamental, implying that the waves generated at the back of the crystal will be out of phase with those from the front.

The phase-matching condition for the general case in which a beam of wave vector $\boldsymbol{k}$ is generated by mixing two beams with wave vectors $\boldsymbol{k}_1$ and $\boldsymbol{k}_2$ can be written in the form:

$$\boldsymbol{k} = \boldsymbol{k}_1 + \boldsymbol{k}_2. \tag{2.68}$$

This corresponds to momentum conservation in the nonlinear process, as indicated in Fig. 2.8. In the case of frequency doubling, eqn 2.68 reduces to:

$$\boldsymbol{k}^{2\omega} = 2\boldsymbol{k}^{\omega}. \tag{2.69}$$

It is apparent from eqn 2.23 that this condition is satisfied when:

$$n^{2\omega} = n^{\omega}. \tag{2.70}$$

It is possible to satisfy eqn 2.70 in a dispersive medium by making use of the birefringence of the nonlinear crystal. (See Section 2.1.4.)

The variation of the refractive index of a medium with frequency is called **dispersion**. The dispersion is described as 'normal' when the refractive index increases with frequency.

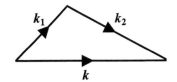

Fig. 2.8 The phase-matching condition for second-order nonlinearities is achieved when momentum conservation occurs in the nonlinear process.

For a positive uniaxial crystal with $n_e > n_o$, the relative polarizations for type I phase matching have to be reversed.

The derivation of eqn 2.71 may be found, for example, in Meschede (2004) or Yariv (1997).

For example, consider a negative uniaxial nonlinear crystal with $n_e < n_o$. With normal dispersion we have $n^{2\omega} > n^{\omega}$, and so we can achieve phase matching by propagating the second-harmonic waves as extraordinary rays and the fundamental waves as ordinary rays. In these circumstances, it can be shown that eqn 2.70 is satisfied when the optic axis of the crystal is set at an angle θ with respect to the propagation direction, where:

$$\frac{1}{(n_o^{\omega})^2} = \frac{\sin^2 \theta}{(n_e^{2\omega})^2} + \frac{\cos^2 \theta}{(n_o^{2\omega})^2} . \tag{2.71}$$

This type of phase matching is called **type I phase matching**. Another type of phase matching is also possible in which one of the fields at frequency ω is propagated as an ordinary ray, and the other as an extraordinary ray. This second type of phase matching is called **type II phase matching**. The critical angle for type II phase matching is different from that given in eqn 2.71.

Further reading

The principles of classical optics are covered in numerous texts, for example: Brooker (2003), Hecht (2002), or Smith and King (2000). There are also many texts available on electromagnetism, for example: Bleaney and Bleaney (1976) or Lorrain et al. (2000). The subject of nonlinear optics is covered in detail in Butcher and Cotter (1990), Shen (1984), or Yariv (1997).

Exercises

(2.1) A plane electromagnetic wave with a constant amplitude $\mathcal{E}$ and wave vector $\boldsymbol{k}$ propagating through an isotropic dielectric medium with relative permittivity ϵ_r may be written in the form:

$$\boldsymbol{\mathcal{E}}(z, t) = \boldsymbol{\mathcal{E}}_0 e^{i(\boldsymbol{k} \cdot \boldsymbol{r} - \omega t)} .$$

Show that the wave is transverse provided that $\epsilon_r \neq 0$. Discuss what might happen when $\epsilon_r = 0$.

(2.2) A linearly polarized laser beam propagating in air has an intensity of 10^6 W m^{-2}. Calculate the amplitudes of the electric and magnetic fields within the electromagnetic wave.

(2.3) Consider two electromagnetic waves of the same frequency propagating along the z-axis, one of which is linearly polarized along the x-axis, and the other along the y-axis. Let the amplitudes of the waves be $\mathcal{E}_{x0}$ and $\mathcal{E}_{y0}$, respectively, and the

phase of the wave polarized along the y-axis relative to that along the x-axis be $\Delta\phi$, so that we can write the components of the electric field as:

$$\mathcal{E}_x(z, t) = \mathcal{E}_{x0} \sin(kz - \omega t) ,$$
$$\mathcal{E}_y(z, t) = \mathcal{E}_{y0} \sin(kz - \omega t + \Delta\phi) ,$$
$$\mathcal{E}_z(z, t) = 0.$$

Describe the resulting polarization for the following cases:

(a) $\mathcal{E}_{x0} = \mathcal{E}_{y0}$, $\Delta\phi = 0$;

(b) $\mathcal{E}_{x0} = \sqrt{3}\mathcal{E}_{y0}$, $\Delta\phi = 0$;

(c) $\mathcal{E}_{x0} = \mathcal{E}_{y0}$, $\Delta\phi = +\pi/2$;

(d) $\mathcal{E}_{x0} = \mathcal{E}_{y0}$, $\Delta\phi = -\pi/2$;

(e) $\mathcal{E}_{x0} = \sqrt{3}\mathcal{E}_{y0}$, $\Delta\phi = +\pi/2$;

(f) $\mathcal{E}_{x0} = \mathcal{E}_{y0}$, $\Delta\phi = +\pi/4$.

(2.4) In the proposed 'LISA' gravity wave detection experiment,[1] a laser beam of wavelength 1064 nm will be collimated with a telescope of diameter 30 cm and fired towards a satellite at a distance of 5×10^6 km. Given that the power of the laser is 1 W, estimate the power collected on the distant satellite through a telescope of the same diameter.

(2.5) Calculate the coherence time for the 589.0 nm line of a sodium lamp operating at $100\,^\circ$C in the following two cases:

  (a) the line is Doppler-broadened;

  (b) the line is pressure-broadened with a full width at half maximum of 5 GHz.

(2.6) Compare the maximum path differences that will give rise to fringes in a Michelson interferometer when using:

  (a) the 546.1 nm line from a Doppler-broadened mercury lamp operating at $150\,^\circ$C,

  (b) the light from a stabilized He–Ne laser with a linewidth of 1 MHz operating at 632.8 nm.

(2.7) A second-order nonlinear crystal has a refractive index of 1.6 and a nonlinear susceptibility of $\sim 10^{-12}$ m V^{-1}. Estimate the optical intensity at which the magnitude of the second-order nonlinear polarization is the same as that of the linear polarization.

(2.8) A signal beam at 1400 nm is generated by parametric down conversion using a pump beam of wavelength 800 nm. What is the wavelength of the idler beam?

(2.9) A beam at 600 nm is to be produced by frequency doubling of the idler beam generated by parametric down conversion using a pump wavelength of 532 nm.

  (a) What is the wavelength of the signal beam?

  (b) What wavelength would be generated by difference-frequency mixing of the signal and idler?

(2.10) A potassium dihydrogen phosphate (KDP) crystal is used for frequency doubling of the radiation from a Nd : YAG laser operating at 1064 nm. Calculate the type I phase matching angle of the crystal if the relevant refractive indices for KDP are: $n_o(1064\,\text{nm}) = 1.494$, $n_o(532\,\text{nm}) = 1.512$, and $n_e(532\,\text{nm}) = 1.471$.

(2.11) (a) Explain why the second-order nonlinear susceptibility of a material that possesses inversion symmetry (i.e. is invariant under the transformation $r \to -r$) has to be zero.

  (b) Isotropic media such as gases, liquids, and glasses may be treated as materials with inversion symmetry on account of their lack of long-range order. It is therefore the case that $\chi^{(2)} = 0$ for these materials, and the dominant nonlinear effects arise from the third-order nonlinear susceptibility. One important consequence of third-order nonlinear effects is that the refractive index depends on the intensity. The **nonlinear refractive index** n_2 is defined according to:

  $$n(I) = n_0 + n_2 I,$$

  where n_0 is the linear refractive index and I is the optical intensity. On the assumption that $n_2 I \ll n_0$, show that:

  $$n_2 = \frac{1}{n_0^2 c \epsilon_0} \chi^{(3)}.$$

[1] 'LISA' is short for 'Laser Interferometer Space Antenna'. Further details of the principles of the experiment are given in Example 7.2 and Exercise 7.9.

3

Quantum mechanics

3.1 Formalism of quantum
    mechanics   26

3.2 Quantized states in
    atoms   35

3.3 The harmonic oscillator   41

3.4 The Stern–Gerlach
    experiment   43

3.5 The band theory of
    solids   45

Further reading   46

Exercises   46

The attribution of a *wave* function to *particles* with non-zero rest mass like electrons is called **first quantization**. The reverse procedure, namely the attribution of particle-like properties to wave fields (e.g. the electromagnetic field), is called **second quantization**. This chapter, in common with most introductory treatments of quantum mechanics, deals only with first quantization. The formalism of second quantization uses the number representation described in Chapter 8.

The subject of quantum optics deals with the application of quantum theory to optical phenomena. It is therefore appropriate to give a brief review of the main results of quantum mechanics in the introductory part of this book. The chapter begins with an overview of the general formalism of quantum mechanics, and then gives a brief summary of the quantum theory of atoms and harmonic oscillators. A discussion of the Stern–Gerlach experiment is then given, and the chapter concludes with a resumé of the band theory of solids. A short bibliography is provided for the benefit of readers who are unfamiliar with any of these topics.

3.1 Formalism of quantum mechanics

Quantum theory represents a fusion of two conflicting classical notions, namely wave and particle behaviour. On the one hand, we have to explain particle-like behaviour for phenomena that we usually consider as waves (e.g. light), and on the other, we have to explain wave-like phenomena associated with particles (e.g. electrons). An example of the former is the momentum exchange between electrons and light in the Compton effect, while an example of the latter is the diffraction of electrons by a crystal.

The basic formalism of quantum mechanics incorporates these two notions by assigning a **wave function** to particle-like objects such as electrons. The task then boils down to calculating the wave function and understanding how to find the values of important physically measurable quantities from it. In the subsections that follow, we give a short summary of how this is done, starting with the Schrödinger equation.

3.1.1 The Schrödinger equation

The physical state of a particle within a quantum system is determined by its wave function Ψ. This wave function is a function of both the position $\boldsymbol{r}$ and time t, and is defined so that the probability of finding the particle within a volume increment dV is given by:

$$\mathcal{P}(\boldsymbol{r},t)\,dV = |\Psi(\boldsymbol{r},t)|^2\,dV. \tag{3.1}$$

The equation of motion of the wave function is given by the **Schrödinger equation**:

$$\hat{H}\Psi(\boldsymbol{r},t) = i\hbar\frac{\partial}{\partial t}\Psi(\boldsymbol{r},t), \tag{3.2}$$

where $\hat{H}$ is the **Hamiltonian** operator. The Hamiltonian represents the total energy of the system:

$$\hat{H} = \hat{T} + \hat{V}, \tag{3.3}$$

where $\hat{T}$ and $\hat{V}$ are the kinetic and potential energy operators, respectively. The position operator is given by

$$\hat{r} = r, \tag{3.4}$$

while the momentum operator is:

$$\hat{p} = -i\hbar\nabla. \tag{3.5}$$

This means that the Hamiltonian operator can be rewritten in a more practical form for a single particle with kinetic energy $p^2/2m$:

$$\hat{H} = \frac{\hat{p}^2}{2m} + \hat{V}(r) = -\frac{\hbar^2}{2m}\nabla^2 + \hat{V}(r). \tag{3.6}$$

We may look for a solution in which the time and spatial parts of the wave function separate by writing:

$$\Psi(r, t) = \psi(r)\,\Theta(t). \tag{3.7}$$

On inserting this into eqn 3.2, we find that if

$$\Theta(t) = \exp(-iEt/\hbar), \tag{3.8}$$

where E is a separation variable, then the spatial part of the wave function satisfies the **time-independent Schrödinger equation**:

$$\hat{H}\psi(r) \equiv -\frac{\hbar^2}{2m}\nabla^2\psi(r) + \hat{V}(r)\psi(r) = E\psi(r). \tag{3.9}$$

The explicit time dependence of the wave function is therefore given by:

$$\Psi(r, t) = \psi(r)\exp(-iEt/\hbar). \tag{3.10}$$

The separation variable E that appears here gives the total energy of the system.

In one-dimensional systems, the position, momentum, and Hamiltonian operators simplify, respectively, to:

$$\hat{x} = x, \tag{3.11}$$

$$\hat{p}_x = -i\hbar\frac{\partial}{\partial x}, \tag{3.12}$$

$$\hat{H} = -\frac{\hbar^2}{2m}\frac{\partial^2}{\partial x^2} + \hat{V}(x), \tag{3.13}$$

and the one-dimensional time-independent Schrödinger equation becomes:

$$-\frac{\hbar^2}{2m}\frac{d^2\psi(x)}{dx^2} + \hat{V}(x)\psi(x) = E\psi(x). \tag{3.14}$$

In quantum mechanics, measurable quantities like the energy, momentum, position, or spin, are represented by **operators**. They are distinguished from the results of the measurements (which are only numbers) by the hat (^) symbol.

The wave functions that satisfy the time-independent Schrödinger equation are called the eigenfunctions of the Hamiltonian. Each function is labelled by a quantum number (or set of numbers) n, so that we can write:

$$\hat{H}\psi_n(\boldsymbol{r}) = E_n\psi_n(\boldsymbol{r}), \tag{3.15}$$

where E_n is the energy of the nth state. These eigenfunctions correspond to the quantized states of a system with Hamiltonian $\hat{H}$. The energy levels and states of a system are thus found by specifying the potential energy operator $\hat{V}$ in eqn 3.6 and then solving eqn 3.15 to find the set of functions $\psi_n(\boldsymbol{r})$ and their corresponding energies E_n.

3.1.2 Properties of wave functions

It is readily verified that if a wave function $\Psi(\boldsymbol{r},t)$ satisfies the Schrödinger equation, then any scalar multiple of $\Psi(\boldsymbol{r},t)$ also satisfies it with the same energy E. However, the probabilistic definition of the wave function given in eqn 3.1, and the fact that the particle must be somewhere, requires that:

$$\int |\Psi(\boldsymbol{r},t)|^2 \, \mathrm{d}^3\boldsymbol{r} = 1. \tag{3.16}$$

This property is called wave-function **normalization**, and serves as a condition to find the correct scalar to pre-multiply the functional part of $\Psi(\boldsymbol{r},t)$. Furthermore, with wave functions of the form given in eqn 3.10, the time dependence is eliminated on taking the square of the modulus, and so the normalization condition simplifies to:

$$\int |\psi(\boldsymbol{r})|^2 \, \mathrm{d}^3\boldsymbol{r} = 1, \tag{3.17}$$

for three-dimensional systems, and to:

$$\int_{-\infty}^{+\infty} |\psi(x)|^2 \, \mathrm{d}x = 1, \tag{3.18}$$

in one-dimensional systems.

Two wave functions $\psi(\boldsymbol{r})$ and $\varphi(\boldsymbol{r})$ are said to be **orthogonal** if:

$$\int \psi^*(\boldsymbol{r})\varphi(\boldsymbol{r}) \, \mathrm{d}^3\boldsymbol{r} = 0. \tag{3.19}$$

A wave function is said to be in a **superposition state** if it can be written as a non-factorizable linear combination of two or more orthogonal wave functions:

$$\Psi(\boldsymbol{r},t) = c_1\Psi_1(\boldsymbol{r},t) + c_2\Psi_2(\boldsymbol{r},t) + \cdots. \tag{3.20}$$

Superposition states are very important in quantum computing. (See Chapter 13.)

In mathematics, a function $u(x)$ is said to be an eigenfunction of a differential operator $F(x)$ if it satisfies the equation:

$$F(x)u(x) = \lambda u(x),$$

where λ is a number. The value of λ, which can be real or complex, is called the eigenvalue.

The word 'orthogonal' is usually first encountered in vector analysis. In that context, vectors $\boldsymbol{a}$ and $\boldsymbol{b}$ are said to be orthogonal if their scalar product is zero:

$$\boldsymbol{a} \cdot \boldsymbol{b} = 0.$$

In practice, this means that the vectors are at right angles to each other. In N-dimensional vector spaces where $N > 3$ (e.g. Hilbert space: see below), the definition of orthogonality based on the scalar product is still valid, even though it is not possible to give an equivalent simple conceptual interpretation.

The eigenfunctions of the Hamiltonian are orthogonal to each other and are normalized so that:

$$\int \psi_n^* \psi_{n'} \, \mathrm{d}^3 \boldsymbol{r} = \delta_{nn'}, \tag{3.21}$$

where $\delta_{nn'}$ is the **Kronecker delta function** defined by:

$$\delta_{nn'} = 1 \quad \text{if } n = n',$$
$$= 0 \quad \text{if } n \neq n'. \tag{3.22}$$

This property is called **orthonormality**.

The eigenfunctions of the Hamiltonian form a complete basis so that, at a given time (say $t = 0$), an arbitrary wave function Ψ can always be written in the form:

$$\Psi(\boldsymbol{r}, 0) = \sum_n c_n \psi_n(\boldsymbol{r}), \tag{3.23}$$

where c_n is a complex number. The orthonormality of the eigenfunctions implies that the normalization condition in eqn 3.17 is satisfied when (see Exercise 3.1):

$$\sum_n |c_n|^2 = 1. \tag{3.24}$$

At subsequent times, the wave function evolves according to:

$$\Psi(\boldsymbol{r}, t) = \sum_n c_n \psi_n(\boldsymbol{r}) \exp(-\mathrm{i}E_n t/\hbar), \tag{3.25}$$

where the exponential factors account for the time dependence of the individual eigenfunctions.

The orthonormality of the eigenfunctions readily lends itself to a geometric interpretation. We make an analogy between the eigenfunctions and the basis vectors $\{\boldsymbol{a}_1, \boldsymbol{a}_2, \ldots, \boldsymbol{a}_N\}$ of an N-dimensional vector space. These vectors have the property that:

$$\boldsymbol{a}_i \cdot \boldsymbol{a}_j = \delta_{ij}, \tag{3.26}$$

so that an arbitrary unit vector $\boldsymbol{\alpha}$ can be written in the form:

$$\boldsymbol{\alpha} = \sum_i c_i \boldsymbol{a}_i, \tag{3.27}$$

where

$$\sum_i |c_i|^2 = 1. \tag{3.28}$$

The formal similarity to eqns 3.21–3.24 explains why the analogy is valid. The vector space that the eigenfunctions span is called **Hilbert space**. The Hilbert space of a particular system has the same number of dimensions as the number of eigenfunctions of the Hamiltonian. In many instances, this will mean that the Hilbert space has an infinite number of dimensions.

The Kronecker delta function should not be confused with the **Dirac delta function** $\delta(x)$, which is defined by:

$$\int_{-\infty}^{+\infty} \delta(x) \, \mathrm{d}x = 1,$$

with $\delta(x) = 0$ everywhere except at $x = 0$.

Since we cannot visualize vector spaces with more than three dimensions, the concepts of Hilbert space are first understood by considering a system with just two or three eigenfunctions, and then generalizing.

The action of a quantum mechanical operator $\hat{O}$ on a system in a state ψ will, in general, have the effect of changing the state to a new one ψ'. In Hilbert space, these wave functions are represented by unit vectors pointing in different directions, and so the action of the operator can be perceived as performing a rotation to the vector. In mathematics, it is possible to represent rotation operations on vectors by matrices. This means that quantum mechanical operators can be represented by matrices, and this forms the conceptual basis for the **matrix representation** of quantum mechanics.

3.1.3   Measurements and expectation values

We have already pointed out that physically measurable quantities are represented by operators in quantum mechanics. Each operator $\hat{O}$ has its own set of eigenfunctions and eigenvalues, which are found by solving the corresponding eigenvalue equation:

$$\hat{O}\varphi_i = O_i\varphi_i. \tag{3.29}$$

As was the case for the Hamiltonian operator, the eigenfunctions $\{\varphi_i\}$ form a complete orthonormal basis, so that an arbitrary state ψ can always be expressed at a specific time as a sum according to:

$$\psi = \sum_i c_i\varphi_i, \tag{3.30}$$

where the coefficients $\{c_i\}$ are, in general, complex numbers.

Equation 3.29 is interpreted as meaning that if we make a measurement of the observable property represented by the operator $\hat{O}$ on a system prepared in the state φ_i, the result O_i will be obtained. In other words, if the particle enters the apparatus in one of the eigenstates of $\hat{O}$ (e.g. φ_i), the result will be equal to the corresponding eigenvalue (i.e. O_i). This is true no matter how many times the measurement is made.

It is one of the fundamental postulates of quantum mechanics that the result of a measurement of an observable property represented by the operator $\hat{O}$ is *always* equal to one of the eigenvalues of $\hat{O}$. The act of measurement 'collapses the wave function' in such a way that, if the particle enters the apparatus in an arbitrary state ψ, it emerges in the state with the eigenfunction corresponding to the result obtained. This means that if we obtain the result O_i, the particle will emerge with the wave function φ_i. Subsequent measurements will therefore always give the same result O_i.

The probability for obtaining the result O_i for a particle that enters the apparatus in an arbitrary state ψ is found by expanding the wave function over the eigenstates of $\hat{O}$ as in eqn 3.30. It is then apparent that the result O_i will be obtained with a probability equal to $|c_i|^2$. If the experiment is repeated many times on an ensemble of particles each prepared in the same state ψ, the average of the results will be equal to (see Exercise 3.2):

$$\langle \hat{O} \rangle = \int \psi^* \hat{O} \psi \, \mathrm{d}^3\boldsymbol{r}. \tag{3.31}$$

'Wave function collapse' is a central feature of the Copenhagen interpretation of quantum mechanics, and has been the subject of much debate over the years. The interpretation of quantum measurements is still a controversial topic. See, for example, J. S. Bell, *Physics World* **3**(8), 33 (1990) or A. J. Leggett, *Science* **307**, 871 (2005).

This average result is called the **expectation value** of the operator. The spread of the results about the expectation value can be obtained from the mean square variation (the **variance**):

$$(\Delta O)^2 = \int \psi^*(\hat{O} - \langle \hat{O} \rangle)^2 \psi \, \mathrm{d}^3 \boldsymbol{r}$$

$$= \langle \hat{O}^2 \rangle - \langle \hat{O} \rangle^2. \tag{3.32}$$

The variance represents the average deviation from the mean value, and can be understood as the uncertainty in the quantity that is being measured.

An important implication of the collapse of the wave function associated with the measurement process is that the act of measurement generally changes the state of the system. Therefore, in general it is not possible to measure a property and leave the system undisturbed in the process. Measurements on quantum systems are therefore *invasive*. The invasiveness of the measurement process is the fundamental principle underlying the security of quantum cryptography systems. (See Chapter 12.)

3.1.4   Commutators and the uncertainty principle

The **commutator** of two quantum mechanical operators $\hat{A}$ and $\hat{B}$ is defined by

$$[\hat{A}, \hat{B}] \equiv \hat{A}\hat{B} - \hat{B}\hat{A}. \tag{3.33}$$

The operators are said to commute if $[\hat{A}, \hat{B}] = 0$. When this occurs, the measurements corresponding to the properties represented by $\hat{A}$ and $\hat{B}$ do not interfere with each other, and it is possible to know their respective values simultaneously with complete accuracy. On the other hand, if the two operators do not commute, it will not be possible to measure their values with arbitrary accuracy at the same time. A measurement of one of the observables will, in general, change the result obtained in a subsequent measurement of the complementary observable.

One very important commutator is that of the position and momentum, namely $[\hat{x}, \hat{p}_x]$. We can work this out by operating on an arbitrary wave function as follows:

$$[\hat{x}, \hat{p}_x]\psi = (\hat{x}\hat{p}_x - \hat{p}_x\hat{x})\psi. \tag{3.34}$$

Then by inserting the definitions of the corresponding operators given in eqns 3.11 and 3.12 we find:

$$[\hat{x}, \hat{p}_x]\psi = -\mathrm{i}\hbar \left(x\frac{\partial \psi}{\partial x} - \frac{\partial(x\psi)}{\partial x} \right)$$

$$= \mathrm{i}\hbar\psi. \tag{3.35}$$

Hence we conclude that:

$$[\hat{x}, \hat{p}_x] = \mathrm{i}\hbar. \tag{3.36}$$

This shows that the position and momentum operators do not commute. A measurement of the position therefore adversely affects the value obtained in a subsequent measurement of the momentum, and vice versa.

The fact that the measurement of one property can interfere with the result for another property leads to uncertainty relationships between the corresponding variances. The most general version of the uncertainty principle states that:

The derivation of eqn 3.37 may be found, for example, in Gasiorowicz (1996).

$$(\Delta A)^2(\Delta B)^2 \geq \left|[\hat{A}, \hat{B}]\right|^2 /4, \tag{3.37}$$

where $(\Delta A)^2$ and $(\Delta B)^2$ are the variances of the measured values as given in eqn 3.32. This sets a minimum limit to the error with which the two properties can be measured, and shows that a very precise measurement of one variable increases the uncertainty of the other, and vice versa.

The general uncertainty relationship given in eqn 3.37 can be used to determine the product of the uncertainties of the position and momentum. On substituting from eqn 3.36, we then find that:

$$(\Delta x)^2(\Delta p_x)^2 \geq |i\hbar|^2/4, \tag{3.38}$$

which implies:

We shall see in Chapter 7 that the Heisenberg uncertainty principle is of central importance in determining the accuracy with which the amplitude and phase of a light field can be measured.

$$\Delta x \Delta p_x \geq \hbar/2. \tag{3.39}$$

This result is called the **Heisenberg uncertainty principle**. It quantifies the way in which a precise measurement of the position increases the uncertainty of a measurement of the momentum, and vice versa.

3.1.5   Angular momentum

Quantum mechanics admits of two different types of angular momentum. The first is called **orbital** and is the quantum mechanical counterpart of classical angular momentum. The second is called **spin** and has no classical counterpart.

There is a general convention in atomic physics that lower case letters refer to single electrons, while upper case letters refer to the equivalent resultant quantities for several electrons in the LS coupling regime. (See, for example, eqns 3.69 and 3.70.) The discussion of angular momentum given in this section refers to a single electron for simplicity, and hence all the quantities are in lower case, except in eqn 3.51. The argument can easily be generalized to multi-electron systems, or to other types of particles.

The definition of the operator for the orbital angular momentum $\hat{l}$ follows the classical one, with:

$$\hat{l} = \hat{r} \times \hat{p}. \tag{3.40}$$

Three related operators, namely $\hat{l}_x$, $\hat{l}_y$, and $\hat{l}_z$, are defined to represent the components along the three Cartesian axes. The operator for the squared magnitude of the angular momentum, namely $\hat{l}^2$, commutes with $\hat{l}_z$:

$$[\hat{l}^2, \hat{l}_z] = 0, \tag{3.41}$$

but the operators for the components do not commute with each other:

$$[\hat{l}_x, \hat{l}_y] = i\hbar\hat{l}_z, \quad \text{and cyclic permutations.} \tag{3.42}$$

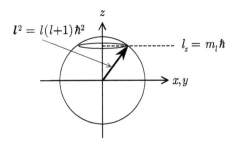

$l^2 = l(l+1)\hbar^2$

$l_z = m_l \hbar$

x, y

Fig. 3.1 Graphical representation of the orbital angular momentum. The angular momentum is represented as a vector of length $\sqrt{l(l+1)}\hbar$ with z-component $m_l\hbar$. The tip of the vector is to be thought of as randomly distributed around the circle centred on the z-axis with $l_z = m_l\hbar$.

Therefore, we can simultaneously know the precise values of the square of the orbital angular momentum and one of its components, but not the other two, except in isolated cases, e.g. when $l = 0$ (see below). This can be given a geometric interpretation as shown in Fig. 3.1.

Since $\hat{l}^2$ and $\hat{l}_z$ commute with each other, it is possible to find wave functions that are simultaneously eigenfunctions of both operators. On writing these functions as Y_{l,m_l}, where l and m_l are quantum numbers, we then find:

$$\hat{l}^2 \, Y_{l,m_l} = l(l+1)\hbar^2 \, Y_{l,m_l}, \tag{3.43}$$

and

$$\hat{l}_z \, Y_{l,m_l} = m_l\hbar \, Y_{l,m_l}. \tag{3.44}$$

The functions Y_{l,m_l} are called **spherical harmonics** and take the form:

$$Y_{l,m_l}(\theta, \varphi) = C_l^{m_l} \, P_l^{m_l}(\cos\theta) \, e^{im_l\varphi}, \tag{3.45}$$

where θ and φ are the polar and azimuthal angles of the spherical coordinate system, $C_l^{m_l}$ is a normalization constant, and $P_l^{m_l}(\cos\theta)$ is the associated Legendre function. These functions are discussed further in Section 3.2. At this stage we simply state that l is called the **orbital quantum number** and can take any positive integer value from 0 upwards, while m_l is called the **magnetic quantum number** and can take integer values from $-l$ to $+l$.

The operators and wave functions for the spin angular momentum s are defined in analogy with eqns 3.41–3.44. There are four operators: $\hat{s}^2$ for the magnitude, and $\hat{s}_x$, $\hat{s}_y$, and $\hat{s}_z$ for the components. The commutators are given by:

$$[\hat{s}^2, \hat{s}_z] = 0, \tag{3.46}$$

and

$$[\hat{s}_x, \hat{s}_y] = i\hbar\hat{s}_z, \tag{3.47}$$

and cyclic permutations.

The fact that $\hat{s}^2$ and $\hat{s}_z$ commute means that simultaneous eigenfunctions can exist. Writing these functions as χ, we then have:

$$\hat{s}^2 \chi = s(s+1)\hbar^2\chi, \tag{3.48}$$

$$\hat{s}_z\chi = m_s\hbar\chi, \tag{3.49}$$

The spin operators cannot be represented as functions of spatial coordinates and time, and neither can their eigenfunctions. Instead, a matrix representation must be used. The matrices that represent the components of the spin for $s = 1/2$ along the axes are called the Pauli spin matrices.

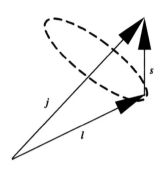

Fig. 3.2 The total angular momentum is found from the resultant of the orbital and spin angular momenta.

We are using upper case letters here because we might be referring to a multi-particle system.

where s and m_s are the quantum numbers for the spin magnitude and its z component, respectively, with m_s taking values from $-s$ to $+s$ in integer steps. Experiments on electrons, protons, and neutrons indicate that they each have $s = 1/2$, and therefore $m_s = \pm 1/2$. The m_s states of $\pm 1/2$ are often called spin 'up' and 'down', respectively, which is an allusion to the sign of the z-component of the spin vector.

When a particle (e.g. an electron in an atom) possesses both orbital and spin angular momentum, the total angular momentum $\boldsymbol{j}$ is defined as the resultant (see Fig. 3.2):

$$\boldsymbol{j} = \boldsymbol{l} + \boldsymbol{s}. \tag{3.50}$$

In this case, the rule is that the quantum number j associated with $\boldsymbol{j}^2$ can take any integer value from $|l - s|$ to $(l + s)$. For each allowed value of j, the components along the z-axis have magnitude $m_j \hbar$, where m_j can take values in integer steps from $-j$ to $+j$.

The rule applied to the total angular momentum of a single particle is an example of a more general rule for the addition of quantum-mechanical angular momenta. This rule is as follows. Suppose $\boldsymbol{J}$ is the resultant of two angular momenta $\boldsymbol{J}_1$ and $\boldsymbol{J}_2$ according to:

$$\boldsymbol{J} = \boldsymbol{J}_1 + \boldsymbol{J}_2. \tag{3.51}$$

If the quantum numbers corresponding to $\boldsymbol{J}$, $\boldsymbol{J}_1$, and $\boldsymbol{J}_2$ are J, J_1, and J_2, respectively, then J can take all the values in integer steps from $|J_1 - J_2|$ up to $(J_1 + J_2)$. For each value of J, the quantum number for the z-component, namely M_J, can take values in integer steps from $-J$ to $+J$. The general rule can be used, for example, for finding the resultant spin and orbital angular momentum of multi-electron atoms, and then to find the total angular momentum of the whole atom.

3.1.6   Dirac notation

We mentioned at the end of Section 3.1.2 that there are two equivalent representations of quantum mechanical systems. On the one hand, we have wave functions that evolve according to the Schrödinger equation, and on the other, we have state vectors in Hilbert space that are governed by matrix mechanics. A convenient shorthand notation was invented by Dirac to represent the state vectors of Hilbert space and their properties. Since this notation has proven to be so useful, it is now widely applied by analogy also to represent the wave functions in the Schrödinger picture, and this is the policy adopted throughout this book.

In the Dirac notation scheme, wave functions can be represented by **ket** vectors:

$$\psi \equiv |\psi\rangle. \tag{3.52}$$

The words 'bra' and 'ket' are derived from the two halves of the word 'bracket'.

The corresponding complex conjugate is represented by a **bra** vector:

$$\psi^* \equiv \langle\psi|. \tag{3.53}$$

Eigenfunctions are usually identified just by the quantum numbers that define them:

$$\psi_{lmn...} \equiv |l, m, n, \ldots\rangle. \tag{3.54}$$

The closing of the 'bra-ket' (i.e. bracket) implies integration. We can therefore represent **overlap integrals** according to:

$$\langle\psi|\varphi\rangle \equiv \int \psi^*\varphi \, \mathrm{d}^3\boldsymbol{r}. \tag{3.55}$$

Similarly, the expectation values defined in eqn 3.31 can be written:

$$\langle\psi|\hat{O}|\psi\rangle \equiv \int \psi^*\hat{O}\psi \, \mathrm{d}^3\boldsymbol{r}. \tag{3.56}$$

Finally, **matrix elements** can be defined according to:

$$\langle n|\hat{O}|m\rangle \equiv \int \psi_n^*\hat{O}\psi_m \, \mathrm{d}^3\boldsymbol{r}. \tag{3.57}$$

Although the use of Dirac notation does not actually make the calculations any simpler, its compact form does make it very convenient for writing equations. That is why it is extensively employed in many quantum mechanics texts.

> In Hilbert space, the closing of the bracket is equivalent to projecting one state vector onto the other. The projection will be zero if the vectors are orthogonal. This is one of the reasons why wave functions with zero overlap are called 'orthogonal'.

3.2 Quantized states in atoms

In many examples throughout this book, we shall be dealing with the interaction of light beams with quantized states of atoms. It is therefore important to give a brief review of the way in which the quantum states of atoms are classified, and the properties that are associated with these states.

3.2.1 The gross structure

The quantized states of atoms are calculated by solving the Schrödinger equation for the atom to find the wave functions and energies. The procedure usually adopted is to start by considering the interactions between the electrons and the nucleus, together with the Coulomb repulsion between the electrons. The energy-level scheme that is obtained in this way is called the **gross structure** of the atom.

The starting model for understanding the gross structure is the hydrogen atom. This has one electron orbiting around the nucleus, and is the simplest atom that we can consider. Owing to the spherical symmetry, it is convenient to work in spherical polar coordinates (r, θ, φ). The time-independent Schrödinger equation is then given by:

$$\left(-\frac{\hbar^2}{2m}\boldsymbol{\nabla}^2 - \frac{Ze^2}{4\pi\epsilon_0 r}\right)\psi(r, \theta, \varphi) = E\,\psi(r, \theta, \varphi), \tag{3.58}$$

where the second term represents the Coulomb interaction between the electron and the nucleus which is assumed to have charge $+Ze$, e being

> For hydrogen itself where the nucleus just consists of a proton, we obviously take $Z = 1$ and $m_N = m_p$. The theory, however, applies to other hydrogenic atoms such as He$^+$, Li^{2+}, Be^{3+}, ... that possess only one electron orbiting the nucleus.

the modulus of the electron charge. The mass m that appears here is the **reduced mass** given by:

$$\frac{1}{m} = \frac{1}{m_0} + \frac{1}{m_N},\tag{3.59}$$

where m_0 is the mass of the electron and m_N is the mass of the nucleus. When the wave function is written in the form:

$$\psi(r, \theta, \varphi) = R(r)\,Y(\theta, \varphi),\tag{3.60}$$

solutions are found with $Y(\theta, \varphi)$ equal to one of the spherical harmonic functions defined in eqns 3.43–3.45. A list of the functional forms of the first few spherical harmonics is given in Table 3.1.

The energy spectrum of the hydrogen atom is determined by the **principal quantum number** n:

$$E_n = -\frac{me^4}{2\hbar^2} \frac{1}{(4\pi\epsilon_0)^2} \frac{Z^2}{n^2},\tag{3.61}$$

where n can take any integer value from 1 to ∞. The energy is often written more simply in the form:

$$E_n = -\frac{m}{m_0} \frac{Z^2}{n^2} R_\infty hc,\tag{3.62}$$

Table 3.1 Spherical harmonic functions.

l	m_l	$Y_{l,m_l}(\theta, \varphi)$
0	0	$(1/4\pi)^{1/2}$
1	0	$(3/4\pi)^{1/2}\cos\theta$
1	± 1	$\mp(3/8\pi)^{1/2}\sin\theta\,\mathrm{e}^{\pm i\varphi}$
2	0	$(5/16\pi)^{1/2}(3\cos^2\theta - 1)$
2	± 1	$\mp(15/8\pi)^{1/2}\sin\theta\cos\theta\,\mathrm{e}^{\pm i\varphi}$
2	± 2	$(15/32\pi)^{1/2}\sin^2\theta\,\mathrm{e}^{\pm 2i\varphi}$

where $R_\infty = 1.0974 \times 10^5$ cm^{-1} is the **Rydberg constant**, and $(R_\infty hc) = 13.61$ eV is equal to the binding energy for the $n = 1$ ground state of a hydrogen atom with a singly charged, infinitely heavy nucleus.

For a given value of n, the orbital quantum number l can take integer values from 0 to $(n - 1)$. For each pair of values of n and l, the radial wave function $R_{nl}(r)$ is of the form:

$$R_{nl}(r) = C_{nl}\,(r/a)^l\,F(r/a)\,\mathrm{e}^{-r/a},\tag{3.63}$$

where C_{nl} is a normalization constant, and $F(x)$ is a function related to the associated Laguerre polynomials. The length parameter a that enters here is given by:

$$a = \frac{n}{Z} \frac{m_0}{m} a_0,\tag{3.64}$$

where a_0 is the **Bohr radius** (5.29×10^{-11} m) given by:

$$a_0 = \frac{4\pi\epsilon_0\hbar^2}{m_0 e^2}.\tag{3.65}$$

Atomic energies are often quoted in **wave-number** (cm^{-1}) units. The conversion factors between wave number, energy, and frequency units are given in the table on the inside of the book jacket.

Table 3.2 lists the functional form of $R_{nl}(r)$ for the first six possibilities.

The Hamiltonian for an N-electron atom with nuclear charge $+Ze$ can be written in the form:

$$\hat{H} = \sum_{i=1}^{N}\left(-\frac{\hbar^2}{2m}\boldsymbol{\nabla}_i^2 - \frac{Ze^2}{4\pi\epsilon_0 r_i}\right) + \sum_{i>j}^{N}\frac{e^2}{4\pi\epsilon_0 r_{ij}}.\tag{3.66}$$

The subscripts i and j refer to individual electrons and $r_{ij} = |\boldsymbol{r}_i - \boldsymbol{r}_j|$. The first summation accounts for the kinetic energy of the electrons and

Table 3.2 Radial wave functions of a hydrogenic atom with a nuclear charge of $+Ze$ and an infinitely heavy nucleus. a_0 is the Bohr radius defined in eqn 3.65 (5.29×10^{-11} m). The wave functions are normalized so that $\int_{r=0}^{\infty} R_{nl}^* R_{nl}\, r^2\, \mathrm{d}r = 1$.

n	l	$R_{nl}(r)$
1	0	$(Z/a_0)^{3/2}\, 2 \exp(-Zr/a_0)$
2	0	$(Z/2a_0)^{3/2}\, 2\,(1 - Zr/2a_0) \exp(-Zr/2a_0)$
2	1	$(Z/2a_0)^{3/2}\, (2/\sqrt{3})\,(Zr/2a_0)\, \exp(-Zr/2a_0)$
3	0	$(Z/3a_0)^{3/2}\, 2\,[1 - (2Zr/3a_0) + (2/3)(Zr/3a_0)^2]\, \exp(-Zr/3a_0)$
3	1	$(Z/3a_0)^{3/2}\, (4\sqrt{2}/3)(Zr/3a_0)(1 - Zr/6a_0)\, \exp(-Zr/3a_0)$
3	2	$(Z/3a_0)^{3/2}\, (2\sqrt{2}/3\sqrt{5})\,(Zr/3a_0)^2\, \exp(-Zr/3a_0)$

their Coulomb interaction with the nucleus, while the second accounts for the electron–electron repulsion. It is not possible to find an exact solution to the Schrödinger equation with a Hamiltonian of the form given by eqn 3.66 because the electron–electron repulsion term is comparable in magnitude to the first summation. The description of multi-electron atoms therefore usually starts with the **central field approximation** in which we rewrite the Hamiltonian of eqn 3.66 in the form:

$$\hat{H} = \sum_{i=1}^{N} \left(-\frac{\hbar^2}{2m} \nabla_i^2 + V_{\text{central}}(r_i) \right) + V_{\text{residual}}, \tag{3.67}$$

A field is described as 'central' if the potential energy has spherical symmetry about the origin, so that $V(\boldsymbol{r})$ only depends on r. Further details about the central field approximation may be found in Foot (2005) or Woodgate (1980).

where V_{central} is the central field and V_{residual} is the **residual electrostatic interaction**. In this approximation, we split the potential that a particular electron experiences due to the nucleus and the other electrons into the central term $V_{\text{central}}(r_i)$ and a non-central term. The non-central terms are lumped together into the residual electrostatic interaction, and it is hoped that this term in the Hamiltonian will be small compared to the central one. In fact, this is a good approximation for most atoms, because closed subshells are spherically symmetric.

The reason why the central field approximation is useful is that the spherical harmonic wave functions discussed in Section 3.1.5 are eigenfunctions of all central potentials, not just the nuclear Coulomb field. This means that the description of multi-electron atoms can be based on the states of the hydrogen atom, albeit with different forms of the radial wave function $R_{nl}(r_i)$.

Within the central field approximation, each electron is specified by four quantum numbers, namely $\{n, l, m_l, m_s\}$. (We do not need to specify the spin quantum number s because it is always equal to $1/2$.) Since electrons are indistinguishable fermions, the N-electron wave function must be antisymmetric under every interchange of the electron labels.

Table 3.3 Spectroscopic notation used to designate orbital states.

l	0	1	2	3	4	$\cdots$
Notation	s	p	d	f	g	$\cdots$

Table 3.4 Electronic configurations of the ground states of the first 11 elements of the periodic table. Z is the atomic number.

Element	Z	Configuration
H	1	$1s^1$
He	2	$1s^2$
Li	3	$1s^2\,2s^1$
Be	4	$1s^2\,2s^2$
B	5	$1s^2\,2s^2\,2p^1$
C	6	$1s^2\,2s^2\,2p^2$
N	7	$1s^2\,2s^2\,2p^3$
O	8	$1s^2\,2s^2\,2p^4$
F	9	$1s^2\,2s^2\,2p^5$
Ne	10	$1s^2\,2s^2\,2p^6$
Na	11	$1s^2\,2s^2\,2p^6\,3s^1$

LS coupling is also called **Russell–Saunders coupling**. The magnitude of spin–orbit interactions generally increases with Z, and so it is possible that the LS coupling approximation is no longer valid in atoms with large Z. In the extreme case where the spin–orbit interaction is the dominant perturbation, a different type of angular-momentum coupling occurs, called jj coupling. In this scheme the spin and orbital angular momenta of the individual electrons are added together first with:

$$\boldsymbol{j}_i = \boldsymbol{l}_i + \boldsymbol{s}_i,$$

and then the total angular momentum is found by summing the $\boldsymbol{j}$s of the individual electrons according to $\boldsymbol{J} = \sum_i \boldsymbol{j}_i$. In fact, only a small number of atoms exhibit pure jj coupling. See Foot (2005) or Woodgate (1980) for further details.

This is achieved by writing the wave function as a **Slater determinant**:

$$\Psi = \frac{1}{\sqrt{N!}} \begin{vmatrix} \psi_\alpha(1) & \psi_\alpha(2) & \cdots & \psi_\alpha(N) \\ \psi_\beta(1) & \psi_\beta(2) & \cdots & \psi_\beta(N) \\ \vdots & \vdots & \ddots & \vdots \\ \psi_\nu(1) & \psi_\nu(2) & \cdots & \psi_\nu(N) \end{vmatrix}, \qquad (3.68)$$

where $\{\alpha, \beta, \ldots, \nu\}$ each represent a set of quantum numbers $\{n, l, m_l, m_s\}$ for the individual electrons, and $\{1, 2, \ldots, N\}$ are the electron labels. Since the determinant is zero if any two rows are equal, it must be the case that each electron in the atom has a unique set of quantum numbers. This conclusion is frequently called the **Pauli exclusion principle**.

The **electronic configuration** of the atom is determined by the quantum states occupied by the electrons. To a first approximation, the energy of the states depends only on n and l, and increases with both quantum numbers, depending most strongly on the value of n. The states are therefore labelled by n and l, following the notation given in Table 3.3. Each of these nl states is called a **shell**. 's' shells have $m_l = 0$ and $m_s = \pm 1/2$, and can therefore hold two electrons. 'p' shells have $m_l = -1$, 0, or $+1$ and $m_s = \pm 1/2$, and can therefore hold six electrons, etc. The configurations of the ground states of the first 11 elements are given in Table 3.4. The configurations of the remaining elements are given in the periodic table on the inside of the jacket of the book. Filled shells have no net orbital or spin angular momentum because there are equal numbers of positive and negative m_l and m_s values. The electrons in the outermost shell are called **valence electrons**.

A particularly important class of elements in quantum optics experiments is the **alkali metals** from group 1A of the periodic table. These have one valence electron in an 's' shell outside filled inner shells. Since there is a large jump in the energy on moving from one shell to the next, the alkalis behave as if they are quasi-one-electron atoms. However, in contrast to true one-electron atoms (i.e. hydrogenic systems), the gross energy of the single valence electron depends on l as well as n.

In multi-electron atoms with more than one valence electron, the spins and orbital angular momenta can combine in several different ways to form their resultants. The way this occurs depends on the hierarchy of interactions that have been ignored in the central field approximation. The two most important of these are the residual electrostatic interaction discussed above and the **spin–orbit interaction**. In many atoms the residual electrostatic interaction is the dominant perturbation, and this leads to LS **coupling** in which the total spin and orbital momenta are determined from:

$$\boldsymbol{L} = \sum_i \boldsymbol{l}_i, \qquad (3.69)$$

$$\boldsymbol{S} = \sum_i \boldsymbol{s}_i, \qquad (3.70)$$

where the lower and upper case vectors refer to individual electrons and the resultants, respectively, and the vector additions are performed according to the rules given in Section 3.1.5. Once the different possible values of L and S have been determined, it is then possible to work out the total angular momentum according to eqn 3.51. The end result is that for each electronic configuration of the valence electrons, we obtain a set of states labelled by the quantum numbers L, S, and J. The LS states are called **atomic terms**, and the states of different J corresponding to a particular LS term are called **levels**. The levels are written in spectroscopic notation as:

$$|L, S, J\rangle \equiv {}^{(2S+1)}\mathrm{L}_J, \qquad (3.71)$$

where the capital roman letter L indicates the value of L according to the convention given in Table 3.3. The factor $(2S+1)$ indicates the spin multiplicity: there are $(2S+1)$ M_S states available for each value of S.

In alkali atoms, there is just a single valence electron, and the value of S is always equal to 1/2. The value of L varies according to the shell of the electron. In the ground state, the electron is in the ns shell and therefore has $L = 0$. This just gives one possibility for J, namely 1/2, and so the ground state is a ${}^2\mathrm{S}_{1/2}$ term. In the excited states, the electron can occupy shells with higher values of l and two values of J are allowed, namely $L - 1/2$ and $L + 1/2$. These two J states have a small splitting caused by the spin–orbit interaction. (See below.)

In atoms with two valence electrons, there are two possible values of S, namely 0 and 1. Terms with $S = 0$ and $S = 1$ are called **singlets** and **triplets**, respectively. The energies of singlet and triplet terms with the same values of L and J differ owing to the exchange interaction, which originates from the residual electrostatic interaction.

Alkali metals have only one valence electron, and it should therefore be appropriate to use lower case lettering to denote the levels. However, atomic physicists tend to extend the spectroscopic notation given in eqn 3.71 to one-electron atoms, including hydrogen, even though it is not strictly consistent with the other conventions of notation.

3.2.2   Fine and hyperfine structure

The gross structure of the atom is calculated by including only the principal Coulomb interactions within the atom in the Hamiltonian. There are other smaller interactions which cause energy shifts to the gross structure terms, giving rise to **fine structure** in the optical spectra. In general, the fine-structure energy shifts are smaller than the gross structure energies by a factor of $\sim Z^2\alpha^2$, where $\alpha = e^2/4\pi\epsilon_0\hbar c \approx 1/137$ is the **fine-structure constant**.

In the LS-coupling limit, the most important of the fine-structure interactions is the **spin–orbit interaction**. This can be understood in simple terms as an interaction between the magnetic field generated by the orbital motion of the electrons and the magnetic dipole associated with the spin. The energy shift due to the spin–orbit interaction can be calculated by finding the expectation value of the effective Hamiltonian:

$$\hat{H}_{\mathrm{so}} = C\boldsymbol{L} \cdot \boldsymbol{S}, \qquad (3.72)$$

where the value of C depends on the electronic configuration. For specific values of L and S, we can use the fact that:

$$\boldsymbol{J}^2 = (\boldsymbol{L} + \boldsymbol{S})^2, \tag{3.73}$$

and therefore that:

$$\langle \boldsymbol{L} \cdot \boldsymbol{S} \rangle = \frac{1}{2} \langle \boldsymbol{J}^2 - \boldsymbol{L}^2 - \boldsymbol{S}^2 \rangle,$$

$$= \frac{\hbar^2}{2} [J(J+1) - L(L+1) - S(S+1)]. \tag{3.74}$$

The energy shifts are therefore of the form:

$$\Delta E_{\text{so}} = C' \, [J(J+1) - L(L+1) - S(S+1)], \tag{3.75}$$

which implies that the different J terms obtained from the same values of L and S have different energies. For example, in the first excited state of sodium, the valence electron is promoted from the 3s to the 3p shell. This gives two terms, namely $^2\text{P}_{3/2}$ and $^2\text{P}_{1/2}$, which are separated by $17 \, \text{cm}^{-1}$ (2.1×10^{-3} eV). The fine-structure splitting should be compared to the $\sim 17\,000 \, \text{cm}^{-1}$ (~ 2.1 eV) energy difference between the 3p and 3s levels. (See Fig. 3.3.)

In high-resolution spectroscopy it is also necessary to consider the interaction between the electrons and the nuclear spin ($\boldsymbol{I}$). The angular momentum of the electrons creates a magnetic field at the nucleus which is proportional to $\boldsymbol{J}$, and this interacts with the magnetic dipole of the nucleus which is proportional to $\boldsymbol{I}$. The small energy shifts that result give rise to **hyperfine structure** in the spectra, with:

$$\Delta E_{\text{HFS}} = A(J) \, \langle \boldsymbol{I} \cdot \boldsymbol{J} \rangle. \tag{3.76}$$

We now have to define the total angular momentum of the atom with the nuclear spin included as $\boldsymbol{F}$, where:

$$\boldsymbol{F} = \boldsymbol{I} + \boldsymbol{J}, \tag{3.77}$$

in the weak field limit. By following a procedure similar to eqns 3.73–3.74, we then readily find that:

$$\Delta E_{\text{HFS}} = A(J) \frac{\hbar^2}{2} [F(F+1) - J(J+1) - I(I+1)]. \tag{3.78}$$

The hyperfine interactions are smaller again than the fine structure, with splittings typically in the range $\sim 0.1 \, \text{cm}^{-1}$ ($\sim 10^{-5}$ eV). For example, in hydrogen the nucleus consists of just a single proton, and we therefore have $I = 1/2$. For the 1s $^2\text{S}_{1/2}$ ground state, we then have $F = 0$ or 1. These two hyperfine levels are split by $0.0475 \, \text{cm}^{-1}$ (5.9×10^{-6} eV). (See Fig. 3.4.) Transitions between these levels occur at $1420 \, \text{MHz}$ ($\lambda = 21 \, \text{cm}$), and are very important in radio astronomy.

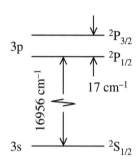

Fig. 3.3 Fine structure of the 3p state in sodium.

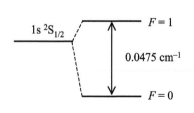

Fig. 3.4 Hyperfine structure of the 1s ground state of hydrogen.

3.2.3   The Zeeman effect

Each atomic level with quantum numbers L, S, and J consists of $(2J + 1)$ degenerate M_J states. This degeneracy can be lifted by applying a magnetic field, and the splitting of the atomic levels into magnetic sublevels is called the **Zeeman effect**.

The energy of an atom in a magnetic field $\boldsymbol{B}$ applied along the z-axis is given by

$$\Delta E = -\boldsymbol{\mu} \cdot \boldsymbol{B} = -\mu_z B_z, \tag{3.79}$$

where $\boldsymbol{\mu}$ and μ_z are the magnetic dipole of the atom and its z-component, respectively. μ_z is given by

$$\mu_z = -g_J \mu_{\mathrm{B}} M_J, \tag{3.80}$$

where g_J is the Landé g-factor:

$$g_J = 1 + \frac{J(J+1) + S(S+1) - L(L+1)}{2J(J+1)}. \tag{3.81}$$

We therefore find that:

$$\Delta E = g_J \mu_{\mathrm{B}} B_z M_J. \tag{3.82}$$

This splits the otherwise degenerate M_J states into a manifold of sublevels with separations that increase linearly with the field. Note that $g_J = 1$ when $S = 0$ (pure orbital angular momentum: $J = L$), while $g_J = 2$ when $L = 0$ (pure spin angular momentum: $J = S$).

Figure 3.5 illustrates the Zeeman splitting of the $^2\mathrm{P}_{3/2}$ and $^2\mathrm{P}_{1/2}$ levels of an np state of an alkali atom, such as, for example, the 3p state of sodium shown in Fig. 3.3. These two levels have Landé g-factors of 4/3 and 2/3 respectively.

In high-resolution spectroscopy, the Zeeman splitting of the hyperfine levels must also be considered. In this case, the energy shift at weak magnetic fields is given by

$$\Delta E = g_F \mu_{\mathrm{B}} B_z M_F, \tag{3.83}$$

with

$$g_F \approx g_J \frac{F(F+1) + J(J+1) - I(I+1)}{2F(F+1)}. \tag{3.84}$$

The Zeeman effect therefore yields $2F + 1$ equidistant components. At higher field strengths, the field breaks the hyperfine coupling of $\boldsymbol{J}$ to $\boldsymbol{I}$ given by eqn 3.77, and the conventional Zeeman effect of eqn 3.82 is observed.

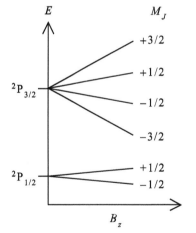

Fig. 3.5 Zeeman splitting of the $^2\mathrm{P}_{3/2}$ and $^2\mathrm{P}_{1/2}$ levels of an alkali atom.

3.3   The harmonic oscillator

The motion of many wave-like and other periodic systems in nature are well described in terms of the formalism of simple harmonic oscillators.

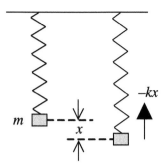

Fig. 3.6 A mass m dangling from a spring of spring constant k experiences a restoring force of $-kx$ when displaced by a distance x from the equilibrium position.

This is particularly so in quantum optics, because the quantum theory of light starts from the quantum harmonic oscillator.

Let us consider a typical example of a harmonic oscillator, namely an oscillating mass dangling from a spring, as shown in Fig. 3.6. The classical equations of motion for the mass are:

$$m\frac{d^2x}{dt^2} = -kx, \tag{3.85}$$

$$p_x = m\frac{dx}{dt}, \tag{3.86}$$

where m is the mass, x is the displacement from the equilibrium position, k is the spring constant, and p_x is the linear momentum. The solutions are of the form:

$$x(t) = x_0 \sin \omega t, \tag{3.87}$$

$$p_x(t) = p_0 \cos \omega t, \tag{3.88}$$

where $p_0 = m\omega x_0$, and

$$\omega = \sqrt{\frac{k}{m}}. \tag{3.89}$$

The potential energy stored in the spring is given by:

$$V(x) = \frac{1}{2}kx^2 = \frac{1}{2}m\omega^2 x^2. \tag{3.90}$$

The quantization of the oscillator is achieved by using this potential in the time-independent Schrödinger equation (eqn 3.14) to find the wave functions and energies of the system. Hence we must solve:

$$-\frac{\hbar^2}{2m}\frac{d^2\psi(x)}{dx^2} + \frac{1}{2}m\omega^2 x^2 \psi(x) = E\psi(x). \tag{3.91}$$

The solutions are of the form:

$$\psi_n(x) = u_n(x)\exp(-m\omega x^2/2\hbar), \tag{3.92}$$

with energy:

$$E_n = \left(n + \frac{1}{2}\right)\hbar\omega, \tag{3.93}$$

where n is an integer ≥ 0. The functions $u_n(x)$ are related to the Hermite polynomials, and are tabulated in Table 3.5. The energy-level spectrum is sketched in Fig. 3.7.

In the quantum theory of light, an important quantity is the uncertainty principle between the position and momentum of the oscillator. On evaluating the uncertainties according to eqn 3.32 for the wave functions given in eqn 3.92, we find for the ground state with $n = 0$ (see Exercise 3.10):

$$\Delta x \Delta p_x = \hbar/2, \tag{3.94}$$

Table 3.5 The polynomial part $u_n(x)$ of the eigenfunctions of the simple harmonic oscillator as defined in eqn 3.92. $a = (\hbar/m\omega)^{1/2}$ and has the dimensions of length, while $H_n(x)$ is the Hermite polynomial of order n. Note that the prefactor of $u_n(x)$ must scale as $1/\sqrt{a}$ in order to ensure correct normalization.

n	$u_n(x)$
0	$(1/a\sqrt{\pi})^{1/2}$
1	$(1/2a\sqrt{\pi})^{1/2}\, 2(x/a)$
2	$(1/8a\sqrt{\pi})^{1/2}\, [2 - 4(x/a)^2]$
3	$(1/48a\sqrt{\pi})^{1/2}\, [12(x/a) - 8(x/a)^3]$
$\vdots$	
n	$(1/n!2^n a\sqrt{\pi})^{1/2}\, H_n(x/a)$

and more generally for the nth level (see Exercise 3.11):

$$\Delta x \Delta p_x = \left(n + \frac{1}{2}\right) \hbar. \tag{3.95}$$

Equation 3.94 shows that the quantum uncertainty of the ground state is at the minimum level permitted by the Heisenberg uncertainty principle given in eqn. 3.39.

Another important aspect of the harmonic oscillator is the $1/2$ that appears in eqn 3.93. This implies that the energy is non-zero even when the system is in the lowest possible level with $n = 0$. The finite energy of the oscillator in the ground state is said to be caused by the **zero-point motion** of the particle. In quantum optics, this zero-point motion is attributed to the vacuum field.

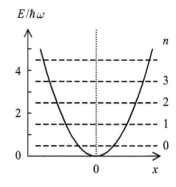

Fig. 3.7 Potential energy and quantized levels of the simple harmonic oscillator.

3.4   The Stern–Gerlach experiment

The **Stern–Gerlach experiment** studies the deflection of atomic beams by a magnet. The force experienced by a magnetic dipole in a non-uniform magnetic field pointing along the z-axis is given by:

$$F_z = \mu_z \frac{\mathrm{d}B}{\mathrm{d}z}, \tag{3.96}$$

where μ_z is the z-component of the dipole. With μ_z given by eqn 3.80, we expect different forces for each M_J state, giving rise to $(2J + 1)$ distinct deflections when the experiment is repeated many times on an ensemble of identical atoms.

In the original version of the experiment, Stern and Gerlach used silver (Ag) atoms and found that two distinct deflections were observed, as indicated in Fig. 3.8. The ground state of Ag is a 5s $^2S_{1/2}$ term, with $L = 0$ and $J = S = 1/2$. In the classical picture, no deflection would be expected because the atom has no orbital angular momentum and hence no classical magnetic dipole. The observation of two deflections in the experiment indicted that the atom possessed a magnetic dipole even when $L = 0$, and was highly instrumental in the discovery of electron spin.

Stern–Gerlach experiments are a useful way to illustrate the general principles of non-commuting operators in quantum mechanics. Consider first the experiment shown in Fig. 3.9(a). The apparatus consists of a beam of spin-1/2 particles and three Stern–Gerlach magnets. The beam enters the first magnet, and the deflected particles are fed separately into identical magnets. It is found that those particles that are deflected up by the first magnet are always deflected up by the second magnet, and vice versa for those deflected down.

The experiment shown in Fig. 3.9(a) can be understood by realizing that the magnet correlates the motion of the particle with its spin. Before entering the first magnet, the particles can be either spin up or spin down

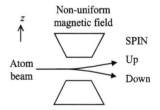

Fig. 3.8 The Stern–Gerlach experiment. A beam of atoms with $J = 1/2$ is deflected in two discrete ways by a non-uniform magnetic field.

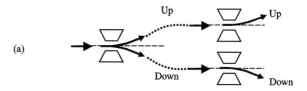

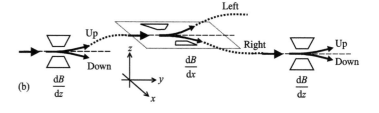

Fig. 3.9 (a) The spin-1/2 particles deflected from a Stern–Gerlach magnet as in Fig. 3.8 are always deflected in the same way by subsequent Stern–Gerlach measurements. (b) Multiple Stern–Gerlach measurements with orthogonal axes. The first and third magnets produce non-uniform magnetic fields along the z-axis, while the middle magnet produces the field along the x-axis.

Probabilities are proportional to $|\psi^2|$, and so the factor of $1/\sqrt{2}$ in ψ is consistent with a 50:50 probability split for the up and down deflections.

with 50:50 probability. After the magnet, the wave function becomes:

$$\psi = \frac{1}{\sqrt{2}}(\psi_{\mathrm{up}}(\boldsymbol{r})|\!\uparrow\rangle + \psi_{\mathrm{down}}(\boldsymbol{r})|\!\downarrow\rangle), \qquad (3.97)$$

where $\psi_{\mathrm{up}}(\boldsymbol{r})$ and $\psi_{\mathrm{down}}(\boldsymbol{r})$ denote the spatial wave functions for the up and down paths, respectively, while $|\!\uparrow\rangle$ and $|\!\downarrow\rangle$ indicate the eigenstates of the $\hat{S}_z$ operator. The atoms that enter the second magnet are always in the $|\!\uparrow\rangle$ state, and a final measurement after the second magnet will therefore always give the spin-up result.

Consider now the experiment shown in Fig. 3.9(b). We again have three Stern–Gerlach magnets, but this time they are arranged consecutively. The axes of the first and third magnets are aligned along the z-axis, but the middle magnet has been rotated by 90° so that the non-uniform field points along the x-axis. The effect of the middle magnet can be understood by rotating the coordinate axes by 90°, in which case it becomes apparent that there will be two possible deflections that we shall call 'left' and 'right'. An additional subtlety is that the second magnet is positioned so that the beam from only one of the output directions of the first magnet enters it, and likewise for the third magnet with respect to the second one. In the scenario depicted in Fig. 3.9(b), the second magnet receives the spin up particles from the first magnet, while the third magnet receives the spin right particles from the second.

Since the $\hat{S}_x$ and $\hat{S}_z$ operators do not commute (see eqn 3.47), it is not possible for the particles to be in simultaneous eigenstates of both $\hat{S}_z$ and $\hat{S}_x$. Therefore, the spin up particles that enter the second magnet will be deflected left or right with a random 50:50 probability. The particles that then enter the third magnet are now eigenstates of $\hat{S}_x$ rather than $\hat{S}_z$, and therefore they are again divided equally between the up and down outputs.

In quantum optics, we shall encounter situations which are formally identical to those illustrated in Fig. 3.9 but involving photon polarization rather than spin. Table 3.6 summarizes the equivalence. In the optical case, the photons pass through a polarizing beam splitter, and the equivalents of up and down spin states are vertically and horizontally

Table 3.6 Equivalence between the spin states of particles with $S = 1/2$ and photon polarization states. In the case of photons, the double-ended arrows represent the direction of the linear polarization.

	Spin state	Photon polarization
Deflection apparatus	Stern–Gerlach magnet	Polarizing beam splitter
Eigenstate basis 1	$\lvert\uparrow\rangle, \lvert\downarrow\rangle$	$\lvert\updownarrow\rangle, \lvert\leftrightarrow\rangle$
Eigenstate basis 2	$\lvert\leftarrow\rangle, \lvert\rightarrow\rangle$	$\lvert\nearrow\rangle, \lvert\searrow\rangle$

polarized photon states, written respectively as $\lvert\updownarrow\rangle$ and $\lvert\leftrightarrow\rangle$. The photon states with polarization angles of $\pm45°$ (i.e. $\lvert\nearrow\rangle$ and $\lvert\searrow\rangle$) are then equivalent to the $\lvert\leftarrow\rangle$ and $\lvert\rightarrow\rangle$ eigenstates of Stern–Gerlach magnets with their axes along the x-direction. The principles behind quantum cryptography (see Chapter 12) are exactly the same as those of performing Stern–Gerlach experiments with the magnet axes at different angles.

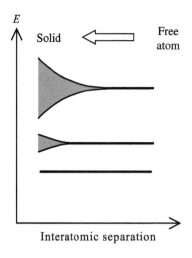

Fig. 3.10 Schematic illustration of the formation of electronic bands in a solid from the condensation of free atoms. As the atoms are brought closer together to form the solid, their outer orbitals begin to overlap with each other. These overlapping orbitals interact strongly, and broad energy bands are formed. The inner core orbitals do not overlap and so remain discrete even in the solid state.

3.5 The band theory of solids

The electronic states of crystals are described by the band theory of solids. The atoms in a solid are packed very close to each other, with the interatomic separation approximately equal to the size of the atoms. Hence the outer orbitals of the atoms overlap and interact strongly with each other. This broadens the discrete levels of the free atoms into bands, as illustrated schematically in Fig. 3.10. Each band can contain $2N$ electrons per unit volume, where N is the number of atoms per unit volume. The factor of two arises from the two possible values of the spin for each available electron space wave function.

The most important class of solid that we shall need to consider within this book is the **semiconductor**. The archetypal semiconductors silicon and germanium come from group IVB of the periodic table, and therefore have four valence electrons per atom. This is also true of III–V compound semiconductors like GaAs, because the group III and group V atoms contribute a total of eight valence electrons, giving an average of four per atom when the covalent bond is formed.

Figure 3.11 shows a generic energy-level diagram of a semiconductor. At absolute zero temperature, the bands can be thought of as filled up, starting from the lowest, until all the electrons are accounted for. With an even number of valence electrons per atom, there are exactly the right number of electrons to completely fill whole bands, and the last band to be filled is called the **valence band**. There is then a gap E_g in energy called the **band gap** to the next band, called the **conduction band**, which is unfilled. As the temperature is increased from absolute zero, a certain number of electrons can be excited from the valence band to the conduction band, leaving empty states called **holes** in the valence band. These empty states are equivalent to the absence of an electron, and behave like particles with a charge of $+e$.

The electrons in the conduction band and the holes in the valence band behave like free particles but with a different mass from that of genuinely

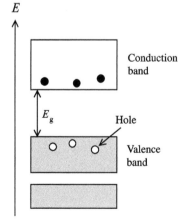

Fig. 3.11 Schematic generic energy-level diagram for a semiconductor. The valence bands are filled with electrons, leaving a gap of energy E_g to the unfilled conduction band. Electrons can be excited across the gap to the conduction band, leaving positively charged empty states called holes in the valence band.

free electrons. This modified mass is called the **effective mass** (m^*), and the values for electrons and holes are usually different. Electrical current can be carried both by the electrons in the conduction band and by the holes in the valence band. At room temperature, there is typically a much smaller number of these free carriers than in a metal, which explains why the materials are called *semi*conductors.

The populations of the free electrons and holes can be controlled by introducing impurities into the crystal. When atoms from group V of the periodic table are added (**n-type doping**), the impurities have an extra valence electron which produces extra free electrons in the conduction band. Similarly, the inclusion of impurities from group III (**p-type doping**) causes a deficiency of electrons, thereby increasing the number of holes.

Table 3.7 gives a list of the band-gap energies of a number of important semiconductors. The band gaps are classified as to whether they are **direct** or **indirect**. In a direct-gap semiconductor, the maximum of the valence band and the minimum of the conduction band occur for the same value of the electron wave vector. This means that an electron in the conduction band can recombine directly with a hole in the valence band by emitting a photon. By contrast, in an indirect-gap semiconductor, the conduction band minimum and valence band maximum occur for different electron wave vectors, and the electrons can only recombine indirectly with holes in a process that involves a quantized vibration (phonon). For this reason, indirect-gap semiconductors like silicon and germanium have very low optical transition probabilities, and are therefore of less interest in quantum optics. Accordingly, the examples of the use of semiconductors in this book will normally involve direct-gap materials like GaAs or InAs. A brief summary of the main features of the optical properties of direct-gap semiconductors is given in Section 4.6.

Table 3.7 Band gap data for a number of common semiconductors. E_g is the band gap at 300 K, and the i/d label indicates whether the gap is indirect or direct.

Semiconductor	E_g (eV)	Type
InAs	0.35	d
Ge	0.66	i
Si	1.12	i
InP	1.34	d
GaAs	1.42	d
CdSe	1.8	d
AlAs	2.15	i
GaP	2.27	i
CdS	2.5	d
ZnSe	2.8	d
ZnO	3.4	d
GaN	3.44	d

Further reading

There are many excellent texts available on quantum mechanics, for example: Cohen-Tannoudji *et al.* (1987), Gasiorowicz (1996), Sakurai (1994), or Schiff (1969). For further details on atomic physics, see Foot (2005), Haken and Wolf (2000) or Woodgate (1980). Explanations of the band theory of solids may be found in Kittel (1996) or Singleton (2001).

Exercises

(3.1) Use the orthonormality condition of eigenfunction wave functions to prove eqn 3.24.

(3.2) Expand an arbitrary wave function over the eigenfunctions of the operator to prove eqn 3.31.

(3.3) By using the definition in the first line of eqn 3.32, prove the second.

(3.4) Insert the definitions of the $\boldsymbol{r}$ and $\boldsymbol{p}$ operators into eqn 3.40 to prove eqn 3.42.

(3.5) The $\boldsymbol{\nabla}^2$ operator in spherical polar coordinates is given by:

$$\boldsymbol{\nabla}^2 = \frac{1}{r^2}\frac{\partial}{\partial r}\left(r^2\frac{\partial}{\partial r}\right) + \frac{1}{r^2\sin\theta}\frac{\partial}{\partial\theta}\left(\sin\theta\frac{\partial}{\partial\theta}\right)$$
$$+ \frac{1}{r^2\sin^2\theta}\frac{\partial^2}{\partial\varphi^2}.$$

Verify that the wave function

$$\psi(r,\theta,\varphi) = \left(1/\sqrt{\pi}a_0^{3/2}\right)\exp(-r/a_0)$$

is an eigenfunction of the hydrogen Hamiltonian with $m = m_0$ if a_0 is given by eqn 3.65, and find its energy. Verify also that the wave function is correctly normalized.

(3.6) An excited state of helium has an electronic configuration of (1s,2p). Write down all the atomic levels that are possible for this configuration.

(3.7) The interval rule of hyperfine structure states that the energy splitting between two levels in a hyperfine multiplet is proportional to the larger of the F-values of the levels spanning the gap.

(a) Explain the origin of the rule.

(b) Deduce a similar rule for the splitting of fine-structure levels.

(c) The 3p $^2\mathrm{P}_{3/2}$ level of sodium consists of four hyperfine levels with splittings of 60 MHz, 36 MHz and 17 MHz.
(i) Explain why it must be the case that $I \geq 3/2$.
(ii) Find the value of I that fits best to the observed splittings.

(3.8) Find the splitting of the Zeeman sublevels from a $^3\mathrm{D}_2$ level in a magnetic field of 0.5 T, expressing your answer in electron volt and wave-number units.

(3.9) Verify that the wave function $\psi(x) = C\exp(-m\omega x^2/2\hbar)$ is an eigenfunction of the simple harmonic oscillator with energy $\frac{1}{2}\hbar\omega$. Find the value of the normalization constant C.

(3.10) Use eqn 3.32 with the definitions of $\hat{x}$ and $\hat{p}_x$ to prove eqn 3.94 for the $n = 0$ state of the simple harmonic oscillator.

(3.11) (a) By comparison with a classical oscillator, explain why the expectation values of x and p_x are both zero for the quantum harmonic oscillator. Hence show that

$$\Delta x\Delta p_x = \left(\langle\hat{x}^2\rangle\langle\hat{p}_x^2\rangle\right)^{1/2}$$

for a quantum harmonic oscillator.

(b) Given that the average kinetic and potential energies of a harmonic oscillator are identical, prove eqn 3.95.

(3.12) A stream of photons from an unpolarized source passes first through a polarizing beam splitter (PBS) and then through a second PBS with its optic axis at an angle θ to the vertical. The second beam splitter is positioned so that only the vertically polarized output from the first beam splitter is incident on it. Single-photon counting detectors are placed at the two outputs of the second beam splitter. On the assumption that the detectors are perfectly efficient, and that there are no losses in the optics, calculate the probability per photon of recording an event on each detector.

4 Radiative transitions in atoms

4.1 Einstein coefficients   48

4.2 Radiative transition
    rates   51

4.3 Selection rules   54

4.4 The width and shape of
    spectral lines   56

4.5 Line broadening in
    solids   58

4.6 Optical properties of
    semiconductors   59

4.7 Lasers   61

Further reading   69

Exercises   69

This chapter gives an overview of the theory of optical absorption and emission in atoms at the level appropriate to introductory quantum and atomic-physics texts. The aim is to provide an introduction to the more advanced treatment of these topics given later in the book, especially in Chapters 9 and 10. It is assumed that the reader is reasonably familiar with the concepts, and so the material is only developed in summary form. Readers who are less well acquainted with these topics may find it helpful to refer to the bibliography for further details.

The chapter begins with a discussion of the Einstein coefficients in order to introduce the concepts of absorption and emission, and the connection between them. We then move on to discuss the calculation of the transition rates in atoms by quantum mechanics, and the selection rules that follow from them. Next we describe the mechanisms that affect the shape of the spectral lines, and the difference between the optical spectra of free atoms and solids, especially semiconductors. Finally, we conclude with a brief discussion of the principal features of lasers.

4.1 Einstein coefficients

The quantum theory of radiation assumes that light is emitted or absorbed whenever an atom makes a jump between two quantum states. These two processes are illustrated in Fig. 4.1. Absorption occurs when the atom jumps to a higher level, while emission corresponds to the process in which a photon is emitted as the atom drops down to a lower level. Conservation of energy requires that the angular frequency ω of the photon satisfies:

$$\hbar\omega = E_2 - E_1, \tag{4.1}$$

where E_2 is the energy of the upper level and E_1 is the energy of the lower level. In Section 4.2 we explain how quantum mechanics enables us to calculate the emission and absorption rates. At this stage we restrict ourselves to a phenomenological analysis based on the **Einstein coefficients** for the transition.

The radiative process by which an electron in an upper level drops to a lower level as shown in Fig. 4.1(a) is called **spontaneous emission**. This is because the atoms in the excited state have a natural (i.e. spontaneous) tendency to de-excite and lose their excess energy. Each type of atom

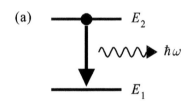

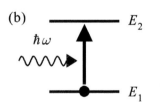

Fig. 4.1 Optical transitions between two states in an atom: (a) spontaneous emission, (b) absorption.

has a characteristic spontaneous-emission spectrum determined by its energy levels according to eqn 4.1.

The rate at which spontaneous emission occurs is governed by the Einstein A coefficient for the transition. This gives the probability per unit time that the electron in the upper level will drop to the lower level by emitting a photon. The photon emission rate is therefore proportional to the number of atoms in the excited state and to the A coefficient for the transition. We thus write down the following rate equation for $N_2(t)$, the number of atoms in the excited state:

$$\frac{\mathrm{d}N_2}{\mathrm{d}t} = -A_{21}N_2. \tag{4.2}$$

The subscript '21' on the A coefficient in eqn 4.2 makes it plain that the transition starts at level 2 and ends at level 1.

Equation 4.2 can be solved for $N_2(t)$ to give:

$$N_2(t) = N_2(0)\,\exp(-A_{21}t) \equiv N_2(0)\exp(-t/\tau), \tag{4.3}$$

where

$$\tau = \frac{1}{A_{21}}. \tag{4.4}$$

τ is the **radiative lifetime** of the excited state. Equation 4.3 shows that the number of atoms in the excited state decays exponentially with a time constant τ due to spontaneous emission. The value of τ for a transition at optical frequencies can range from about a nanosecond to several milliseconds, according to the type of radiative process that occurs.

The process of **absorption** is illustrated in Fig. 4.1(b). The atom is promoted from the lower level to the excited state by absorbing the required energy from a photon. Unlike emission, it is not a spontaneous process. The electron cannot jump to the excited state unless it receives the energy required from an incoming photon. Following Einstein's treatment, we write the rate of absorption transitions per unit time as:

$$\frac{\mathrm{d}N_1}{\mathrm{d}t} = -B_{12}^{\omega}N_1 u(\omega), \tag{4.5}$$

where $N_1(t)$ is the number of atoms in level 1 at time t, B_{12}^{ω} is the Einstein B coefficient for the transition, and $u(\omega)$ is the spectral energy density of the electromagnetic field in $\mathrm{J\,m^{-3}(rad/s)^{-1}}$ at angular frequency ω. By writing $u(\omega)$ we are explicitly stating that only the part of the spectrum of the incoming radiation at angular frequencies around ω, where $\hbar\omega = E_2 - E_1$, can induce the absorption transitions. Equation 4.5 may be considered to be the definition of the Einstein B coefficient.

The processes of absorption and spontaneous emission that we have described above are fairly intuitive. Einstein realized that the analysis was not complete, and introduced a third type of transition called **stimulated emission**. In this process, the incoming photon field can

The selection rules that govern whether a particular transition is fast or slow are discussed in Section 4.3.

Throughout this book, we choose to work mainly with the *angular* frequency ω of the light rather than its actual frequency ν. The two are, of course, simply related by:

$$\omega = 2\pi\nu,$$

which means that all of the formulae that are derived can be quickly converted between the two conventions. A case in point is the definition of the Einstein B coefficient in eqn 4.5, which could equally well have been defined in terms of $u(\nu)$ rather than $u(\omega)$. The two B coefficients differ by a factor of 2π, with

$$B_{12}^{\omega} = 2\pi B_{12}^{\nu}.$$

We shall see in Section 10.3.2 how spontaneous emission can in fact be considered as a stimulated-emission process instigated by the ever-present zero-point fluctuations of the electromagnetic field.

stimulate downward emission transitions as well as upward absorption transitions. The stimulated-emission rate is governed by a second Einstein B coefficient, namely B_{21}. The subscript is now essential to distinguish the B coefficients for the two distinct processes of absorption and stimulated emission.

In analogy with eqn 4.5, we write the rate of stimulated emission transitions by the following rate equation:

$$\frac{dN_2}{dt} = -B_{21}^{\omega} N_2 u(\omega). \tag{4.6}$$

Stimulated emission is a coherent quantum-mechanical effect in which the photons emitted are in phase with the photons that induce the transition.

The three Einstein coefficients introduced above are not independent parameters: they are all related to each other. If we know one of them, we can work out the other two. To see how this works, we follow Einstein's analysis.

We imagine that we have a gas of N atoms inside a box with black walls at temperature T. We assume that the atoms only interact with the black-body radiation filling the cavity and not directly with each other. The black-body radiation will induce both absorption and stimulated-emission transitions, while spontaneous-emission transitions will also be occurring at a rate determined by the Einstein A coefficient. The three types of transition are indicated in Fig. 4.2. If we leave the atoms for long enough, they will come to thermal equilibrium with the black-body radiation. In these steady-state conditions, the rate of upward transitions due to absorption must exactly balance the rate of downward transitions due to spontaneous and stimulated emission. From eqns 4.2–4.6, we must therefore have:

$$B_{12}^{\omega} N_1 u(\omega) = A_{21} N_2 + B_{21}^{\omega} N_2 u(\omega). \tag{4.7}$$

Since the atoms are in thermal equilibrium with the radiation field at temperature T, the distribution of the atoms among the various energy levels will be governed by the laws of thermal physics. The ratio of N_2 to N_1 will therefore be given by Boltzmann's law:

$$\frac{N_2}{N_1} = \frac{g_2}{g_1} \exp\left(-\frac{\hbar\omega}{k_B T}\right), \tag{4.8}$$

where g_1 and g_2 are the degeneracies of levels 1 and 2, respectively. Now the energy spectrum of a black-body source is given by the Planck

Stimulated emission is the basis of laser operation, as will be discussed in Section 4.7.

The state of equilibrium between the atoms and the radiation occurs whether or not level 1 is the ground state and whether or not transitions take place to and from other levels. The principle of **detailed balance** guarantees that eqn 4.7 must hold regardless.

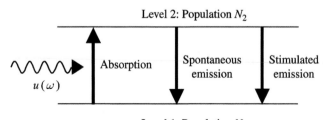

Fig. 4.2 Absorption, spontaneous emission and stimulated emission transitions between two levels of an atom in the presence of electromagnetic radiation with spectral energy density $u(\omega)$.

formula:

$$u(\omega) = \frac{\hbar\omega^3}{\pi^2 c^3} \frac{1}{\exp\left(\hbar\omega/k_{\mathrm{B}}T\right) - 1}. \tag{4.9}$$

The only way that eqns 4.7–4.9 can be consistent with each other at all temperatures is if:

$$g_1 B_{12}^\omega = g_2 B_{21}^\omega, \tag{4.10}$$

and

$$A_{21} = \frac{\hbar\omega^3}{\pi^2 c^3} B_{21}^\omega. \tag{4.11}$$

If the atom is embedded within an optical medium with a refractive index n, we replace c by c/n in eqn 4.9 and hence 4.11 to account for the reduced velocity of light.

Equation 4.10 tells us that the probabilities for stimulated absorption and emission are the same apart from the degeneracy factors. Furthermore, the interrelationship of the Einstein coefficients tells us that transitions that have a high absorption probability will also have a high emission probability, both for spontaneous processes and stimulated ones.

We shall see in the next section that the Einstein coefficients for a particular transition are determined by the wave functions of the initial and final levels, and hence are intrinsic properties of the atoms. This means that the relationships between the Einstein coefficients given in eqns 4.10 and 4.11 apply in all cases, even though they were derived for the special case when the atoms are in equilibrium with black-body radiation at a specific temperature. This is very useful, because we then only need to know one of the coefficients to work out the other two. For example, we can measure the radiative lifetime to determine A_{21} using eqn 4.4, and then work out the B coefficients using eqns 4.11 and 4.10.

4.2   Radiative transition rates

The calculation of radiative transition rates by quantum mechanics is based on time-dependent perturbation theory. The light–matter interaction is described by transition probabilities, which can be calculated for the case of spontaneous emission by using **Fermi's golden rule**. According to this rule, the transition rate is given by:

$$W_{1\rightarrow 2} = \frac{2\pi}{\hbar}|M_{12}|^2 g(\hbar\omega), \tag{4.12}$$

where M_{12} is the **matrix element** for the transition, and $g(\hbar\omega)$ is the **density of states**.

Note that $g(\hbar\omega)$ refers to the density of *final* states.

Let us first consider the density of states factor that appears in the golden rule. The density of states is defined so that $g(\hbar\omega)\mathrm{d}E$ is the number of final states per unit volume that fall within the energy range E to $E + \mathrm{d}E$, where $E = \hbar\omega$. In the standard case of transitions between quantized levels in an atom, the initial and final electron states are *discrete*. In this case, the density of final states factor that enters eqn 4.12 is the density of *photon* states.

In some situations it will be necessary to consider the density of electron states as well as the density of photon states. Two obvious examples are the transitions from an initial discrete atomic level to a continuum of levels (e.g. above the ionization threshold) and the transitions between two continuous electron energy bands in solid-state physics. (See Section 4.6.)

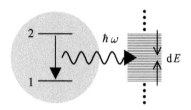

Fig. 4.3 Optical transitions between discrete atomic states involving photon emission into a continuum of states.

The classical theory of radiative emission and absorption treats the atoms as oscillating electric dipoles. Absorption occurs when electromagnetic waves at the natural resonant frequency force oscillations that transfer energy from the light to the atoms. Emission also occurs at the natural frequency of the oscillator, and can be understood by applying the theory of Hertzian aerials. The link between the classical and quantum theories can be made by relating the transition dipole moment in eqn 4.20 with the classical electric dipole of the electron.

In considering spontaneous radiative emission by an atom, we shall usually be interested in the situation where the photons are emitted into free space. In this case, the photons are emitted into a continuum of states, as illustrated schematically in Fig. 4.3. The density of photon modes is proportional to ω^2 in free space. (See eqn C.11 in Appendix C.) This factor of ω^2, together with a third factor of ω to account for the photon energy, normally appears in the spontaneous-emission probability. (See eqns 4.11 and eqn 4.23.) Note, however, that the photon density of states can be modified by making the atoms emit into an optical cavity or into a photonic crystal. This modification of the photon density of states can have a profound effect on the radiative emission rate, as we shall consider in Chapter 10.

Now let us consider the matrix element that appears in Fermi's golden rule. This is given by:

$$M_{12} = \langle 2|H'|1 \rangle = \int \psi_2^*(\boldsymbol{r})H'(\boldsymbol{r})\psi_1(\boldsymbol{r})\mathrm{d}^3\boldsymbol{r}, \qquad (4.13)$$

where H' is the perturbation caused by the light, $\boldsymbol{r}$ is the position vector of the electron, and $\psi_1(\boldsymbol{r})$ and $\psi_2(\boldsymbol{r})$ are the wave functions of the initial and final states.

It is convenient to adopt a semi-classical approach in which the atoms are treated as quantum-mechanical objects but the light is treated classically. There are a number of different types of interaction that can be considered between the light and the atom, and this gives rise to a classification of the radiative transitions according to the scheme shown in Table 4.1. It is apparent from this table that the transition rate decreases by several orders of magnitude each time the multipolarity increases (i.e. dipole → quadrupole → octupole) and also that the magnetic interactions are weaker than the equivalent electric ones by a similar factor. In what follows, we concentrate on the **electric dipole** (E1) interaction, which is the strongest of the different possible types of transition by several orders of magnitude.

The perturbation to the atom in an E1 transition is caused by the interaction between the electric field amplitude $\boldsymbol{\mathcal{E}}_0$ of the light and the electric dipole $\boldsymbol{p}$ of the atom:

$$H' = -\boldsymbol{p} \cdot \boldsymbol{\mathcal{E}}_0. \qquad (4.14)$$

Table 4.1 Classification of radiative transitions. The figures quoted for the Einstein coefficients and radiative lifetimes should be considered only as order of magnitude values for transitions at frequencies around the visible spectral region.

Transition	Notation	Einstein A coefficient	Radiative lifetime	Parity change
Electric dipole	E1	10^7–10^9 s^{-1}	1–100 ns	yes
Magnetic dipole	M1	10^3–10^5 s^{-1}	0.01–1 ms	no
Electric quadrupole	E2	10^3–10^5 s^{-1}	0.01–1 ms	no
Magnetic quadrupole	M2	0.1–10 s^{-1}	0.1–10 s	yes
Electric octupole	E3	0.1–10 s^{-1}	0.1–10 s	yes
⋮				

At optical frequencies, we assume that only the electrons are light enough to respond, and so we write:

$$\boldsymbol{p} = -e\boldsymbol{r}. \tag{4.15}$$

The perturbation is then given by:

$$H' = e(x\mathcal{E}_x + y\mathcal{E}_y + z\mathcal{E}_z), \tag{4.16}$$

where $\mathcal{E}_x$ is the component of the field amplitude along the x-axis, etc. Since atoms are small compared to the wavelength of light, the amplitude of the electric field will not vary significantly over the dimensions of an atom. We can therefore take $\mathcal{E}_x$, $\mathcal{E}_y$, and $\mathcal{E}_z$ in eqn 4.16 to be constants in the calculation of the integrals in eqn 4.13 to obtain:

$$M_{12} = e\mathcal{E}_x \int \psi_2^* x \psi_1 \, \mathrm{d}^3 \boldsymbol{r} \quad x\text{-polarized light,}$$

$$M_{12} = e\mathcal{E}_y \int \psi_2^* y \psi_1 \, \mathrm{d}^3 \boldsymbol{r} \quad y\text{-polarized light,} \tag{4.17}$$

$$M_{12} = e\mathcal{E}_z \int \psi_2^* z \psi_1 \, \mathrm{d}^3 \boldsymbol{r} \quad z\text{-polarized light.}$$

These matrix elements can be written in the more succinct form:

$$M_{12} = -\boldsymbol{\mu}_{12} \cdot \boldsymbol{\mathcal{E}}_0, \tag{4.18}$$

where

$$\boldsymbol{\mu}_{12} = -e\big(\langle 2|x|1 \rangle \hat{\mathbf{i}} + \langle 2|y|1 \rangle \hat{\mathbf{j}} + \langle 2|z|1 \rangle \hat{\mathbf{k}} \big) \tag{4.19}$$

is the electric **dipole moment** of the transition. For the case of light polarized along the x-axis (and equivalently for the y or z-polarizations), this simplifies to:

$$\boldsymbol{\mu}_{12} = -e\langle 2|x|1 \rangle \equiv -e \int \psi_2^* x \psi_1 \mathrm{d}^3 \boldsymbol{r}. \tag{4.20}$$

The dipole moment is thus the key parameter that determines the transition rate for the electric-dipole process.

The result given in eqn 4.17 allows us to evaluate the matrix elements for particular transitions if the wave functions of the initial and final states are known. We can then use Fermi's golden rule to calculate the transition rate per atom, which can be equated with the transition probability $B_{12}^\omega u(\omega)$ in eqn 4.5. Since the energy density is proportional to $\mathcal{E}^2$, we can eliminate the electric field amplitude from the transition rate, and deduce the Einstein coefficients. The final results for transitions between non-degenerate discrete atomic levels by absorption or emission of unpolarized light of angular frequency ω are:

$$B_{12}^\omega = \frac{\pi}{3\epsilon_0 \hbar^2} |\boldsymbol{\mu}_{12}|^2, \tag{4.21}$$

and

$$A_{21} = \frac{\omega^3}{3\pi\epsilon_0 \hbar c^3} |\boldsymbol{\mu}_{12}|^2. \tag{4.22}$$

Here, as elsewhere in the book, we take e to represent the magnitude of the electron charge, so that the electron charge itself is equal to $-e$.

The energy density per unit volume of an electromagnetic wave is given by:

$$U = \frac{1}{2}\left(\epsilon_0 \boldsymbol{\mathcal{E}} \cdot \boldsymbol{\mathcal{E}} + \frac{1}{\mu_0} \boldsymbol{B} \cdot \boldsymbol{B} \right),$$

where $\boldsymbol{\mathcal{E}}$ and $\boldsymbol{B}$ are the electric and magnetic fields, respectively. The electric and magnetic terms are equal, and so U is proportional to $\mathcal{E}^2$. However, U is not the same as the *spectral* energy density $u(\omega)$ that appears in eqn 4.5. The elimination of $\mathcal{E}^2$ from the transition rate therefore requires careful consideration of the spectral width of the light beam and the atomic transition line. (See, e.g. Cohen-Tannoudji 1987.) Equation 4.21 is derived from first principles in Section 9.4.

When the levels are degenerate, we must modify eqns 4.21 and 4.22 to allow for the different transition pathways. For example, if we consider the transitions at angular frequency ω_{ji} between atomic levels with quantum numbers j and i, each of which consists of a manifold of degenerate levels labelled by additional quantum numbers m_j and m_i, then eqn 4.22 is modified to:

In solid-state physics, the summation over discrete levels is replaced by the joint density of states for the initial and final electron bands.

$$A_{ji} = \frac{e^2 \omega_{ji}^3}{3\pi\epsilon_0 \hbar c^3} \frac{1}{g_j} \sum_{m_j, m_i} |\langle j, m_j | \boldsymbol{r} | i, m_i \rangle|^2 , \qquad (4.23)$$

where g_j is the degeneracy of the upper state.

The dipole moment is directly related to the **oscillator strength** f_{ij} of the transition according to:

$$f_{ij} = \frac{2m\omega_{ji}}{3\hbar} |\langle j | \boldsymbol{r} | i \rangle|^2 \equiv \frac{2m\omega_{ji}}{3\hbar e^2} |\mu_{ij}|^2 . \qquad (4.24)$$

The oscillator strength was introduced before quantum theory was developed to explain how some atomic absorption and emission lines are stronger than others. With the hindsight of quantum mechanics, it is easy to understand that this is simply caused by the different dipole moments for the transitions.

4.3   Selection rules

The electric-dipole matrix element given in eqn 4.18 can be easily evaluated for simple atoms with known wave functions. This leads to the notion of electric-dipole **selection rules**. These are rules about the quantum numbers of the initial and final states. If the states do not satisfy the selection rules, then the electric-dipole transition rate will be zero.

Transitions that obey the electric-dipole selection rules are called **allowed** transitions, while those which do not are called **forbidden** transitions. E1-allowed transitions have high transition probabilities, and therefore have short radiative lifetimes, typically in the range 1–100 ns. (See Table 4.1.) Forbidden transitions, by contrast, are much slower. The different time-scales for allowed and forbidden transitions lead to another general classification of the spontaneous emission as **fluorescence** and **phosphorescence**, respectively. Fluorescence is a 'prompt' process in which the photon is emitted within a few nanoseconds after the atom has been excited, while phosphorescence gives rise to 'delayed' emission which persists for a substantial time.

The electric-dipole selection rules for a single electron in a hydrogenic system with quantum numbers l, m, s, and m_s are summarized in Table 4.2. The origin of these rules is as follows:

- The parity change rule follows from the fact that the electric-dipole operator is proportional to $\boldsymbol{r}$, which is an odd function.

Table 4.2 Electric-dipole selection rules for single-electron atoms. The z-axis is usually defined by the direction of an applied static magnetic or electric field. The rule on Δm for circular polarization applies to absorption. The sign is reversed for emission.

Quantum number	Selection rule	Polarization
Parity	Changes	
l	$\Delta l = \pm 1$	
m	$\Delta m = +1$	Circular: σ^+
	$\Delta m = -1$	Circular: σ^-
	$\Delta m = 0$	Linear: $\parallel z$
	$\Delta m = \pm 1$	Linear: $\parallel (x, y)$
s	$\Delta s = 0$	
m_s	$\Delta m_s = 0$	

- The rule for Δl derives from the properties of the spherical harmonic functions and is consistent with the parity rule because the wave functions have parity $(-1)^l$.

- The rules on Δm can be understood by realizing that σ^+ and σ^- circularly polarized photons carry angular momenta of $+\hbar$ and $-\hbar$, respectively, along the z-axis, and hence m must change by one unit to conserve angular momentum. For linearly polarized light along the z-axis, the photons carry no z-component of momentum, implying $\Delta m = 0$, while x or y-polarized light can be considered as an equal combination of σ^+ and σ^- photons, giving $\Delta m = \pm 1$.

- The spin selection rules follow from the fact that the photon does not interact with the electron spin, and so the spin quantum numbers never change in the transition.

The σ^+ and σ^- polarizations refer to light in which the electric field vector rotates positively or negatively around the z-axis, respectively. For light travelling in the $+z$ direction, σ^+ polarization thus corresponds to positive circular polarization, and negative circular for the σ^- case. The relationship between positive/negative and left/right circular polarizations is explained in Section 2.1.4.

These selection rules can be generalized to many-electron atoms with quantum numbers (L, S, J) as follows:

(1) the parity of the wave function must change;

(2) $\Delta l = \pm 1$ for the changing electron;

(3) $\Delta L = 0, \pm 1$, but $L = 0 \to 0$ is forbidden;

(4) $\Delta J = 0, \pm 1$, but $J = 0 \to 0$ is forbidden;

(5) $\Delta S = 0$.

The parity rule follows from the odd parity of the dipole operator. The rule on l applies the single-electron rule to the individual electron that makes the jump in the transition. The rules on L and J follow from the fact that the photon carries one unit of angular momentum. The final rule is a consequence of the fact that the photon does not interact with the spin.

The selection rules for higher-order transitions are different. For example, magnetic-dipole and electric-quadrupole transitions can take place between states of the same parity. This can allow an atom in an

$J = 0 \rightarrow 0$ transitions are strictly forbidden for single-photon transitions

excited state with no possibility of decay by electric-dipole transitions to relax to the ground state. In extreme cases it may happen that all standard types of single-photon radiative transitions are forbidden. In this case, the excited state is said to be **metastable**, and the atom must de-excite by transferring its energy to other atoms in collisions, or by some other low-probability mechanism that we have not considered here, such as multi-photon emission.

4.4   The width and shape of spectral lines

4.4.1   The spectral lineshape function

The radiation emitted in atomic transitions is not perfectly monochromatic. The shape of the emission line is described by the **spectral lineshape function** $g_\omega(\omega)$. This is a function that peaks at the line centre defined by

$$\hbar\omega_0 = (E_2 - E_1), \qquad (4.25)$$

and is normalized so that:

$$\int_0^\infty g_\omega(\omega)\,\mathrm{d}\omega = 1. \qquad (4.26)$$

An equivalent lineshape function $g_\nu(\nu)$ can be defined in terms of frequency rather than angular frequency. The two functions are related to each other through

$$g_\nu(\nu) = 2\pi g_\omega(\omega).$$

The linewidths are contrariwise related:

$$\Delta\nu = \Delta\omega/2\pi.$$

The most important parameter of the lineshape function is the **full width at half maximum** (FWHM) $\Delta\omega$, which quantifies the width of the spectral line.

In the following subsections we shall briefly review the main line-broadening mechanisms that can occur in gases, namely:

- lifetime (natural) broadening,
- collisional (pressure) broadening,
- Doppler broadening.

Then in Section 4.5 we shall see how these processes are adapted when considering the emission spectra of atoms or ions embedded in solid-state hosts.

Lifetime and collisional broadening are examples of homogeneous broadening mechanisms, while Doppler broadening is an example of an inhomogeneous one.

Before going into the details, it is important to point out a general classification of the broadening mechanisms as either **homogeneous** or **inhomogeneous**. In the former case, all the individual atoms behave in the same way, and produce the same spectrum, while in the latter, the individual atoms behave differently and contribute to different parts of the spectrum. This distinction is important for a number of reasons, but here we concentrate on just one, namely that the spectral lineshapes are different. We shall now see that homogeneous mechanisms generally give rise to **Lorentzian** lineshapes, while inhomogeneous processes tend to produce **Gaussian** spectral lines.

4.4.2   Lifetime broadening

Light is emitted when an electron in an excited state drops to a lower level by spontaneous emission, as shown in Fig. 4.1(a). The rate at which this occurs is determined by the Einstein A coefficient, which in turn determines the radiative lifetime τ, as discussed in Section 4.1.

The finite lifetime of the excited state leads to a broadening of the spectral line in accordance with the energy–time uncertainty principle:

$$\Delta E \Delta t \gtrsim \hbar. \tag{4.27}$$

On setting $\Delta t = \tau$, we then deduce that the amount of broadening in angular-frequency units must satisfy:

$$\Delta \omega = \frac{\Delta E}{\hbar} \gtrsim \frac{1}{\tau}. \tag{4.28}$$

Since this broadening mechanism is intrinsic to the transition, it is alternatively called **natural broadening** or simply **radiative broadening**.

The detailed form of the lifetime broadening can be deduced by taking the Fourier transform of a burst of light that decays exponentially with time constant τ. (See Exercise 4.4.) This gives the following spectral lineshape function:

$$g_\omega(\omega) = \frac{\Delta \omega}{2\pi} \frac{1}{(\omega - \omega_0)^2 + (\Delta \omega / 2)^2}, \tag{4.29}$$

where the FWHM is given by:

$$\Delta \omega_{\text{lifetime}} = \frac{1}{\tau}. \tag{4.30}$$

The spectrum described by eqn 4.29 is called a **Lorentzian** lineshape, and is plotted in Fig. 4.4. Note that the rigorous result in eqn 4.30 agrees with the approximate one of eqn 4.28 from the uncertainty principle.

4.4.3   Collisional (pressure) broadening

The atoms in a gas frequently collide with each other and with the walls of the containing vessel. This interrupts the process of light emission and can shorten the effective lifetime of the excited state. If the mean time between collisions $\tau_{\text{collision}}$ is shorter than the radiative lifetime, then we need to replace τ by $\tau_{\text{collision}}$ in eqn 4.30, thereby giving rise to additional line broadening.

A simple analysis based on the kinetic theory of gases gives the following result for $\tau_{\text{collision}}$:

$$\tau_{\text{collision}} \sim \frac{1}{\sigma_s P} \left(\frac{\pi m k_B T}{8} \right)^{1/2}, \tag{4.31}$$

where σ_s is the collision cross-section, and P is the pressure. It is apparent that $\tau_{\text{collision}}^{-1}$, and hence $\Delta \omega$, are proportional to P. Collisional broadening is therefore also called **pressure broadening**. At standard temperature and pressure (STP) we typically find $\tau_{\text{collision}} \sim 10^{-10}$ s, (see Exercise 4.6) which is much shorter than typical radiative lifetimes, and gives a linewidth from eqn 4.30 of $\sim 10^{10}$ rad s^{-1}.

In conventional atomic discharge tubes, we reduce the effects of collisional broadening by working at low pressures. We see from eqn 4.31 that this increases $\tau_{\text{collision}}$, and hence reduces the linewidth. This is why we tend to use 'low pressure' discharge lamps for spectroscopy.

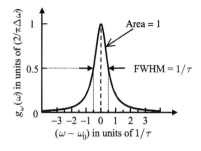

Fig. 4.4 The Lorentzian lineshape function. The form is given in eqn 4.29. The function peaks at the line centre ω_0 and has a FWHM of $1/\tau$. The function is normalized so that the total area is unity.

The derivation of eqn 4.31 is outlined in Exercise 4.5. The hard-sphere model on which the concept of the collision cross-section is based is only an approximation that breaks down under detailed scrutiny. Moreover, the averaging process in the kinetic theory involves a number of assumptions that are not always valid, and the result assumes ideal-gas behaviour. For these reasons, the result in eqn 4.31 should only be considered as a rough order-of-magnitude estimate. See Corney (1977) for a more detailed discussion.

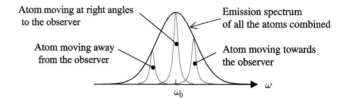

Fig. 4.5 The Doppler broadening mechanism. The thermal motion of the atoms causes their lab-frame frequencies to be shifted by the Doppler effect.

4.4.4   Doppler broadening

Doppler broadening originates from the random motion of the atoms in the gas. The random thermal motion of the atoms gives rise to Doppler shifts in the observed frequencies, which then causes line broadening, as illustrated in Fig. 4.5.

The broadening caused by the Doppler mechanism can be quantified by considering the light emitted by an atom moving with velocity component v_x towards the observer. If the transition frequency in the rest frame of the atom is ω_0, the observed frequency will be Doppler shifted to:

If the atom is moving away from the observer, v_x will be negative, which gives a negative frequency shift.

$$\omega = \omega_0 \left(1 + \frac{v_x}{c}\right). \tag{4.32}$$

The number of atoms with velocity between v_x and $v_x + dv_x$, namely $N(v_x)dv_x$, is given by the **Maxwell–Boltzmann distribution**:

$$N(v_x) = N_0 \left(\frac{2k_\mathrm{B}T}{\pi m}\right)^{1/2} \exp\left(-\frac{mv_x^2}{2k_\mathrm{B}T}\right), \tag{4.33}$$

where T is the temperature, N_0 is the total number of atoms, and m is their mass. By combining eqns 4.32 and 4.33, we find the normalized **Gaussian lineshape** function:

$$g_\omega(\omega) = \frac{c}{\omega_0}\sqrt{\frac{m}{2\pi k_\mathrm{B}T}} \exp\left(-\frac{mc^2(\omega - \omega_0)^2}{2k_\mathrm{B}T\omega_0^2}\right), \tag{4.34}$$

with a FWHM given by:

$$\Delta\omega_\mathrm{Doppler} = 2\omega_0 \left(\frac{(2\ln 2)k_\mathrm{B}T}{mc^2}\right)^{1/2} = \frac{4\pi}{\lambda}\left(\frac{(2\ln 2)k_\mathrm{B}T}{m}\right)^{1/2}. \tag{4.35}$$

The Doppler linewidth in a gas at STP is usually much larger than the natural linewidth. For example, the Doppler linewidth of the 589.0 nm line of sodium at 300 K works out to be 1.3 GHz, which is about two orders of magnitude larger than the 10 MHz natural broadening due to the radiative lifetime of 16 ns. The dominant broadening mechanism in low-pressure gases at room temperature is therefore usually Doppler broadening, and the lineshape is closer to Gaussian than Lorentzian.

4.5   Line broadening in solids

In many instances we will be interested in the emission spectra of atoms embedded within crystalline or amorphous solids. The spectra will be subject to lifetime broadening as in gases, since this is a fundamental property of radiative emission. However, the atoms are locked into their positions within the solid and do not move about freely as in a gas. This means that neither pressure nor Doppler broadening is relevant in solids. On the other hand, the emission and absorption lines can be broadened by other mechanisms, as we discuss below.

In some cases it may be possible for the atoms to de-excite from the upper level to the lower level by making a **non-radiative transition**. One way this could happen is to drop to the lower level by emitting phonons (i.e. heat) instead of photons. To allow for this possibility, we must rewrite eqn 4.2 in the following form:

$$\frac{\mathrm{d}N_2}{\mathrm{d}t} = -A_{21}N_2 - \frac{N_2}{\tau_{\mathrm{NR}}} = -\left(A_{21} + \frac{1}{\tau_{\mathrm{NR}}}\right)N_2, \qquad (4.36)$$

where τ_{NR} is the non-radiative relaxation time. This shows that non-radiative transitions shorten the lifetime of the excited state according to:

$$\frac{1}{\tau} = A_{21} + \frac{1}{\tau_{\mathrm{NR}}}. \qquad (4.37)$$

We thus expect additional homogeneous lifetime broadening according to eqn 4.30. The phonon emission times in solids are often very fast, and can cause substantial broadening of the emission lines. This is the solid-state equivalent of collisional broadening.

Another factor that may cause line broadening is the inhomogeneity of the host medium, for example, when the atoms are doped into a glass. If the environments in which the atoms find themselves are not entirely uniform, the emission spectrum will be affected through the interaction between the atoms and their different local environments. This inhomogeneous broadening mechanism is called **environmental broadening**.

An example of the effects of environmental broadening is the difference between the emission spectra of Nd^{3+} ions doped into a crystalline host such as yttrium aluminium garnet (YAG) and that of the same ions doped into a glass. The 1064 nm transition of a Nd:YAG crystal at room temperature is homogenously broadened to around 120 GHz by phonon emission. The equivalent transition in Nd:glass is 40–60 times broader owing to the effects of the inhomogeneity of the glass medium on the emission frequency of the Nd^{3+} ions.

4.6   Optical properties of semiconductors

Semiconductor materials—especially quantum wells and quantum dots—are important for a number of recent demonstrations of quantum optical effects. In this section we give a brief summary of the main optical properties of bulk semiconductors. A discussion of the optical properties of low-dimensional semiconductor structures may be found in Appendix D.

We have seen in Section 3.5 that the electronic states of semiconductors are broadened into bands. The optical properties of semiconductors are therefore determined by transitions between energy bands rather than between discrete levels. These transitions are generally called **interband transitions**. Figure 4.6 illustrates the interband transitions that occur between the conduction and valence bands of a semiconductor. Figures 4.6(a) and (b) illustrate the processes of interband absorption and spontaneous emission, respectively.

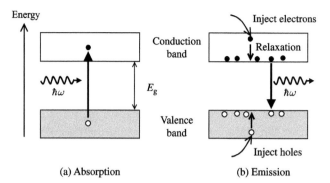

Fig. 4.6 Interband transitions in a semiconductor: (a) absorption; (b) emission. E_g is the band-gap energy.

The spontaneous emission from a semiconductor is generally called **luminescence**. This can be further classified as either **electroluminescence** or **photoluminescence** depending on whether the emission is excited electrically or optically.

In the absorption process illustrated in Fig. 4.6(a), an electron is promoted from the valence band to the conduction band, leaving a hole in the valence band. The transitions can take place over a continuous range of photon energies determined by the lower and upper energy limits of the bands. An absorption band is therefore observed, with a threshold at the band-gap energy E_g.

The interband luminescence process shown schematically in Fig. 4.6(b) is more complicated. For emission to be possible, it is necessary that there should be an electron in the conduction band and an unoccupied level (i.e. a hole) in the valence band. These electrons and holes are typically injected into their respective bands either from an electrical current or by previous optical excitation. The electrons that are injected relax very rapidly to the bottom of the conduction band by emission of phonons. Similarly, the injected holes relax very rapidly to the top of the valence band. (Hole energies are measured downwards from the top of the valence band.) The radiative transitions therefore take place at energies very close to the band-gap energy E_g. The width of the emission line is determined by the thermal spread of the charge carriers within their bands or by inhomogeneous effects. As a rule of thumb, the linewidth at temperature T in energy units is of order $k_B T$ unless this energy is smaller than the inhomogeneous broadening, in which case the latter determines the linewidth.

Strong interband transitions can occur when the transitions are allowed by the electric-dipole selection rules, and when the semiconductor has a **direct band gap**. In the case of a semiconductor with an **indirect band gap**, a phonon must be absorbed or emitted whenever the electron jumps between the bands, and this substantially reduces the transition probability. Many III–V compound semiconductors like GaAs exhibit very strong interband transitions because they have direct band gaps and E1 transitions are allowed between the conduction and valence bands. By contrast, the elemental semiconductors silicon and germanium have weaker transition probabilities because their band gaps are indirect.

The fact that the electrons and holes relax within their bands before emission occurs means that there is a qualitative difference between the emission and absorption spectra of a semiconductor. This contrasts with atomic spectra, where the absorption and emission lines both occur at the same energy. In a semiconductor, the band gap E_g represents the threshold for absorption to occur, whereas it corresponds to the transition energy in the case of emission. This point is illustrated in Fig. 4.7, which shows the absorption and emission spectra of the direct-gap III–V compound semiconductor GaN at 4 K. In the absorption spectrum, a threshold is observed at the band gap (3.50 eV) and a continuous absorption band is observed for photon energies that exceed this threshold. The emission spectrum, by contrast, consists of a single line at an energy close to E_g.

The broadened peak identified by the arrow in the absorption spectrum shown in Fig. 4.7(a) is caused by the formation of **excitons**. Excitons are bound electron and hole pairs held together by their mutual Coulomb interaction. Excitons are important in quantum optical experiments because they behave like two-level atoms to a certain level of approximation. Throughout this book, we shall refer to a number of quantum optical effects relating to excitons in low-dimensional semiconductor structures. Further details about their optical properties may be found in Appendix D.

Excitons are observed at a photon energy of $E_g - E_X$, where E_X is the exciton binding energy. The value of E_X is typically of order 0.01 eV. See Exercise 4.10.

4.7   Lasers

The word 'laser' is an acronym that stands for 'Light Amplification by Stimulated Emission of Radiation'. Laser operation was first demonstrated in 1960, and since then, lasers have become essential tools in nonlinear and quantum optics. In this section we give a brief review of the physical principles that underly laser operation, and then give a short description of the main properties of the lasers that are commonly used in the laboratory.

4.7.1   Laser oscillation

Figure 4.8 shows a schematic diagram of a typical laser oscillator. The laser consists of a **gain medium** and two end mirrors called the **output coupler** and **high reflector** with reflectivities of R_1 and R_2, respectively. Light bounces between the two end mirrors and is amplified each time it passes through the gain medium. If the amplification in the gain medium is sufficient to balance the losses during a round trip, then oscillation can occur and the laser will operate. The output of the laser emerges through the output coupler, which has a partially transmitting coating.

The light amplification that occurs within the gain medium is quantified by the **gain coefficient** $\gamma(\omega)$ defined by:

$$\frac{dI}{dz} = \gamma(\omega)I(z), \qquad (4.38)$$

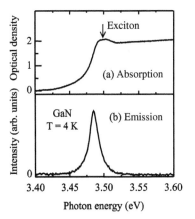

Fig. 4.7 (a) Absorption and (b) emission spectra of a GaN crystal of thickness 0.5 μm at 4 K. In part (a), the optical density is directly proportional to the absorption coefficient. (Unpublished data from K. S. Kyhm and R. A. Taylor.)

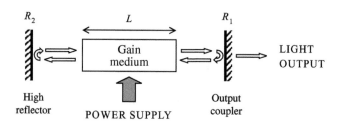

Fig. 4.8 Schematic diagram of a laser oscillator.

where I is the optical intensity, ω is the angular frequency of the light, and z is the direction of propagation of the beam. Integration of eqn 4.38 yields:

$$I(z) = I_0 e^{\gamma z}, \qquad (4.39)$$

which shows that the light intensity grows exponentially inside the gain medium, in the absence of gain saturation (see below).

Let us consider the case in which the light beam is close to resonance with an atomic transition of angular frequency ω_0. The beam will trigger both absorption and stimulated-emission transitions as shown in Fig. 4.2. For amplification to occur, we require that the stimulated-emission rate should exceed the absorption rate, so that the number of photons in the beam increases as it propagates through the gain medium. From eqns 4.5 and 4.6 we see that this occurs when:

$$B_{21}^\omega N_2 u(\omega) > B_{12}^\omega N_1 u(\omega), \qquad (4.40)$$

which, on substituting from eqn 4.10, implies:

$$N_2 > \frac{g_2}{g_1} N_1. \qquad (4.41)$$

In thermal equilibrium, the ratio of N_2 to N_1 is given by the Boltzmann formula of eqn 4.8. This means that it is never possible to satisfy eqn 4.41, and the light intensity decays as it propagates because the absorption rate exceeds the stimulated-emission rate. Equation 4.41 can therefore only be satisfied in non-equilibrium conditions called **population inversion**. Population inversion is normally achieved by pumping energy into the medium to excite a large number of atoms to the excited state. The energy is derived from an external power source, as indicated schematically in Fig. 4.8.

The population-inversion density ΔN can be defined as:

$$\Delta N = N_2 - \frac{g_2}{g_1} N_1. \qquad (4.42)$$

The gain coefficient that is achieved for an inversion density ΔN is given by (see Exercise 4.11):

$$\gamma(\omega) = \frac{\lambda^2}{4n^2\tau} \Delta N g_\omega(\omega), \qquad (4.43)$$

where λ is the vacuum wavelength, n is the refractive index of the gain medium, τ is the radiative lifetime of the upper level, and $g_\omega(\omega)$ is the

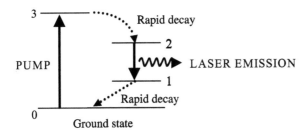

Fig. 4.9 Population inversion mechanism in a four-level laser.

spectral lineshape function defined in eqn 4.26. This shows that the gain is directly proportional to the inversion density and also to the transition probability via $1/\tau \equiv A_{21}$.

The population inversion required for laser oscillation is usually obtained by 'pumping' atoms to a higher level. Figure 4.9 illustrates the general scheme for obtaining population inversion in a **four-level laser**. Atoms are pumped from the ground state to level 3 from where they decay rapidly to level 2, creating population inversion with respect to level 1. The pumping process to level 3 can be optical (e.g. from a flash lamp or another laser) or electrical. The decay rate from level 1 back to the ground state must be fast to prevent atoms accumulating in that level and destroying the population inversion. In the case of flash-lamp pumping, it is convenient if level 3 is in fact a broad band, so that a large fraction of the lamp's output energy can be harnessed.

In normal operation the population inversion will be proportional to the pumping rate $\mathcal{R}$, which in turn is proportional to the power supplied by the pump source. The variation of the gain in the medium with the pumping rate will then be linear at first, as sketched in Fig. 4.10. However, a situation is eventually reached when the gain is sufficient to initiate laser operation. This is called the **laser threshold**. At threshold, the laser begins to emit light, and the gain coefficient (and hence the population inversion) gets clamped at the threshold value. (See Fig. 4.10.)

The value of the gain coefficient at the threshold can be calculated by considering the amount of amplification required to maintain laser oscillation. In general, this is a rather complicated calculation, because the population inversion will often vary throughout the gain medium. Moreover, **gain saturation** occurs as the photon density inside the cavity increases. The analysis below is therefore only valid for a uniform gain medium in the weak-saturation limit.

In stable oscillation conditions, the increase of the intensity due to the gain must exactly balance the losses due to the imperfect reflectivity of the end mirrors and any other losses that may be present within the cavity. On following the beam through a round-trip of the cavity shown in Fig. 4.8, we see that the oscillation condition can be written:

$$R_1 R_2 \xi e^{2\gamma L} = 1, \tag{4.44}$$

where L is the length of the gain medium and ξ is a factor that accounts for other losses such as scattering and absorption in the optics. The

Population inversion can also be achieved in **three-level laser** schemes in which level 1 and the ground state coincide. In general, three-level lasers have higher thresholds than four-level lasers, because the lower level is initially occupied, and it is therefore necessary to pump more than half of the atoms out of the ground state to achieve population inversion.

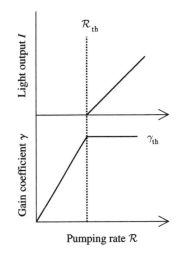

Fig. 4.10 Idealized variation of the gain coefficient and light output with the pumping rate $\mathcal{R}$ in a laser with a threshold at $\mathcal{R}_{\text{th}}$.

factor of two in the exponential allows for the fact that the light passes through the gain medium twice during a round trip.

The oscillation condition in eqn 4.44 can be rewritten as:

$$\gamma = -\frac{1}{2L}\ln(R_1 R_2) - \frac{1}{2L}\ln\xi. \qquad (4.45)$$

This defines the threshold gain γ_{th} required to make the laser oscillate. This gain will be achieved for a certain pumping rate $\mathcal{R}_{\text{th}}$. For pumping rates larger than $\mathcal{R}_{\text{th}}$, the gain cannot increase further since it is clamped by the oscillation condition. The extra energy of the pumping source thus goes into generating the light output, which increases linearly with $(\mathcal{R} - \mathcal{R}_{\text{th}})$ for $\mathcal{R} > \mathcal{R}_{\text{th}}$ in this simplified model, as shown in Fig. 4.10.

In an ideal laser in which the losses are low and the high reflector has near perfect reflectivity, the value of $\mathcal{R}_{\text{th}}$ is determined by the transmission of the output coupler. A low value of $(1 - R_1)$ will give a low threshold, but also a low power output, because very little of the energy oscillating inside the cavity can escape. Conversely, a higher value of $(1 - R_1)$ increases the threshold, but also increases the output coupling efficiency, so that higher powers can in principle be obtained. In practice, the choice of the value of output coupler is often determined by the amount of power available from the pumping source.

4.7.2   Laser modes

The cavity is an essential part of a laser, providing the positive feedback that turns an amplifier into an oscillator. Furthermore, it has a profound effect on the properties of the beam that emerges from the output coupler. In this section we briefly discuss the mode structure of the laser light that is determined by the cavity, starting with the spatial properties of the beam.

Many lasers employ slightly curved mirrors rather than plane mirrors. The general conclusions of the discussion of the beam modes given here are not affected by this detail.

Consider the beam emerging through the output coupler of a laser as shown in Fig. 4.11. The fact that the light rays have to bounce repeatedly between the cavity mirrors leads to one of the most obvious properties of laser beams, namely that they are highly *directional*. In ideal circumstances, the beam will have only a very small divergence determined by the design of the cavity.

The variation of the electric field amplitude through a cross-sectional slice of the beam is determined by the **transverse mode** structure. The

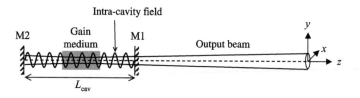

Fig. 4.11 Schematic representation of the output beam from a laser propagating in the z-direction. M1 and M2 are the output coupler and high reflector mirrors, respectively. Also shown is the intra-cavity electric field, which has nodes at the end mirrors. L_{cav} is the length of the cavity.

modes are labelled by two integers m and n. If the beam is propagating in the z-direction as shown in Fig. 4.11, the (x, y)-dependence of the field amplitude for a particular mode is given by:

$$\mathcal{E}_{mn}(x,y) = \mathcal{E}_0 H_m(\sqrt{2}x/w)H_n(\sqrt{2}y/w)\exp\left(-\frac{x^2+y^2}{w^2}\right), \quad (4.46)$$

where H_m and H_n are the Hermite polynomials that we encountered previously in the harmonic oscillator. (See Table 3.5.) The first three Hermite polynomials are given by:

$$H_0(u) = 1,$$
$$H_1(u) = 2u,$$
$$H_2(u) = 4u^2 - 2. \quad (4.47)$$

The parameter w that appears in eqn 4.46 determines the width of the beam and is called the beam spot size.

In general, the cross-section of the beam may be described by any of the transverse modes, or by a superposition of several of them. However, it is normal to try to operate the laser on the 00 mode, which has a Gaussian field distribution:

$$\mathcal{E}_{00}(x,y) = \mathcal{E}_0 \exp[-(x^2+y^2)/w^2] \equiv \mathcal{E}_0 \exp(-r^2/w^2), \quad (4.48)$$

where $r = \sqrt{x^2+y^2}$ is the radial distance from the centre of the beam. The 00 mode is the closest approximation to an idealized ray of light that can be found in nature. It has the smallest divergence of all the modes and can be focused to the smallest size.

The suppression of the transverse modes with $m \geq 1$ or $n \geq 1$ is usually achieved by placing an appropriate aperture within the cavity. The size of the aperture is chosen so that it clips the higher modes but not the smaller 00 mode. The higher modes thus experience a severe loss that prevents them from oscillating.

Now let us consider the **longitudinal mode** structure of the laser, which relates to the variation of the electric field with z, where the z-axis lies along the cavity axis. The light bouncing repeatedly around the cavity must have nodes (field zeros) at the mirrors because they have high reflectivities. The intra-cavity field is therefore a standing wave, with an integer number of half wavelengths inside the cavity, as shown in Fig 4.11. If the length of the cavity is L_{cav}, the standing wave condition can be written as:

$$L_{\text{cav}} = \text{integer} \times \frac{\lambda}{2} = \text{integer} \times \frac{\pi c}{n_{\text{cav}}\omega}, \quad (4.49)$$

where n_{cav} is the average refractive index within the cavity. This can be rearranged to give the allowed angular frequencies of the cavity modes:

$$\omega_{\text{mode}} = \text{integer} \times \frac{\pi c}{n_{\text{cav}}L_{\text{cav}}}, \quad (4.50)$$

which implies that the modes are separated in angular frequency by:

$$\Delta\omega_{\text{mode}} = \frac{\pi c}{n_{\text{cav}}L_{\text{cav}}}. \quad (4.51)$$

The separation of the longitudinal modes in frequency units ($\Delta\nu_{\text{mode}}$) is equal to $\Delta\omega_{\text{mode}}/2\pi = c/2n_{\text{cav}}L_{\text{cav}}$.

Thus the longitudinal-mode spacing is larger in shorter cavities.

The longitudinal-mode structure, together with the properties of the gain medium, determine the emission spectrum of the laser. For a given

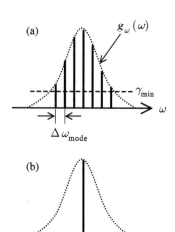

Fig. 4.12 (a) Multi-mode and (b) single-mode operation of a laser with longitudinal mode spacing $\Delta\omega_{\text{mode}}$. The dotted line represents the spectral lineshape function $g_\omega(\omega)$. In (a), the dashed line labelled γ_{min} represents the minimum gain required to overcome the cavity losses. The modes with gain values larger than γ_{min} can oscillate. The figure is drawn for the case where the laser is operating well above threshold, so that most of the modes within the spectral line can oscillate.

The time-averaged output spectrum of a mode-locked laser would appear similar to that of the multi-mode laser shown in Fig. 4.12(a). The difference is that the phases of the longitudinal modes in a mode-locked laser are all locked together. This contrasts with a multi-mode laser, in which the modes oscillate independently of each other and have random relative phases.

mode to oscillate, its frequency must lie within the linewidth of the laser transition as determined by its lineshape function $g_\omega(\omega)$. It will normally be the case that the mode spacing is much smaller than the linewidth, and so there will be many modes that satisfy this condition, as illustrated in Fig. 4.12(a). If the line broadening is inhomogeneous, as with Doppler-broadened lines in a gas laser, the different atoms that contribute to different parts of the spectrum can support lasing on any of the modes that have sufficient gain to overcome the cavity losses, and the laser will oscillate on many modes simultaneously. This type of operation is called **multi-mode**. In the case where the laser is operating well above threshold, most of the modes that fall within the spectral line of the transition will have enough gain to oscillate. In this situation, the spectral width of the laser spectrum is roughly the same as that of the equivalent line in a discharge lamp. Since the modes are effectively independent of each other in an inhomogeneous gain medium, their relative optical phases are random.

Figure 4.12(b) illustrates the **single-mode** operation of the laser in which only one longitudinal mode is oscillating. Single-mode operation is typically achieved by introducing a frequency-selective element such as a Fabry–Perot etalon into the cavity. The etalon introduces a frequency-selective loss into the cavity, thereby picking out the single longitudinal mode with the lowest loss. In this mode of operation, the linewidth of the laser is very narrow, being determined by the properties of the cavity rather than the atomic transition. It is not uncommon to achieve linewidths in the MHz range or less by this method, leading to coherence lengths of hundreds of metres. (See eqns 2.40 and 2.41.) Furthermore, by tuning the etalon, the frequency of the laser can be scanned through the lineshape function, thereby generating tunable narrow-band emission. Such lasers are routinely used for high-resolution laser spectroscopy.

A third important mode of operation of the laser is called **mode-locked**. In this case, the laser operates on as many longitudinal modes as the gain medium can support, but the phases of all the modes are locked together. The temporal properties of the output beam can be found by taking the Fourier transform of a comb of fields with a regular frequency separation given by eqn 4.51 and with their amplitude modulated by the gain spectrum of the laser transition. The regular frequency spacing of the modes leads to a regular train of pulses separated in time by $2n_{\text{cav}}L_{\text{cav}}/c$. The duration Δt of the pulses is determined by the spectral width of the gain according to the time–bandwidth product:

$$\Delta\omega\Delta t \sim 1, \tag{4.52}$$

where $\Delta\omega$ is the spectral width of the gain medium. Shorter pulses are therefore generated by gain media with a very broad spectral range. Dye lasers and Ti:sapphire lasers have very broad gain bandwidths, and can be used to generate pulses in the femtosecond time range.

The pulse structure of a mode-locked laser can be understood by realizing that the time for light to travel around the cavity is equal to $2n_{\text{cav}}L_{\text{cav}}/c$. It is then apparent that mode-locked operation implies that

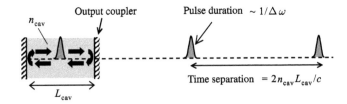

Fig. 4.13 Mode-locked laser operation. The pulses are separated by the cavity round-trip time, namely $2n_{cav}L_{cav}/c$, where n_{cav} and L_{cav} are the average refractive index and length of the cavity, respectively. The duration of the pulses is determined by the gain bandwidth $\Delta\omega$ of the gain medium.

there is just a single pulse circulating within the cavity, so that an output pulse is emitted each time it hits the output coupler, as illustrated in Fig. 4.13. The single-pulse operation is typically achieved by introducing a time-dependent loss modulator into the cavity. This both suppresses continuous operation and favours the mode of operation with a pulse passing through the modulator at the time when its loss is smallest.

4.7.3   Laser properties

Laser light has many attractive features which can be adapted to the needs of particular experiments or applications. The properties of the light are determined mainly by:

- the gain medium that is employed,
- the design of the cavity,
- the mode of operation.

The gain medium primarily determines the wavelengths that are generated. It also affects whether the laser can be operated continuously or only in pulses. The cavity design determines the transverse and longitudinal mode structure, while the choice of the operational mode determines the linewidth and pulse width, as appropriate.

Two features that are common to all types of lasers are the directionality of the beam and its spectral brightness. This is apparent when comparing to black-body or atomic-discharge lamps. The lamps emit in all directions, which means that the intensity in any particular direction is rather small. Furthermore, in the case of the black-body source, the energy is distributed over a very broad spectrum. Thus the red spot generated by a 1 mW He:Ne laser is many orders of magnitude brighter than the filtered red spot of the same area obtained by collecting light from a 100 W tungsten filament lamp. Other useful properties such as monochromaticity, long coherence length, and short pulse emission depend upon the mode of operation, according to Table 4.3.

Table 4.4 lists some of the more common lasers that are used in the laboratory and in industry. The lasers are generally classified according to the chemical phase of the gain medium, that is, solid, liquid, or gas. Most lasers can be operated continuously (i.e. in 'continuous wave' or 'CW' mode), but some (e.g. ruby) generally only operate in pulsed mode.

Table 4.3 Properties of the light emitted by the three different types of laser discussed in Section 4.7.2.

	Multimode	Single-mode	Mode-locked
Beam directionality	✓	✓	✓
High spectral brightness	✓	✓	✓
Highly monochromatic emission		✓	
Very long coherence length		✓	
Ultrashort pulse duration			✓

Table 4.4 Common lasers and their main emission wavelengths. When the laser can operate on several different lines, only the strongest wavelengths are listed. In the case of semiconductor lasers, the individual devices only operate at one wavelength, and different wavelengths are obtained by using different crystals.

Laser	Type	Wavelength(s) (nm)	Notes
CO_2	Gas	10 600	
Nd:YAG	Solid state	1064	
Nd:glass	Solid state	1054	
Semiconductor	Solid state	670, 800, 1300, 1550	Nominal values only
Ti:sapphire	Solid state	700–1100	Tunable
Ruby	Solid state	694.3	
Krypton ion	Gas	676.4	
He:Ne	Gas	632.8	
Rhodamine 6G dye	Liquid	550–650	Tunable
Argon ion	Gas	488.0, 514.5	
He:Cd	Gas	325.0, 441.6	

Others (e.g. CO_2 or Nd:YAG) can be operated either way. Note that in the case of solid-state lasers, the emission wavelength can change slightly when the active atoms are doped into different host materials. Thus the principal emission line of the Nd^{3+} ion is 1064 nm in an yttrium aluminium garnet (YAG) crystal, but 1054 nm in phosphate glass. Note also that the different wavelengths listed for semiconductor diode lasers cannot be obtained from a single device. Each individual device operates at a single wavelength determined by the band gap of the active material, and the different wavelengths are obtained by using separate crystals with differing band gaps.

It is apparent from Table 4.4 that it is possible to obtain laser radiation at many different wavelengths. Moreover, the range of available wavelengths can be extended further by using techniques of nonlinear optics. (See Table 2.2.) For example, it is very common to use frequency-doubling techniques to generate 532 nm radiation from a Nd:YAG laser, and sometimes to repeat the process to produce fourth-harmonic radiation at 266 nm.

The wavelength of semiconductor lasers can also be tuned, although only over a rather limited range.

The two broadly tunable lasers listed in Table 4.4, namely the dye and Ti:sapphire lasers, have found widespread application in quantum optics. When operating on a single longitudinal mode, they give very narrow emission lines that can be tuned to resonance with atomic transitions. Alternatively, their broad gain bandwidth permits ultrashort pulse generation for investigating dynamical processes on very fast time scales.

Further reading

The basic principles of radiative transitions are covered in most standard quantum-mechanics textbooks, for example: Gasiorowicz (1996) or Schiff (1969). More detailed information on atomic selection rules and transition rates may be found in atomic-physics texts such as Corney (1977), Foot (2005), Haken and Wolf (2000), or Woodgate (1980). The optical properties of solids are described at length in Fox (2001), while laser physics is covered in Silfvast (1996), Svelto (1998), or Yariv (1997).

Exercises

(4.1) In the Einstein analysis we assume that the light radiation has a broad spectrum compared to the transition line. Let us now consider the contrary situation in which the spectral width of the light beam is much smaller than the linewidth of the transition. This is the kind of situation that occurs when a narrow-band laser beam interacts with an atom, either inside a laser cavity or externally.

(a) Explain why it is appropriate to write the spectral energy intensity of the beam as:

$$u(\omega') = u_\omega \delta(\omega' - \omega),$$

where ω is the angular frequency of the beam, u_ω is its energy density in $\mathrm{J\,m}^{-3}$, and $\delta(x)$ is the Dirac delta function.

(b) Let us assume that the frequency dependence of the absorption probability follows the spectral lineshape function $g_\omega(\omega)$. This implies that the Einstein B coefficients will also vary with frequency. Explain why it is appropriate to write the frequency dependence of the

Einstein B_{12} coefficient as:[1]

$$B_{12}(\omega') = \frac{g_2}{g_1} \frac{\pi^2 c^3}{\hbar n^3 \omega'^3} \frac{1}{\tau} g_\omega(\omega'),$$

where g_1 and g_2 are the lower and upper level degeneracies, n is the refractive index of the medium, and τ is the radiative lifetime of the upper level.

(c) Hence show that the total absorption rate defined as

$$W_{12} = N_1 \int_0^\infty B_{12}(\omega') u(\omega') \, \mathrm{d}\omega'$$

is given by:

$$W_{12} = N_1 \frac{g_2}{g_1} \frac{\pi^2 c^3}{\hbar n^3 \omega^3 \tau} u_\omega g_\omega(\omega).$$

(d) Repeat the argument to show that the total stimulated-emission rate is given by:

$$W_{21} = N_2 \frac{\pi^2 c^3}{\hbar n^3 \omega^3 \tau} u_\omega g_\omega(\omega).$$

---

[1] Spectroscopists often take a slightly different approach to that adopted in this exercise and define a unique value of B_{12} for the transition. The variation of the absorption probability with ω is then achieved by including the lineshape function explicitly in the calculation of the transition rate. The end result is the same.

(4.2) Show that the parities of the initial and final states involved in E1 transitions must be different.

(4.3) The wave functions of the hydrogen atom may be written in the form (cf. eqns 3.60 and 3.45):

$$\psi(r,\theta,\varphi) = F(r,\theta)\exp(im_l\varphi),$$

where m_l is the magnetic quantum number. Consider an E1 transition between an initial state with magnetic quantum number m to a final state with magnetic quantum number m'. By considering the integral over φ, show that the matrix element is zero unless:

(a) $m' = m$ for $\hat{\mathbf{z}}$-polarized light;

(b) $m' = m+1$ for σ^+ light (polarization $\hat{\mathbf{x}}+i\hat{\mathbf{y}}$);

(c) $m' = m-1$ for σ^- light (polarization $\hat{\mathbf{x}}-i\hat{\mathbf{y}}$);

(d) $m' = m\pm1$ for $\hat{\mathbf{x}}$- or $\hat{\mathbf{y}}$-polarized light.

(4.4) Consider an atom emitting a burst of light of angular frequency ω_0 with an exponentially decaying intensity $I(t) = I(0)\exp(-t/\tau)$ for $t \geq 0$.

(a) Explain why the time-dependent electric field can be taken in the form:

$$t < 0: \quad \mathcal{E}(t) = 0,$$
$$t \geq 0: \quad \mathcal{E}(t) = \mathcal{E}_0\cos\omega_0 t\, e^{-t/2\tau}.$$

(b) By taking the Fourier transform of the electric field, namely:

$$\mathcal{E}(\omega) = \frac{1}{\sqrt{2\pi}}\int_{-\infty}^{+\infty}\mathcal{E}(t)e^{i\omega t}\,dt,$$

and making the assumption that $\omega_0 \gg 1/\tau$, show that the emission spectrum is given by:

$$I(\omega) \propto |\mathcal{E}(\omega)|^2 \propto \frac{1}{(\omega-\omega_0)^2 + (1/2\tau)^2}.$$

(c) Hence show that the normalized spectral lineshape function is given by eqn 4.29.

(4.5) The detailed calculation of collision times and mean free paths in a gas is rather complicated. In this exercise we give a highly simplified treatment that will suffice to calculate the order of magnitude of the linewidth in pressure broadening.

(a) Consider a single molecule with a collision cross-section σ_s moving through a gas of stationary molecules with N/V molecules per unit volume. Show that the mean free path between collisions is given by:

$$L = \frac{1}{(N/V)\sigma_s}.$$

(b) Explain why the probability that a molecule's speed lies in the range c to $c+dc$ is given by:

$$\mathcal{P}(c)dc \propto \exp(-mc^2/2k_BT)4\pi c^2\,dc.$$

Hence show that the average speed of the molecules is given by:

$$\bar{c} = \sqrt{\frac{8k_BT}{\pi m}}.$$

(c) The average time between collisions in a gas is given by:

$$\tau_{\text{collision}} \sim L/\bar{c}.$$

Use this result to derive eqn 4.31 for an ideal gas.

(4.6) The collision cross-section is an effective area which determines whether two atoms will collide or not. If we assume that it is approximately equal to the cross-sectional area of the atom, estimate the value of $\tau_{\text{collision}}$ for sodium under STP conditions. (The relative atomic mass of the sodium atom is 23.0, and its radius is $\sim$0.2 nm.)

(4.7) Mercury has a relative atomic mass of 200.6 and an atomic radius of 0.17 nm. The transition at 546.1 nm has an Einstein A coefficient of 4.9×10^7 s^{-1}. Calculate the natural, Doppler, and collisional linewidths in frequency units (Hz) for this transition in:

(a) a low-pressure lamp operating at 250 °C with a pressure of 10^{-4} atmospheres,

(b) a high-pressure lamp operating at 500 °C with a pressure of 1 atmosphere.

(4.8) The 692.9 nm line of neon (relative atomic mass 20.18) has an Einstein A coefficient of 1.7×10^7 s^{-1}. Find the temperature at which the natural and Doppler linewidths would be the same in a low-pressure lamp.

(4.9) Consider a spectral line of centre frequency ω_0, FWHM $\Delta\omega$, and spectral lineshape function $g_\omega(\omega)$. Explain why the value of the lineshape function at the line centre must be given by:

$$g_\omega(\omega_0) = C/\Delta\omega,$$

where C is a numerical constant of order unity. Evaluate C for Lorentzian and Gaussian lines.

(4.10) An exciton may be considered as a hydrogen-like atom in which an electron in the conduction band and a hole in the valence band are bound together by their mutual Coulomb interaction. By treating the electron and hole as free particles with masses of m_e^* and m_h^*, respectively, and considering the semiconductor as a dielectric medium with a relative permittivity of ϵ_r, explain why the binding energy is given by:

$$E_X = \frac{\mu}{m_0} \frac{1}{\epsilon_r^2} (R_\infty hc),$$

where μ is the reduced mass of the exciton and R_∞ is the Rydberg constant. Evaluate E_X for GaAs where $m_e^* = 0.067m_0$, $m_h^* = 0.2m_0$, and $\epsilon_r = 12.8$.

(4.11) Consider an atom interacting with a monochromatic beam of light with angular frequency ω, where ω is close to a transition frequency ω_0.

(a) Use the results of Exercise (4.1) to show that the net rate of downward transitions (defined as the stimulated-emission rate less the absorption rate) is given by:

$$W_{21}^{net} = \frac{\pi^2 c^3}{\hbar n^3 \omega^3 \tau} u_\omega g_\omega(\omega) \Delta N,$$

where $\Delta N = N_2 - (g_2/g_1)N_1$, u_ω is the energy per unit volume of the beam, and $g_\omega(\omega)$ is the spectral lineshape function.

(b) Show that optical intensity is given by $I = u_\omega c/n$.

(c) Consider a unit area of beam propagating in the $+z$-direction through the medium. Show that the incremental increase in the intensity dI in a length element dz is given by:

$$dI = W_{21}^{net} \hbar\omega \, dz.$$

(d) Hence show that the gain coefficient is given by eqn 4.43.

(4.12) Calculate the fraction of the energy of a 00-mode laser beam with beam radius w within a distance w from the beam centre.

(4.13) A helium–neon laser consists of a laser tube of length 0.3 m with mirrors bonded to the end of the tube. The output coupler has a reflectivity of 99%. The laser operates on the 632.8 nm transition of neon (relative atomic mass 20.18), which has an Einstein A coefficient of 3.4×10^6 s^{-1}. The tube runs at 200 °C and the laser transition is Doppler-broadened. On the assumption that the only loss in the cavity is through the output coupler, that the average refractive index is equal to unity, and that the laser operates at the line centre, calculate:

(a) the gain coefficient in the laser tube;

(b) the population inversion density.

(4.14) A mode-locked Ti:sapphire laser has a gain bandwidth of 100 nm centred at 800 nm. The cavity length is 2 m, and the average refractive index is effectively unity because the laser crystal is much shorter than the cavity.

(a) Calculate the pulse repetition frequency.

(b) Estimate the duration of the shortest pulses that can be produced by the laser.

(c) Estimate the number of longitudinal modes that will be oscillating when the laser is producing the shortest possible pulses.

Part II

Photons

Introduction to Part II

We begin our discussion of quantum optics by considering the intrinsic properties of light that manifest its quantum nature. We start in Chapter 5 by considering the classification of the light according to the type of photon statistics that can occur, namely sub-Poissonian, Poissonian, or super-Poissonian. Then, in Chapter 6, we look at the work of Hanbury Brown and Twiss which leads to the concept of second-order correlation functions. This will allow us to introduce another classification of light according to whether the photon streams are antibunched, coherent, or bunched. We shall see that both sub-Poissonian photon statistics and photon antibunching are pure quantum effects with no classical counterpart.

In Chapter 7 we study the properties of coherent states, which are the quantum equivalents to classical electromagnetic waves. We shall again discover new states called squeezed states that have no classical equivalent. Finally, in Chapter 8 we give a brief introduction to the quantum theory of light. This will allow us to understand the properties of photon number states and see how they relate to the experimental results described in the previous three chapters.

The four chapters that comprise Part II of this book assume a reasonable familiarity with classical optics. A brief summary of the relevant background theory may be found in Chapter 2.

Photon statistics

The task of quantum optics is to study the consequences of considering a beam of light as a stream of photons rather than as a classical wave. It turns out that the differences are rather subtle, and we have to look quite hard to see significant departures from the predictions of the classical theories. In this chapter we shall approach the subject from the perspective of the statistical properties of the photon stream. We shall study the three different types of photon statistics that can occur, namely: Poissonian, super-Poissonian, and sub-Poissonian. A key result that emerges is that the observation of Poissonian and super-Poissonian statistics in photodetection experiments is consistent with the classical theory of light, but not sub-Poissonian statistics. Hence the observation of sub-Poissonian photon statistics constitutes direct confirmation of the photon nature of light. Unfortunately, it transpires that sub-Poissonian light is very sensitive to optical losses and inefficient detection. This explains why it has only been observed relatively recently, following the development of high efficiency detectors.

The chapter concludes with a consideration of some of the practical consequences of the photon statistics. This will lead us to discuss the origin of shot noise in photodetectors, and to consider how it can be reduced by using light with sub-Poissonian statistics.

5.1	Introduction	75
5.2	Photon-counting statistics	76
5.3	Coherent light: Poissonian photon statistics	78
5.4	Classification of light by photon statistics	82
5.5	Super-Poissonian light	83
5.6	Sub-Poissonian light	87
5.7	Degradation of photon statistics by losses	88
5.8	Theory of photodetection	89
5.9	Shot noise in photodiodes	94
5.10	Observation of sub-Poissonian photon statistics	99
Further reading		103
Exercises		103

5.1 Introduction

We can introduce our discussion of photon statistics by considering the detection of a light beam by a photon counter as illustrated in Fig. 5.1. The photon counter consists of a very sensitive light detector such as a photomultiplier tube (PMT) or avalanche photodiode (APD) connected to an electronic counter. The detector produces short voltage pulses in response to the light beam and the counter registers the number of pulses that are emitted within a certain time interval set by the user. Photon counters thus operate in a very similar way to the Geiger counters used to count the particles emitted by the decay of radioactive nuclei.

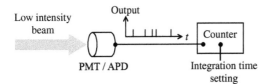

Fig. 5.1 Detection of a faint light beam by a PMT or APD and pulse counting electronics.

The analogy between a photon counter and a Geiger counter makes it apparent that the number of counts that we might expect to observe in a given time interval would not be constant. When using a Geiger counter, the count rate fluctuates about the average value due to the intrinsically random nature of the radioactive decay process. The same thing happens with a photon counter: the average count rate is determined by the intensity of the light beam, but the actual count rate fluctuates from measurement to measurement. It is these fluctuations in the count rate that are our concern here.

At first sight it might appear that the fact that the detector emits individual pulses is clear and conclusive evidence that the impinging light beam consists of a stream of discrete energy packets that we generally call 'photons'. The fluctuations in the count rate would then give information about the statistical properties of the incoming photon stream. Unfortunately, the argument is not quite that simple. It has been a long-standing issue in optical physics whether the individual events registered by photon counters are necessarily related to the photon statistics, or whether they are just an artefact of the detection process. This means that we have to distinguish carefully between:

(1) the statistical nature of the photodetection process;
(2) the intrinsic photon statistics of the light beam.

If we were approaching this subject from a historical perspective, it would make sense to look at the theory of photodetection first in order to avoid the danger of jumping to false conclusions. From a conceptual point of view, however, it is more interesting to examine the intrinsic statistical nature of the light first, and then return to consider how this relates to the results of photodetection experiments. It is this second approach that we adopt here.

In adopting an approach that starts from the photons, we are anticipating the final result that some experiments can *only* be explained if we attribute the photocount fluctuations to the underlying photon statistics. It must be emphasized, however, that the actual number of experiments that fall into this category is rather small. In Section 5.8.1 we shall show that *most* of the results obtained in photon-counting experiments can be explained by **semi-classical** models in which we treat the light classically but quantize the photoelectric effect in the detector. At the same time, this semi-classical approach tells us where to look for effects that *cannot* be explained by the classical theories of light. This second type of experiment is particularly interesting because it gives a clear proof of the quantum nature of light.

5.2 Photon-counting statistics

Let us consider the outcome of a photon-counting experiment as illustrated in Fig. 5.1. The basic function of the experiment is to count the number of photons that strike the detector in a user-specified time interval T. We start with the simplest case and consider the detection of a perfectly coherent monochromatic beam of angular frequency ω and constant intensity I. In the quantum picture of light, we consider the beam to consist of a stream of photons. The **photon flux** Φ is defined as the average number of photons passing through a cross-section of the beam in unit time. Φ is easily calculated by dividing the energy flux by the energy of the individual photons:

$$\Phi = \frac{IA}{\hbar\omega} \equiv \frac{P}{\hbar\omega} \text{ photons s}^{-1}, \tag{5.1}$$

where A is the area of the beam and P is the power.

Photon-counting detectors are specified by their **quantum efficiency** η, which is defined as the ratio of the number of photocounts to the number of incident photons. The average number of counts registered by the detector in a counting time T is thus given by:

$$N(T) = \eta\Phi T = \frac{\eta PT}{\hbar\omega}. \tag{5.2}$$

The corresponding average **count rate** $\mathcal{R}$ is given by:

$$\mathcal{R} = \frac{N}{T} = \eta\Phi = \frac{\eta P}{\hbar\omega} \text{ counts s}^{-1}. \tag{5.3}$$

The maximum count rate that can be registered with a photon-counting system is usually determined by the fact that the detectors need a certain amount of time to recover after each detection event, which means that a 'dead time' of ~ 1 µs must typically elapse between successive counts. This sets a practical upper limit on $\mathcal{R}$ of around 10^6 counts s^{-1}. With typical values of η for modern detectors of 10% or more, eqn 5.3 shows that photon counters are only useful for analysing the properties of very faint light beams with optical powers of $\sim 10^{-12}$ W or less. The detection of light beams with higher power levels is done by a different method and will be discussed in Section 5.9.

The photon flux given in eqn 5.1 and the detector count rate given by eqn 5.3 both represent the *average* properties of the beam. A beam of light with a well-defined average photon flux will nevertheless show photon number fluctuations at short time intervals. This is a consequence of the inherent 'graininess' of the beam caused by chopping it up into photons. We can see this more clearly with the aid of a simple example.

Consider a beam of light of photon energy 2.0 eV with an average power of 1 nW. Such a beam could be obtained by taking a He:Ne laser operating at 633 nm with a power of 1 mW and attenuating it by a factor 10^6 by using appropriate filters. The average photon flux from eqn 5.1 is:

$$\Phi = \frac{10^{-9}}{2.0 \times (1.6 \times 10^{-19})} = 3.1 \times 10^9 \text{ photons s}^{-1}.$$

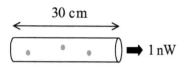

Fig. 5.2 A 30 cm section of a beam light at 633 nm with a power of 1 nW contains three photons on average.

Since the velocity of light is 3×10^8 m s^{-1}, a segment of the beam with a length of 3×10^8 m would contain 3.1×10^9 photons on average. On a smaller scale, we would expect an average of 31 photons within a 3 m segment of beam. In still smaller segments, the average photon number becomes fractional. For example, a 1-ns count time corresponds to a 30 cm segment of beam, and contains an average of 3.1 photons. Now photons are discrete energy packets, and the actual number of photons has to be an integer. We must therefore have an integer number of photons in each beam segment, as illustrated in Fig. 5.2. In the next section we shall show that if we assume that the photons are equally likely to be at any point within the beam, then we find random fluctuations above and below the mean value. If we were to look at 30 such beam segments, we might therefore find a sequence of photon counts that looks something like:

$$1, 6, 3, 1, 2, 2, 4, 4, 2, 3, 4, 3, 1, 3, 6, 5, 0, 4, 1, 1, 6, 2, 2, 6, 4, 1, 4, 3, 4, 6.$$

Statistical analysis of this sequence, which is based on uniform random numbers, gives a sum of 95, a mean of 3.16, and a standard deviation of 1.81. The statistical fluctuations arise from the fact that we do not know exactly where the photons are within the beam.

If we make the length of the segments even smaller, the fluctuations become even more apparent. For example, in a 3 cm segment of beam corresponding to a time interval of 100 ps, the average photon number falls to 0.31. The majority of beam segments are now empty, and a sequence of 10 beam segments equivalent to any one of the 30 cm segments considered above might look like:

$$1, 0, 0, 1, 0, 0, 0, 0, 0, 1.$$

This sequence has a sum of 3, a mean of 0.3, and a standard deviation of 0.46. It is apparent that the shorter the time interval, the more difficult it becomes to know where the photons are. Thus if we split the 30 cm beam segment shown in Fig. 5.2 into 1000 intervals of 0.3 mm length and 1 ps duration, we would find that typically only three intervals contain photons, and 997 are empty. We have no way of predicting which three of these 1000 segments contain the photons.

These examples show that although the average photon flux can have a well-defined value, the photon number on short time-scales fluctuates due to the discrete nature of the photons. These fluctuations are described by the **photon statistics** of the light. In the following sections, we shall investigate the statistical nature of various types of light, starting with the simplest case, namely a perfectly stable monochromatic light source.

5.3 Coherent light: Poissonian photon statistics

In classical physics, light is considered to be an electromagnetic wave. The most stable type of light that we can imagine is a perfectly coherent light beam which has constant angular frequency ω, phase ϕ, and amplitude $\mathcal{E}_0$:

$$\mathcal{E}(x,t) = \mathcal{E}_0 \sin(kx - \omega t + \phi), \tag{5.4}$$

where $\mathcal{E}(x,t)$ is the electric field of the light wave and $k = \omega/c$ in free space. The beam emitted by an ideal single-mode laser operating well above threshold is a reasonably good approximation to such a field. The intensity I of the beam is proportional to the square of the amplitude (cf. eqn 2.28), and is constant if $\mathcal{E}_0$ and ϕ are independent of time. There will therefore be no intensity fluctuations and the average photon flux defined by eqn 5.1 will be constant in time.

The intensity is understood here to be determined by the average value of $\mathcal{E}(t)^2$ during an optical cycle.

It might be thought that a beam of light with a time-invariant average photon flux would consist of a stream of photons with regular time intervals between them. This is not in fact the case. We have seen above that there must be statistical fluctuations on short time-scales due to the discrete nature of the photons. We shall now show that perfectly coherent light with a constant intensity has **Poissonian** photon statistics.

Consider a light beam of constant power P. The average number of photons within a beam segment of length L is given by

$$\bar{n} = \Phi L/c, \tag{5.5}$$

where Φ is the photon flux given by eqn 5.1. We assume that L is large enough that $\bar{n}$ takes a well-defined integer value. We now subdivide the beam segment into N subsegments of length L/N. N is assumed to be sufficiently large that there is only a very small probability $p = \bar{n}/N$ of finding a photon within any particular subsegment, and a negligibly small probability of finding two or more photons.

We have assumed that p is the same for each subsegment because the intensity is identical at all points within the beam.

We now ask: what is the probability $\mathcal{P}(n)$ of finding n photons within a beam of length L containing N subsegments? The answer is given by the probability of finding n subsegments containing one photon and $(N-n)$ containing no photons, in any possible order. This probability is given by the binomial distribution:

$$\mathcal{P}(n) = \frac{N!}{n!(N-n)!} \, p^n \, (1-p)^{N-n}, \tag{5.6}$$

which, with $p = \bar{n}/N$, gives

$$\mathcal{P}(n) = \frac{N!}{n!(N-n)!} \left(\frac{\bar{n}}{N}\right)^n \left(1 - \frac{\bar{n}}{N}\right)^{N-n}. \tag{5.7}$$

We now take the limit as $N \to \infty$. To do this, we first rearrange eqn 5.7 in the following form:

$$\mathcal{P}(n) = \frac{1}{n!} \left(\frac{N!}{(N-n)!N^n}\right) \bar{n}^n \left(1 - \frac{\bar{n}}{N}\right)^{N-n}. \tag{5.8}$$

Now by using Stirling's formula:

$$\lim_{N \to \infty} [\ln N!] = N \ln N - N, \tag{5.9}$$

we can see that

$$\lim_{N \to \infty} \left[\ln \left(\frac{N!}{(N-n)!N^n} \right) \right] = 0.$$

Hence:

$$\lim_{N \to \infty} \left[\frac{N!}{(N-n)!N^n} \right] = 1. \qquad (5.10)$$

Furthermore, by applying the binomial theorem and comparing the result for the limit $N \to \infty$ to the series expansion of $\exp(-\overline{n})$, we can see that:

$$\left(1 - \frac{\overline{n}}{N} \right)^{N-n} = 1 - (N-n)\frac{\overline{n}}{N} + \frac{1}{2!}(N-n)(N-n-1)\left(\frac{\overline{n}}{N} \right)^2 - \cdots$$

$$\to 1 - \overline{n} + \frac{\overline{n}^2}{2!} - \cdots$$

$$= \exp(-\overline{n}). \qquad (5.11)$$

On using these two limits in eqn 5.8, we find

$$\lim_{N \to \infty} [\mathcal{P}(n)] = \frac{1}{n!} \cdot 1 \cdot \overline{n}^n \cdot \exp(-\overline{n}). \qquad (5.12)$$

We thus conclude that the photon statistics for a coherent light wave with constant intensity are given by:

A summary of the mathematical properties of Poisson distributions may be found in Appendix A.

$$\mathcal{P}(n) = \frac{\overline{n}^n}{n!}e^{-\overline{n}}, \quad n = 0, 1, 2, \cdots. \qquad (5.13)$$

This equation describes a **Poisson distribution**.

Poissonian statistics generally apply to random processes that can only return integer values. We have already mentioned one of the standard examples of Poissonian statistics, namely the number of counts from a Geiger tube pointing at a radioactive source. In this case, the number of counts is always an integer, and the average count value $\overline{n}$ is determined by the half life of the source, the amount of material present, and the time interval set by the user. The actual count values fluctuate above and below the mean value due to the random nature of the radioactive decay, and the probability for registering n counts is given by the Poissonian formula in eqn 5.13. A similar situation applies to the count rate of a photon-counting system detecting individual photons from a light beam with constant intensity. In this second case, the randomness originates from chopping the continuous beam into discrete energy packets with an equal probability of finding the energy packet within any given time subinterval.

Poisson distributions are uniquely characterized by their mean value $\overline{n}$. Representative distributions for $\overline{n} = 0.1$, 1, 5, and 10 are shown in Fig. 5.3. It is apparent that the distribution peaks at $\overline{n}$ and gets broader as $\overline{n}$ increases. The fluctuations of a statistical distribution about its mean value are usually quantified in terms of the **variance**. The variance is equal to the square of the **standard deviation** Δn and is defined by:

$$\mathrm{Var}(n) \equiv (\Delta n)^2 = \sum_{n=0}^{\infty}(n-\overline{n})^2 \mathcal{P}(n). \qquad (5.14)$$

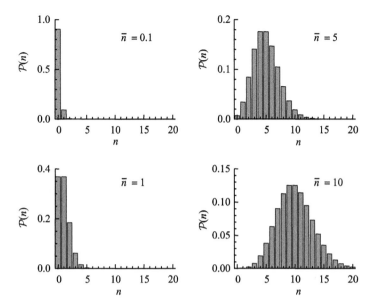

Fig. 5.3 Poisson distributions for mean values of 0.1, 1, 5, and 10. Note that the vertical axis scale changes between each figure.

It is a well-known result for Poisson statistics that the variance is equal to the mean value $\bar{n}$ (see eqn A.10):

$$(\Delta n)^2 = \bar{n}. \tag{5.15}$$

The standard deviation for the fluctuations of the photon number above and below the mean value is therefore given by:

$$\Delta n = \sqrt{\bar{n}}. \tag{5.16}$$

This shows that the relative size of the fluctuations decreases as $\bar{n}$ gets larger. If $\bar{n} = 1$, we have $\Delta n = 1$ so that $\Delta n / \bar{n} = 1$. On the other hand, if $\bar{n} = 100$, we have $\Delta n = 10$, and $\Delta n / \bar{n} = 0.1$.

Example 5.1   An attenuated beam from an argon laser operating at 514 nm (2.41 eV) with a power of 0.1 pW is detected with a photon-counting system of quantum efficiency 20% with the time interval set at 0.1 s. Calculate (a) the mean count value, and (b) the standard deviation in the count number.

Solution
(a) We first calculate the photon flux from eqn 5.1. This gives

$$\Phi = \frac{10^{-13} \text{ W}}{2.41 \text{ eV}} = 2.59 \times 10^5 \text{ photon s}^{-1}.$$

The average photon count is then given by eqn 5.2:

$$N = 0.2 \times (2.59 \times 10^5) \times 0.1 = 5180.$$

(b) We assume that the detected counts have Poissonian statistics with a standard deviation given by eqn 5.16. With $\bar{n} \equiv N = 5180$, we then find:

$$\Delta n = \sqrt{5180} = 72.$$

5.4 Classification of light by photon statistics

In the previous section, we considered the photon statistics of a perfectly coherent light beam with constant optical power P. We saw that the statistics are described by a Poisson distribution with photon number fluctuations that satisfy eqn 5.16. From a classical perspective, a perfectly coherent beam of constant intensity is the most stable type of light that can be envisaged. This therefore provides a bench mark for classifying other types of light according to the standard deviation of their photon number distributions. In general, there are three possibilities:

- **sub-Poissonian statistics**: $\Delta n < \sqrt{\bar{n}}$,
- **Poissonian statistics**: $\Delta n = \sqrt{\bar{n}}$,
- **super-Poissonian statistics**: $\Delta n > \sqrt{\bar{n}}$.

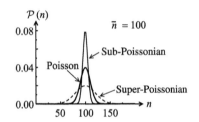

Fig. 5.4 Comparison of the photon statistics for light with a Poisson distribution, and those for sub-Poissonian and super-Poissonian light. The distributions have been drawn with the same mean photon number $\bar{n} = 100$. The discrete nature of the distributions is not apparent in this figure due to the large value of $\bar{n}$.

The difference between the three different types of statistics is illustrated in Fig. 5.4. This figure compares the photon number distributions of super-Poissonian and sub-Poissonian light to that of a Poisson distribution with the same mean photon number. We see that distributions of super-Poissonian and sub-Poissonian light are, respectively, broader or narrower than the Poisson distribution.

It is not difficult to think of types of light which would be expected to have super-Poissonian statistics. If there are any classical fluctuations in the intensity, then we would expect to observe larger photon number fluctuations than for the case with a constant intensity. Since a perfectly stable intensity gives Poissonian statistics, it follows that all classical light beams with time-varying light intensities will have super-Poissonian photon number distributions. In the next section we shall see that the thermal light from a black-body source and the partially coherent light from a discharge lamp fall into this category. These types of light are clearly 'noisier' than perfectly coherent light in both the classical sense that they have larger variations in the intensity, and also in the quantum sense that they have larger photon number fluctuations.

Sub-Poissonian light, by contrast, has a narrower distribution than the Poissonian case and is therefore 'quieter' than perfectly coherent light. Now we have already emphasized that a perfectly coherent beam is the most stable form of light that can be envisaged in classical optics. It is therefore apparent that sub-Poissonian light has no classical counterpart, and is therefore the first example of **non-classical light** that we have met. Needless to say, the observation of sub-Poissonian light is quite difficult, which explains why it is not normally discussed in standard optics texts.

Table 5.1 gives a summary of the classification of light according to the criteria established in this section.

Table 5.1 Classification of light according to the photon statistics. $I(t)$ is the time dependence of the optical intensity.

Photon statistics	Classical equivalents	$I(t)$	Δn
Super-Poissonian	Partially coherent (chaotic), incoherent, or thermal light	Time-varying	$> \sqrt{\bar{n}}$
Poissonian	Perfectly coherent light	Constant	$\sqrt{\bar{n}}$
Sub-Poissonian	None (non-classical)	Constant	$< \sqrt{\bar{n}}$

5.5 Super-Poissonian light

In this section we shall consider two examples of super-Poissonian statistics, namely thermal light and chaotic light. We have seen above that super-Poissonian light is defined by the relation:

$$\Delta n > \sqrt{\bar{n}}. \qquad (5.17)$$

We have also mentioned that super-Poissonian photon statistics have a classical interpretation in terms of fluctuations in the light intensity. It is always easier to make an unstable light source than a stable one, and therefore the observation of super-Poissonian statistics is commonplace. At the same time, it is important to understand the properties of super-Poissonian sources since they are frequently used in the laboratory.

In the next chapter we shall see that super-Poissonian statistics can be related to **photon bunching**. In this present chapter, we concentrate on the statistical classification of the sources as determined by a photon-counting experiment.

5.5.1 Thermal light

The electromagnetic radiation emitted by a hot body is generally called **thermal light** or **black-body radiation**. The properties of thermal light are conventionally understood by applying the laws of statistical mechanics to the radiation within an enclosed cavity at a temperature T. The radiation pattern consists of a continuous spectrum of oscillating modes, with the energy density within the angular frequency range ω to $\omega + \mathrm{d}\omega$ given by Planck's law:

$$\rho(\omega, T) \, \mathrm{d}\omega = \frac{\hbar \omega^3}{\pi^2 c^3} \, \frac{1}{\exp(\hbar\omega/k_\mathrm{B}T) - 1} \, \mathrm{d}\omega. \qquad (5.18)$$

The derivation of eqn 5.18 requires that the energy of the radiation should be quantized. We can consider each individual mode as a harmonic oscillator of angular frequency ω, and write down the quantized energy as (cf. eqn 3.93):

$$E_n = (n + \tfrac{1}{2})\hbar\omega, \qquad (5.19)$$

where n is an integer ≥ 0.

We consider a single radiation mode within the cavity at angular frequency ω. The probability that there will be n photons in the mode is given by Boltzmann's law:

$$\mathcal{P}_\omega(n) = \frac{\exp(-E_n/k_\mathrm{B}T)}{\sum_{n=0}^{\infty} \exp(-E_n/k_\mathrm{B}T)}. \qquad (5.20)$$

In quantum optics, we interpret eqn 5.19 as meaning that there are n photons excited at angular frequency ω in the particular mode.

The subscript on $\mathcal{P}(n)$ makes it plain that the probability refers specifically to a single mode at angular frequency ω.

On substituting E_n from eqn 5.19, we find:

$$\mathcal{P}_\omega(n) = \frac{\exp(-n\hbar\omega/k_BT)}{\sum_{n=0}^{\infty}\exp(-n\hbar\omega/k_BT)}, \tag{5.21}$$

which is of the form:

$$\mathcal{P}_\omega(n) = \frac{x^n}{\sum_{n=0}^{\infty}x^n}, \tag{5.22}$$

where

$$x = \exp(-\hbar\omega/k_BT). \tag{5.23}$$

The general result for the summation of a geometric progression is:

$$\sum_{i=1}^{k}r^{i-1} \equiv \sum_{j=0}^{k-1}r^j = \frac{1-r^k}{1-r}, \tag{5.24}$$

which implies:

$$\sum_{n=0}^{\infty}x^n = \frac{1}{1-x}, \tag{5.25}$$

since $x < 1$. We therefore find:

$$\mathcal{P}_\omega(n) = x^n(1-x)$$
$$\equiv \left(1 - \exp(-\hbar\omega/k_BT)\right)\exp(-n\hbar\omega/k_BT). \tag{5.26}$$

The mean photon number is given by:

$$\bar{n} = \sum_{n=0}^{\infty}n\,\mathcal{P}_\omega(n)$$

$$= \sum_{n=0}^{\infty}nx^n(1-x)$$

$$= (1-x)x\frac{\mathrm{d}}{\mathrm{d}x}\left(\sum_{n=0}^{\infty}x^n\right)$$

$$= (1-x)x\frac{\mathrm{d}}{\mathrm{d}x}\left(\frac{1}{1-x}\right)$$

$$= (1-x)x\frac{1}{(1-x)^2}$$

$$= \frac{x}{(1-x)}, \tag{5.27}$$

which, on substitution from eqn 5.23, gives the Planck formula:

$$\bar{n} = \frac{1}{\exp(\hbar\omega/k_BT) - 1}. \tag{5.28}$$

Equation 5.27 implies that $x = \overline{n}/(\overline{n}+1)$, and we are thus able to rewrite the probability given in eqn 5.26 in terms of $\overline{n}$ as:

$$P_\omega(n) = \frac{1}{\overline{n}+1}\left(\frac{\overline{n}}{\overline{n}+1}\right)^n. \tag{5.29}$$

This distribution is called the **Bose–Einstein distribution**. From eqn 5.26 we can see that $P_\omega(n)$ is always largest for $n = 0$, and decreases exponentially for increasing n.

Figure 5.5 compares the photon statistics for a single mode of thermal light with the Bose–Einstein distribution to those of a Poisson distribution with the same value of $\overline{n}$. It is apparent that the distribution of photon numbers for thermal light is much broader than for Poissonian light. This is hardly surprising, given the nature of thermal energy fluctuations. The variance of the Bose–Einstein distribution can be found by inserting $P_\omega(n)$ from eqn 5.29 into eqn 5.14, giving (see Exercise 5.3):

$$(\Delta n)^2 = \overline{n} + \overline{n}^2. \tag{5.30}$$

This shows that the variance of the Bose–Einstein distribution is always larger than that of a Poisson distribution (cf. eqn 5.15), and that thermal light therefore falls into the category of super-Poissonian light defined by eqn 5.17. For example, if $\overline{n} = 10$ as in Fig. 5.5, we have $\Delta n = 3.2$ for the Poisson distribution, but $\Delta n = 10.5$ for thermal light.

It should be stressed that the Bose–Einstein distribution only applies to a *single mode* of the radiation field. In reality, black-body radiation consists of a continuum of modes, and in most experiments we have to consider the properties of multi-mode thermal light. It can be shown that the photon number variance of N_m thermal modes of similar frequency is given by:

$$(\Delta n)^2 = \overline{n} + \frac{\overline{n}^2}{N_m}, \tag{5.31}$$

which reduces to the result for a Poisson distribution when N_m is large. In practice, it is very difficult to measure a single mode of the thermal field, and the statistics measured in most experiments with thermal light will therefore be Poissonian. (See Exercise 5.5.)

The single-mode variance given in eqn 5.30 can be interpreted in an interesting way if we refer to Einstein's analysis of the energy fluctuations of back-body radiation originally given in 1909. According to statistical mechanics, the magnitude of the energy fluctuations from the mean value at thermal equilibrium $\langle E \rangle$ is given by:

$$\langle \Delta E^2 \rangle = k_B T^2 \frac{\partial \langle E \rangle}{\partial T}. \tag{5.32}$$

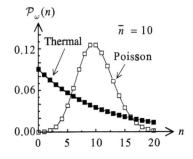

Fig. 5.5 Comparison of the photon statistics for a single mode of a thermal source with $\overline{n} = 10$ and a Poisson distribution with the same value of $\overline{n}$.

The derivation of eqn 5.31 may be found, for example, in Mandel and Wolf (1995, §13.3.2). It should also be pointed out that if the detection time interval is long, the intensity fluctuations will be averaged out, and Poisson counting statistics will be obtained even for single-mode thermal light. This point will be explained further in the next subsection.

See Pais (1982, §21a).

If we apply this result to the energy fluctuations of black-body radiation in the angular frequency range ω to $\omega + \mathrm{d}\omega$, we obtain:

$$\langle \Delta E^2 \rangle \, \mathrm{d}\omega = k_\mathrm{B} T^2 \, \frac{\partial}{\partial T} (V \rho \, \mathrm{d}\omega)$$

$$= k_\mathrm{B} T^2 V \, \mathrm{d}\omega \, \frac{\partial \rho}{\partial T}$$

$$= \left(\rho \hbar \omega + \frac{\pi^2 c^3}{\omega^2} \rho^2 \right) V \, \mathrm{d}\omega, \qquad (5.33)$$

where V is the volume of the cavity, and ρ is the spectral energy density given by the Planck formula (eqn 5.18). These energy fluctuations can be connected to the photon number fluctuations per mode by writing:

$$\langle \Delta E^2 \rangle \, \mathrm{d}\omega = \text{density of states} \times \text{energy fluctuations per mode} \times \text{volume}$$

$$= \frac{\omega^2}{\pi^2 c^3} \, \mathrm{d}\omega \times \left\langle \left(\Delta(n \hbar \omega) \right)^2 \right\rangle \times V$$

$$= \frac{\omega^2}{\pi^2 c^3} \, (\Delta n)^2 \, (\hbar \omega)^2 \, V \, \mathrm{d}\omega, \qquad (5.34)$$

where we made use of eqn C.11 in Appendix C for the density of states. On comparing eqns 5.33 and 5.34 we find that:

$$(\Delta n)^2 = \frac{\pi^2 c^3}{\hbar \omega^3} \rho + \left(\frac{\pi^2 c^3}{\hbar \omega^3} \rho \right)^2. \qquad (5.35)$$

Then, by using eqn 5.28 to rewrite eqn 5.18 as:

$$\rho = \frac{\hbar \omega^3}{\pi^2 c^3} \, \bar{n}, \qquad (5.36)$$

we find as before (cf. eqn 5.30):

$$(\Delta n)^2 = \bar{n} + \bar{n}^2. \qquad (5.37)$$

The same argument applies to the first and second terms in eqn 5.37: the first term which gives the Poissonian statistics arises from the quantization of the light, while the second is caused by classical intensity fluctuations.

Einstein realized that the first term in eqn 5.33 originates from the particle nature of the light, while the second originates from the thermal fluctuations of the energy of the electromagnetic radiation. The latter term is therefore classical in origin, and is called the **wave noise**. The first term, by contrast, originates from the quantization of the energy of the electromagnetic radiation, that is, the photon nature of light.

5.5.2   Chaotic (partially coherent) light

Note that the use of the term 'chaotic' for partially coherent light predates chaos theory. Chaos theory has nothing to do with chaotic light.

The light from a single spectral line of a discharge lamp is generally called **chaotic light**. Chaotic light has partial coherence, with classical intensity fluctuations on a time-scale determined by the coherence time τ_c. (See Section 2.3.) These intensity fluctuations will obviously give rise to greater fluctuations in the photon number than for a source with a constant power (i.e. a perfectly coherent source).

It can be shown that the fluctuations in the photocount rate when a chaotic source is incident on a detector are given by:

$$(\Delta n)^2 = \langle W(T) \rangle + \langle \Delta W(T)^2 \rangle, \qquad (5.38)$$

where W represents the count rate in the detection time interval T:

$$W(T) = \int_t^{t+T} \eta \Phi(t') \, \mathrm{d}t', \qquad (5.39)$$

η being the detection efficiency and $\Phi(t)$ the instantaneous photon flux given by eqn 5.1. The mean count rate $\langle W(t) \rangle$ is of course equal to $\bar{n}$. If there were no intensity fluctuations so that $\Phi(t)$ was constant, then eqn 5.38 would just revert to the Poissonian case with $(\Delta n)^2 = \bar{n}$.

In chaotic light, the photon flux is *not* constant due to the fluctuations in the intensity of the light on time-scales of the order of the coherence time. These intensity fluctuations will be significant if the measurement time T is comparable to, or less than, the coherence time τ_c. The second term in eqn 5.38 would then be non-zero, implying that chaotic light, when measured on short time-scales, is super-Poissonian. On the other hand, when $T \gg \tau_c$, the intensity fluctuations on time-scales of order τ_c will not be noticed, and the intensity may be taken as effectively constant. In this case we again revert to the Poissonian formula.

The two terms in eqn 5.38 can be interpreted as originating, respectively, from the Poissonian statistics associated with the particle nature of light and the classical power fluctuations in the source. The classical fluctuations due to the time-varying intensity of the source are often called wave noise in analogy with the case of black-body radiation considered in the previous subsection.

See, for example, Mandel and Wolf (1995, §9.7), or Goodman (1985, §9.2).

See Loudon (2000, §3.9), especially eqns 3.9.22 and 3.9.23.

5.6   Sub-Poissonian light

Sub-Poissonian light is defined by the relation:

$$\Delta n < \sqrt{\bar{n}}. \qquad (5.40)$$

It is apparent from Fig. 5.4 that sub-Poissonian light has a narrower photon number distribution than for Poissonian statistics. We have seen in Section 5.3 that a perfectly coherent beam with constant intensity has Poissonian photon statistics. We thus conclude that sub-Poissonian light is somehow more stable than perfectly coherent light, which has been our paradigm up to this point. In fact, sub-Poissonian light has no classical equivalent. Therefore, the observation of sub-Poissonian statistics is a clear signature of the quantum nature of light.

Even though there is no direct classical counterpart of sub-Poissonian light, it is easy to conceive of conditions that would give rise to sub-Poissonian statistics. Let us consider the properties of a beam of light in which the time intervals Δt between the photons are identical, as

In the next chapter, we shall see that sub-Poissonian photon statistics are often associated with the observation of another purely quantum optical effect, namely photon **antibunching**.

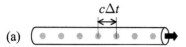

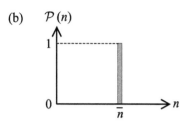

Fig. 5.6 (a) A beam of light containing a stream of photons with a fixed time spacing Δt between them. (b) Photon-counting statistics for such a beam.

illustrated schematically in Fig. 5.6(a). The photocount obtained for such a beam in a time T would be the integer value determined by:

$$N = \text{Int}\left(\eta \frac{T}{\Delta t}\right), \tag{5.41}$$

which would be exactly the same for every measurement. The experimenter would therefore obtain the histogram shown in Fig. 5.6(b), with $\bar{n} = N$ given by eqn 5.41. This is highly sub-Poissonian, and has $\Delta n = 0$.

Photon streams of the type shown in Fig. 5.6(a) with $\Delta n = 0$ are called **photon number states**. Further details of photon number states will be given in Chapter 8. Photon number states are the purest form of sub-Poissonian light. Other types of sub-Poissonian light can be conceived in which the time intervals between the photons in the beam are not exactly the same, but are still more regular than the random time intervals appropriate to a beam with Poissonian statistics. Such types of light are fairly easy to generate in the laboratory, although their detection is quite problematic, and will be discussed in Section 5.10.

5.7   Degradation of photon statistics by losses

It should be apparent from the discussion in the previous section that light with sub-Poissonian statistics is particularly interesting. In Section 5.10 we shall discuss how such light might be observed in the laboratory. Before we do this, we need to cover an important issue related to optical losses.

Fig. 5.7 (a) The effect of a lossy medium with transmission $\mathcal{T}$ on a beam of light can be modelled as a beam splitter with splitting ratio $\mathcal{T}:(1-\mathcal{T})$ as shown in (b). The beam splitting process is probabilistic at the level of the individual photons, and so the incoming photon stream splits randomly towards the two outputs with a probability set by the transmission : reflection ratio (50 : 50 in this case) as shown in part (c).

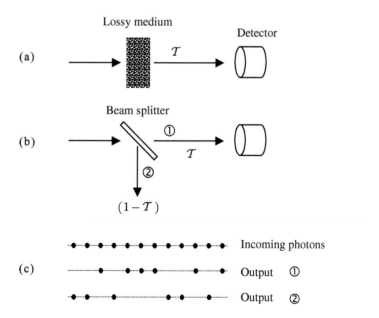

Let us suppose that we have a beam of light that passes through a lossy medium and is then detected as shown in Fig. 5.7(a). If the transmission of the medium is T, then we can model the losses as a beam splitter with splitting ratio $T : (1 - T)$, as indicated in Fig. 5.7(b). The beam splitter separates the photons into two streams going towards its two output ports, so that only a fraction T of the incoming photons impinge on the detector and are registered as counts. Now the beam splitting process occurs randomly at the level of individual photons, with weighted probabilities for the two output paths of $T : (1 - T)$, respectively. We therefore see that the lossy medium randomly selects photons from the incoming beam with probability T. It is well known that the distribution obtained by **random sampling** of a given set of data is more random than the original distribution. This point is illustrated in Fig 5.7(c), which presents the case of a regular stream of photons split with $50 : 50$ probability towards two output ports. It is apparent that the regularity of the time intervals in the photon stream going to the detector is reduced compared to the incoming photon stream. Thus the random sampling nature of optical losses degrades the regularity of the photon flux, and would eventually make the time intervals completely random for low values of T.

The beam splitter model of optical losses is a convenient way to consider the many different factors that reduce the efficiency of photon-counting experiments. These factors include:

(1) inefficient collection optics, whereby only a fraction of the light emitted from the source is collected;

(2) losses in the optical components due to absorption, scattering, or reflections from the surfaces;

(3) inefficiency in the detection process due to using detectors with imperfect quantum efficiency.

All of these processes are equivalent to random sampling of the photons. The first one randomly selects photons from the source. The second randomly deletes photons from the beam. The third randomly selects a subset of photons to be detected. The first two degrade the photon statistics themselves, while the third degrades the correlation between the photon statistics and the photoelectron statistics.

This argument unfortunately shows that sub-Poissonian light is very fragile: all forms of loss and inefficiency will tend to degrade the statistics to the Poissonian (random) case. This means that we must be very careful to avoid optical losses and use very high-efficiency detectors to observe large quantum effects in the photon statistics.

5.8   Theory of photodetection

Now that we are familiar with the different types of photon statistics that can occur, it is appropriate to consider the relationship between the counting statistics registered by the detector and the underlying

The relationship between the photon and detection statistics has been a contentious issue in quantum optics, and has only been definitively resolved in relatively recent times.

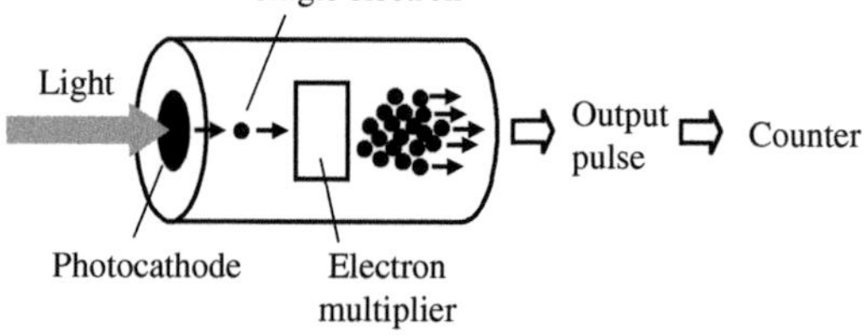

Fig. 5.8 Schematic diagram of the operation of a single-photon counting photomultiplier. The incident light ejects single electrons from the photocathode, which then trigger the release of an avalanche of electrons in the multiplier region, thereby generating an output pulse large enough to be detected by electronics.

photon statistics of the light beam. We start by outlining the semi-classical theory of photodetection in which the light is assumed to consist of classical electromagnetic waves. This will enable us to highlight the critical results that prove that the light beam really does consist of a stream of photons. We shall then give the results for the full quantum theory of photodetection, and hence see the conditions under which non-classical effects can be observed.

5.8.1 Semi-classical theory of photodetection

Single-photon avalanche photodiode (SPAD) detectors are now also commonly used for single-photon counting experiments.

Let us consider a photon-counting detector such as a photomultiplier illuminated by a faint light beam, as shown in Fig. 5.8. The light interacts with the atoms in the photocathode of the detector and liberates individual electrons by the photoelectric effect. These single photoelectrons then trigger the release of many more electrons in the multiplier region of the tube, thereby generating a current pulse of sufficient magnitude to be detected with an electronic counter. The pulses counted thus correspond to the release of individual electrons from the photocathode.

In the following we assume that the light beam is a classical electromagnetic wave of intensity I. The atoms in the photocathode are assumed to be quantized so that the photoelectrons are ejected in a probabilistic fashion after the absorption of a quantum of energy from the light beam. The statistical nature of the timing between the output pulses can then be explained by the making the following three assumptions about the photodetection process:

1. The probability of the emission of a photoelectron in a short time interval Δt is proportional to the intensity I, the area A illuminated, and the time interval Δt.

2. If Δt is sufficiently small, the probability of emitting two photoelectrons is negligibly small.

3. Photoemission events registered in different time intervals are statistically independent of each other.

See, for example, Goodman (1985).

From assumption (1) we can write the probability of observing one photoemission event in the time interval $t \to t + \Delta t$ as:

$$\mathcal{P}(1; t, t + \Delta t) = \xi I(t)\Delta t, \qquad (5.42)$$

where ξ is proportional to the area illuminated and is equal to the emission probability per unit time per unit intensity. Assumption (2) then implies that the probability of observing no events in the same time interval is given by:

$$P(0; t, t + \Delta t) = 1 - P(1; t, t + \Delta t) = 1 - \xi I(t)\Delta t. \quad (5.43)$$

We now ask: what is the probability of obtaining n events in the time interval $0 \to t + \Delta t$? If the events are statistically independent (assumption 3) and Δt is small, then there are only two ways of achieving this result: n events in the time interval $0 \to t$ and none in time interval $t \to t + \Delta t$, or $n - 1$ events in time interval $0 \to t$ and one in time interval $t \to t + \Delta t$. We must therefore have:

$$P(n; 0, t + \Delta t) = P(n; 0, t)P(0; t, t + \Delta t) + P(n - 1; 0, t)P(1; t, t + \Delta t)$$
$$= P(n; 0, t)[1 - \xi I(t)\Delta t] + P(n - 1; 0, t)\xi I(t)\Delta t. \quad (5.44)$$

On writing $P(n; 0, t) \equiv P_n(t)$ and rearranging, we find:

$$\frac{P_n(t + \Delta t) - P_n(t)}{\Delta t} = \xi I(t)[P_{n-1}(t) - P_n(t)], \quad (5.45)$$

which, on taking the limit $\Delta t \to 0$, gives:

$$\frac{dP_n(t)}{dt} = \xi I(t)[P_{n-1}(t) - P_n(t)]. \quad (5.46)$$

The general solution to this recursion relation, with the boundary condition $P_0(0) = 1$, is:

$$P_n(t) = \frac{\left[\int_0^t \xi I(t')\, dt'\right]^n}{n!} \exp\left(-\int_0^t \xi I(t')\, dt'\right). \quad (5.47)$$

The derivation of eqn 5.47 is beyond the scope of this book. We can, however, show that the solution is correct for the simplest case in which $I(t)$ is constant, that is independent of t. With $\xi I = \text{constant} \equiv C$, eqn 5.46 reduces to:

Constant intensity corresponds to perfectly coherent light.

$$\frac{dP_n(t)}{dt} + CP_n(t) = CP_{n-1}(t). \quad (5.48)$$

For $n = 0$ it must be the case that $P_{n-1}(t) = 0$ because we cannot have a negative count value. The first recursion relation is therefore of the form:

$$\frac{dP_0(t)}{dt} = -CP_0(t), \quad (5.49)$$

which, with the boundary condition $P_0(0) = 1$, has the solution:

$$P_0(t) = \exp(-Ct). \quad (5.50)$$

For $n \geq 1$ we multiply eqn 5.48 by the integrating factor e^{Ct} to obtain:

$$\frac{d}{dt}\left(e^{Ct}P_n(t)\right) = Ce^{Ct}P_{n-1}(t), \quad (5.51)$$

which, on integrating, gives:

$$\mathcal{P}_n(t) = e^{-Ct} \int_0^t C e^{Ct'} \mathcal{P}_{n-1}(t') \, dt'. \qquad (5.52)$$

With $\mathcal{P}_0(t)$ given by eqn 5.50, we can then solve recursively to obtain:

$$\mathcal{P}_1(t) = e^{-Ct} \int_0^t C e^{Ct'} \mathcal{P}_0(t') \, dt' = (Ct) \, e^{-Ct},$$

$$\mathcal{P}_2(t) = e^{-Ct} \int_0^t C e^{Ct'} \mathcal{P}_1(t') \, dt' = \frac{(Ct)^2}{2} \, e^{-Ct},$$

$$\mathcal{P}_3(t) = e^{-Ct} \int_0^t C e^{Ct'} \mathcal{P}_2(t') \, dt' = \frac{(Ct)^3}{3!} \, e^{-Ct},$$

$$\vdots$$

$$\mathcal{P}_n(t) = e^{-Ct} \int_0^t C e^{Ct'} \mathcal{P}_{n-1}(t') \, dt' = \frac{(Ct)^n}{n!} \, e^{-Ct}. \qquad (5.53)$$

Since $\int_0^t \xi I(t') dt' = \xi I t = Ct$ if $I(t)$ is constant, it is evident that eqn 5.53 is consistent with eqn 5.47.

We can cast eqn 5.53 into a more familiar form if we notice that eqn 5.42 implies that the event probability per unit time is equal to $\xi I(t)$. Therefore, if $I(t)$ is constant, the mean count rate $\overline{n}$ for the time interval $0 \to t$ is just given by:

$$\overline{n} = \xi I t \equiv Ct. \qquad (5.54)$$

Hence we can rewrite eqn 5.53 as

$$\mathcal{P}_n(t) = \frac{\overline{n}^n}{n!} e^{-\overline{n}}, \qquad (5.55)$$

which shows that we obtain a Poisson distribution when $I(t)$ is constant. (cf. eqn 5.13.)

Equation 5.55 demonstrates that we can explain the Poissonian photocount statistics observed when detecting light with a time-independent intensity without invoking the concept of photons at all. All that we require is that the emission of photoelectrons is a probabilistic process triggered by the absorption of a quantum of energy from the light beam. Hence the analysis of the photocount statistics does not necessarily tell us anything about the underlying photon statistics. At the same time, it is clear that sub-Poissonian statistics are not possible within a semi-classical theory. This follows because we obtain the Poissonian formula if the intensity is constant, and if the intensity varies with time, it can be shown that we obtain the super-Poissonian result given in eqn 5.38. Hence the observation of *sub*-Poissonian photocount statistics constitutes a clear demonstration that the semi-classical approach is inadequate. In Section 5.10 we shall describe experimental work that gives direct evidence of sub-Poissonian photodetection statistics. These experiments can *only* be explained by the full quantum treatment of light detection, and definitively establish the quantum nature of light.

5.8.2 Quantum theory of photodetection

The aim of the quantum theory of photodetection is to relate the photocount statistics observed in a particular experiment to those of the incoming photons. The derivation of the final result is beyond the scope of this work, and at this level we can only quote the conclusion and discuss its implications at a qualitative level.

As usual, we consider the photocount statistics measured in a time interval of T. We are interested in the relationship between the variance in the photocount number $(\Delta N)^2$ and the corresponding variance $(\Delta n)^2$ in the number of photons impinging on the detector in the same time interval. This relationship is given by:

$$(\Delta N)^2 = \eta^2 (\Delta n)^2 + \eta(1-\eta)\overline{n}, \qquad (5.56)$$

where η is the quantum efficiency of the detector, defined previously as the ratio of the average photocount number $\overline{N}$ to the mean photon number $\overline{n}$ incident on the detector in the same time interval (cf. eqn 5.2):

$$\eta = \frac{\overline{N}}{\overline{n}}. \qquad (5.57)$$

Several important conclusions follow from eqn 5.56.

1. If $\eta = 1$, we have $\Delta N = \Delta n$ and the photocount fluctuations faithfully reproduce the fluctuations of the incident photon stream.

2. If the incident light has Poissonian statistics with $(\Delta n)^2 = \overline{n}$, then $(\Delta N)^2 = \eta\overline{n} \equiv \overline{N}$ for all values of η. In other words, the photocount statistics always give a Poisson distribution.

3. If $\eta \ll 1$, the photocount fluctuations tend to the Poissonian result with $(\Delta N)^2 = \eta\overline{n} \equiv \overline{N}$ irrespective of the underlying photon statistics.

The conclusion is obvious: if we want to measure the photon statistics we need high efficiency detectors. If we have such detectors, the photocount statistics give a true measure of the incoming photon statistics, with a fidelity that increases as the efficiency of the detector increases.

It is apparent from the comments above that the quantum efficiency of the detector is the critical parameter that determines the relationship between the photoelectron and photon statistics. We can understand why this should be so by reference to Fig. 5.7(b). An imperfect detector of efficiency η is equivalent to a perfect detector of 100% efficiency with a beam splitter of transmission η in front of it. As discussed in Section 5.7, the random sampling nature of the beam-splitting process gradually randomizes the statistics, irrespective of the original statistics of the incoming photons. In the limit of very low efficiencies, the time intervals between the photoelectrons would become completely random, and the counting statistics would be Poissonian for all possible incoming distributions.

The difficulty in producing single-photon detectors with high quantum efficiencies is one of the reasons why it is difficult to observe sub-Poissonian statistics in the laboratory. With low efficiency detectors,

Students who wish to pursue the quantum theory of photodetection in more detail are referred to the more advanced texts. See, for example, Mandel and Wolf (1995, Chapter 14), or Loudon (2000, §6.10).

See, for example, Loudon (2000, eqn 6.10.8).

the photocount statistics will always be random (i.e. Poissonian), irrespective of the incoming photon distribution. Nowadays, however, single-photon detectors with quantum efficiencies in excess of 50% are available for some wavelengths. Furthermore, by using a different detection strategy employing high-intensity beams and photodiode detection, quantum efficiencies approaching 90% can be obtained. This is the topic of the next section.

5.9   Shot noise in photodiodes

Photodiodes, in contrast to the detectors used for single-photon counting (i.e. photomultipliers or *avalanche* photodiodes), do not contain electron multiplication regions. Hence there is a one-to-one relationship between the number of photoelectrons generated in the active region of the photodiode and the number of photons incident on the detector. This should not be misinterpreted to imply that it is the *same* electron that was excited to the conduction band by the photon that flows in the external circuit: the correspondence between individual electrons and photons is lost by the myriad of electron–electron scattering processes that occur in the semiconductor material and in the wires. However, the charge flow is conserved throughout the circuit, and it is this that determines the current that is measured by the detection electronics.

Up to this point, we have been thinking exclusively about the detection of light beams by single-photon counting methods as sketched in Fig. 5.1. As mentioned in Section 5.2, this method is only appropriate for very weak beams with a flux of $\sim 10^6$ photons s^{-1} or less. In many cases we shall be dealing with light beams of much higher photon flux. For example, a He:Ne laser beam with a power of 1 mW at 633 nm has, from eqn 5.1, a flux of 3.3×10^{15} photons s^{-1}. No detector can respond fast enough to register the individual photons in this case, and we would completely saturate the output of a single-photon counting detector such as a photomultiplier. A different detection strategy must therefore be used.

The normal method used to detect high flux light beams is to employ **photodiode** detectors. Photodiodes are semiconductor devices that generate electrons in an external circuit when photons excite electrons from the valence band to the conduction band. A key parameter of a photodiode is its quantum efficiency η, which is defined in this context as the ratio of the number of photoelectrons generated in the external circuit to the number of photons incident. Hence the current generated in the external circuit for an incident photon flux Φ, namely the **photocurrent** i, is given by:

$$i = \eta e \Phi \equiv \eta e \frac{P}{\hbar \omega}, \tag{5.58}$$

Well-designed photodiodes can have quantum efficiencies in excess of 90%.

where e is the modulus of the charge of the electron, P is the power of the beam, and ω is its angular frequency (cf. eqn 5.1). The ratio $i/P = \eta e/\hbar \omega$ is called the **responsivity** of the photodiode and has the units of A W^{-1}. The value of η can therefore be worked out from the measured responsivity at the detection wavelength.

A spectrum analyser is an electronic device that displays the Fourier transform of the time-dependent voltage at its input.

Figure 5.9(a) gives a schematic representation of the detection of a high-intensity light beam with a photodiode (PD) detector. The light is incident on the detector, and the time dependence of the resulting photocurrent is displayed on an oscilloscope after appropriate amplification. Alternatively, the photocurrent is analysed in the frequency domain by using a spectrum analyser. Figure 5.9(b) gives a simplified circuit diagram for the detection system. The photocurrent $i(t)$ flows through a load resistor R_L, thereby generating a time-dependent voltage $i(t)R_L$. The capacitor C blocks the DC component of the voltage, and the AC part is then amplified to produce the output voltage $V(t)$. Measurement

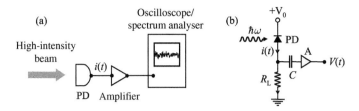

Fig. 5.9 (a) Detection of a high-intensity light beam with a photodiode (PD) detector. The time dependence of the photocurrent fluctuations relates to the photon statistics of the incoming beam. (b) Simplified diagram for the detector circuit. The diode is reverse-biased with a voltage V_0. The photocurrent $i(t)$ generated in the detector flows through a load resistor R_L, and the AC voltage across R_L is amplified to produce a time-dependent output voltage $V(t)$. The capacitor C blocks the DC voltage across R_L from saturating the amplifier A.

of the DC voltage across R_L permits the average photocurrent $\langle i \rangle$ to be determined.

The principle behind using photodiode detectors to study the statistical properties of light is that the photocurrent generated by the beam will fluctuate because of the underlying fluctuations in the impinging photon number. These photon number fluctuations will be reflected in the photocurrent fluctuations with a fidelity determined by η. The fluctuations manifest themselves as **noise** in the photocurrent, as illustrated in Fig 5.10(a). The time-varying photocurrent $i(t)$ can be broken into a time-independent average current $\langle i \rangle$ and a time-varying fluctuation $\Delta i(t)$ according to:

$$i(t) = \langle i \rangle + \Delta i(t). \tag{5.59}$$

The average value of $\Delta i(t)$ must, of course, be zero, but the average of the square of Δi, namely $\langle (\Delta i(t))^2 \rangle$, will not be zero. Since the photocurrent flows through the load resistor R_L, which then generates energy at the rate of $i^2 R_L$, it is convenient to analyse the fluctuations in terms of a time-varying **noise power** according to:

$$P_{\text{noise}}(t) = (\Delta i(t))^2 R_L. \tag{5.60}$$

The Fourier transform of this noise power can be displayed on a spectrum analyser after suitable amplification. Figure 5.10(b) shows the type of noise power spectrum that might typically be obtained.

Let us consider what happens if we illuminate the photodiode with the light from a single-mode laser operating high above threshold. Such light is nearly perfectly coherent, and is expected to have Poissonian photon statistics, in which the photon number fluctuations obey eqn 5.15. The photoelectron statistics will therefore also follow a Poisson distribution with:

$$(\Delta N)^2 = \langle N \rangle. \tag{5.61}$$

Since $i(t)$ is proportional to the number of photoelectrons generated per second, it follows that the photocurrent variance Δi will satisfy:

$$(\Delta i)^2 \propto \langle i \rangle. \tag{5.62}$$

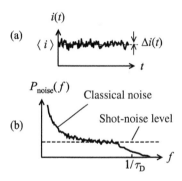

Fig. 5.10 (a) Time-varying photocurrent resulting from the detection of a high-intensity light beam with a photodiode detector as in Fig. 5.9. $\langle i \rangle$ represent the average photocurrent, while Δi represents the fluctuation from the mean value. (b) Fourier transform of $(\Delta i(t))^2$ showing the typical dependence of the photocurrent noise on frequency f. It is assumed that the photodiode used to detect the light has a response time of τ_D, and that the light source has excess classical noise at low frequencies.

The term 'shot noise' was originally used to describe the random spread of the pellets from a shot gun. Electrical shot noise was extensively studied in the days of vacuum tube electronics. The current in a vacuum tube is ultimately determined by the random thermal emission of electrons from the hot cathode, and thus exhibits Poissonian statistics. Simple ohmic circuits with a battery and a resistor, by contrast, do not usually exhibit shot noise. Instead, they have Johnson noise, which arises from the thermal fluctuations of the current.

On taking the Fourier transform of $i(t)$ and then measuring the variance of the current fluctuations within a frequency bandwidth Δf, we find:

$$(\Delta i)^2 = 2e\Delta f \langle i \rangle. \tag{5.63}$$

The corresponding noise power is given from eqn 5.60 as:

$$P_{\text{noise}}(f) = 2eR_{\text{L}}\Delta f \langle i \rangle. \tag{5.64}$$

The fluctuations described by eqns 5.63 and 5.64 are called **shot noise**.

Two characteristic features of shot noise are apparent from eqns 5.63 and 5.64:

- The variance of the current fluctuations (or equivalently, the noise power) is directly proportional to the average value of the current.

- The noise spectrum is 'white', that is, independent of frequency.

The second characteristic is a consequence of the random timing between the arrival of the photons in a beam with Poissonian statistics. The 'whiteness' of the noise, is, of course, subject to the response time τ_{D} of the photodiode, which means that in practice the shot noise can only be detected up to a maximum frequency of $\sim (1/\tau_{\text{D}})$. This point is illustrated in the schematic representation of the noise power spectrum shown in Fig. 5.10(b).

In the semi-classical theory of photodetection, the shot noise level corresponding to Poissonian photoelectron statistics is the fundamental detection limit. Hence the shot noise power level is often called the **quantum limit** or **standard quantum limit** of detection. For a similar reason, shot noise is often called **quantum noise**.

All light sources will show some classical intensity fluctuations due to noise in the electrical drive current, and lasers are subject to additional classical noise due to mechanical vibrations in the cavity mirrors. These classical noise sources tend to produce intensity fluctuations at fairly low frequencies, and so the noise spectrum tends to be well above the shot noise level in the low-frequency limit. However, at high frequencies, the classical noise sources are no longer present, and we are left with only the fundamental noise caused by the photon statistics. Hence a typical spectrum will show a noise level well above the shot noise limit at low frequencies, but should eventually reach the shot noise limit at high frequencies as shown in Fig. 5.10(b). Shot noise is present at all frequencies and the high frequency roll-off shown in Fig. 5.10(b) only reflects the frequency limit imposed by the detector response time.

Figure 5.11 shows the noise spectrum measured for a Nd:YAG laser operating at 1064 nm. The noise power is specified in 'dBm' units, which is a logarithmic scale defined by:

The decibel scale itself gives a logarithmic representation of a ratio r as $10 \times \log_{10} r$.

$$\text{Power in dBm units} = 10 \times \log_{10} \left(\frac{\text{Power}}{1 \text{ mW}} \right). \tag{5.65}$$

The data clearly show that the laser exhibits classical noise at low frequencies, but ultimately reaches the shot-noise limit at around 15 MHz.

The linear relationship between the shot noise and the photocurrent predicted by eqn. 5.64 is demonstrated by the data plotted in Fig. 5.12. This shows the high-frequency photocurrent noise power of a Ti:sapphire laser as a function of the optical power incident on the detector. The data show a linear increase in the noise power with the optical power.

Since the average photocurrent is directly proportional to the average power via eqn 5.58, the results clearly demonstrate the linear relationship between the shot noise and the average current.

The low-frequency classical noise that is apparent in Fig. 5.11 can, in principle, be removed. Two ways in which this might be done are shown in Fig. 5.13, namely the '**noise eater**' and the **balanced detector**. Consider first the noise eater shown in Fig. 5.13(a). The figure shows an intensity stabilization scheme in which a signal proportional to the laser output is fed back to the power supply. Negative feedback is used, so that the output of the power supply is reduced to compensate for fluctuations of the laser power above the average value, and vice versa. Alternatively, a modulator could be placed after the laser with a negative input proportional to the laser output, so that high-intensity fluctuations get attenuated more. These schemes can compensate (to a greater or lesser extent) for the classical power fluctuations in the laser output, but can do nothing about the shot noise, which is intrinsic to the light and cannot be removed by any classical stabilization methods. The best that such a stabilization scheme can hope to achieve is to remove all the excess classical noise and bring the output noise level down to the shot-noise level.

Now consider the balanced detector scheme shown in Fig. 5.13(b). The output of the laser is split into two beams of equal intensity, which are then detected with two identical photodiodes D1 and D2, generating photocurrents i_1 and i_2, respectively. The outputs of the photodiodes are connected so that the subtracted signal $(i_1 - i_2)$ can be detected. From a classical perspective, the two currents should be identical, so that the output signal is zero. If an absorbing sample S is introduced into the path leading to D2, i_2 will decrease and a positive signal will result. In this way, it is possible to measure very small absorption levels from, say, thin film samples. Alternatively, it is possible to detect a very weak modulation signal applied to one of the beams after the beam splitter.

The key point about the balanced detector scheme is that it usually gives much better signal-to-noise ratios than a single detector. If only a single detector were to be used, the small change in the intensity caused by the presence of the sample might be lost in the laser noise. With balanced detectors, by contrast, the classical noise is exactly cancelled (at least in principle) by the subtraction of the photocurrents, and much smaller changes in the intensity should be resolvable. Note, however, that the balanced detection scheme *cannot* remove the shot noise. From the perspective of the photons, the 50 : 50 beam splitting process is random, and therefore any noise associated with the photon nature of the light cannot be cancelled. Since the quantum nature of the light gives rise to shot noise, the output of the balanced detectors with no sample present will correspond to the shot-noise level.

It should by now be clear that shot noise is very important in optical science and telecommunications because it sets a practical limit to the signal-to-noise ratios that can be obtained in normal circumstances. For example, we can encode information onto a laser beam by modulating its intensity at a particular frequency. The information is retrieved by analysing the time-dependence of the photocurrent generated in the

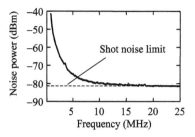

Fig. 5.11 Laser intensity noise spectrum measured for a Nd:YAG laser operating at 1064 nm with a fast detector of reponsivity 0.7 A W^{-1}. The detection bandwidth was 100 kHz, and the optical power and average photocurrent were 66 mW and 46 mA, respectively. (After D.J. Ottaway *et al.*, *Appl. Phys. B* **71**, 163 (2000), reproduced with permission of Springer Science and Business Media.)

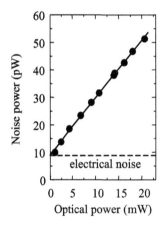

Fig. 5.12 Power dependence of the amplified photocurrent noise at 50 MHz within a 3 MHz bandwidth measured for a Ti:sapphire laser operating at 930 nm. The offset at zero power is caused by the ever-present electrical noise in the detector circuit. This noise is uncorrelated with the photocurrent noise, and so the two noise sources just add together, leading to the constant offset observed in the data. (Data by the author.)

We shall come across balanced detectors again in Section 7.8. In that discussion, we shall analyse the balanced detector from a different perspective and assign the shot noise output with no sample present to the vacuum modes that enter the unused input port of the 50 : 50 beam splitter.

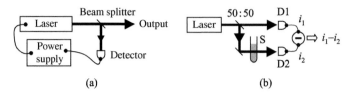

Fig. 5.13 (a) Noise eater scheme for stabilizing the power output of a laser. The laser power is monitored by sending a portion of the output to a detector from a beam-splitter. The detected signal is then used to control the power supply to the laser in a negative feedback loop. (b) Balanced detection scheme for cancelling classical noise. The beam is split into two equal parts by a 50:50 beam splitter, which are then incident on identical detectors D1 and D2. The output is equal to the difference of the photocurrents i_1 and i_2 from D1 and D2. When a sample S is inserted into the path to D2, the intensities on the detectors are no longer balanced, which then gives rise to a positive output.

receiving circuit. The size of the detected photocurrent signal must be larger than the photocurrent fluctuations due to the laser noise, and a glance at Fig. 5.11 suggests that the best strategy is to work at high frequencies where the laser noise is smallest. At these frequencies the laser noise is determined by the photon statistics set by the shot-noise limit. Hence the shot noise imposes a basic limit on the minimum signal-to-noise ratio that can be achieved. The only way to beat the shot noise limit is to use non-classical light sources with sub-Poissonian photon statistics. (See Section 5.10 below.)

The presence of shot noise in the photocurrent generated by the detection of light raises similar questions about its origin as arise with the observation of Poissonian statistics in a photon-counting detector. In analogy to the discussion of eqn 5.56 for photon-counting statistics, the photoelectron statistics from a photodiode will always show a Poisson distribution if the detector is inefficient. Moreover, we would also expect to observe shot noise after the detection of a purely classical light wave of constant intensity due to the probabilistic nature of the photodetection process at the microscopic level. In the next section we shall see that noise levels below the shot limit have been obtained in a number of experiments using sub-Poissonian light and high-efficiency detectors. This is not understandable within the semi-classical approach in which the noise originates in the photodetection process, and establishes that the shot noise in a high-efficiency photodiode can originate from the light, not the detector.

Example 5.2   A 10 mW He:Ne laser operating at 632.8 nm is detected with a photodiode of responsivity of 0.43 A W^{-1} via a load resistor of 50 Ω. Calculate:

(a) the quantum efficiency of the detector,

(b) the average photocurrent,

(c) the root-mean-square (r.m.s.) current fluctuations within a bandwidth of 100 kHz,

(d) the noise power measured in dBm units after amplification by an amplifier with a power gain of 20 dB in the same 100 kHz bandwidth.

Solution

(a) The quantum efficiency is worked out from the responsivity via eqn 5.58:

$$\eta = \frac{\hbar\omega}{e} \times \frac{i}{P} = 1.96 \text{ V} \times 0.43 \text{ A W}^{-1} = 84\%.$$

(b) The photocurrent is worked out from the responsivity:

$$i(\text{A}) = \text{responsivity (A W)}^{-1} \times \text{power (W)} = 0.43 \times 0.01 = 4.3 \text{ mA}.$$

(c) The variance of the current fluctuations within the bandwidth Δf is given by eqn 5.63. Hence the r.m.s. fluctuation for the photocurrent worked out above is given by:

$$\Delta i_{\text{r.m.s.}} = \sqrt{2e\Delta f \langle i \rangle} = (2e \times 10^5 \times 0.0043)^{1/2} = 12 \text{ nA}.$$

(d) The noise power in the load resistor is given by eqn 5.64 as:

$$P_{\text{noise}} = 2e \times 50 \times 10^5 \times 0.0043 = 6.9 \times 10^{-15} \text{ W}.$$

An amplification factor of +20 dB implies a power gain of $10^{(+20/10)} = 100$. Hence the amplified shot noise power would be 6.9×10^{-13} W, which, from eqn 5.65, is equivalent to -91.6 dBm.

5.10 Observation of sub-Poissonian photon statistics

The demonstration of sub-Poissonian photon statistics depends on two key aspects:

- the discovery of light sources with sub-Poissonian statistics;
- the development of detectors with high quantum efficiencies.

In practice, the second point has been a severe limitation, because, as eqn 5.56 demonstrates, there is no hope of demonstrating sub-Poissonian photoelectron statistics with a detector of low quantum efficiency. Fortunately, high-efficiency detectors are now readily available for many wavelengths. This has led to an increasing number of demonstrations of sub-Poissonian light. In the following two subsections we concentrate on the methods for the direct generation of sub-Poissonian light from light sources driven by electrical power supplies.

Other methods to generate sub-Poissonian light will be covered in Chapters 6–7. Sections 6.6 and 6.7 describe the generation of antibunched light, while Section 7.9.2 describes the generation of amplitude squeezed light by nonlinear optics. Note, however, that antibunched light is not necessarily sub-Poissonian: see the discussion in Section 6.5.3.

5.10.1 Sub-Poissonian counting statistics

Figure 5.14 shows an experimental scheme for generating sub-Poissonian light at 253.7 nm. The experiment works on the principle that the time taken by the atoms to emit a photon is short compared to the time-scales of the fluctuations in the electrical current used to excite the atoms.

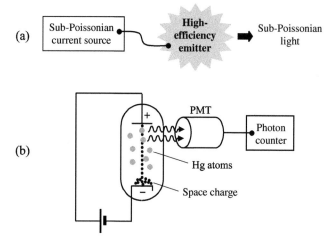

Fig. 5.14 (a) General scheme for generating sub-Poissonian light by driving a high-efficiency light emitter with a current source with sub-Poissonian electron statistics. (b) Experimental scheme for generating sub-Poissonian ultraviolet light at 253.7 nm from Hg atoms in a Franck–Hertz tube. The tube was operated in the space-charge-limited mode in which the electron statistics were sub-Poissonian. The photons were detected with a photomultiplier (PMT) and a photon counter. (After M.C. Teich and B. E. A. Saleh, *J. Opt. Soc. Am. B* **2**, 275 (1985).)

This implies that the statistical properties of the photons emitted in a discharge tube are closely related to the statistical properties of the electrons that comprise the current. It is intuitively obvious that if the electron flow is completely regular, then the photon flux is also regular, with equal time spacing between the photons. This point is summarized schematically in Fig. 5.14(a). Such a stream of photons is highly sub-Poissonian and contrasts with the usual (Poissonian) case in which the time spacing is random. The efficiency of the emission process needs to be high for the method to work well. If it is not, only a random subset of the electrons generate photons, and, as discussed in Section 5.7, such random sampling eventually randomizes the statistics, whatever the properties of the original photon distribution.

The experimental arrangement used to generate the sub-Poissonian light is illustrated schematically in Fig. 5.14(b). The light source consists of a Franck–Hertz tube filled with mercury (Hg) atoms. These atoms emit photons at 4.887 eV (253.7 nm) after excitation by electrons with sufficient energy to generate the photon. The electrons that comprise the anode current in the discharge tube are generated by thermal emission from the cathode and their energy is determined by the voltage between the anode and cathode. The statistics of the electrons generated by thermal emission would normally be random (i.e. Poissonian). However, it is well known that at the relatively small tube voltages required to initiate the mercury emission (i.e. 4.887 V), a space charge develops around the cathode. The presence of the space charge tends to regularize the electron flow, so that the statistics of the electrons in the anode current are sub-Poissonian. The photons emitted when the electrons collide with the mercury atoms are thus expected to have sub-Poissonian statistics.

The light measured in the experiment shown in Fig. 5.14 was found to have a variance smaller than the Poissonian value by 0.16%. The reason why the measured effect was so small was that the overall efficiency for conversion of electrons in the anode current to photoelectrons in the PMT was only 0.25%. This low efficiency was caused by a product of factors, including: the inefficiency of the electron–atom excitation process (25%), the imperfect photon collection efficiency (10%), the imperfect transmission of the optics (83%), and the imperfect quantum efficiency of the detector (15%). Although the light generated was only very slightly sub-Poissonian, the experiment was a clear proof of principle and paved the way for the experiments described in the next subsection which produce much larger effects.

It is instructive to consider briefly why it is so much easier to generate a sub-Poissonian current in an electrical circuit than it is to generate sub-Poissonian light. It is obvious that the negatively charged electrons repel each other. Furthermore, electrons are fermions and it is not possible for two of them to occupy the same quantum state. Both of these effects tend to keep the electrons apart, and hence to produce regular streams, randomized only by relatively small thermal fluctuations. Neither of these two regularizing mechanisms works for photons, which are neutral bosons, and can bunch together or spread themselves randomly with ease.

5.10.2   Sub-shot-noise photocurrent

The principle for generating sub-Poissonian light shown in Fig. 5.14(a) can readily be extended to solid-state emitters such as light-emitting diodes (LEDs) or laser diodes (LDs), which have much higher efficiencies than discharge tubes. Figure 5.15(a) shows a scheme for generating sub-Poissonian light from an LED and detecting it with a photodiode

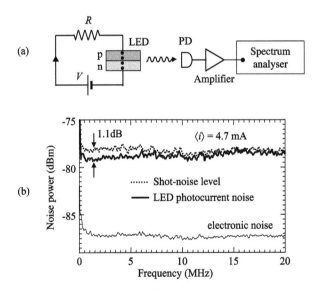

Fig. 5.15 (a) Generation of sub-Poissonian light from a high-efficiency LED and detection with a photodiode (PD). (b) Amplified photocurrent noise power spectrum measured for an AlGaAs LED emitting at 875 nm and measured with a photodiode of quantum efficiency 90%. The average photocurrent detected was 4.7 mA, and the detection bandwidth was 30 kHz. The curve shown by the dotted line corresponds to the calibrated shot-noise limit for the same current of 4.7 mA. The amplifier noise was about 9 dB below the shot-noise level, as shown by the lower curve in the graph. (After F. Wölfl *et al.*, *J. Mod. Opt.* **45**, 1147 (1998). © Taylor and Francis, reproduced with permission.)

detector. In this case the LED is driven by a battery with a series resistor R in the drive circuit. The purpose of the resistor is to control the current that flows, and in these circumstances the current fluctuations are determined by the thermal (Johnson) noise in the resistor. Provided the voltage dropped across the resistor is greater than $2k_BT/e$, where T is the temperature, then the fluctuations in the drive current are below the shot noise level. (See Exercise 5.12.) With $2k_BT/e \sim 50$ mV at room temperature, this condition is easily achieved, and the drive current is then strongly sub-Poissonian. If the LED has high efficiency, then the photon statistics should reflect the sub-Poissonian character of the drive current.

The experiments to demonstrate sub-shot-noise photocurrents have to be calibrated very carefully when the optical power level on the detector is high. This is because the photodiode response tends to saturate at high powers, and this can lead to erroneous measurements of the photocurrent noise.

Figure 5.15(b) show typical results obtained for a commercial AlGaAs LED operating at 875 nm. The photocurrent noise is observed to lie approximately 1.1 dB (21%) below the shot-noise level at frequencies of around 1 MHz. At higher frequencies the photocurrent noise tends to the shot-noise level due to the inability of the LED to follow the drive current at frequencies above its modulation response limit (~ 5 MHz). The observation of photocurrent noise below the shot-noise level clearly indicates that the photon statistics emitted by the LED are sub-Poissonian.

It is convenient to quantify the shot-noise reduction in terms of the **Fano factor** F_{Fano} defined by:

$$F_{\text{Fano}} = \frac{\text{measured noise}}{\text{shot noise limit}}. \tag{5.66}$$

See H.A. Bachor *et al.*, *Appl. Phys. B* **55**, 258 (1992).

The Fano factor for the data shown in Fig. 5.15(b) is thus 0.79 at ~ 1 MHz. If the total efficiency of the system in converting drive electrons from the battery into photoelectrons in the detector circuit is η_{total}, then the measured Fano factor is expected to be:

$$F_{\text{Fano}} = \eta_{\text{total}} F_{\text{dr}} + (1 - \eta_{\text{total}}), \tag{5.67}$$

where F_{dr} is the noise level of the drive current relative to the shot noise level. With $F_{\text{dr}} = 1$, we find $F_{\text{Fano}} = 1$ for all values of η_{total}, but for a strongly sub-Poissonian drive current, we have $F_{\text{dr}} \to 0$ and $F_{\text{Fano}} \to (1 - \eta_{\text{total}})$. The Fano factor of 0.79 deduced from the data in Fig. 5.15(b) agreed closely with the total conversion efficiency deduced from the product of the LED emission efficiency, the photon collection efficiency, and the detector quantum efficiency.

The principle shown in Fig. 5.15 is also applicable to semiconductor laser diodes. The use of such lasers in optical experiments has been shown to result in signal-to-noise ratios substantially better than the shot-noise level. Laser diodes offer a number of advantages over LEDs in the context of sub-shot-noise light generation. They usually have higher emission efficiencies and also emit into preferred directions, making the photon collection more efficient. Furthermore, they have large gain bandwidths, leading to the generation of sub-shot-noise light up to very high frequencies. The down-side is that laser diodes are far more sensitive to other noise sources than LEDs. In particular, they are very sensitive to instabilities in the power distribution between the longitudinal modes of the cavity. This means that most laser diodes show noise levels well above the shot-noise limit at all frequencies. In practice, the generation of sub-shot-noise light from laser diodes usually requires single-mode lasers with very high modal purities, often incorporating mode stabilization techniques employing external cavities.

Further reading

More advanced treatments of photon statistics may be found, for example, in Mandel and Wolf (1995) or Loudon (2000). Both texts give a thorough treatment of the semi-classical and quantum theories of photoelectric detection, while the semi-classical approach is also well described in Goodman (1985). Bachor and Ralph (2004) give a detailed explanation of the experimental techniques required to measure sub-Poissonian light, while Yamamoto and Imamoglu (1999) describe the theory of sub-Poissonian light generation by LEDs and laser diodes.

Introductory review articles on sub-Poissonian light have been given by Teich and Saleh (1990) and Rarity (1994). Undergraduate experiments to measure photon-counting statistics and to demonstrate sub-Poissonian light from an LED are described respectively by Koczyk *et al.* (1996) and Funk and Beck (1997).

Exercises

(5.1) A light beam of wavelength 633 nm and power 0.01 pW is detected with a photon-counting system of quantum efficiency 30% with a time interval of 10 ms. Calculate:

  (a) the count rate;

  (b) the average count value;

  (c) the standard deviation in the count value.

(5.2) Calculate the probability of obtaining a count value in the range 48–52 in a Poisson distribution with an average value of 50. Compare the exact probability to that obtained by approximating the Poisson distribution to a Gaussian (normal) distribution and calculating the probability that the count value lies between 47.5 and 52.5. Try to repeat the exercise for a mean value of 100 and a range from 95 to 105.

(5.3) Prove eqn 5.30.

(5.4) Calculate the mean photon number per mode at 500 nm from a tungsten lamp source operating at 2000 K, and also the temperature required to achieve $\bar{n} = 1$ at this wavelength. What is the equivalent temperature for a wavelength of 10 μm?

(5.5) In an experiment to measure the photon statistics of thermal light, the radiation from a black-body source is filtered with an interference filter of bandwidth 0.1 nm centered at 500 nm, and allowed to fall on a photon-counting detector. Calculate the number of modes incident on the detector, and hence discuss the type of statistics that would be expected.

(5.6) A pulsed diode laser operating at 800 nm emits 10^8 pulses per second. The average power measured on a slow response power meter is 1 mW. On the assumption that the laser light has Poissonian photon statistics, calculate the mean photon number and its standard deviation per pulse.

(5.7) The laser beam described in the previous question is attenuated by a factor 10^9. For the attenuated beam, calculate:

    (a) the mean photon number per pulse;

    (b) the fraction of the pulses containing one photon;

    (c) the fraction of the pulses containing more than one photon.

(5.8) A beam with a photon flux of $1\,000$ photons s^{-1} is incident on a detector with a quantum efficiency of 20%. If the time interval of the counter is set to 10 s, calculate the average and standard deviation of the photocount number for the following scenarios:

    (a) the light has Poissonian statistics;

    (b) the light has super-Poissonian statistics with $\Delta n = 2 \times \Delta n_{\mathrm{Poisson}}$;

    (c) the light is in a photon number state.

(5.9) A 10 mW He:Ne laser beam at 632.8 nm is incident on a photodiode with a quantum efficiency of 90%. Calculate the noise power per unit bandwidth when the photocurrent generated by the laser flows through a 50 Ω resistor.

(5.10) Estimate the quantum efficiency of the detector used for the data shown in Fig. 5.11.

(5.11) The photodiode used for the data shown in Fig. 5.12 had a responsivity of 0.40 $\mathrm{A\,W}^{-1}$ at the laser wavelength. Estimate the power gain of the amplifier in dB units, on the assumption that the input impedance of the amplifier was 50 Ω.

(5.12) Consider the current flowing through a resistor R at temperature T in an ohmic circuit. The current fluctuations in a frequency band Δf are given by the Johnson noise formula:

$$\langle (\Delta i)^2 \rangle = 4k_{\mathrm{B}}T\Delta f/R.$$

Show that the Johnson noise is smaller than the shot noise for the same average current value provided that the voltage dropped across the resistor is greater than $2k_{\mathrm{B}}T/e$, and evaluate this voltage for $T = 300\,\mathrm{K}$.

(5.13) The quantum efficiency of an LED is defined as the ratio of the number of photons emitted to the number of electrons flowing through the device. An LED emitting light at 800 nm is driven by a 9 V battery through a resistor with $R = 1000$ Ω. The LED has a quantum efficiency of 40%, and 80% of the photons emitted are focussed onto a photodiode detector with a quantum efficiency of 90%.

    (a) Calculate the average drive current, given that the voltage drop across the LED is approximately equivalent to the photon energy in eV in normal operating conditions.

    (b) Use the result in the previous exercise to calculate the Fano factor of the drive current for $T = 293\mathrm{K}$.

    (c) Calculate the average photocurrent in the detection circuit.

    (d) Calculate the Fano factor of the photocurrent.

    (e) Compare the photocurrent noise power in a 50 Ω load resistor with the shot-noise level for a bandwidth of 10 kHz.

(5.14) A laser pulse of energy 1 pJ and wavelength 800 nm is transmitted down an optical fibre with a loss of $3\,\mathrm{dB\,km}^{-1}$. Calculate the maximum distance that the pulses can propagate before the probability that a pulse contains no photons exceeds 10^{-9}. Discuss the implications of this result for data communications.

Photon antibunching

6

In the previous chapter we studied how light beams can be classified according to their photon statistics. We saw that the observation of Poissonian and super-Poissonian statistics could be explained by classical wave theory, but not sub-Poissonian statistics. Hence sub-Poissonian statistics is a clear signature of the photon nature of light. In this chapter we shall look at a different way of quantifying light according to the second-order correlation function $g^{(2)}(\tau)$. This will lead to an alternative threefold classification in which the light is described as *antibunched*, *coherent*, or *bunched*. We shall see that antibunched light is only possible in the photon interpretation, and is thus another clear signature of the quantum nature of light.

We begin with a classical description of the time-dependent intensity fluctuations in a light beam. These effects were first investigated in detail by R. Hanbury Brown and R. Q. Twiss in the 1950s, and their work has subsequently proven to be of central importance in the development of modern quantum optics. The Hanbury Brown–Twiss (HBT) experiments naturally led to the concept of the second-order correlation function, and we shall study the values that $g^{(2)}(\tau)$ can take for different types of classical light. We shall then see that the quantum theory of light can predict values of $g^{(2)}(\tau)$ that are completely impossible for classical light waves. The light that exhibits these non-classical results is described as being *antibunched*, and is particularly interesting in quantum optics. We conclude with a discussion of the experimental demonstrations of photon antibunching, and the practical application of antibunched light in *single-photon sources*.

6.1 Introduction: the intensity interferometer	**105**
6.2 Hanbury Brown–Twiss experiments and classical intensity fluctuations	**108**
6.3 The second-order correlation function $g^{(2)}(\tau)$	**111**
6.4 Hanbury Brown–Twiss experiments with photons	**113**
6.5 Photon bunching and antibunching	**115**
6.6 Experimental demonstrations of photon antibunching	**117**
6.7 Single-photon sources	**120**
Further reading	**123**
Exercises	**123**

This chapter assumes a reasonable familiarity with the coherence properties of light. A short summary of the main points may be found in Section 2.3.

6.1 Introduction: the intensity interferometer

Hanbury Brown and Twiss were astronomers who had a particular interest in measuring the diameters of stars. To this end, they developed the **intensity interferometer** while working at the Jodrell Bank telescope in England. Their interferometer was designed to be an improvement on the Michelson stellar interferometer which was originally implemented on the 2.5 m telescope at Mount Wilson near Los Angeles in the 1920s.

Figure 6.1(a) gives a schematic diagram of the Michelson stellar interferometer. The light from a bright star is collected by two mirrors M_1 and M_2 that are separated by a distance d. The light from each mirror is

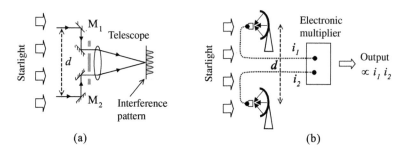

Fig. 6.1 (a) The Michelson stellar interferometer. Coherent light striking the two collection mirrors M_1 and M_2 produces an interference pattern in the focal plane of the telescope. (b) The Hanbury Brown–Twiss (HBT) stellar intensity interferometer. The light recorded on two separated detectors generates photocurrents i_1 and i_2, which are then correlated with each other with an electronic multiplier.

directed through separate slits into the telescope. If the light collected by the two mirrors is coherent, then an interference pattern will be formed in the focal plane of the telescope. On the other hand, if the light is incoherent, no interference pattern will be formed, and the intensities will just add together. The experiment consists in varying d and studying the visibility of the fringes that are observed in the focal plane. An analysis of the variation of the fringe visibility with d enables the angular size of the star to be measured. The actual diameter can then be determined if the star's distance from the earth is known.

We can understand how the Michelson stellar interferometer works by realizing that we are referring to the **spatial coherence** of the starlight, and not its temporal coherence. The spatial coherence is determined by the spread of angles within which light arrives at the interferometer. In an interference experiment, the light arriving at a particular angle generates its own set of bright and dark fringes, but displaced from each other by a distance depending on the angle. (See Exercise 6.1.) If we are not careful, we will have bright fringes for one angle where dark fringes for another angle occur, and vice versa. This would have the effect of washing out the visibility of the fringes, and so it is apparent that interference patterns are only observed when the spread of angles from the source is carefully controlled.

The angular spread $\delta\theta_s$ of an extended source such as a large star or galaxy is given by:

$$\delta\theta_s = D/L, \tag{6.1}$$

where D is its diameter and L the distance from the earth. This needs to be compared to the diffraction-limited angular resolution of the stellar interferometer $\delta\theta_r$ given by:

$$\delta\theta_r = 1.22\lambda/d, \tag{6.2}$$

where λ is the wavelength of the light and d is the mirror separation. Since d can be larger than the diameter of the telescope optics, the angular resolution is improved compared to the original instrument.

Light from a point source has no angular spread and therefore has perfect spatial coherence. An extended source, on the other hand, delivers light within a finite angular range, and therefore yields only partial spatial coherence.

The discussion of the Michelson stellar interferometer given here is somewhat simplified. In particular, the factor of 1.22 in eqn 6.2 is not immediately obvious, given that we are dealing with the diffraction pattern from two mirrors arranged in a line, rather than from a circular aperture. See Brooker (2003) for more details.

On the other hand, the light collection efficiency is worse, because the collection mirrors are usually relatively small. The Michelson stellar interferometer thus improves the angular resolution at the expense of light collection efficiency, and is therefore only useful for observing bright objects.

Let us suppose that we point the stellar interferometer at a small bright star, which acts like a point source in this context with $\delta\theta_s \ll \lambda/d$, for all practical values of d. In these conditions the instrument will not be able to resolve the different angles from the source, and interference fringes will be observed throughout. Now suppose we point the interferometer at an extended source such as a large star or a galaxy so that $\delta\theta_s > 1.22\lambda/d$ for some practical value of d. For $d > 1.22\lambda/\delta\theta_s$, the interferometer will be able to resolve the spread of angles from the source, and the light will be spatially incoherent, so that no interference fringes will be observed. Thus by varying d and recording the fringe visibility, we can determine $\delta\theta_s$, and thus deduce D from eqn 6.1 if L is known.

In the original experiments performed at Mount Wilson in the 1920s, the maximum practical value of d was about 6 m. The angular resolution $\delta\theta_r$ was therefore about 10^{-7} radians for wavelengths in the middle of the visible spectral region with $\lambda \sim 500$ nm. This was sufficient to determine the size of red giants like Betelgeuse in the Orion constellation, which has $\delta\theta_s = 2.2 \times 10^{-7}$ radians. In fact, this was how red giants were discovered.

It is apparent from eqn 6.2 that we need to increase d to improve the angular resolution $\delta\theta_r$ of the stellar interferometer. However, as d get larger, it becomes more and more difficult to hold the collection mirrors steady enough to observe interference fringes. To get around this problem, Hanbury Brown and Twiss proposed the much simpler arrangement shown in Fig. 6.1(b). In their experiment, they used two separated searchlight mirrors to collect light from a chosen star and focused it directly onto separate photomultipliers. This got around the need to form an interference pattern, and made the experiment much easier to perform.

The operating principles of the intensity interferometer will be studied at length in the subsequent sections of this chapter. At this stage we just need to point out that the interferometer measures the correlations between the photocurrents i_1 and i_2 generated by the starlight that impinges on the two photomultipliers. This is done by an electronic multiplying circuit, so that the output of the experiment is proportional to the time average of $i_1 i_2$. This in turn is proportional to $I_1 I_2$, where I_1 and I_2 are the light intensities incident on the two detectors. When d is small, the two detectors collect light from the same area of the source, and hence $I_1(t)$ and $I_2(t)$ will be the same. On the other hand, when d is large, the detectors can differentiate between the light arriving with different angles from the source, so that $I_1(t) \neq I_2(t)$, and the average of $I_1(t)I_2(t)$ will be different. The output of the detector will thus depend on d, and this provides another way of determining the angular spread of the star.

See R. Hanbury Brown and R. Q. Twiss, *Nature* **178**, 1046 (1957). The comment made above about the Michelson stellar interferometer sacrificing sensitivity for resolution applies even more strongly to the intensity interferometer. The latter is typically several orders of magnitude less sensitive.

Hanbury Brown and Twiss carried out a series of experiments to test the stellar interferometer during the winter of 1955–6. They demonstrated the validity of their method by obtaining $\delta\theta_s = 3.3 \times 10^{-8}$ radians for the star Sirius, which was in good agreement with the value determined by other methods. Hanbury Brown then moved to the clearer skies of Australia, where he set up a larger version of the interferometer with d values up to 188 m. The angular resolution of this improved instrument was 2×10^{-9} radians, which led to the measurement of the diameters of several hundreds of the brighter stars for the first time.

In the context of this present work on quantum optics, the interest in the HBT experiments is in the interpretation of the results. We have mentioned above that the interferometer measures *correlations* between the light intensities recorded by two separated photodetectors. This raises many conceptual difficulties. If each individual photodetection event is a statistical quantum process, how can separated events be correlated with each other?

The conceptual difficulty can be resolved by taking a semi-classical approach, such as the one taken in Section 5.8.1, in which the light is treated classically, and quantum theory is only introduced in the photodetection process itself. This approach was enough to satisfy the original critics of the HBT experiments. However, it transpires that if we really treat the light as a quantum object, then the objections raised are perfectly valid. In fact, the quantum theory of light predicts results in HBT experiments that are impossible for classical light. The aim of this chapter is to explain how these quantum optical effects can be observed and to describe the sources that produce them. Before we do this, however, we first review the classical theory of the HBT experiments.

The original experiments of Hanbury Brown and Twiss did not reveal any quantum optical effects because single-photon detectors with high quantum efficiencies were not available and they were looking at the thermal light from stars and galaxies. It was not until the 1970s that detectors of sufficient efficiency and new types of light sources were available to demonstrate the purely quantum optical effects.

6.2   Hanbury Brown–Twiss experiments and classical intensity fluctuations

The original HBT experiment is described in *Nature*, **177**, 27 (1956). Hanbury Brown and Twiss subsequently gave more detailed accounts in *Proc. Roy. Soc. A* **242**, 300 (1957) and **243**, 291 (1958). AC coupling was used so that the large DC photocurrent from the detectors and the electrical $1/f$ noise did not saturate the high-gain amplifiers. *RC* filters were used to block the low frequencies, and only the fluctuations in the frequency range 3–27 MHz were amplified.

Hanbury Brown and Twiss realized that their stellar interferometer was raising conceptual difficulties, and so they decided to test the principles of their experiment in the laboratory with the simpler arrangement shown in Fig. 6.2. In this experiment the 435.8 nm line from a mercury discharge lamp was split by a half-silvered mirror and then detected by two small photomultipliers PMT1 and PMT2, generating photocurrents i_1 and i_2, respectively. These photocurrents were then fed into AC-coupled amplifiers, which gave outputs proportional to the fluctuations in the photocurrents, namely Δi_1 and Δi_2. One of these was passed through an electronic time delay generator set to a value τ. Finally, the two signals were connected to a multiplier–integrator unit which multiplied them together and averaged them over a long time. The final output signal was thus proportional to $\langle \Delta i_1(t) \Delta i_2(t + \tau) \rangle$, where the symbol $\langle \cdots \rangle$ indicates the time average. Since the photocurrents were proportional to the impinging light intensities, it is apparent that the

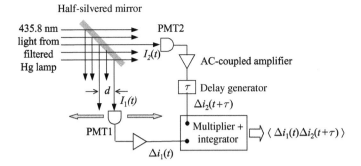

Fig. 6.2 Schematic representation of the Hanbury Brown–Twiss (HBT) intensity correlation experiment. The light from a mercury lamp was filtered so that only the 435.8 nm emission line impinged on a half-silvered mirror. Two photomultipliers tubes PMT1 and PMT2 detected the reflected and transmitted light intensities $I_1(t)$ and $I_2(t)$, respectively. The photocurrent signals generated by the detectors were filtered and amplified, and one of them was delayed by a time τ. The two amplified photocurrent fluctuation signals $\Delta i_1(t)$ and $\Delta i_2(t+\tau)$ were then fed into an electronic multiplier–integrator, giving an output proportional to $\langle \Delta i_1(t)\Delta i_2(t+\tau)\rangle$. PMT1 was placed on a translation stage, so that the two detectors could register light separated by a distance d. In this way, the spatial coherence of the source could be investigated. (After R. Hanbury Brown and R.Q. Twiss, *Nature*, **177**, 27 (1956).)

output was in fact proportional to $\langle \Delta I_1(t)\Delta I_2(t + \tau)\rangle$, where $I_1(t)$ and $I_2(t)$ were the light intensities incident on the respective detectors, and ΔI_1 and ΔI_2 were their fluctuations.

The light emitted by a mercury lamp originates from many different atoms. This leads to fluctuations in the light intensity on time-scales comparable to the coherence time, τ_c. These light intensity fluctuations originate from fluctuations in the numbers of atoms emitting at a given time, and also from jumps and discontinuities in the phase emitted by the individual atoms. The partially coherent light emitted from such a source is called **chaotic** to emphasize the underlying randomness of the emission process at the microscopic level.

See Section 5.5.2 for further discussion of chaotic light.

Figure 6.3 shows a computer simulation of the time dependence of the intensity of the light emitted by a chaotic source with a coherence time of τ_c. It is apparent that the intensity fluctuates wildly above and below the average value $\langle I \rangle$ on time-scales comparable to τ_c. These intensity fluctuations are caused by the addition of the randomly phased light from the millions of light-emitting atoms in the source. We suppose that each atom emits at the same frequency, but that the phase of the light from the individual atoms is constantly changing due to the random collisions.

A typical collision-broadened discharge lamp will have a spectral width $\Delta\nu \sim 10^9$ Hz, so that from eqn 2.41 we expect $\tau_c \sim 1$ ns.

The principle behind the HBT experiments is that the intensity fluctuations of a beam of light are related to its coherence. If the light impinging on the two detectors is coherent, then the intensity fluctuations will be *correlated* with each other. Thus by measuring the correlations of the intensity fluctuations, we can deduce the coherence

Fig. 6.3 Computer simulation of the time dependence of the light intensity emitted by a chaotic source with a coherence time of τ_c and average intensity $\langle I \rangle$. (After A.J. Bain and A. Squire, *Opt. Commun.* **135**, 157 (1997), © Elsevier, reproduced with permission.)

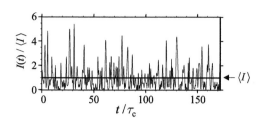

We have been assuming throughout this discussion that the detector can respond to the fast fluctuations in the light intensity on time-scales comparable to the coherence time τ_c. This requires very fast detectors that were not available in the 1950s. If the response time of the detector τ_D is longer than τ_c, it can be shown that the signal at $\tau = 0$ is reduced by a factor (τ_c/τ_D). See, for example, Loudon (2000, §3.8), or Mandel and Wolf (1995, §14.7.1.) Therefore, when $\tau_D > \tau_c$, we still expect the experiment to work and the output to fall to zero on a time-scale $\sim \tau_c$, although it becomes more difficult to observe the effect due to the smaller size of the signal.

properties of the light. This is much easier than setting up interference experiments, and gives us other insights as well.

Consider the results of the HBT experiment shown in Fig. 6.2 with d set at zero. We adjust the beam splitter so that the average intensity $\langle I(t) \rangle$ impinging on the detectors is identical. From a classical perspective, we can write the time-varying light intensity on the detectors as:

$$I_1(t) = I_2(t) \equiv I(t) = \langle I \rangle + \Delta I(t). \tag{6.3}$$

where $2I(t)$ is the intensity incident on the beam splitter and $\Delta I(t)$ is the fluctuation from the mean intensity $\langle I \rangle$. With identical intensities on the detectors, the output of the HBT experiment is proportional to $\langle \Delta I(t) \Delta I(t+\tau) \rangle$.

Let us suppose that we set the time delay τ to be zero. The output is then:

$$\langle \Delta I(t) \Delta I(t+\tau) \rangle_{\tau=0} = \langle \Delta I(t)^2 \rangle. \tag{6.4}$$

Although $\langle \Delta I(t) \rangle$ is equal to zero by definition, $\langle \Delta I(t)^2 \rangle$ will be non-zero due to the intensity fluctuations in the chaotic light from the discharge lamp. Hence there will be a non-zero output for $\tau = 0$. On the other hand, if we make $\tau \gg \tau_c$, the intensity fluctuations will be completely uncorrelated with each other, so that $\Delta I(t) \Delta I(t+\tau)$ randomly changes sign with time and averages to zero:

$$\langle \Delta I(t) \Delta I(t+\tau) \rangle_{\tau \gg \tau_c} = 0. \tag{6.5}$$

The output therefore falls to zero for values of $\tau \gg \tau_c$. Hence by measuring the output as a function of τ, we can determine τ_c directly.

In their original experiments, Hanbury Brown and Twiss set $\tau = 0$ and varied d. As d increased, the spatial coherence of the light impinging on the two detectors decreased. Hence the correlations between ΔI_1 and ΔI_2 eventually vanished for large values of d, and the output fell to zero. Their method therefore provided a way to determine the spatial coherence of the source through the decreased intensity correlations at large d values. The stellar intensity interferometer works by the same principle.

6.3   The second-order correlation function $g^{(2)}(\tau)$

In the previous section we considered how the results of the HBT experiments can be explained classically in terms of intensity correlations. In order to analyse these results in a quantifiable way, it is helpful to introduce the **second-order correlation function** of the light defined by:

$$g^{(2)}(\tau) = \frac{\langle \mathcal{E}^*(t)\mathcal{E}^*(t+\tau)\mathcal{E}(t+\tau)\mathcal{E}(t)\rangle}{\langle \mathcal{E}^*(t)\mathcal{E}(t)\rangle\langle \mathcal{E}^*(t+\tau)\mathcal{E}(t+\tau)\rangle} = \frac{\langle I(t)I(t+\tau)\rangle}{\langle I(t)\rangle\langle I(t+\tau)\rangle}, \quad (6.6)$$

where $\mathcal{E}(t)$ and $I(t)$ are the electric field and intensity of the light beam at time t. The $\langle\cdots\rangle$ symbols again indicate the time average computed by integrating over a long time period.

Let us consider a source with constant average intensity such that $\langle I(t)\rangle = \langle I(t+\tau)\rangle$. We shall also assume from now on that we are testing the spatially coherent light from a small area of the source. In these circumstances the second-order correlation function investigates the *temporal* coherence of the source.

We have seen above that the time-scale of the intensity fluctuations is determined by the coherence time τ_c of the source. If $\tau \gg \tau_c$, the intensity fluctuations at times t and $t+\tau$ will be completely uncorrelated with each other. On writing

$$I(t) = \langle I\rangle + \Delta I(t) \quad (6.7)$$

as before, with $\langle \Delta I(t)\rangle = 0$, we then have from eqn 6.5 that:

$$\langle I(t)I(t+\tau)\rangle_{\tau \gg \tau_c} = \left\langle \left(\langle I\rangle + \Delta I(t)\right)\left(\langle I\rangle + \Delta I(t+\tau)\right)\right\rangle$$

$$= \langle I\rangle^2 + \langle I\rangle\langle \Delta I(t)\rangle + \langle I\rangle\langle \Delta I(t+\tau)\rangle$$

$$+ \langle \Delta I(t)\Delta I(t+\tau)\rangle$$

$$= \langle I\rangle^2. \quad (6.8)$$

It is therefore apparent that:

$$g^{(2)}(\tau \gg \tau_c) = \frac{\langle I(t)I(t+\tau)\rangle}{\langle I(t)\rangle^2} = \frac{\langle I(t)\rangle^2}{\langle I(t)\rangle^2} = 1. \quad (6.9)$$

On the other hand, if $\tau \ll \tau_c$, there will be correlations between the fluctuations at the two times. In particular, if $\tau = 0$, we have

$$g^{(2)}(0) = \frac{\langle I(t)^2\rangle}{\langle I(t)\rangle^2}. \quad (6.10)$$

It can be shown that for any conceivable time dependence of $I(t)$, it will always be the case that

$$g^{(2)}(0) \geq 1, \quad (6.11)$$

and

$$g^{(2)}(0) \geq g^{(2)}(\tau). \quad (6.12)$$

The second-order correlation function $g^{(2)}(\tau)$ is the intensity analogue of the first-order correlation function $g^{(1)}(\tau)$ that determines the visibility of interference fringes. (See Section 2.3.) By comparing eqns 2.42 and 6.6, we can see that $g^{(1)}(\tau)$ quantifies the way in which the electric field fluctuates in time, whereas $g^{(2)}(\tau)$ quantifies the intensity fluctuations. In classical optics texts, $g^{(2)}(\tau)$ is often called the **degree of second-order coherence**.

These results can be proven rigorously (see Exercises 6.3 and 6.4), but we can also give a simple intuitive explanation of why they must apply.

Consider first a perfectly coherent monochromatic source with a time-independent intensity I_0. In this case, it is trivial to see that:

$$g^{(2)}(\tau) = \frac{\langle I(t)I(t+\tau)\rangle}{\langle I(t)\rangle^2} = \frac{I_0^2}{I_0^2} = 1, \tag{6.13}$$

for all values of τ, because I_0 is a constant. Next, recall that we expect from eqn 6.9 that $g^{(2)}(\tau) = 1$ for all large values of τ. Finally consider any source with a time-varying intensity. It is apparent that $\langle I(t)^2\rangle > \langle I(t)\rangle^2$ because there are equal intensity fluctuations above and below the average, and the squaring process exaggerates the fluctuations above the mean value. (See Example 6.1 below.) On using this fact in eqn 6.10, we see that we must have $g^{(2)}(0) > 1$. Putting it all together, we realize that, for any source with a time-varying intensity, we expect $g^{(2)}(\tau)$ to decrease with τ, reaching the value of unity for large τ. In the special case where $I(t)$ does not vary with time, we expect a constant value of $g^{(2)}(\tau) = 1$. These conclusions concur with the rigorous results given in eqns 6.11 and 6.12.

It is instructive to consider the explicit forms of the second-order correlation function for the various forms of light that we usually consider in classical optics. We have already seen that perfectly coherent light has $g^{(2)}(\tau) = 1$ for all τ. The values of $g^{(2)}(\tau)$ for the chaotic light from an atomic discharge lamp can be calculated by assuming simple models of the source. If the spectral line is Doppler-broadened with a Gaussian lineshape, the second-order correlation function is given by:

$$g^{(2)}(\tau) = 1 + \exp\left[-\pi(\tau/\tau_c)^2\right]. \tag{6.14}$$

This function is plotted in Fig. 6.4 and compared to that of perfectly coherent light. Similarly, a lifetime-broadened source with a Lorentzian lineshape has a $g^{(2)}$ function given by:

$$g^{(2)}(\tau) = 1 + \exp\left(-2|\tau|/\tau_0\right), \tag{6.15}$$

where τ_0 is the radiative lifetime for the spectral transition, or the collision time, as appropriate.

The main properties of the second-order correlation function $g^{(2)}(\tau)$ are listed in Table 6.1. These properties are derived by assuming that the

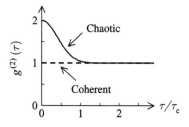

Fig. 6.4 Second-order correlation function $g^{(2)}(\tau)$ for chaotic and perfectly coherent light plotted on the same time-scale. The chaotic light is assumed to be Doppler-broadened with a coherence time of τ_c.

The derivations of eqns 6.14 and 6.15 may be found, for example, in Loudon (2000, §3.7). Note that both Doppler and Lorentzian-broadened chaotic light have $g^{(2)}(0) = 2$, with the value of $g^{(2)}(\tau)$ decreasing towards unity for $\tau \gg \tau_c$. Hence they both satisfy the general conditions set out in eqns 6.11 and 6.12.

Table 6.1 Properties of the second-order correlation function $g^{(2)}(\tau)$ for classical light.

Light source	Property	Comment
All classical light	$g^{(2)}(0) \geq 1$	$g^{(2)}(0) = 1$ when $I(t) = $ constant
	$g^{(2)}(0) \geq g^{(2)}(\tau)$	
Perfectly coherent light	$g^{(2)}(\tau) = 1$	Applies for all τ
Gaussian chaotic light	$g^{(2)}(\tau) = 1 + \exp\left[-\pi(\tau/\tau_c)^2\right]$	$\tau_c = $ coherence time
Lorentzian chaotic light	$g^{(2)}(\tau) = 1 + \exp\left(-2|\tau|/\tau_0\right)$	$\tau_0 = $ lifetime

light consists of classical electromagnetic waves. In the sections below
we shall reconsider the HBT experiments with photons incident on the
beam splitter rather than classical light waves. We shall find that the
two conditions given in eqns 6.11 and 6.12 do not necessarily have to be
obeyed. In particular, we shall see that it is possible to have light with
$g^{(2)}(0) < 1$, in violation of the classical result given in eqn 6.11. The
observation of $g^{(2)}(0) < 1$ is thus a conclusive signature of the quantum
nature of light.

Example 6.1   Evaluate $g^{(2)}(0)$ for a monochromatic light wave with
a sinusoidal intensity modulation such that $I(t) = I_0(1 + A \sin \omega t)$ with
$|A| \le 1$.

Solution
We compute $g^{(2)}(0)$ from eqn 6.10 according to:

$$g^{(2)}(0) = \frac{\langle I(t)^2 \rangle}{\langle I(t) \rangle^2} = \frac{\langle I_0^2(1 + A \sin \omega t)^2 \rangle}{I_0^2} = \langle (1 + A \sin \omega t)^2 \rangle,$$

where we used $\langle I(t) \rangle = I_0 \langle (1 + A \sin \omega t) \rangle = I_0$ since $\langle \sin \omega t \rangle = 0$. We
compute the time average by taking the integral over a long time interval
T, with $T \gg 1/\omega$:

$$g^{(2)}(0) = (1/T) \int_0^T (1 + A \sin \omega t)^2 \, \mathrm{d}t$$

$$= (1/T) \int_0^T (1 + 2A \sin \omega t + A^2 \sin^2 \omega t) \, \mathrm{d}t.$$

On using $2 \sin^2 x = (1 - \cos 2x)$, and with both $\sin \omega t$ and $\cos 2\omega t$
averaging to zero, this then gives:

$$g^{(2)}(0) = 1 + \frac{A^2}{2T} \int_0^T (1 - \cos 2\omega t) \mathrm{d}t = 1 + A^2/2.$$

$g^{(2)}(0)$ is therefore always greater than unity, and its maximum value is
equal to 1.5 for $|A| = 1$.

6.4   Hanbury Brown–Twiss experiments with photons

It is now time to re-examine the Hanbury Brown–Twiss (HBT) experiment in the quantum picture of light. Figure 6.5(a) illustrates the
experimental arrangement for a HBT experiment configured with single-
photon counting detectors. A stream of photons is incident on a $50:50$
beam splitter, and is divided equally between the two output ports. The
photons impinge on the detectors and the resulting output pulses are
fed into an electronic counter/timer. The counter/timer records the time
that elapses between the pulses from D1 and D2, while simultaneously

The quantum theory of the HBT
experiment will be given in Section 8.5.
We restrict ourselves here to a qualitative understanding of the experiments
and the classification of light according
to the second-order correlation function that naturally emerges from the
analysis.

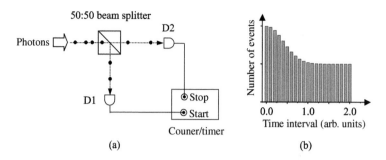

Fig. 6.5 (a) Hanbury Brown–Twiss (HBT) experiment with a photon stream incident on the beam splitter. The pulses from the single-photon counting detectors D1 and D2 are fed into the start and stop inputs of an electronic counter/timer. The counter/timer both counts the number of pulses from each detector and also records the time that elapses between the pulses at the start and stop inputs. (b) Typical results of such an experiment. The results are presented as a histogram showing the number of events recorded within a particular time interval. In this case the histogram shows the results that would be obtained for a bunched photon stream.

counting the number of pulses at each input. The results of the experiment are typically presented as a histogram, as shown in Fig. 6.5(b). The histogram displays the number of events that are registered at each value of the time τ between the start and stop pulses.

The correct normalization of $g^{(2)}(\tau)$ is very important for establishing non-classical results with $g^{(2)}(0) < 1$. The counter/timer arrangement shown in Fig. 6.5 produces a histogram that is proportional to $g^{(2)}(\tau)$ but does not give its exact value. From an experimental point of view, the normalization can be performed by assuming that $g^{(2)}(\tau) = 1$ for very long time delays. Alternatively, the non-classical source can be replaced by a Poissonian source of the same average intensity and the coincidence rates compared. Single-mode laser light can serve as a convenient Poissonian calibration source.

In Section 6.3 we discussed the $g^{(2)}(\tau)$ function classically in terms of intensity correlations. Since the number of counts registered on a photon-counting detector is proportional to the intensity (cf. eqn 5.2), we can rewrite the classical definition of $g^{(2)}(\tau)$ given in eqn 6.6 as:

$$g^{(2)}(\tau) = \frac{\langle n_1(t) n_2(t+\tau) \rangle}{\langle n_1(t) \rangle \langle n_2(t+\tau) \rangle}, \tag{6.16}$$

where $n_i(t)$ is the number of counts registered on detector i at time t. This shows that $g^{(2)}(\tau)$ is dependent on the simultaneous probability of counting photons at time t on D1 and at time $t+\tau$ on D2. In other words, $g^{(2)}(\tau)$ is proportional to the conditional probability of detecting a second photon at time $t = \tau$, given that we detected one at $t = 0$. This is exactly what the histogram from the HBT experiment with photon-counting detectors records. Hence the results of the HBT experiment also give a direct measure of the second-order correlation function $g^{(2)}(\tau)$ in the photon interpretation of light.

A moment's thought makes us realize that completely different results are possible with photons at the input port of the beam splitter than with a classical electromagnetic wave. Let us suppose that the incoming light consists of a stream of photons with long time intervals between successive photons. The photons then impinge on the beam splitter one by one and are randomly directed to either D1 or D2 with equal proba-

We are, of course, assuming here that the detectors have unity quantum efficiencies. Less perfect detectors would reduce the overall count rate, but would not affect the essential gist of the argument.

bility. There is therefore a 50% probability that a given photon will be detected by D1 and trigger the timer to start recording. The generation of a start pulse in D1 implies that there is a zero probability of obtaining a stop pulse from D2 from this photon. Hence the timer will record no events at $\tau = 0$. Now consider the next photon that impinges on the beam splitter. This will go to D2 with probability 50%, and if it does so, it will stop the timer and record an event. If the photon goes to D1, then nothing happens and we have to wait again until the next photon arrives to get a chance of having a stop pulse. The process proceeds until a stop pulse is eventually achieved. This might happen with the first or second or any subsequent photon, but never at $\tau = 0$. We therefore have a situation where we expect no events at $\tau = 0$, but some events for larger values of τ, which clearly contravenes the classical result given in eqns 6.11 and 6.12. We thus immediately see that the experiment with photons can give results that are not possible in the classical theory of light.

The observation of the non-classical result with $g^{(2)}(0) = 0$ arose from the fact that the photon stream consisted of individual photons with long time intervals between them. Let us now consider a different scenario in which the photons arrive in bunches. Half of the photons are split towards D1 and the other half towards D2. These two subdivided bunches strike the detectors at the same time and there will be a high probability that both detectors register simultaneously. There will therefore be a large number of events near $\tau = 0$. On the other hand, as τ increases, the probability for getting a stop pulse after a start pulse has been registered decreases, and so the number of events recorded drops. We thus have a situation with many events near $\tau = 0$ and fewer at later times, which is fully compatible with the classical results in eqns 6.11 and 6.12.

This simple discussion should make it apparent that sometimes the photon picture concurs with the classical results and sometimes it does not. The key point relates to the time intervals between the photons in the light beam; that is, whether the photons come in bunches or whether they are regularly spread out. This naturally leads us to the concepts of bunched and antibunched light, which is the subject of the next section.

6.5   Photon bunching and antibunching

In Section 5.4 we introduced a threefold classification of light according to whether the statistics were sub-Poissonian, Poissonian, or super–Poissonian. We now make a different threefold classification according to the second-order correlation function $g^{(2)}(\tau)$. The classification is based on the value of $g^{(2)}(0)$ and proceeds as follows:

- **bunched light**: $g^{(2)}(0) > 1$,
- **coherent light**: $g^{(2)}(0) = 1$,
- **antibunched light**: $g^{(2)}(0) < 1$.

This point is summarized in Table 6.2. A comparison of Tables 6.1 and 6.2 makes us realize that bunched and coherent light are compatible with the classical results, but not antibunched light. Antibunched light has no classical counterpart and is thus a purely quantum optical phenomenon.

Figure 6.6 is a simplistic attempt to illustrate the difference between the three different types of light in terms of the photons streams. The

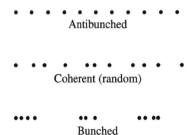

Fig. 6.6 Comparison of the photon streams for antibunched light, coherent light, and bunched light. For the case of coherent light, the Poissonian photon statistics correspond to random time intervals between the photons.

Table 6.2 Classification of light according to the photon time intervals. Antibunched light is a purely quantum state with no classical equivalent: classical light must have $g^{(2)}(0) \geq 1$.

Classical description	Photon stream	$g^{(2)}(0)$
Chaotic	Bunched	>1
Coherent	Random	1
None	Antibunched	<1

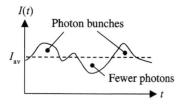

Fig. 6.7 Link between the classical intensity fluctuations about the average intensity I_{av} and photon bunching in a chaotic source. The photon bunches coincide with the high-intensity fluctuations.

reference point is the case where the time intervals between the photons are random. On either side of this we have the case where the photons spread out with regular time intervals between them, or where they clump together in bunches. These three cases correspond to coherent, antibunched, and bunched light, respectively. In what follows below, we explore the properties of each of these three types of light in more detail, starting with coherent light.

6.5.1   Coherent light

We have seen in Section 6.3 that perfectly coherent light has $g^{(2)}(\tau) = 1$ for all values of τ including $\tau = 0$. It thus provides a convenient reference for classifying other types of light.

In Section 5.3 we found that perfectly coherent light has Poissonian photon statistics, with random time intervals between the photons. This implies that the probability of obtaining a stop pulse is the same for all values of τ. We can thus interpret the fact that coherent light has $g^{(2)}(\tau) = 1$ for all values of τ (cf. eqn 6.13 and Fig. 6.4) as a manifestation of the randomness of the Poissonian photon statistics.

6.5.2   Bunched light

The tendency for photons to bunch together may be considered to be a manifestation of the fact that they are bosons.

Bunched light is defined as light with $g^{(2)}(0) > 1$. As the name suggests, it consists of a stream of photons with the photons all clumped together in bunches. This means that if we detect a photon at time $t = 0$, there is a higher probability of detecting another photon at short times than at long times. Hence we expect $g^{(2)}(\tau)$ to be larger for small values of τ than for longer ones, so that $g^{(2)}(0) > g^{(2)}(\infty)$.

We have seen in Section 6.3 that classical light must satisfy eqns 6.11 and 6.12. It is apparent that bunched light satisfies these conditions and is therefore consistent with a classical interpretation. It is also apparent from Table 6.1 that chaotic light (whether Gaussian or Lorentzian) also satisfies these conditions. The chaotic light from a discharge lamp is therefore bunched.

The link between photon bunching and chaotic light is illustrated schematically in Fig. 6.7, which shows the classical fluctuations in the light intensity as a function of time. Since the photon number is proportional to the instantaneous intensity, there will be more photons in the time intervals that correspond to high-intensity fluctuations and fewer in the low-intensity fluctuations. The photon bunches will therefore coincide with the high-intensity fluctuations.

6.5.3   Antibunched light

In antibunched light the photons come out with regular gaps between them, rather than with a random spacing. This is illustrated schematically in Fig. 6.6. If the flow of photons is regular, then there will be long time intervals between observing photon counting events. In this case, the probability of getting a photon on D2 after detecting one on D1 is small for small values of τ and then increases with τ. Hence antibunched light has

$$g^{(2)}(0) < g^{(2)}(\tau),$$
$$g^{(2)}(0) < 1. \tag{6.17}$$

This is in violation of eqns 6.11 and 6.12 which apply to classical light. Hence the observation of photon antibunching is a purely quantum effect with no classical counterpart. The $g^{(2)}(\tau)$ functions for two possible forms of antibunched light are sketched schematically in Fig. 6.8. The key point is that $g^{(2)}(0)$ is less than unity.

In Section 5.6 we studied the properties sub–Poissonian light and concluded that it, like antibunched light, is also a clear signature of the quantum nature of light. The question then arises whether photon antibunching and sub-Poissonian photon statistics are different manifestations of the same quantum optical phenomenon. This point has been considered by Zou and Mandel and the answer is negative. At the same time, it is apparent that a regular photon stream such as that illustrated in Fig. 6.6 will have sub-Poissonian photon statistics. Thus although the two phenomena are not identical, it will frequently be the case that nonclassical light will show both photon antibunching and sub-Poissonian photon statistics at the same time.

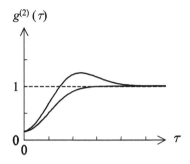

Fig. 6.8 Second-order correlation function $g^{(2)}(\tau)$ for two possible forms of antibunched light.

See X. T. Zou, and L., Mandel, *Phys. Rev A* **41**, 475 (1990).

6.6   Experimental demonstrations of photon antibunching

We have seen above that the observation of photon antibunching is a clear proof of the quantum nature of light. The first successful demonstration of photon antibunching was made by Kimble *et al.* in 1977 using the light emitted by sodium atoms. The basic principle of an antibunching experiment is to isolate an *individual* emitting species (i.e. an individual atom, molecule, quantum dot, or colour centre) and regulate the rate at which the photons are emitted by fluorescence. This is done by shining a laser onto the emissive species to excite it, and then waiting

See H. J. Kimble, M. Dagenais, and L. Mandel, *Phys. Rev. Lett.* **39**, 691 (1977).

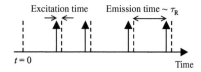

Excitation time   Emission time ~ τ_R

$t = 0$ Time

Fig. 6.9 Schematic representation of the photon emission sequence from a single atom excited by an intense laser. The dashed lines indicate the times at which the atom is promoted to the excited state, while the arrows indicate the photon emission events. τ_R is the radiative lifetime of the excited state.

Note that it is important to filter the light so that only a single transition wavelength is detected. See the discussion of Fig. 6.12(b) below.

for the photon to be emitted. Once a photon has been emitted, it will take a time approximately equal to the radiative lifetime of the transition, namely τ_R, before the next photon can be emitted. This leaves long time gaps between the photons, and so we have antibunched light.

We can understand this process in more detail by referring to Fig. 6.9, which shows a schematic representation of the photon emission sequence from a single atom. Let us suppose that the atom is promoted to an excited state at time $t = 0$, as indicated by the dashed line. The emission probability of the transition dictates that the average time to emit the photon is equal to τ_R. Once the photon has been emitted, the atom can be re-excited by the laser, which will only require a short amount of time if a high-power laser is used. The atom can then emit another photon after a time $\sim \tau_R$, at which point the excitation–emission cycle can start again. Since spontaneous emission is a probabilistic process, the emission time will not be the same for each cycle, which means that the stream of photons will not be exactly regular. However, it is clear that the probability for the emission of two photons with a time separation $\ll \tau_R$ will be very small. There will therefore be very few events when both the start and stop detectors of the HBT correlator in Fig. 6.5 fire simultaneously, and so we shall have $g^{(2)}(0) \approx 0$.

At this point we might legitimately ask why we do not observe the same antibunching effects from a conventional light source such as a discharge lamp. The point is that the antibunching effects are only observed if we look at the light from a *single* atom. The excitation–emission cycle shown in Fig. 6.9 is taking place for each individual atom in a discharge

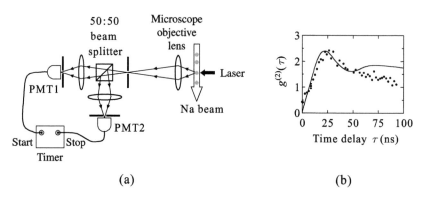

(a)   (b)

Fig. 6.10 (a) Schematic representation of the apparatus used to observe photon antibunching from the $3^2P_{3/2} \to 3^2S_{1/2}$ transition at 589.0 nm in atomic sodium. The sodium atomic beam was excited with a resonant laser and the fluorescence from one or two atoms only was collected with a microscope objective lens. The light beam was then split by a 50 : 50 beam splitter and detected with two photomultiplier tubes (PMT1 and PMT2) in a HBT arrangement. (b) Second-order correlation function $g^{(2)}(\tau)$ extracted from the data. The solid line is a theoretical fit calculated for a single atom. (After H.J. Kimble, M. Dagenais, and L. Mandel, *Phys. Rev. Lett.*, **39**, 691 (1977), © American Physical Society, reproduced with permission.)

lamp, but the light that is emitted originates from millions of atoms. The excitation and emission processes for the different atoms are statistically independent, and so they all emit at different times. This produces the photon bunches that are observed in the light emitted from a large number of atoms in a discharge lamp.

Figure 6.10(a) gives a schematic diagram of the apparatus used by Kimble *et al.* in 1977 to observe photon antibunching from a sodium atom. The sodium atomic beam was excited with a laser and the fluorescence from the $3^2P_{3/2} \rightarrow 3^2S_{1/2}$ transition at 589.0 nm was collected with a microscope objective lens. By using a very dilute beam, it was possible to arrange that, on average, no more than one or two atoms were able to contribute to the collected fluorescence at the same time. The fluorescence was then divided by a 50:50 beam splitter and detected by two photomultipliers in a HBT arrangement. The results obtained are shown in Fig. 6.10(b). Very few events were recorded near $\tau = 0$, and then $g^{(2)}(\tau)$ increased on a time-scale comparable to the radiative lifetime, namely 16 ns. At large time delays $g^{(2)}(\tau)$ decayed towards the asymptotic value of unity. The measured value of $g^{(2)}(0)$ was 0.4, which was a clear indication that the light was antibunched.

From a theoretical standpoint, it was expected that $g^{(2)}(0)$ should be zero if only a single atom was being observed. (See solid curve in Fig. 6.10(b).) The reason why the experimental value was larger was related to the experimental difficulty in arranging that only one atom should be in the field of view of the collecting lens at any one time. In practice, there were sometimes two or more, and this increased the value of $g^{(2)}(0)$ because of the possibility that two photons originating from different atoms should impinge on the beam splitter at the same time, and subsequently produce an event at $\tau = 0$ if the two photons go to different detectors. (See Exercise 6.11.)

In the years following Kimble *et al.*'s work, much progress has been made in atomic antibunching experiments. For example, antibunching has been demonstrated from a one-atom laser in the strong coupling regime of cavity quantum electrodynamics (QED). Moreover, antibunching has also been observed from many other types of light emitters, including a number of solid-state sources, such as:

- fluorescent dye molecules doped in a glass or crystal;
- semiconductor quantum dots;
- colour centres in diamonds.

As an example, Fig. 6.11 shows the $g^{(2)}(\tau)$ function measured for an individual semiconductor quantum dot at cryogenic temperatures. The sample consisted of an InAs quantum dot embedded within a GaAs microdisk as shown in Fig. 6.11(a). The purpose of the microdisk was to increase the collection efficiency of the photons emitted by the quantum dot. The quantum dot was excited with continuous light at 760 nm and the photons emitted across the band gap of the quantum dot at 937.7 nm

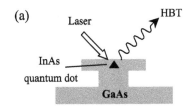

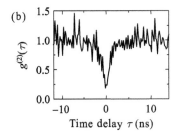

Fig. 6.11 (a) Excitation of an individual InAs quantum dot embedded within a GaAs microdisk structure using a continuous Ti:sapphire laser at 760 nm. The photons emitted from the quantum dot at 937.7 nm were detected with a HBT arrangement similar to the one shown in Fig. 6.5. (b) Second-order correlation function $g^{(2)}(\tau)$ measured for the quantum dots at 4 K. (After P. Michler *et al.*, *Science* **290**, 2282 (2000), ©AAAS, reprinted with permission.)

See Section 10.4 for a discussion of strong coupling effects in cavity QED, and Appendix D for an overview of the properties of semiconductor quantum dots. The strongly coupled one-atom laser is reported by J. McKeever *et al.* in *Nature* **425**, 268 (2003). For further information on antibunched light from solid-state sources, see, for example: Th. Basché *et al.*, *Phys. Rev. Lett.* **69**, 1516 (1992); P. Michler *et al.*, *Nature* **406**, 968 (2000); A. Beveratos *et al.*, *Phys. Rev. A*, **64**, 061802 (2001).

were detected with a HBT correlator. The results obtained at 4 K are shown in Fig. 6.11(b). The observed $g^{(2)}(0)$ value of ~ 0.2 is a clear signature of photon antibunching.

In this experiment the main reason why $g^{(2)}(0)$ was not zero was related to the finite response time of the detector, namely 0.42 ns. With a radiative lifetime of 2.2 ns, there was a significant probability of emission by 0.42 ns, and this contributed to the signal recorded at $\tau = 0$ because the detectors could not discriminate between these two times. (See Exercise 6.9.) Note that this is a different situation from the sodium experiment shown in Fig. 6.10, where the lifetime was much longer. At the same time, the quantum dots were fixed in the crystal lattice and so there was no chance that more than one emissive species could contribute to the fluorescence, as was the case for the atomic beam.

6.7   Single-photon sources

Single-photon sources are also required for the scheme proposed in Knill *et al.* (2001) to perform efficient quantum computation with linear optics. This scheme contrasts with the work described in Chapter 13 in which the quantum information is stored in atoms or ions and the photons are only used to manipulate the quantum bits.

An application of the techniques for the generation of antibunched light is the development of a triggered **single-photon source**. As explained in Section 12.5, these sources are needed to improve the security in quantum cryptography experiments. The basic idea of a single-photon source is that the source should emit exactly one photon in response to a trigger pulse, which can be either electrical or optical. The operating principle is shown in Fig. 6.12. The source consists of a single emissive species (say an atom), and the trigger pulse excites the atom to an upper excited state, as shown in Fig. 6.12(a). The atom then emits a cascade of photons as it relaxes to the ground state. Since the photons have different wavelengths, it is possible to select the photon from a particular transition by filtering the fluorescence. There will only ever be one photon emitted from a specific transition in each cascade.

Consider now the timing of the photons emitted by this process. An intense trigger pulse will rapidly promote an electron to the excited state,

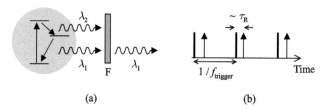

(a)   (b)

Fig. 6.12 Excitation–emission cycle from a single atom in response to trigger pulses. (a) The atom emits a cascade of photons of different wavelengths as it relaxes, but by using a suitable filter (F), only one of them is selected. (b) Schematic representation of the photon emission sequence. Trigger pulses are indicated by the thick lines, while the arrows indicate the photon emission events, which occur at a time roughly equal to the radiative lifetime (τ_R) after the trigger pulses. When the pulse separation $1/f_{\text{trigger}}$ is significantly longer than τ_R, the photon stream is controlled by the trigger pulse sequence.

and the atom will emit exactly one photon after a time roughly equal to the radiative lifetime τ_R, as shown schematically in Fig. 6.12(b). No more photons can be emitted until the next trigger pulse arrives, when the process repeats itself. The time separation of the trigger pulses is determined by the frequency f_{trigger} at which the trigger source operates. If the time separation between the pulses is significantly longer than τ_R, the trigger pulses control the separation of the photons in the fluorescence. We thus have a source that emits exactly one photon of a particular wavelength whenever a trigger pulse is applied.

The easiest way to make a triggered single-photon source is to use an optical trigger from a suitable laser. However, in the long run it will be important to develop electrically triggered devices. Figure 6.13 illustrates one such implementation incorporating a single quantum dot as the light-emitting species. The device consisted of a GaAs light emitting diode (LED) with a layer of InAs quantum dots inserted within the active region. The quantum dots were excited by a programmed sequence of

Experiments describing a molecular single-photon source are reported by B. Lounis and W.E. Moerner, *Nature* **407**, 491 (2000). The equivalent experiments for colour centres in diamond are described in C. Kurtsiefer *et al.*, *Phys. Rev. Lett.*, **85**, 290 (2000). The first two results on quantum dot single-photon sources are described in P. Michler *et al.*, *Science* **290**, 2282 (2000) and C. Santori *et al.*, *Phys. Rev. Lett.* **86** 1502 (2001). All of these experiments use optical trigger pulses. The electrically driven single-photon source discussed here was reported by Z. Yuan *et al.*, *Science* **295**, 102–5 (2002).

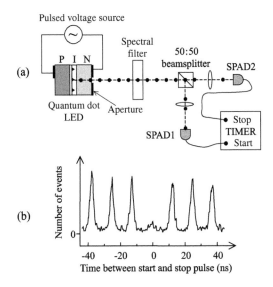

Fig. 6.13 An electrically driven triggered single-photon source. (a) Schematic representation of the experiment. The source consisted of a quantum dot LED driven by a pulsed voltage source. A single layer of InAs quantum dots was inserted within the intrinsic (I) region of a GaAs P-I-N diode. The quantum dots emitted light pulses in response to the pulsed voltage source. The photons were transmitted through the transparent N-type GaAs layer above the quantum dots and then through a small aperture in the metallic top contact. This aperture was small enough that it selected the light from only a few quantum dots, which emitted at different wavelengths due to their differing sizes. A spectrometer was then used as a spectral filter to select the emission at 889.3 nm from an individual quantum dot. The statistics of these selected photons were measured using a HBT arrangement with fast SPADs as the detectors. (b) Results obtained for a pulse repetition rate of 80 MHz with the device at 5 K. The count rate showed peaks corresponding to the pulse train period of 12.5 ns. The absence of a peak at zero time interval indicates the low probability that the quantum dot emitted two photons of the same wavelength at the same time. (After Z. Yuan *et al.*, Science **295**, 102 (2002), ©AAAS, reprinted with permission.)

current pulses produced by a pulsed voltage source. The current pulse injected electrons and holes into the device, and the quantum dots then emitted a light pulse in response to each trigger pulse. An aperture in the top contact ensured that the light from only a few of the InAs quantum dots was collected. The emission wavelength of a quantum dot depends on its size, which varies from dot to dot due to statistical fluctuations related to the crystal growth. Hence the wavelength varied slightly from dot to dot, which allowed the light emitted from a particular emission line of an individual quantum dot to be selected by using a spectrometer as a spectral filter. In these circumstances, we expect the light to be antibunched, as demonstrated previously in Fig. 6.11.

Figure 6.13(b) presents the results of the HBT experiment performed on the filtered light emitted from the device using fast single-photon avalanche photodiodes (SPADs) as the detectors. These results can be understood as follows. Let us suppose a photon strikes SPAD1 and generates a trigger pulse to start the timer. The timer will then measure the time that elapses before another photon strikes SPAD2 and generates the stop signal. This second photon may have come from the same light pulse as the first one, or from a different one. In the former case, we will record an event near $\tau = 0$. In the latter case, we will record an event near $\tau = m/f_{\text{trigger}}$, where m is an integer and $f_{\text{trigger}} = 80$ MHz is the frequency of the trigger pulse sequence. Hence the histogram of events will show peaks separated by 12.5 ns in these experiments. The key feature of the results is the very small number of events recorded near $\tau = 0$. This indicates that the source is emitting only one photon in each pulse, because there would have to be at least two photons in the pulse in order to register events at $\tau = 0$. In other words, we must have achieved a single-photon light source.

The results shown in Fig. 6.13 represent a substantial step towards the development of a convenient source for generating single photons on demand. At the present time, the main experimental difficulties that have to be overcome before these single-photon sources find more widespread applications is the low overall quantum efficiency and the operating temperature, which was 5 K for the data presented in Fig. 6.13.

An elegant experiment demonstrating the wave–particle duality of light using a single-photon source is shown schematically in Fig. 6.14. The light from a quantum dot single-photon source was divided equally with a 50 : 50 beam splitter and sent either to a Michelson interferometer or to a HBT experiment. The data from the HBT experiment was collected simultaneously with the fringe pattern from the interferometer. Clear interference fringes demonstrating the wave nature of light were observed at the same time as antibunching, which is a purely photon (i.e. particle) effect. Although it is clear that the individual photons go either to the interferometer or to the HBT experiment, it is very unlikely that the presence of one piece of apparatus can affect the results of the other. Therefore, the simultaneous observation of fringes and antibunching during a data collection run is a good demonstration of the wave–particle duality of light.

The principles of the Michelson interferometer are explained in Section 2.2.2.

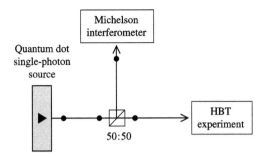

Fig. 6.14 Demonstration of the wave–particle duality of light using a quantum dot single-photon source. A 50:50 beam splitter sends half the photons to a Michelson interferometer and the other half to a HBT experiment. Interference fringes were observed at the same time as antibunching. Details of the experiment may be found in Zwiller *et al.*, *Phys. Rev. B* **69**, 165307 (2004).

Further reading

Introductory descriptions of the stellar intensity interferometer may be found in many standard optics texts, for example Brooker (2003), Hecht (2002), or Smith and King (2000). A more detailed discussion may be found in Hanbury Brown (1974).

The classical interpretation of the second-order correlation function is covered very rigorously in Mandel and Wolf (1995), while many useful insights are also to be found in Loudon (2000). Both of these texts develop the equivalent quantum theory in depth. Teich and Saleh (1990) give an introduction to antibunched light, and a more detailed account may be found in Teich and Saleh (1988). Thorn (2004) describes an undergraduate experiment to demonstrate photon antibunching.

An introductory account of single-photon sources has been given by Grangier and Abram (2003), and a collection of papers on the sources and their applications may be found in Grangier *et al.* (2004). There have now been many reports of the generation of antibunched light by semiconductor quantum dots. A review may be found in Michler (2003), while Petroff (2001) gives details of the techniques used to grow the dots.

Exercises

(6.1) Consider the fringe pattern from light of wavelength λ produced in a Young's double-slit experiment from a source of finite size D. Let the distance from the source to the slits be L ($L \gg D$) and the separation of the slits be d. Show that the dark fringes from the centre of the source coincide with the bright fringes from the edges when $D/L = \lambda/d$.

(6.2) In the HBT experiment shown in Fig. 6.2, a high-pass filter was inserted between the detector and the amplifier to remove the DC component of the photocurrent and the low-frequency electrical noise.

(a) Design a simple circuit employing a capacitor and a resistor to act as the filter.

(b) Calculate the value of the resistor to use for a cut-off frequency of 1 MHz if the capacitor has a capacitance of 1 nF.

(6.3) In this exercise we will prove eqn 6.11 by following the approach in Loudon (2000).

(a) By considering the quantity $(x(t_1) - x(t_2))^2$, where $x(t)$ is a real number, prove Cauchy's inequality:

$$x(t_1)^2 + x(t_2)^2 \geq 2x(t_1)x(t_2).$$

(b) Cauchy's inequality applied to $I(t)$ gives:

$$I(t_1)^2 + I(t_2)^2 \geq 2I(t_1)I(t_2).$$

Consider the quantity:

$$\left(\sum_{i=1}^N I(t_i)\right)^2 = (I(t_1) + I(t_2) + \cdots + I(t_N))^2$$

$$= \sum_{i=1}^N \sum_{j=1}^N I(t_i)I(t_j).$$

Apply Cauchy's inequality to each of the cross-terms to show that:

$$\left(\sum_{i=1}^N I(t_i)\right)^2 \leq N \sum_{i=1}^N I(t_i)^2.$$

(c) We define the average of the intensity as the mean of a large number of measurements made at different times according to:

$$\langle I(t)\rangle = \frac{1}{N}\sum_{i=1}^N I(t_i).$$

The average of $I(t)^2$ is defined in the same way:

$$\langle I(t)^2\rangle = \frac{1}{N}\sum_{i=1}^N I(t_i)^2.$$

Substitute these definitions into the result of part (b) to show that:

$$\langle I(t)\rangle^2 \leq \langle I(t)^2\rangle.$$

Hence derive eqn 6.11.

(6.4) The purpose of this exercise is to establish eqn 6.12 by a method similar to the one used in the previous exercise.

(a) Use Cauchy's inequality to show that:

$$\sum_{i=1}^N I(t_i)I(t_i + \tau) \leq \frac{1}{2}\sum_{i=1}^N (I(t_i)^2 + I(t_i + \tau)^2).$$

(b) Explain why, in any stationary light source (i.e. one in which the averages do not vary with time), we expect:

$$\sum_i I(t_i)^2 = \sum_i I(t_i + \tau)^2.$$

(c) The argument of part (b) allows us to rewrite the result of part (a) in the form:

$$\sum_{i=1}^N I(t_i)I(t_i + \tau) \leq \sum_{i=1}^N I(t_i)^2.$$

Following the definition of time averages given in part (c) of the previous exercise, we can write:

$$\langle I(t)I(t + \tau)\rangle = \frac{1}{N}\sum_{i=1}^N I(t_i)I(t_i + \tau).$$

Use this definition to show that:

$$\langle I(t)I(t + \tau)\rangle \leq \langle I(t)^2\rangle.$$

Hence derive eqn 6.12.

(6.5) Show that, in terms of the intensity fluctuations defined by $\Delta I(t) = (I(t) - \langle I(t)\rangle)$, the second-order correlation function $g^{(2)}(\tau)$ may be written in the form:

$$g^{(2)}(\tau) = 1 + \frac{\langle \Delta I(t)\Delta I(t + \tau)\rangle}{\langle I(t)\rangle\langle I(t + \tau)\rangle}.$$

Hence prove that $g^{(2)}(0) \geq 1$ for classical light.

(6.6) Calculate the values of $g^{(2)}(0)$ for a monochromatic light wave with a square wave intensity modulation of $\pm 20\%$.

(6.7) The 632.8 nm line of a neon discharge lamp is Doppler-broadened with a linewidth of 1.5 GHz. Sketch the second-order correlation function $g^{(2)}(\tau)$ for τ in the range 0–1 ns.

(6.8) The 546.1 nm line of a pressure-broadened mercury lamp has a line width of 0.001 nm. Sketch the second-order correlation function $g^{(2)}(\tau)$ for τ in the range 0–1 ns.

(6.9) (a) A single atom is irradiated with a powerful beam from a continuous wave laser which can promote it to an excited state with a radiative lifetime of τ_R. On the assumption that the excitation time is negligibly small, calculate the probability that the atom emits two photons in a time T.

(b) In a Hanbury Brown–Twiss (HBT) experiment, single-photon counting detectors with a response time of τ_D are connected to the start and stop inputs of a timer. The finite response time of the detectors implies that two events separated in time by $\leq \tau_D$ will be registered as simultaneous. Use this fact, together with the result of part (a), to estimate the value of $g^{(2)}(0)$ that would be expected in a HBT experiment using detectors of response times τ_D on a single atom with a radiative lifetime of τ_R.

(6.10) Discuss the dependence of $g^{(2)}(\tau)$ on the power of the exciting laser for an antibunching experiment such as that shown in Fig. 6.11 when using detectors of response time τ_D as in the previous question.

(6.11) A source emits a regular train of pulses, each containing exactly two photons. What value of $g^{(2)}(0)$ would be expected?

(6.12) A quantum dot with a radiative lifetime of 1 ns is used to make a single-photon source. What is the maximum photon bit rate that can be achieved?

(6.13) A source emits a train of single photons with exactly regular time intervals between them. Sketch the $g^{(2)}(\tau)$ function that would be expected:

(a) when the time interval between the photons is very much larger than the response time τ_D of the detector;

(b) when the time interval is very much smaller than τ_D.

7 Coherent states and squeezed light

7.1 Light waves as classical
    harmonic oscillators   126
7.2 Phasor diagrams and
    field quadratures   129
7.3 Light as a quantum
    harmonic oscillator   131
7.4 The vacuum field   132
7.5 Coherent states   134
7.6 Shot noise and
    number–phase
    uncertainty   135
7.7 Squeezed states   138
7.8 Detection of squeezed
    light   139
7.9 Generation of squeezed
    states   142
7.10 Quantum noise in
     amplifiers   146

Further reading   148
Exercises   148

In the preceding two chapters we have explored the consequences of quantizing the energy of a light beam in terms of the number of photons. This enabled us to classify light according to either the photon statistics or the second-order correlation function. In this chapter we shall consider the effects of quantizing the electric and magnetic fields that comprise the light. We shall make use of our knowledge of the quantum harmonic oscillator, and adopt a predominantly intuitive approach, leaving the rigorous mathematics to the next chapter.

We begin by explaining the connection between light and the harmonic oscillator at both the classical and quantum-mechanical level. This will lead us to discuss the properties of the *vacuum field* that corresponds to the zero-point fluctuations of the quantized light field, and those of *coherent states* which are the quantum-mechanical equivalents of classical electromagnetic waves. We shall see that this leads to a new type of uncertainty principle, namely *number–phase uncertainty*, and that this can give us an alternative way to understand the shot noise observed in optical detectors. Finally, we shall describe the properties of another class of non-classical light, namely *squeezed states*, and discuss the methods used to generate them in the laboratory.

7.1 Light waves as classical harmonic oscillators

The connection between light and the harmonic oscillator is, in one sense, completely obvious: light is a wave, and all wave phenomena can be related to harmonic oscillators. The link can be formalized by establishing the equations of motion for the light wave and showing that they are equivalent to those of a harmonic oscillator of mass m and angular frequency ω, namely:

$$p_x = m\dot{x} \qquad (7.1)$$

and

$$m\ddot{x} = \dot{p}_x = -m\omega^2 x, \qquad (7.2)$$

where x is the displacement and p_x is the linear momentum. The solutions can be written in the form:

$$x(t) = x_0 \sin \omega t, \tag{7.3}$$

$$p(t) = p_0 \cos \omega t, \tag{7.4}$$

where

$$p_0 = m\omega x_0. \tag{7.5}$$

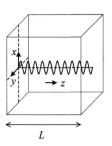

Fig. 7.1 Electric field of an electromagnetic wave polarized in the x-direction enclosed within an empty cavity of dimension L.

It will also be important to show that the energy of the light wave can be written in an equivalent form to that of the mechanical oscillator, namely:

$$E_{\text{SHO}} = \frac{p_x^2}{2m} + \frac{1}{2}m\omega^2 x^2. \tag{7.6}$$

Our task here is thus to find the equivalents of the position and linear momentum for the electromagnetic wave.

Let us consider a linearly polarized electromagnetic wave of wavelength λ enclosed within an empty cavity of dimension L as illustrated in Fig. 7.1. We assume that the light is polarized along the x-axis, and that the direction of the wave is along the z-axis. We can then write down the electric field in the following form:

$$\mathcal{E}_x(z,t) = \mathcal{E}_0 \sin kz \sin \omega t, \tag{7.7}$$

We have assumed that we have defined $t = 0$ so that we do not have to include a phase factor ϕ at this stage.

where $\mathcal{E}_0$ is the amplitude, $k = 2\pi/\lambda$ is the wave vector, and ω is the angular frequency. With the electric field polarized along the x-axis, the magnetic field will be along the y-axis. Writing this field as $B_y(z,t)$, the fourth Maxwell equation (eqn 2.12) with $\boldsymbol{j} = 0$, $\boldsymbol{B} = \mu_0 \boldsymbol{H}$, and $\boldsymbol{D} = \epsilon_0 \boldsymbol{\mathcal{E}}$ then reads:

$$-\frac{\partial B_y}{\partial z} = \epsilon_0 \mu_0 \frac{\partial \mathcal{E}_x}{\partial t}. \tag{7.8}$$

This implies:

$$B_y(z,t) = B_0 \cos kz \cos \omega t, \tag{7.9}$$

with

$$B_0 = \mathcal{E}_0/c, \tag{7.10}$$

since $\omega = ck$. It is apparent from eqns 7.7 and 7.9 that the electric and magnetic fields are 90° out of phase with each other, in exact analogy to $x(t)$ and $p(t)$ in the mechanical oscillator. (cf. eqns 7.3 and 7.4.)

The energy of the wave in the cavity can be found by integrating the energy density, namely:

$$U = \frac{1}{2}\left(\epsilon_0 \mathcal{E}^2 + \frac{1}{\mu_0}B^2\right), \tag{7.11}$$

This 90° phase shift between the electric and magnetic fields occurs because we are considering the standing waves in a cavity. The fields of travelling waves are in phase.

over the mode volume V. If we take the mode area to be A, the electric field energy for the spatially varying field given by eqn 7.7 is equal to:

$$E_{\text{electric}} = \frac{1}{2}\epsilon_0 A \int_0^L \mathcal{E}_0^2 \sin^2 kz \sin^2 \omega t \, dz$$

$$= \frac{1}{4}\epsilon_0 A \mathcal{E}_0^2 \sin^2 \omega t \int_0^L (1 - \cos 2kz) \, dz$$

$$= \frac{1}{4}\epsilon_0 V \mathcal{E}_0^2 \sin^2 \omega t, \qquad (7.12)$$

where we have made use of the identity $2\sin^2 \theta = 1 - \cos 2\theta$ in the second line, and equated $AL = V$ in the third. We also used the fact that we have a standing wave in the cavity, with nodes at $z = 0$ and $z = L$, so that:

The precise boundary conditions at the cavity walls are unimportant, because they potentially add only a term that varies as $1/L$ when the integration over z is performed, and this can be neglected for sufficiently large L. The insensitivity to the boundary conditions is important because we do not want any of the essential results to depend on the presence of the cavity.

$$\sin kL = 0, \qquad (7.13)$$

and therefore

$$\int_0^L \cos 2kz \, dz = \sin 2kL/2k = \sin kL \cos kL/k = 0. \qquad (7.14)$$

The magnetic field energy is likewise given by:

$$E_{\text{magnetic}} = \frac{1}{2\mu_0} A \int_0^L B_0^2 \cos^2 kz \cos^2 \omega t \, dz$$

$$= \frac{1}{4\mu_0} A B_0^2 \cos^2 \omega t \int_0^L (1 + \cos 2kz) \, dz$$

$$= \frac{1}{4\mu_0} V B_0^2 \cos^2 \omega t. \qquad (7.15)$$

The total energy is thus:

$$E = \frac{V}{4}\left(\epsilon_0 \mathcal{E}_0^2 \sin^2 \omega t + \frac{B_0^2}{\mu_0}\cos^2 \omega t\right), \qquad (7.16)$$

which shows that the energy oscillates back and forth between the electric and magnetic fields.

We now introduce two new coordinates $q(t)$ and $p(t)$ defined as follows:

The use of the generalized coordinates q and p instead of x and p_x avoids the need to introduce a mass m in the expression for the energy of the electromagnetic wave: see eqn 7.23.

$$q(t) = \left(\frac{\epsilon_0 V}{2\omega^2}\right)^{1/2} \mathcal{E}_0 \sin \omega t, \qquad (7.17)$$

$$p(t) = \left(\frac{V}{2\mu_0}\right)^{1/2} B_0 \cos \omega t \equiv \left(\frac{\epsilon_0 V}{2}\right)^{1/2} \mathcal{E}_0 \cos \omega t, \qquad (7.18)$$

where we made use of eqn 7.10 to substitute for B_0 in the definition of $p(t)$. It becomes clear that $q(t)$ and $p(t)$ are equivalent to the position and momentum of the electromagnetic harmonic oscillator, respectively, by noting that eqns 7.17 and 7.18 imply that

$$p = \dot{q}, \qquad (7.19)$$

and

$$\dot{p} = -\omega^2 q. \tag{7.20}$$

These can be made identical to the standard equations of motion of a harmonic oscillator given in eqns 7.1 and 7.2, by making the substitutions:

$$q(t) = \sqrt{m}\, x(t) \tag{7.21}$$

$$p(t) = (1/\sqrt{m})p_x(t). \tag{7.22}$$

On substituting $q(t)$ and $p(t)$ into eqn 7.16, we can rewrite the energy as:

$$E = \frac{1}{2}(p^2 + \omega^2 q^2). \tag{7.23}$$

Note again that this can be recast into its more familiar form given in eqn 7.6 by using eqns 7.21 and 7.22.

Equations 7.19, 7.20, and 7.23 together show that the $q(t)$ and $p(t)$ variables introduced in eqns 7.17 and 7.18 act like the position and momentum of the electromagnetic oscillator. Furthermore, since $q(t) \propto \mathcal{E}_x(t)$ and $p(t) \propto B_y(t)$, it is apparent that we can consider these variables to be equivalent to the electric and magnetic fields of the wave, respectively. We can then understand the oscillation of the energy back and forth between the electric and magnetic fields (cf. eqn 7.16) as equivalent to the oscillation between the potential and kinetic energies of a mechanical oscillator.

The oscillation of the energy between the electric and magnetic fields does not occur in travelling waves as the fields are in phase.

7.2 Phasor diagrams and field quadratures

The discussion of quantized light waves in subsequent sections makes frequent references to **phasor diagrams** and the electric field **quadratures**. It is therefore convenient to study these concepts first with reference to classical light waves.

Let us consider again a plane-polarized classical monochromatic wave within a cavity as shown in Fig. 7.1. In writing the field in eqn 7.7, a specific choice of the optical phase was made. In general, we ought to write:

$$\mathcal{E}_x(z,t) = \mathcal{E}_0 \sin kz \sin(\omega t + \phi), \tag{7.24}$$

where $\mathcal{E}_0$, k, and ω have the same meaning as in eqn 7.7, and ϕ is a phase factor that depends on how we define $t = 0$. By making use of the identity $\sin(\alpha + \beta) = \sin \alpha \cos \beta + \cos \alpha \sin \beta$, we can rewrite the field as:

$$\mathcal{E}_x(z,t) = \mathcal{E}_0 \sin kz (\cos \phi \sin \omega t + \sin \phi \cos \omega t)$$
$$= \mathcal{E}_1 \sin \omega t + \mathcal{E}_2 \cos \omega t, \tag{7.25}$$

where $\mathcal{E}_1 = \mathcal{E}_0 \sin kz \cos \phi$ and $\mathcal{E}_2 = \mathcal{E}_0 \sin kz \sin \phi$. The two amplitudes $\mathcal{E}_1$ and $\mathcal{E}_2$ are called the field quadratures. They correspond to two oscillating electric fields 90° out of phase with each other.

Fig. 7.2 (a) Phasor diagram for a classical wave of amplitude $\mathcal{E}_0$ and phase ϕ. (b) Equivalent phasor diagram in dimensionless quadrature field units. (c) Time dependence of the X_1 field quadrature. The quadrature amplitude X_{10} is related to the electric field amplitude $\mathcal{E}_0$ through $(\epsilon_0 V/4\hbar\omega)^{1/2}\mathcal{E}_0$.

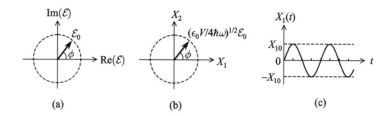

(a)　　　　　　　(b)　　　　　　　(c)

The two field quadratures can be conveniently incorporated into a single expression by using complex arithmetic. At a specific point in space, we write the field amplitude as:

$$\mathcal{E}(z) = \mathcal{E}_0(z)\,e^{i\phi}$$
$$= (\mathcal{E}_0(z)\cos\phi + i\mathcal{E}_0(z)\sin\phi)$$
$$= (\mathcal{E}_1(z) + i\mathcal{E}_2(z)), \qquad (7.26)$$

where $\mathcal{E}_0(z) = \mathcal{E}_0 \sin kz$. The complex field amplitude $(\mathcal{E}_1 + i\mathcal{E}_2)$ can be represented as a vector in the Argand diagram in which the real part of $\mathcal{E}$ corresponds to the x-axis, and the imaginary part to the y-axis, as shown in Fig. 7.2(a). This type of diagram is called a **phasor diagram**. The field is represented by a vector of length $\mathcal{E}_0$ at an angle of ϕ with respect to the x-axis.

In quantum optics, it is convenient to work in units in which the field is *dimensionless*. We therefore redraw the field phasor as a vector of length $(\epsilon_0 V/4\hbar\omega)^{1/2}\mathcal{E}_0$ as shown in Fig. 7.2(b). The axes are labelled as X_1 and X_2, respectively. The X_1 and X_2 quadratures of the field correspond to the sine and cosine parts of the time-dependent electric field, respectively:

$$X_1(t) = \left(\frac{\epsilon_0 V}{4\hbar\omega}\right)^{1/2} \mathcal{E}_0 \sin\omega t,$$

$$X_2(t) = \left(\frac{\epsilon_0 V}{4\hbar\omega}\right)^{1/2} \mathcal{E}_0 \cos\omega t. \qquad (7.27)$$

We can then substitute back into eqn 7.25 to find the time dependence of the electric field in terms of the quadrature amplitudes:

$$\mathcal{E}_x(z,t) = \left(\frac{4\hbar\omega}{\epsilon_0 V}\right)^{1/2} \sin kz\,(\cos\phi\, X_1(t) + \sin\phi\, X_2(t)). \qquad (7.28)$$

The time dependence of the X_1 field quadrature is shown in Fig. 7.2(c).

By comparing eqn 7.27 with eqns 7.17 and 7.18, it becomes apparent that the two field quadratures can be directly related to the generalized position and momentum coordinates $q(t)$ and $p(t)$, respectively, with:

$$X_1(t) = \left(\frac{\omega}{2\hbar}\right)^{1/2} q(t), \qquad (7.29)$$

$$X_2(t) = \left(\frac{1}{2\hbar\omega}\right)^{1/2} p(t). \qquad (7.30)$$

Many texts and papers use an alternative notation with the two field quadratures labelled as X and Y instead of X_1 and X_2. There is clear advantage to doing this as it makes a direct link to the axes of the Argand diagram. On the other hand, this can lead to some confusion, because there is no connection between the X and Y quadratures and the x- and y-components of the electric field vector. *Both* quadratures refer to a single field polarization. Furthermore, many authors associate X_1 and X_2 (or equivalently X and Y) with the cosine and sine quadratures, respectively, rather than with the sine and cosine quadratures as we have done here. This makes no fundamental difference to the analysis, and the author's choice is based on starting from a sine function in eqn 7.24 to avoid the inconvenient minus sign that occurs on expanding $\cos(\omega t + \phi)$.

Many quantum optics texts start with eqns 7.29–7.30, so that the definitions of X_1 and X_2 can apply to any type of harmonic oscillator. The approach taken here is less general, but has the advantage of highlighting the relationship between the quadratures and the electric field for the case of an optical harmonic oscillator.

The connection between the X_1 and X_2 quadratures and the position and momentum coordinates (see eqns 7.29–7.30) provides a formalism to apply the quantum theory of the simple harmonic oscillator to the electromagnetic wave. This link is developed in the following sections.

7.3   Light as a quantum harmonic oscillator

The equivalence between a light wave and a harmonic oscillator established in Section 7.1 means that we can apply our knowledge of the quantized harmonic oscillator to the quantized electromagnetic field states. In particular, we make use of two well-known results for the quantum harmonic oscillator (see Section 3.3):

1. The energy is quantized in units of $\hbar\omega$:

$$E_n = \left(n + \frac{1}{2}\right)\hbar\omega. \tag{7.31}$$

2. The position and momentum must satisfy the Heisenberg uncertainty principle:

$$\Delta x \Delta p_x \geq \frac{\hbar}{2}. \tag{7.32}$$

The first point can be interpreted as saying that we have n photons of angular frequency ω, together with a zero-point energy of $(1/2)\hbar\omega$. The second implies that all harmonic oscillators have quantum uncertainty. We have studied the consequences of the discreteness of the energy in the previous two chapters. We shall now consider the quantum uncertainty associated with electromagnetic harmonic oscillators and the zero-point energy.

Let us consider a single electromagnetic mode at angular frequency ω within a cavity of volume V. We describe the field in terms of the field quadratures $X_1(t)$ and $X_2(t)$ as defined in eqns 7.29–7.30. We define ΔX_1 and ΔX_2 as the uncertainties in the two field amplitudes, respectively. Their product is given by:

$$\Delta X_1 \Delta X_2 = \left(\frac{\omega}{2\hbar}\right)^{1/2} \Delta q \left(\frac{1}{2\hbar\omega}\right)^{1/2} \Delta p$$

$$= \frac{1}{2\hbar}\Delta q \Delta p. \tag{7.33}$$

On recalling that q and p are related to x and p_x through eqns 7.21 and 7.22, respectively, we can recast this in the form:

$$\Delta X_1 \Delta X_2 = \frac{1}{2\hbar}\left(\frac{\Delta x}{\sqrt{m}}\right)(\sqrt{m}\Delta p_x) = \frac{1}{2\hbar}\Delta x \Delta p_x. \tag{7.34}$$

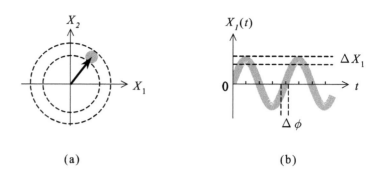

Fig. 7.3 (a) Phasor diagram for a quantized light field. (b) Time dependence of the X_1 quadrature for a quantized light field. These diagrams should be compared to the classical versions given in Fig. 7.2.

(a)   **(b)**

Finally, we introduce the quantum uncertainty of the harmonic oscillator given in eqn 7.32 to obtain:

$$\Delta X_1 \Delta X_2 \geq \frac{1}{4}. \tag{7.35}$$

We therefore conclude that the field quadratures are subject to quantum uncertainty in exact analogy to the quantum uncertainty of the position and momentum of a harmonic oscillator.

The quantum uncertainty in the field quadratures implies that the magnitude and direction of the electric field vector in a phasor diagram must be uncertain to some extent. If we assume that the uncertainties in the field quadratures are the same, then we would expect the phasor diagram to appear as in Fig. 7.3(a). In this figure the shaded circle represents the equal uncertainty in the two quadratures. The electric field phasor can lie anywhere within this uncertainty circle. Figure 7.3(b) shows the corresponding time dependence for the X_1 quadrature. It is apparent that the quantum uncertainty introduces uncertainty into both the amplitude and the phase of the wave.

From what we have seen here we realize that the classical picture of an electromagnetic wave with a perfectly well-defined amplitude and phase is an over-simplification. Quantum theory introduces an intrinsic uncertainty into the amplitude and phase. The consequences of this quantum uncertainty will be explored in the following sections.

7.4   The vacuum field

It is apparent from eqn 7.31 that the energy of the quantum harmonic oscillator is equal to $(1/2)\hbar\omega$, even when no photons are excited. This non-zero energy is usually described in standard quantum mechanics texts as the **zero-point energy** of the oscillator.

In quantum optics it is more normal to consider the zero-point energy as originating from a randomly fluctuating electric field called the **vacuum field**. This field is present everywhere, even in a complete vacuum. Its magnitude $\mathcal{E}_{\text{vac}}$ can be worked out by considering an evacuated optical cavity of volume V at a temperature where the thermal energy is very much less than the oscillator quantum energy. In these conditions there will be negligible thermal excitation of the oscillator, and in the absence of other energy sources, the electromagnetic modes will be in

The thermal energy will be much less than the energy spacing of the quantum levels when $k_B T \ll \hbar\omega$. This condition is easily achieved for optical frequencies because $\hbar\omega/k_B \sim 30\,000$ K for $\lambda \sim 500$ nm.

the $n = 0$ state. The zero-point energy of $(1/2)\hbar\omega$ per mode can then be equated with the electromagnetic energy within the mode volume V. On recalling that the time-averaged energy contributions of the electric and magnetic fields are identical, we may write:

$$2 \times \int \frac{1}{2}\epsilon_0 \mathcal{E}_{\text{vac}}^2 \, dV = \frac{1}{2}\hbar\omega, \tag{7.36}$$

which implies:

$$\mathcal{E}_{\text{vac}} = \left(\frac{\hbar\omega}{2\epsilon_0 V}\right)^{1/2}. \tag{7.37}$$

Equation 7.37 tells us that the magnitude of the vacuum field is largest for small cavities. (See Example 7.1.)

The classical field amplitude $\mathcal{E}_0$ is zero for the vacuum, and so the vacuum state is represented on a phasor diagram as an uncertainty circle centred at the origin as shown in Fig. 7.4. The shaded region of the phasor diagram indicates the random fluctuating field of the vacuum, with an average magnitude in real units given by eqn 7.37. The uncertainties in the two quadratures are identical, and each is equal to the minimum allowed by eqn 7.35. We therefore have:

$$\Delta X_1^{\text{vac}} = \Delta X_2^{\text{vac}} = \frac{1}{2}. \tag{7.38}$$

States like the vacuum field that satisfy eqn 7.35 with the minimum allowed uncertainty are called **minimum uncertainty states**.

The existence of the vacuum field is usually demonstrated by the **Casimir force**. This is an attractive force between two parallel conducting mirrors placed in a vacuum. The force arises from the change in the vacuum energy caused by the presence of the mirror cavity, and its magnitude per unit area is equal to:

$$F_{\text{Casimir}} = \frac{\pi^2 \hbar c}{240 L^4}, \tag{7.39}$$

where L is the separation of the mirrors. Although the force is extremely small (see Exercise 7.5), sensitive measurements have confirmed its existence.

The vacuum field has important consequences for several quantum optical phenomena. One of the best-known examples is the explanation of spontaneous emission as a stimulated emission process triggered by the vacuum field. Another topic in which the vacuum field is important is in considering the strong coupling regime in cavity quantum electrodynamics. (See Section 10.2.) The Lamb shift of atomic energy levels is also attributed to the vacuum field fluctuations.

Example 7.1   Calculate the magnitude of the vacuum field in a cavity of volume:
(a) 1 mm³, and (b) 1 μm³ at 500 nm.

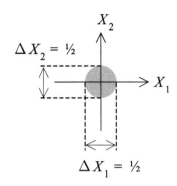

Fig. 7.4 Phasor diagram for the vacuum state. The uncertainties in the two field quadratures are identical, with $\Delta X_1 = \Delta X_2 = 1/2$. Note that this figure is essentially the same as Fig. 7.3(a) except that the uncertainty circle is displaced to the origin to account for the zero classical field of the vacuum.

A derivation of eqn 7.39 may be found in Loudon (2000, §6.12).

Solution

The magnitude of the vacuum field in a cavity of mode volume V is given by eqn 7.37.

(a) With $V = 10^{-9}$ m^3 and $\omega = 3.8 \times 10^{15}$ rad s^{-1}, we find $\mathcal{E}_{\text{vac}} = 4.7$ V m^{-1}.

(b) On repeating the calculation for $V = 10^{-18}$ m^3, we find $\mathcal{E}_{\text{vac}} = 1.5 \times 10^5$ V m^{-1}.

7.5   Coherent states

The theory of coherent states was introduced by Schrödinger in 1926. Their importance in quantum optics was first realized by Glauber in 1963 for which work he received the Nobel prize for physics in 2005. The mathematical definition of coherent states will be given in Section 8.4.

The quantum-mechanical equivalent of a classical monochromatic electromagnetic wave is called a **coherent state**. These states are represented in Dirac notation as $|\alpha\rangle$, where α is a dimensionless complex number. The significance of α can be understood by considering a linearly polarized mode of angular frequency ω enclosed within a cavity of volume V. In this situation, α is defined according to:

$$\alpha = X_1 + \mathrm{i}X_2, \tag{7.40}$$

where X_1 and X_2 are the dimensionless quadratures of the field within the cavity, as defined in eqn 7.27. We can separate α into its amplitude and phase ϕ by writing:

$$\alpha = |\alpha| e^{\mathrm{i}\phi} \tag{7.41}$$

with

$$|\alpha| = \sqrt{X_1^2 + X_2^2}, \tag{7.42}$$

and

$$\begin{aligned} X_1 &= |\alpha| \cos\phi, \\ X_2 &= |\alpha| \sin\phi. \end{aligned} \tag{7.43}$$

These definitions make it apparent that the coherent state $|\alpha\rangle$ can be represented as a phasor of length $|\alpha|$ at angle ϕ as shown in Fig. 7.5.

It can be shown that a coherent state is a minimum uncertainty state so that the equality sign in eqn 7.35 is appropriate. There is no intrinsic preference to either of the two quadratures, and thus their uncertainties must be identical. We therefore have:

$$\Delta X_1 = \Delta X_2 = \frac{1}{2}, \tag{7.44}$$

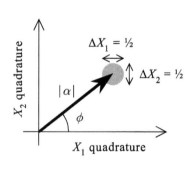

Fig. 7.5 Phasor diagram for the coherent state $|\alpha\rangle$. The length of the phasor is equal to $|\alpha|$, and the angle from the X_1-axis is the optical phase ϕ. The quantum uncertainty is shown by a circle of diameter 1/2 at the end of the phasor.

as for the vacuum state (cf. eqn 7.38.) Coherent states can therefore be considered as displaced vacuum states, with the uncertainty circle of the vacuum displaced from the origin by the field vector $\boldsymbol{\alpha}$ of the coherent state. The shaded circle of diameter 1/2 at the end of the phasor in Fig. 7.5 represents this quantum uncertainty.

A comparison of Fig. 7.5 with the phasor of the classical field in Fig. 7.2(b) makes it apparent that $|\alpha|$ is related to the electric field amplitude $\mathcal{E}_0$ according to:

$$|\alpha| = \sqrt{\frac{\epsilon_0 V}{4\hbar\omega}} \mathcal{E}_0. \tag{7.45}$$

The classical electromagnetic energy due to the mode is given by eqn 7.16. On substituting for B_0 from eqn 7.10 and using $c^2 = 1/\mu_0\epsilon_0$, we find:

$$E_{\text{classical}} = \frac{V}{4}\epsilon_0\mathcal{E}_0^2(\cos^2\omega t + \sin^2\omega t) = \frac{V}{4}\epsilon_0\mathcal{E}_0^2, \tag{7.46}$$

which, on using eqn 7.45, gives:

$$E_{\text{classical}} = \hbar\omega|\alpha|^2. \tag{7.47}$$

We can link this to the quantum theory of the electromagnetic harmonic oscillator by recalling that the excitation energy in the cavity can be written in the form (see eqn 7.31):

$$E_{\text{quantum}} = \bar{n}\,\hbar\omega + \frac{1}{2}\hbar\omega, \tag{7.48}$$

where $\bar{n}$ is the average number of photons excited in the cavity at angular frequency ω. The second term in eqn 7.48 is the zero-point energy associated with the vacuum field fluctuations. We can therefore equate the first term in eqn 7.48 with the classical energy due to $\mathcal{E}_0$. On setting $E_{\text{classical}} = \bar{n}\hbar\omega$ in eqn 7.47, we then find:

$$|\alpha| = \sqrt{\bar{n}}. \tag{7.49}$$

We therefore see that the length of the vector that represents the coherent state $|\alpha\rangle$ in a phasor diagram is equal to $\sqrt{\bar{n}}$.

7.6   Shot noise and number–phase uncertainty

It is apparent from Fig. 7.5 that both the length and angle of the coherent state are uncertain. This contrasts with the phasor of a classical wave shown in Fig. 7.2, in which both quantities are defined with complete precision. In this section, we shall see that the quantum uncertainty causes both shot noise and number–phase uncertainty.

Let us consider the quantum uncertainty of a coherent state with amplitude α as shown in Fig. 7.6. The phasor of the coherent state has an average length of $|\alpha|$ and makes an average angle of ϕ with respect to the X_1-axis. The photon number uncertainty Δn can be worked out by realizing that, if the uncertainty circle has a diameter of $1/2$, then the length of the phasor is uncertain between $(\alpha + 1/4)$ and $(\alpha - 1/4)$. Equation 7.49 tells us that the photon number is equal to $|\alpha|^2$, and we therefore have:

$$\Delta n = (|\alpha| + 1/4)^2 - (|\alpha| - 1/4)^2 = |\alpha| = \sqrt{\bar{n}}. \tag{7.50}$$

Equation 7.47 can also be derived directly from the relationship between α and the field quadratures: see Exercise 7.3. The separation of the mode energy in eqn 7.48 into a part that corresponds to the classical field energy plus the zero-point energy ensures that the vacuum state with zero classical amplitude has the correct energy of $\hbar\omega/2$.

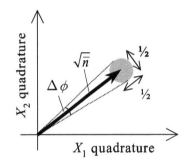

Fig. 7.6 The uncertainty circle of a coherent state $|\alpha\rangle$ introduces both photon number and phase uncertainty. Note that the phase uncertainty $\Delta\phi$ is only well-defined when $|\alpha| = \sqrt{\bar{n}} \gg 1$.

The geometric derivation of eqn 7.50 gives an intuitive understanding of the origin of shot noise. However, a sleight of hand was performed in using n rather than $\bar{n}$ in eqn 7.49. Concerned readers may rest assured that a rigorous derivation of eqn 7.50 can be given using the operator techniques in Chapter 8. (See eqn 8.45.)

This shows that coherent states have Poissonian photon statistics (cf. eqn 5.16). We saw in Section 5.9 that Poissonian photon statistics cause shot noise in optical detection. We thus see that the shot noise observed in optical detection can be thought of as originating from the quantum uncertainty in the light. This point will be developed further in Section 7.8, where we shall see that the shot noise can be attributed to the presence of vacuum modes.

A tutorial review on the phase in quantum optics may be found in Pegg and Barnett (1997).

The evaluation of the uncertainty in the optical phase is more problematic. In fact, in quantum optics the optical phase is not uniquely defined. On the other hand, when $|\alpha| = \sqrt{\bar{n}} \gg 1$, a useful result can be derived. This limit corresponds to fields of large amplitude where the quantum effects are small and the optical phase of the coherent state should equate with the classical phase of the electromagnetic wave. With $|\alpha|$ large, the phase uncertainty $\Delta\phi$ can be worked out from the angle subtended by the uncertainty circle (see Fig. 7.6):

$$\Delta\phi = \frac{\text{uncertainty diameter}}{\alpha} = \frac{1/2}{\sqrt{\bar{n}}}. \tag{7.51}$$

By combining this with eqn 7.50, we can form the **number–phase uncertainty** relationship:

$$\Delta n \Delta\phi \geq \frac{1}{2}. \tag{7.52}$$

The photon number–phase uncertainty relationship given in eqn 7.52 only holds accurately when n is large. When n is small, the definition of ϕ is ambiguous, and the relationship breaks down. Moreover, the number–phase uncertainty can always exceed the minimum value allowed by quantum theory, which is why the '$\geq$' sign is introduced in eqn 7.52.

This shows that it is not possible to know the photon number (i.e. the amplitude) and the phase of a wave with perfect precision at the same time.

It should be apparent from the number–phase uncertainty given in eqn 7.52 why the definition of optical phase in quantum optics is problematic. If $\bar{n}$ is small, then $\Delta\phi$ becomes large. Eventually $\Delta\phi$ approaches its maximum value of 2π, where the phase is totally undefined. The inability to define ϕ in this limit is a manifestation of the inherent difficulty in finding a quantum-mechanical definition of the optical phase.

The phase uncertainty implied by eqn 7.52 is important for high-precision measurements in interferometry. Example 7.2 below illustrates this point for the case of the ultra-high-precision interferometers designed to detect gravity waves.

Example 7.2 Figure 7.7 shows a schematic diagram of the LIGO gravity wave interferometer. The interferometer consists of a Michelson interferometer (see Section 2.2.2) comprising a 50 : 50 beam splitter BS and two end mirrors M1 and M2 each mounted on test masses. Gravity waves are predicted to produce oscillatory tidal distortions such that the displacements δL of the test masses are in opposite directions with respect to the beam splitter. This should produce a shift of the fringe pattern observed at the output, and hence leads to the possibility of detecting the displacements due to the gravity wave.

The LIGO experiment uses a Nd laser operating at 1064 nm with a power output of about 5 W. The experiment contains two additional

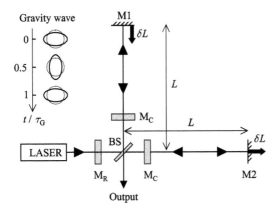

Fig. 7.7 Schematic diagram of the LIGO interferometer. A laser injects photons into a Michelson interferometer of arm length L comprising a 50:50 beam splitter (BS) and two end mirrors M1 and M2. A recycling mirror M_R recycles the power in the interferometer by a factor of 60, while two cavity mirrors M_C increase the effective arm length by a factor of 50. The effect of a gravity wave with period τ_G on a circular object is shown schematically in the top-left corner of the figure. The oscillatory tidal forces cause elliptical distortions, which displace the test masses attached to M1 and M2 in opposite directions, leading to an oscillatory shift in the fringe pattern observed at the output.

features compared to a standard Michelson interferometer, both of which are designed to improve its sensitivity. First, the power recycling mirror M_R recycles the power within the interferometer so that the effective power is 60 times higher, namely 300 W. Second, the two cavity mirrors M_C increase the effective arm length L by a factor of 50 to $50L$.

In this example we shall work out the sensitivity of the interferometer. Calculate:

(a) the phase uncertainty of the light within the cavity;

(b) the minimum displacement δL that can be detected;

(c) the minimum strain that can be detected for $L = 4$ km.

Solution

(a) We first calculate the average photon flux. With a photon energy of 1.17 eV, eqn 5.1 gives:

$$\overline{n} = \frac{300 \text{ W}}{1.17 \text{ eV}} = 1.6 \times 10^{21} \text{ photons s}^{-1}.$$

The uncertainty in the photon number is then given by eqn 7.50 as:

$$\Delta n = \sqrt{1.6 \times 10^{21}} = 4.0 \times 10^{10}.$$

On the assumption that the classical laser noise has been eliminated, the phase uncertainty is finally calculated from the minimum value allowed by eqn 7.52:

$$\Delta\phi = 1/2\Delta n = 1.3 \times 10^{-12} \text{ radians.}$$

LIGO is short for 'Light Interferometer Gravitational wave Observatory'. Gravity waves were predicted by Einstein in 1916 but have yet to be discovered. Gravity waves produce very small oscillatory tidal distortions that change spherical objects into elliptical ones, as shown in the inset to Fig. 7.7. Optimistic theories predict that the gravity waves from large astronomical events might produce strains of 10^{-21}. As this example shows, the LIGO experiment should be sensitive enough to detect such waves, on the assumption that all other sources of noise such as vibrations of the mirrors are eliminated, a condition which is extremely difficult to achieve in practice. The proposal to perform a similar experiment in space is explored in Exercise 7.9.

(b) In order for the fringe shift induced by the displacement δL to be observable, we require:

$$\frac{\delta L}{\lambda} > \frac{\Delta \phi}{2\pi}.$$

With $\Delta \phi = 1.3 \times 10^{-11}$ radians and $\lambda = 1064$ nm, we then find $\delta L = 2.2 \times 10^{-18}$ m.

(c) The strain is equal to the fractional length change divided by the original length. With the cavity enhancement included, this gives a strain h of

$$h = \frac{2.2 \times 10^{-18} \text{ m}}{50 \times 4 \text{ km}} = 1.1 \times 10^{-23}.$$

7.7   Squeezed states

The vacuum and coherent states studied in the previous two sections are both examples of minimum uncertainty states with equal uncertainties in the two quadratures so that:

$$\Delta X_1 = \Delta X_2 = \tfrac{1}{2}. \tag{7.53}$$

The uncertainty product in eqn 7.35 allows for other types of minimum uncertainty states in which the quadrature uncertainties are different. One way in which this can be achieved is to squeeze the uncertainty circle of the vacuum or the coherent state into an ellipse of the same area. Such states are called **quadrature-squeezed states**.

Figure 7.8 illustrates three different types of quadrature-squeezed states. Figure 7.8(a) illustrates the **squeezed-vacuum state**. This is a quantum state of light in which the quadrature uncertainty circle of the vacuum shown in Fig. 7.4 has been squeezed in one direction at the expense of the other to give an ellipse. In this particular example, the X_1 quadrature has been squeezed by a factor of $\sqrt{2}$, so that we have $\Delta X_1 = 0.35$ and $\Delta X_2 = 0.71$. The squeezing of the X_1 quadrature is apparent by comparing the ellipse to the dotted circle which corresponds to the original vacuum state with $\Delta X_1 = \Delta X_2 = 0.5$. (cf. eqn 7.38.)

Figures 7.8(b) and (c) illustrate two other forms of squeezed light in which the uncertainty circle of the coherent state shown in Fig. 7.5 has been squeezed into an ellipse of the same area. In Fig. 7.8(b) the major axis of the ellipse has been aligned with the phasor of the coherent state, so that the phase uncertainty is smaller than that in the original coherent state, while in Fig. 7.8(c) the minor axis has been aligned in order to reduce the amplitude uncertainty. The two states shown in Fig. 7.8(b) and (c) are therefore called **phase-squeezed light** and **amplitude-squeezed light**, respectively.

The use of phase-squeezed light allows interferometric measurements with greater precision than that obtained with a coherent state, as given in eqn 7.51. Similarly, the use of amplitude-squeezed light gives smaller amplitude noise than that of a coherent state. Now we have seen in

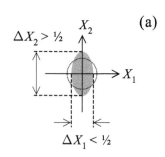

(a)

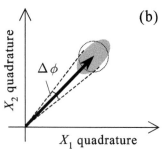

(b)

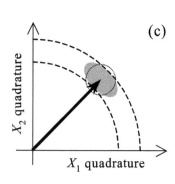

(c)

Fig. 7.8 Quadrature squeezed states. (a) Squeezed vacuum. (b) Phase-squeezed light. (c) Amplitude-squeezed light. The dotted circle in each of the diagrams shows the quadrature uncertainty of the vacuum/coherent states with $\Delta X_1 = \Delta X_2 = 1/2$.

eqn 7.50 that coherent states have Poissonian photon number fluctuations and therefore generate shot noise. Hence amplitude-squeezed light has sub-Poissonian photon statistics and produces a smaller noise level in optical detection than the shot-noise limit. The observation of photodetection noise below the shot-noise limit is thus one of the ways that squeezed states are detected in the laboratory.

The angles of the axes of the ellipses shown in Fig. 7.8 were chosen to illustrate the most important type of quadrature-squeezed states. There are, of course, many other examples of quadrature-squeezed states with axes at different angles. Furthermore, the uncertainty principle requires that there is a minimum *area* for the phasor uncertainty, but does not impose any limit on the *shape* of the uncertainty profile, leading to the possibility of other types of squeezed states. The only requirement on these states is that the uncertainty area in quadrature units must be $\geq 1/4$.

One of the most important types of squeezed states are the **photon number states** that we introduced previously in Section 5.6. These are states of perfectly defined photon number n, which implies $\Delta n = 0$, and, by the same token, a completely undefined phase. (See the discussion of eqn 7.52.) This contrasts with coherent states which have larger photon number fluctuations ($\Delta n = \sqrt{\overline{n}}$), but also have a much better defined phase.

Figure 7.9 shows the phasor diagram for a photon number state. The phasor is a circle of radius $(n + 1/2)^{1/2}$. The length of the phasor is perfectly well defined, and so there is no uncertainty in the electric field amplitude $\mathcal{E}_0$. On the other hand, the phase is totally undefined. The field is therefore a superposition of waves with the same amplitude but with all possible phases. Note that photon number states are not minimum uncertainty states. (See Exercise 7.13.)

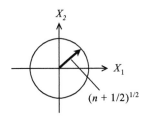

Fig. 7.9 Phasor diagram for a photon number state. The amplitude is perfectly defined, but the phase is completely uncertain. The phasor thus maps out a circle of radius $(n+1/2)^{1/2}$.

Note that the photon number states are not quadrature squeezed states because both ΔX_1 and ΔX_2 are larger than $\frac{1}{2}$.

At first sight, it would seem natural to suppose that the radius of the phasor for a photon number state should be equal to $n^{1/2}$ rather than $(n+1/2)^{1/2}$. The origin for the extra term only becomes apparent from consideration of the eigenvalues of the quadrature operators. See Exercise 8.7.

7.8   Detection of squeezed light

The detection strategy for squeezed light depends on the type of states that have been generated. Most detection schemes employ some form of balanced detection, a concept that was introduced previously in Section 5.9. In that context, we saw that by subtracting the photocurrents from two balanced detectors, we could cancel the classical noise and bring the noise level down to the shot-noise limit. We shall now see that we can actually get below the shot-noise limit when detecting squeezed light with a balanced detector, and in the process we shall also obtain a new perspective on the origin of shot noise in terms of vacuum noise.

7.8.1   Detection of quadrature-squeezed vacuum states

Figure 7.10(a) shows a schematic diagram of a **balanced homodyne detector**, which is the normal method used for the detection

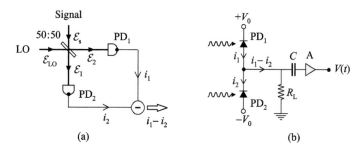

Fig. 7.10 The balanced homodyne detector. (a) The detector consists of a 50 : 50 beam splitter and two photodiodes PD_1 and PD_2 connected together so that their photocurrents i_1 and i_2 are subtracted, giving an output equal to $i_1 - i_2$. The signal field $\mathcal{E}_s$ is incident at one of the input ports of the beam splitter, while the local oscillator (LO) with amplitude $\mathcal{E}_{LO}$ is incident at the other. (b) Possible circuit diagram for the two photodiodes using a power supply of voltage $\pm V_0$. R_L is the load resistor, C is a capacitor, and A is an amplifier. The output voltage $V(t)$ is typically fed into a spectrum analyser.

The terminology of the balanced homodyne detector is borrowed from the engineering theory of heterodyne receivers. In a radio receiver, for example, the incoming signal at the aerial is mixed with a 'local oscillator' which is inside the radio set: that is, 'local' to the receiver, as opposed to the 'remote' oscillator in the transmitter. The receiver is called *homo*dyne or *hetero*dyne depending on whether the local oscillator has the *same* or *different* frequency as the signal's carrier wave. In squeezed-light experiments, it is common to use the *same* laser for both the signal generation and the local oscillator, and it is therefore appropriate to speak of *homodyne* detectors. See Haus (2000) for further details about heterodyne and homodyne detectors.

The effectiveness of the balanced detector in removing classical noise can be seen in the data presented in Fig. 7.14.

of quadrature-squeezed states. The detector consists of a 50 : 50 beam splitter and two photodiodes PD_1 and PD_2. The photodiodes are connected together in such a way that the output is equal to $(i_1 - i_2)$, where i_1 and i_2 are the photocurrents generated by PD_1 and PD_2, respectively. Figure 7.10(b) shows one way in which this can be done. The two diodes are connected in series to power supplies of $\pm V_0$. Provided that the load resistor R_L is relatively small, the DC voltage at the midpoint between PD_1 and PD_2 will be close to zero, and both diodes will be in reverse bias. In this situation, the dark current will be very small, and the difference of the two photocurrents will flow towards the amplifier A. The capacitor C ensures that the DC component of $(i_1 - i_2)$ flows through R_L, so that the output voltage $V(t)$ is the amplified AC component of $(i_1 - i_2)$.

The balanced detector has two input ports. The signal field is fed into one of the input ports of the beam splitter, while a **local oscillator** field is fed into the other. The local oscillator is a large-amplitude light wave with the same frequency as the signal, and is usually derived from the same laser that was used to generate the squeezed light. Let us first consider what happens when there is no input at the signal port of the beam splitter. From a classical perspective, the beam splitter divides the intensity of the local oscillator equally between the two photodiodes. The photocurrents generated by PD_1 and PD_2 will therefore be identical, so that $(i_1 - i_2) = 0$, and all the classical intensity fluctuations in the local oscillator are removed. On the other hand, the Poissonian statistics of the photon beams impinging on PD_1 and PD_2 will generate shot noise in i_1 and i_2. (See Section 5.9.) Since shot noise is random, the photocurrent fluctuations in i_1 and i_2 will be completely *uncorrelated*, and the two noise signals will *add* together at the output. Furthermore, the combined shot noise in i_1 and i_2 must have the same magnitude as that from a single photodiode detecting the whole of the intensity from the local oscillator. This follows from the fact that the shot-noise power scales as the average photocurrent (cf. eqn 5.63), which is itself proportional to the average optical power incident on each detector. The output of the balanced detector with no input at the signal port is therefore equal to the shot noise in the local oscillator, as we saw previously in Section 5.9.

Let us now consider the effect of introducing a signal field $\mathcal{E}_s$ into the other input port of the beam splitter. The output fields $\mathcal{E}_1$ and $\mathcal{E}_2$ are given by:

$$\mathcal{E}_1 = \frac{1}{\sqrt{2}}\left(\mathcal{E}_{LO}e^{i\phi_{LO}} + \mathcal{E}_s\right), \tag{7.54}$$

$$\mathcal{E}_2 = \frac{1}{\sqrt{2}}\left(\mathcal{E}_{LO}e^{i\phi_{LO}} - \mathcal{E}_s\right), \tag{7.55}$$

where $\mathcal{E}_{LO}$ is the amplitude of the local oscillator beam, and ϕ_{LO} is its phase relative to the signal field. Since the local oscillator is a large-amplitude field, it can be treated classically. On the other hand, the signal is a weak field and must therefore be treated quantum mechanically. We therefore split the signal field into its two quadrature components:

$$\mathcal{E}_s = \mathcal{E}_s^{X_1} + i\mathcal{E}_s^{X_2}, \tag{7.56}$$

with the factor of $i \equiv e^{i\pi/2}$ representing the 90° phase shift between the two quadratures. On splitting the output fields into their real and imaginary parts, we find:

$$\mathcal{E}_1 = \frac{1}{\sqrt{2}}\left[(\mathcal{E}_{LO}\cos\phi_{LO} + \mathcal{E}_s^{X_1}) + i(\mathcal{E}_{LO}\sin\phi_{LO} + \mathcal{E}_s^{X_2})\right], \tag{7.57}$$

$$\mathcal{E}_2 = \frac{1}{\sqrt{2}}\left[(\mathcal{E}_{LO}\cos\phi_{LO} - \mathcal{E}_s^{X_1}) + i(\mathcal{E}_{LO}\sin\phi_{LO} - \mathcal{E}_s^{X_2})\right]. \tag{7.58}$$

The photocurrent generated by a detector with a field $\mathcal{E}$ incident is proportional to $|\mathcal{E}|^2 = \mathcal{E}\mathcal{E}^*$. Hence the output of the balanced homodyne detector will be given by:

$$\text{output} \propto i_1 - i_2$$
$$\propto \mathcal{E}_1\mathcal{E}_1^* - \mathcal{E}_2\mathcal{E}_2^*$$
$$\propto 2\mathcal{E}_{LO}\left(\cos\phi_{LO}\mathcal{E}_s^{X_1} + \sin\phi_{LO}\mathcal{E}_s^{X_2}\right). \tag{7.59}$$

The output is therefore phase sensitive. If we pick $\phi_{LO} = 0, \pi, 2\pi, \ldots$, we find the output magnitude proportional to $\mathcal{E}_{LO}\mathcal{E}_s^{X_1}$, whereas for $\phi_{LO} = \pi/2, 3\pi/2, \ldots$, the output is proportional to $\mathcal{E}_{LO}\mathcal{E}_s^{X_2}$. The balanced homodyne detector therefore gives an output proportional to the signal field quadrature that is *in phase* with the local oscillator.

Let us consider again the case with no signal field present. In quantum optics, 'no signal field' means that there are vacuum modes entering at the signal port. The output of the detector is thus proportional to $\mathcal{E}_{LO}\mathcal{E}^{vac}$. We have seen above that this output level is equivalent to the shot noise in the local oscillator. We can thus interpret the shot-noise output with no signal present as a result of homodyning the local oscillator with the vacuum field.

Since the optical power is proportional to the square of the field, a power splitting ratio of 50:50 implies that the fields are reduced by a factor of $\sqrt{2}$ at the output ports. The minus sign in eqn 7.55 is a consequence of the fact that a 50:50 beam splitter always introduces a relative phase shift of π between the two output ports. At the most basic level, this phase shift is a consequence of the need to conserve energy: see Exercise 7.14.

It is worth stressing again that the observation of photocurrent noise below the shot-noise limit is a purely quantum optical effect with no classical counterpart. (See Section 5.8.1.)

Since a vacuum field input at the signal gives a shot-noise output, a squeezed vacuum signal will produce a noise level smaller than the shot-noise limit for the deamplified quadrature, and above the shot-noise level for the amplified quadrature. The observation of phase-dependent noise that goes below the shot-noise level for certain local oscillator phases is thus the signature of a quadrature-squeezed vacuum input.

7.8.2   Detection of amplitude-squeezed light

Amplitude-squeezed light has sub–Poissonian statistics, and is therefore easier to detect than quadrature-squeezed light. In principle, all that has to be done is to shine the light onto a photodiode and look for a noise signal below the shot-noise level. However, due to difficulties in obtaining an accurate calibration of the shot noise level, it is convenient to use a modified balanced detector as shown in Fig. 7.11. The arrangement is basically the same as for the balanced homodyne detector shown in Fig. 7.10 except that only one input port is used, and the photodiodes PD_1 and PD_2 are wired together so that the photocurrents i_1 and i_2 can either be subtracted ('−' output) or added ('+' output) by choice.

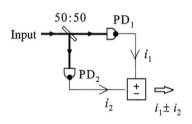

Fig. 7.11 Balanced detector scheme for the detection of amplitude squeezed light. The photocurrents i_1 and i_2 from the two photodiodes PD_1 and PD_2 can either be subtracted ('−' output) or added ('+' output).

When the '−' output is selected, we again have a balanced homodyne detector with no signal input. The output will therefore be at the shot-noise level, as discussed in the previous subsection. On the other hand, with the '+' output, we simply add the photocurrent fluctuations together as if we had detected the radiation with just a single photodiode. If the photon statistics are sub-Poissonian, the noise power for the '+' output is therefore expected to be lower than the shot noise level. Thus by switching between the '+' and '−' outputs, we can obtain a precise measurement of the photocurrent noise power of the input beam relative to the shot noise level.

7.9   Generation of squeezed states

The generation of squeezed states is a large subject, and it is not possible to cover all the possibilities in a text such as this. We therefore concentrate here on two of the most important types of squeezed states, namely quadrature-squeezed vacuum states and amplitude-squeezed light.

The three first experiments demonstrating quadrature squeezing are described in R.E. Slusher *et al.*, *Phys. Rev. Lett.* **55**, 2409 (1985), *ibid.* **56**, 788 (1986), R.M. Shelby *et al.*, *Phys. Rev. Lett.* **57**, 691 (1986), and L.-A. Wu, *et al.*, *Phys. Rev. Lett.*, **57**, 2520 (1986). The first two experiments used *third-order* nonlinear optical media, namely, sodium vapour and an optical fibre, respectively. For brevity, we just consider here the third experiment which used a second-order nonlinearity and produced the largest effect.

7.9.1   Squeezed vacuum states

Quadrature-squeezed vacuum states are generated by techniques of nonlinear optics. Figure 7.12 explains the general principle. The core of the experiment involves a **degenerate parametric amplifier**, consisting of a second-order nonlinear crystal pumped by an intense laser beam at angular frequency $\omega_p = 2\omega$. A weak **signal** beam at angular frequency $\omega_s = \omega$ is also introduced, as shown in Fig. 7.12(a). The nonlinear crystal

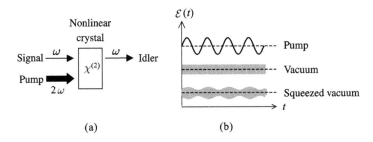

Fig. 7.12 (a) A degenerate parametric amplifier consists of a second-order nonlinear crystal pumped by an intense laser at angular frequency 2ω. The pumped nonlinear crystal acts as a phase-sensitive amplifier for signal modes at angular frequency ω. (b) With no signal input, the nonlinear crystal amplifies and de-amplifies the vacuum modes (middle panel), hence producing quadrature-squeezed vacuum states (bottom panel). Note that the y-axis scales for the classical pump laser field in the top panel and for the quantum fields in the other two panels are completely different.

mixes the signal with the pump and produces an **idler** beam at angular frequency ω_i by difference frequency mixing, where (see Section 2.4.2):

$$\omega_i = \omega_p - \omega_s = 2\omega - \omega = \omega. \tag{7.60}$$

These idler photons then mix again with the pump to produce more signal photons, and so on. In the special case that we are considering here where the signal and idler photons are degenerate, the nonlinear process produces amplification or de-amplification of the signal depending on its phase relative to the pump field. (See Appendix B.)

The effect of the phase-sensitive amplification of the degenerate parametric amplifier is illustrated in Fig. 7.12(b). We assume that there is *no* signal beam present at the input of the crystal. In this case, the signal is taken from the ever-present vacuum modes. The vacuum modes consist of a randomly fluctuating field of average amplitude given by eqn 7.37. On an $\mathcal{E}(t)$ diagram, the vacuum is represented by a fuzzy line of constant magnitude placed symmetrically about the $\mathcal{E} = 0$ axis, as shown in the middle panel of Fig. 7.12(b). The nonlinear process either amplifies or de-amplifies the vacuum depending on its phase. This produces an output field as shown in the bottom panel of the figure. The magnitude of the field is smaller than that of the vacuum for certain phases and therefore can correspond to a quadrature-squeezed vacuum state.

Figure 7.13(a) shows a schematic diagram of an experimental arrangement to generate squeezed vacuum states by degenerate parametric amplification using a second-order nonlinear crystal and a Nd:YAG laser operating at 1064 nm. The parametric amplifier was pumped by second harmonic radiation at 532 nm generated by frequency doubling of the laser using a second-order nonlinear crystal. (See Section 2.4.2.) Vacuum modes at 1064 nm were then amplified or de-amplified depending on their phase relative to the pump beam. A resonant cavity was used to enhance the magnitude of the nonlinear effects. The transmitted pump beam was removed from the output of the parametric amplifier by a

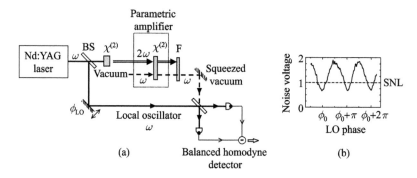

Fig. 7.13 (a) Schematic arrangement for generating quadrature squeezed vacuum states by degenerate parametric amplification. The beam from a Nd:YAG laser operating at angular frequency ω (1064 nm) was split into two powerful beams by a beam splitter (BS). One of the beams was used to generate a pump beam at 2ω (532 nm) by frequency doubling in a second-order nonlinear crystal $\chi^{(2)}$, while the other was used as the local oscillator (LO) for the balanced homodyne detector. Vacuum modes at angular frequency ω incident on another second-order nonlinear crystal within a resonant cavity and pumped by the beam at 2ω experienced parametric amplification, thereby generating squeezed vacuum states. A filter (F) selectively absorbed the transmitted pump beam, allowing the squeezed vacuum states to be fed into the signal port of the balanced homodyne detector. The phase of the local oscillator ϕ_{LO} was adjusted by placing one of the mirrors on the local oscillator path on a piezoelectric transducer and scanning its position over a few wavelengths. (b) Experimental results obtained using a MgO : LiNbO₃ crystal in the parametric amplifier. The noise voltage has been normalized so that the shot noise level (SNL) corresponds to a noise level of unity. (After L.-A. Wu *et al.*, *Phys. Rev. Lett.* **57**, 2520 (1986), ©American Physical Society, reproduced with permission.)

More recent experiments have now demonstrated larger squeezing levels. See, for example: P.K. Lam *et al.*, *J. Opt. B: Quantum Semiclass. Opt.* **1**, 469 (1999). By making use of the higher intensities available from pulsed lasers, quadrature-squeezed vacuum states can also be generated in a simpler experimental configuration that does not require a resonant cavity. See R.E. Slusher, *et al.*, *Phys. Rev. Lett.* **59**, 2566 (1987) and Exercise 7.15.

filter and the squeezed vacuum modes were fed into the signal port of a balanced homodyne detector. The local oscillator was derived from the original laser beam and its phase was scanned by placing one of the mirrors on a piezoelectric transducer.

Figure 7.13(b) presents the results of the experiment performed by Wu *et al.* with an MgO : LiNbO₃ crystal in the parametric amplifier. The output of the balanced homodyne detector has been normalized so that a noise level of unity corresponds to the shot-noise level (SNL). The data show that the noise voltage drops below the SNL for relative local oscillator phases of 0, π, and 2π, while rising above it for $\phi_{LO} = \pi/2$ and $3\pi/2$, indicating that the output of the parametric amplifier is a quadrature-squeezed vacuum state.

In the results shown in Fig. 7.13(b), the minimum r.m.s. noise voltage V_{rms} was about 70% of the SNL. This indicates that the noise power (proportional to V_{rms}^2) is below the SNL by a factor of ~ 2. This result, achieved in 1986, stood as a bench mark for several years.

7.9.2   Amplitude-squeezed light

In Section 5.10 we explained how sub–Poissonian light (i.e. amplitude-squeezed light) can be generated in light-emitting devices driven by an electrical supply with sub-Poissonian electron current statistics. In this section we shall explain how amplitude-squeezed light can be generated by techniques of nonlinear optics. The principle is to use the fact that nonlinear processes such as second-harmonic generation are sensitive to the fluctuations in the instantaneous photon flux. The principle has been demonstrated for a number of different types of nonlinear processes, including second-harmonic generation, two-photon absorption, and self-phase modulation.

Let us consider the process of frequency-doubling in a second-order nonlinear crystal. This process converts two photons from the fundamental beam into one photon in the second-harmonic beam, as shown in Fig. 2.6(b). In a typical experiment, the nonlinear crystal is pumped by a powerful beam at angular frequency ω, and has two output beams at ω and 2ω, as shown in Fig. 2.7(a). The probability for second-harmonic generation is proportional to $\chi^{(2)}\mathcal{E}_p^2$, where $\chi^{(2)}$ is the nonlinear susceptibility and $\mathcal{E}_p$ is the pump field. (See eqn 2.58.) The probability is therefore proportional to the intensity of the pump, which in turn is proportional to its photon flux.

Let us suppose that we use a stabilized laser generating light at the SNL as the input to the frequency doubler. The pump beam is therefore in a coherent state with Poissonian photon statistics, so that the incoming photon stream is random. Within this random photon stream, there will be instances when two photons arrive close together, giving rise to a higher conversion probability. This means that the above-average fluctuations in the photon flux of the pump beam will be selectively filtered out, leaving the output pump beam with a more regular flow than the incoming beam. At the same time, second-harmonic photons are only generated for these high-intensity fluctuations, and so the interval between successive photons will be larger than in the incoming pump beam. The result is that the fluctuations in both the transmitted fundamental and the second-harmonic beam are expected to be smaller than those of the incoming beam. Since the incoming photon stream is Poissonian, both the transmitted fundamental and the frequency-doubled beam are expected to be sub-Poissonian. In other words, we have amplitude squeezing, with photon fluctuations below the shot noise limit.

Figure 7.14(a) shows a scheme for demonstrating amplitude squeezing of the second harmonic of a Nd:YAG laser operating at 1064 nm. The transmitted 1064 nm radiation was selectively removed with a filter (F) and the 532 nm radiation was fed into the input port of a $+/-$ balanced detector as discussed in Section 7.8.2.

Figure 7.14(b) shows the results obtained with a MgO:LiNbO$_3$ frequency doubling crystal inside a resonant cavity. At low frequencies, the laser has classical noise and resonances giving large intensity fluctuations well above the SNL. However, the noise fluctuations for the second-harmonic beam are below the shot-noise level for frequencies above $\sim$15 MHz. The amount of noise reduction deduced from the data was $\sim$30%, after allowing for the inefficiency of the photodiodes.

The simultaneous squeezing of *both* the transmitted fundamental and the second-harmonic beam may appear counter-intuitive. See Loudon (2000, §9.3) for a more detailed explanation.

In more recent amplitude-squeezing experiments, the observed noise reduction has been increased to over 6 dB. See, for example: K. Schneider *et al.*, *Optics Express* **2**, 59 (1998). The noise spectrum in Fig. 7.14(b) can be compared to that shown previously for a different Nd:YAG laser in Fig. 5.11. Both lasers have noise powers above the shot-noise level for frequencies below about 10 MHz. The peaks in the noise spectrum in Fig. 7.14(b) at 6, 9, 11, and 12 MHz are related to classical resonances in the laser.

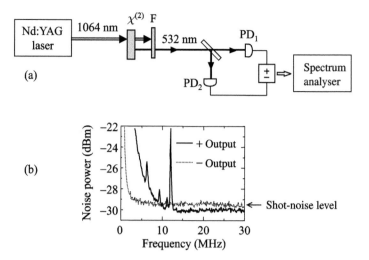

Fig. 7.14 (a) Schematic arrangement for generating amplitude-squeezed light at 532 nm by frequency doubling of the radiation at 1064 nm from a Nd:YAG laser in a second-order nonlinear crystal $\chi^{(2)}$. The second-harmonic beam was separated from the transmitted pump radiation by a filter (F) and was fed into a '+/−' balanced detector as described in Section 7.8.2. The added or subtracted photocurrents from the photodiodes PD_1 and PD_2 were fed into a spectrum analyser, and the AC noise power generated in its $50\,\Omega$ input resistor was recorded. (b) Experimental results. (See eqn 5.65 for a definition of dBm units.) The short-noise level was calibrated from the subtracted ('−') noise power, while the noise level of the second-harmonic beam was determined from the addition ('+') of the photocurrents from PD_1 and PD_2. The sharp increase of the noise power below 5 MHz for the '−' signal is caused by classical amplifier noise. (After R. Paschotta *et al.*, *Phys. Rev. Lett.* **72**, 3807 (1994), ©American Physical Society, reproduced with permission.)

7.10   Quantum noise in amplifiers

Optical amplifiers are important components in telecommunication systems. As the light pulses propagate down the optical fibres, their intensity decays due to absorption and scattering losses, and the signals therefore have to be boosted at regular intervals by amplifiers called repeaters. The issue that we wish to address briefly here is whether the amplification process necessarily adds noise to the signal, as illustrated schematically in Fig. 7.15(a).

Let us first make a few definitions. The gain G of an amplifier is determined by the gain coefficient γ. If the amplifier is operating in the linear regime and has a length L, the total gain will be given by (see eqn 4.39):

$$G = \exp(\gamma L). \tag{7.61}$$

The signal-to-noise ratio of the light beam can be defined classically in terms of the power of the signal compared to that of the fluctuations:

$$\mathrm{SNR} = \frac{(\text{signal amplitude})^2}{(\text{noise amplitude})^2} = \frac{\langle I \rangle^2}{\langle \Delta I^2 \rangle}. \tag{7.62}$$

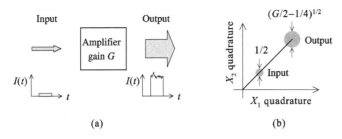

Input   Output

Amplifier
gain G

$I(t)$ t $I(t)$ t

(a)

X_2 quadrature

$(G/2-1/4)^{1/2}$

Output

$1/2$

Input

X_1 quadrature

(b)

Fig. 7.15 (a) Effect of a noisy amplifier with gain G on an input signal. The output pulse is amplified compared to the input pulse, but its amplitude is noisier. (b) Phasor diagram for the output field when the input signal is a coherent state.

The second equality follows from the fact that the maximum signal amplitude that can be modulated onto a beam is equal to its average intensity $\langle I \rangle$. The quantum equivalent to eqn 7.62 is:

$$\text{SNR} = \frac{\overline{n}^2}{(\Delta n)^2}. \tag{7.63}$$

A coherent state with Poisson statistics ($\Delta n = \sqrt{\overline{n}}$) therefore has a signal-to-noise ratio of $\overline{n}$.

The **noise figure** of the amplifier is defined as the ratio of the signal-to-noise ratio of the input beam compared to that of the output. It can be shown that, when the input to the amplifier is a coherent state, the noise figure is given by:

$$\text{Noise figure} \equiv \frac{\text{SNR}_{\text{in}}}{\text{SNR}_{\text{out}}} = 2 - \frac{1}{G}, \tag{7.64}$$

where G is the gain. This implies that the amplifier adds noise for $G > 1$. It is, of course, possible to make bad amplifiers with higher noise figures, but many well-designed travelling-wave optical amplifiers can come close to the theoretical limit set by eqn 7.64.

Two important consequences follow immediately from eqn 7.64:

(1) the amplified states are not minimum uncertainty states;
(2) for large values of G, we can expect a noise figure of two ($+3\,\text{dB}$).

The first point is illustrated in the phasor diagram shown in Fig. 7.15(b): the length of the phasor increases by $\sqrt{G}$, but the area of the uncertainty circle also increases.

An obvious question arises as to why the amplifier adds noise. Let us suppose that we use a *non-degenerate* parametric amplifier with a coherent state as the signal input as shown in Fig. 7.16. The signal beam is amplified as it propagates through the nonlinear crystal, taking energy from the pump beam. The mechanism for the amplification is the difference frequency mixing process, in which idler photons are generated and then mix with the pump to produce more signal photons (cf. eqn 2.63). In this process, the noise of the idler photons gets mixed into the signal. Since there is no idler input, the idler beam starts from vacuum noise. It is the mixing of the vacuum noise of the idler with the signal through the nonlinear interaction that produces the excess noise. A similar argument

The derivation of eqn 7.64 may be found, for example, in Loudon (2000, §7.6), or Abram and Levenson (1994).

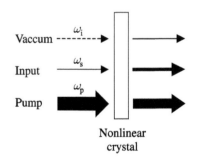

Vaccum ω_i

Input ω_s

Pump ω_p

Nonlinear
crystal

Fig. 7.16 Amplification of a signal input at angular frequency ω_s in a non-degenerate parametric amplifier with a pump at ω_p. The idler input at ω_i is taken to be vacuum noise.

can be made about conventional travelling wave amplifiers. In this second case, some modes of the laser cavity are amplified, but many others are not. Spontaneous emission into these non-lasing modes provides an equivalent noise source that degrades the signal-to-noise ratio of the amplifier.

Up to this point, we have been considering phase-insensitive amplifiers. However, we have seen in Section 7.9.1 that the parametric amplifier acts as a *phase-sensitive* amplifier when the signal and idler are *degenerate*. In this situation, it is possible to amplify one of the phase quadratures at the expense of extra noise in the other one. Such an amplifier is called a **noiseless amplifier**.

Experimental results on noiseless amplifiers are presented in J.A. Levenson *et al.*, *Quantum Semiclass. Opt.* **9**, 221 (1997).

Further reading

The subject of coherent states and squeezed light is covered in greater depth in Gerry and Knight (2005), Haus (2000), Loudon (2000), Mandel and Wolf (1995), or Walls and Milburn (1994). An extensive discussion of the experimental techniques used to generate squeezed light may be found in Bachor and Ralph (2004). Useful information about the award of the 2005 Nobel Prize to Glauber may be found at http://nobelprize.org/physics/.

Introductory reviews on coherent states and squeezed light may be found in Slusher and Yurke (1988), Leuchs (1988), Barnett and Gilson (1988), Teich and Saleh (1990), or Ryan and Fox (1996). A scholarly review is given by Loudon and Knight (1987a), while Haus (2004) presents an overview of the whole subject of quantum noise in optics. Collections of articles on the subject of squeezed light written soon after the first experimental observations may be found in Loudon and Knight (1987b) and Kimble and Walls (1987).

Pegg and Barnett (1997) give a tutorial review on the subject of phase in quantum optics, while Giovannetti *et al.* (2004) give an overview of quantum-enhanced measurements. Reviews on gravity wave interferometers may be found in Corbitt and Mavalvala (2004) or Hough and Rowan (2005). Slusher and Yurke (1990) give a review of the applications of squeezed light in optical communications, while Abram and Levenson (1994) provide a detailed overview of the effects of quantum noise in amplifiers.

Exercises

(7.1) Show that the time-averaged energy in the electric and magnetic fields of an electromagnetic wave are identical.

(7.2) Use the definitions of the field quadratures in eqn 7.27 to verify that $X_1(t)$ and $X_2(t)$ are dimensionless variables.

(7.3) Show that the energy of a classical electromagnetic wave with field quadratures $X_1(t)$ and $X_2(t)$ is given by:

$$E = \hbar\omega(X_1(t)^2 + X_2(t)^2).$$

Hence derive eqn 7.47 for a coherent state.

(7.4) Calculate the volume required to make the vacuum field magnitude equal to $1~\mathrm{V\,m^{-1}}$ for a wavelength of (a) $1~\mu\mathrm{m}$ and (b) $100~\mathrm{nm}$.

(7.5) Calculate the Casimir force between two conducting plates of area $1~\mathrm{cm^2}$ separated by (a) $1~\mathrm{mm}$, (b) $1~\mu\mathrm{m}$.

(7.6) For the coherent states $|\alpha\rangle$ with $\alpha = 5$, calculate:

 (a) the mean photon number;

 (b) the standard deviation in the photon number;

 (c) the quantum uncertainty in the optical phase.

(7.7) A ruby laser operating at 693 nm emits pulses of energy 1 mJ. Calculate the quantum uncertainty in the phase of the laser light.

(7.8) In the LIGO experiment described in Example 7.2, the power recycling mirror increases the power within the cavity by a factor of 60. Calculate the improvement of the sensitivity introduced by the power recycling effect.

(7.9) A proposed space experiment for gravity wave detection will use a standard Michelson interferometer (i.e. no power recycling or cavity enhancement) with a laser operating at 1064 nm. The length of the arms of the interferometer is 5×10^6 km and the power of the beams that form the interference pattern is $\sim 10^{-11}$ W. Calculate the minimum strain that can be detected.[1]

(7.10) Sketch the time dependence of the electric field equivalent to Fig. 7.3(b) for (a) phase-squeezed light, and (b) amplitude-squeezed light.

(7.11) Explain why light with very strong quadrature squeezing will not exhibit amplitude squeezing, no matter how the axes of the uncertainty ellipse are chosen. Explain further why strongly amplitude-squeezed light would have an uncertainty area shaped like a banana.[2]

(7.12) Calculate the electric field amplitude for a photon number state of wavelength 800 nm with $n = 10^6$ in a microcavity of volume $10~\mathrm{mm^3}$.

(7.13) By considering the phasor diagram of a photon number state, calculate the uncertainty $\Delta X_1 \Delta X_2$, and hence show that photon number states are not minimum uncertainty states.

(7.14) Consider a $50:50$ beam splitter with input fields of $\mathcal{E}_1$ and $\mathcal{E}_2$ and output fields of $\mathcal{E}_3$ and $\mathcal{E}_4$ as shown in Fig. 7.17. Let the phase shifts of the fields on transmission and reflection be written ϕ_i^{t} and ϕ_i^{r}, respectively, where $i = 1, 2$. Assume that $\mathcal{E}_1$ and $\mathcal{E}_2$ are real.

 (a) Verify that the output fields must be in the form:

$$\mathcal{E}_3 = \tfrac{1}{\sqrt{2}}\left[\mathcal{E}_1 \exp\left(i\phi_1^{\mathrm{t}}\right) + \mathcal{E}_2 \exp\left(i\phi_2^{\mathrm{r}}\right)\right],$$
$$\mathcal{E}_4 = \tfrac{1}{\sqrt{2}}\left[\mathcal{E}_1 \exp\left(i\phi_1^{\mathrm{r}}\right) + \mathcal{E}_2 \exp\left(i\phi_2^{\mathrm{t}}\right)\right].$$

 (b) By considering conservation of energy, show that:

$$\cos(\phi_2^{\mathrm{r}} - \phi_1^{\mathrm{t}}) + \cos(\phi_1^{\mathrm{r}} - \phi_2^{\mathrm{t}}) = 0.$$

 (c) The phase change on transmission is usually taken to be zero, which implies that:

$$\cos\phi_2^{\mathrm{r}} + \cos\phi_1^{\mathrm{r}} = 0.$$

 Show that this condition is satisfied when there is a relative phase difference of π between the two reflections.[3]

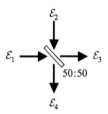

Fig. 7.17 The $50:50$ beam splitter.

---

[1] This exercise is based on the proposed Laser Interferometer Space Antenna (LISA) mission. The absolute sensitivity is smaller than that of the ground-based LIGO experiment described in Example 7.2, but the instrument is sensitive to gravity waves of much lower frequencies, which are more likely to be produced in astronomical events. See Exercise 2.4 for a discussion of the power levels involved.

[2] Such states are sometimes called 'banana states' for obvious reasons.

[3] The exact values of ϕ_i^{r} and ϕ_i^{t} depend on both the polarization of the light and the type of reflective coating on the beam splitter. See Brooker (2003) or Hecht (2002) for further details.

(7.15) Equation B.35 in Appendix B shows that the de-amplification factor for a nonlinear crystal of length L pumped by a laser beam with electric field amplitude $\mathcal{E}_\text{p}$ inside the crystal is $\exp(-\gamma L)$, where $\gamma = \omega \chi^{(2)} \mathcal{E}_\text{p}/2nc$. Calculate the quadrature squeezing expected for 1064 nm vacuum modes in a nonlinear crystal with $\chi^{(2)} = 4 \times 10^{-12}$ m V^{-1}, $n = 1.75$, and $L = 10$ mm, for a pump intensity of 2×10^{10} W m^{-2} at 532 nm.

(7.16) An erbium-doped optical fibre amplifier of length 10 m has a gain coefficient of 0.14 m^{-1}. Calculate the noise figure of the amplifier in decibels, on the assumption that no classical noise is added.

Photon number states

8

In the previous three chapters we have studied several different ways in which the quantum states of light can be classified. In Chapter 5 we looked at the photon statistics and classified the light as being sub-Poissonian, Poissonian, or super-Poissonian. Then in Chapter 6 we studied the second-order correlation function, and classified the light as antibunched, coherent or bunched. Finally, in Chapter 7 we studied coherent states and several forms of squeezed light. In each case, the approach we took was primarily phenomenological, with an emphasis on understanding the basic physical concepts and the experimental results. In this chapter we shall redress the balance somewhat, by giving a brief introduction to the quantum theory of light. This will allow us to revisit some of our main results from a more formal perspective, and should also serve as an introduction to the more advanced texts on quantum optics.

The quantum theory of light is based on the quantum harmonic oscillator. The chapter therefore begins with a review of the operator solution of the simple harmonic oscillator, and the number state representation that follows from it. We shall then look at the properties of coherent states, and conclude with a discussion of the Hanbury Brown–Twiss experiment with quantized light fields incident on the beam splitter.

8.1 Operator solution of the harmonic oscillator   151

8.2 The number state representation   154

8.3 Photon number states   156

8.4 Coherent states   157

8.5 Quantum theory of Hanbury Brown–Twiss experiments   160

Further reading   163

Exercises   163

No new results of major significance will be introduced here, which means that readers with an aversion to endless pages of mathematics can pass over this chapter without detriment to the subject matter developed in the remainder of the book.

8.1 Operator solution of the harmonic oscillator

The potential energy of a one-dimensional harmonic oscillator of mass m and angular frequency ω is given by:

$$V(x) = \frac{1}{2}m\omega^2 x^2. \tag{8.1}$$

The Hamiltonian is therefore of the form:

$$\hat{H} = \frac{\hat{p}_x^2}{2m} + \frac{1}{2}m\omega^2 \hat{x}^2. \tag{8.2}$$

The standard derivation for the quantized energies follows by solving the time-independent Schrödinger equation:

$$\hat{H}\psi(x) = E\psi(x) \tag{8.3}$$

for the eigenfunctions $\psi_n(x)$ and eigenenergies E_n, with $\hat{p}_x$ and $\hat{x}$ defined by eqns 3.12 and 3.11, respectively. This solution is reviewed briefly in

The reason why $\hat{a}$ and $\hat{a}^\dagger$ are called ladder operators will become apparent as we go along.

Section 3.3. We take a different approach here, making use of operator methods to find the solutions.

We define the **ladder operator** $\hat{a}$ and its Hermitian conjugate $\hat{a}^\dagger$ in terms of the position and momentum operators according to:

$$\hat{a} = \frac{1}{(2m\hbar\omega)^{1/2}}\left(m\omega\hat{x} + i\hat{p}_x\right), \tag{8.4}$$

$$\hat{a}^\dagger = \frac{1}{(2m\hbar\omega)^{1/2}}\left(m\omega\hat{x} - i\hat{p}_x\right). \tag{8.5}$$

We can turn these around to find $\hat{x}$ and $\hat{p}_x$ in terms of $\hat{a}$ and $\hat{a}^\dagger$:

$$\hat{x} = \left(\frac{\hbar}{2m\omega}\right)^{1/2}(\hat{a} + \hat{a}^\dagger), \tag{8.6}$$

$$\hat{p}_x = -i\left(\frac{m\hbar\omega}{2}\right)^{1/2}(\hat{a} - \hat{a}^\dagger). \tag{8.7}$$

The first thing we usually work out with quantum-mechanical operators is their commutator brackets. (See Section 3.1.4.) To do this, we need to work out the product operators $\hat{a}\hat{a}^\dagger$ and $\hat{a}^\dagger\hat{a}$. For $\hat{a}\hat{a}^\dagger$ we find from eqns 8.4 and 8.5 that:

$$\hat{a}\hat{a}^\dagger = \frac{1}{2m\hbar\omega}(m\omega\hat{x} + i\hat{p}_x)(m\omega\hat{x} - i\hat{p}_x)$$

$$= \frac{1}{2m\hbar\omega}\left(\hat{p}_x^2 + m^2\omega^2\hat{x}^2 + im\omega(\hat{p}_x\hat{x} - \hat{x}\hat{p}_x)\right)$$

$$= \frac{1}{2m\hbar\omega}\left(\hat{p}_x^2 + m^2\omega^2\hat{x}^2 + im\omega[\hat{p}_x, \hat{x}]\right)$$

$$= \frac{1}{\hbar\omega}\left(\frac{\hat{p}_x^2}{2m} + \frac{1}{2}m\omega^2\hat{x}^2 + \frac{1}{2}\hbar\omega\right), \tag{8.8}$$

where we made use of the standard result for the commutator of $\hat{x}$ and $\hat{p}_x$ in the last line. (See eqn 3.36.) In the same way we can work out that:

$$\hat{a}^\dagger\hat{a} = \frac{1}{\hbar\omega}\left(\frac{\hat{p}_x^2}{2m} + \frac{1}{2}m\omega^2\hat{x}^2 - \frac{1}{2}\hbar\omega\right). \tag{8.9}$$

We thus find the required commutator:

$$[\hat{a}, \hat{a}^\dagger] = \hat{a}\hat{a}^\dagger - \hat{a}^\dagger\hat{a} = 1. \tag{8.10}$$

On comparing Equations 8.2 and 8.9 we further see that the Hamiltonian may be written in terms of the ladder operators as:

$$\hat{H} = \hbar\omega\left(\hat{a}^\dagger\hat{a} + \frac{1}{2}\right). \tag{8.11}$$

We can use this result together with eqn 8.10 to work out that:

$$[\hat{H}, \hat{a}^\dagger] = \hbar\omega\left[(\hat{a}^\dagger\hat{a} + \frac{1}{2}), \hat{a}^\dagger\right]$$

$$= \hbar\omega(\hat{a}^\dagger\hat{a}\hat{a}^\dagger - \hat{a}^\dagger\hat{a}^\dagger\hat{a})$$

$$= \hbar\omega\hat{a}^\dagger(\hat{a}\hat{a}^\dagger - \hat{a}^\dagger\hat{a})$$

$$= \hbar\omega\hat{a}^\dagger[\hat{a}, \hat{a}^\dagger]$$

$$= \hbar\omega\hat{a}^\dagger. \tag{8.12}$$

Similarly, we find that:

$$[\hat{H}, \hat{a}] = -\hbar\omega\hat{a}. \tag{8.13}$$

We can now use these commutators to work out the energy spectrum of $\hat{H}$. Suppose ψ_n is an eigenfunction of $\hat{H}$ with energy E_n:

$$\hat{H}\psi_n = E_n\psi_n. \tag{8.14}$$

We operate on ψ_n with the operator $\hat{H}\hat{a}^\dagger$:

$$\begin{aligned}
\hat{H}\hat{a}^\dagger\psi_n &= (\hat{H}\hat{a}^\dagger - \hat{a}^\dagger\hat{H} + \hat{a}^\dagger\hat{H})\psi_n \\
&= ([\hat{H}, \hat{a}^\dagger] + \hat{a}^\dagger\hat{H})\psi_n \\
&= (\hbar\omega\hat{a}^\dagger + \hat{a}^\dagger E_n)\psi_n \\
&= (\hbar\omega + E_n)\hat{a}^\dagger\psi_n.
\end{aligned} \tag{8.15}$$

This shows that $\hat{a}^\dagger\psi_n$ is also an eigenfunction of $\hat{H}$, with an energy of $(E_n + \hbar\omega)$. Similarly, we find that:

$$\hat{H}\hat{a}\psi_n = (-\hbar\omega + E_n)\hat{a}\psi_n, \tag{8.16}$$

which shows that $\hat{a}\psi_n$ is an eigenfunction of $\hat{H}$ with energy of $(E_n - \hbar\omega)$.

Equations 8.15 and 8.16 indicate that the energy spectrum of the harmonic oscillator consists of a ladder of equally spaced energy levels as shown in Fig. 8.1. This is why $\hat{a}$ and $\hat{a}^\dagger$ are called 'ladder' operators. $\hat{a}^\dagger$ is called the **raising operator**, while $\hat{a}$ is called the **lowering operator**.

The total energy of a quantum harmonic oscillator must always be positive because both the kinetic and potential energy are always positive. Therefore the ladder of levels cannot continue going down in energy indefinitely: it must have a bottom rung, with energy in the range $0 \le E < \hbar\omega$. This bottom rung of the ladder of levels corresponds to the ground state of the system. If its wave function is $\psi_0(x)$, then we must have that:

$$\hat{a}\psi_0 = 0 \tag{8.17}$$

to prevent the energy from going negative. We can substitute for $\hat{a}$ from eqn 8.4 and use eqns 3.12 and 3.11 to rewrite eqn 8.17 in the following form:

$$\frac{\mathrm{d}\psi_0}{\mathrm{d}x} = -\left(\frac{m\omega}{\hbar}\right)x\,\psi_0. \tag{8.18}$$

This is a differential equation, with solution:

$$\psi_0(x) = C\exp\left(-\frac{m\omega x^2}{2\hbar}\right). \tag{8.19}$$

The constant C is determined by the normalization condition (eqn 3.18) and is given by $C = (m\omega/\hbar\pi)^{1/4}$. The energy E_0 of the ground state can be found by direct substitution of $\psi_0(x)$ into eqn 8.14, or more simply by using eqn 8.11:

$$\hat{H}\psi_0 = \hbar\omega\left(\hat{a}^\dagger\hat{a} + \frac{1}{2}\right)\psi_0 = \frac{1}{2}\hbar\omega\psi_0 = E_0\psi_0, \tag{8.20}$$

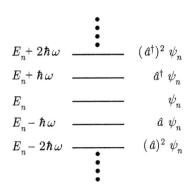

Fig. 8.1 Ladder of equally spaced energy levels generated by repeated action of the ladder operators to the state ψ_n at energy E_n.

where we made use of eqn 8.17. Equation 8.20 implies that:

$$E_0 = \frac{1}{2}\hbar\omega. \tag{8.21}$$

This is the zero-point energy of the harmonic oscillator.

The wave functions of the excited states can be found by repeated application of the raising operator $\hat{a}^\dagger$:

$$\psi_n(x) = C_n(\hat{a}^\dagger)^n\psi_0(x), \tag{8.22}$$

where C_n is a normalization constant. This implies that the energy E_n of the nth level is:

$$E_n = E_0 + n\hbar\omega = \left(n + \frac{1}{2}\right)\hbar\omega. \tag{8.23}$$

We thus find the expected result for the quantized energy of the simple harmonic oscillator (cf. eqn 3.93).

We finally introduce the **number operator** $\hat{n}$ which is defined by the following equation:

$$\hat{n}\psi_n = n\psi_n. \tag{8.24}$$

This operator gives the number of energy quanta excited, and will be interpreted later as the photon number operator. By making use of eqns 8.11 and 8.23, we can rewrite the Schrödinger equation given by eqn 8.14 in the following form:

$$\hbar\omega\left(\hat{a}^\dagger\hat{a} + \frac{1}{2}\right)\psi_n = \left(n + \frac{1}{2}\right)\hbar\omega\,\psi_n. \tag{8.25}$$

On comparing with eqn 8.24, we see that:

$$\hat{n} = \hat{a}^\dagger\hat{a}. \tag{8.26}$$

This is a key result for the number state representation, as we shall see in the following section.

8.2   The number state representation

The ladder operator solution of the harmonic oscillator allows us to construct a set of wave functions determined by an integer quantum number n, with energy increasing in steps of $\hbar\omega$. This naturally leads to the concept of the **number state representation**, in which we represent the system by a series of states labelled by n. These states are called **number states**, and are usually written in Dirac notation as $|n\rangle$. The correspondence between the number states and the states of the harmonic oscillator is shown in Table 8.1.

It is apparent from Table 8.1 that the number state $|n\rangle$ corresponds to the harmonic oscillator eigenstate with n quanta of energy excited above the ground state. The number states are therefore eigenstates of the simple harmonic oscillator Hamiltonian $\hat{H}$ with energies given by:

$$\hat{H}|n\rangle = E_n|n\rangle = \left(n + \frac{1}{2}\right)\hbar\omega|n\rangle. \tag{8.27}$$

Table 8.1 Correspondence between the number states and the wave functions of the harmonic oscillator.

State	Wave function	Energy
$|0\rangle$	$\psi_0(x)$	$(1/2)\,\hbar\omega$
$|1\rangle$	$\psi_1(x)$	$(3/2)\,\hbar\omega$
$|2\rangle$	$\psi_2(x)$	$(5/2)\,\hbar\omega$
$\vdots$	$\vdots$	$\vdots$
$|n\rangle$	$\psi_n(x)$	$\left(n + \frac{1}{2}\right)\hbar\omega$

Since the eigenstates of the Hamiltonian form an orthonormal basis, the number states must satisfy:

$$\langle n|n'\rangle = \delta_{nn'}, \tag{8.28}$$

where $\delta_{nn'}$ is the Kronecker delta function defined in eqn 3.22.

The raising and lowering operators $\hat{a}^\dagger$ and $\hat{a}$ defined for the harmonic oscillator in eqns 8.5 and 8.4, respectively, are very important in the number state representation. Equation 8.15 shows that the application of the raising operator $\hat{a}^\dagger$ to the state ψ_n produces a new eigenstate with energy $E_n + \hbar\omega$. This can be interpreted as saying that the operator $\hat{a}^\dagger$ *creates* one quantum of energy by raising the system from the state $|n\rangle$ to $|n+1\rangle$. We therefore define the **creation operator** according to:

$$\hat{a}^\dagger|n\rangle = (n+1)^{1/2}|n+1\rangle. \tag{8.29}$$

Similarly, the **annihilation operator** $\hat{a}$ is defined by:

$$\hat{a}|n\rangle = n^{1/2}|n-1\rangle. \tag{8.30}$$

The factors $(n + 1)^{1/2}$ and $n^{1/2}$ that appear on the right-hand side of eqns 8.29 and 8.30, respectively, are needed to ensure that the wave functions are all properly normalized.

This follows from eqn 8.16, which shows that $\hat{a}$ destroys one quantum of energy, and thus lowers the system from $|n\rangle$ to $|n-1\rangle$. The annihilation operator $\hat{a}$ is alternatively called the **destruction operator**.

The ground state $|0\rangle$ corresponds to the state in which no quanta are excited. Equation 8.29 implies that the number states can be built up from the ground state by repeated application of the creation operator:

$$|n\rangle = \frac{1}{(n!)^{1/2}}(\hat{a}^\dagger)^n|0\rangle. \tag{8.31}$$

By contrast, if we operate on the ground state with the annihilation operator, we find from eqn 8.30 that:

$$\hat{a}|0\rangle = 0. \tag{8.32}$$

This is consistent with the same result written in eqn 8.17.

The number operator $\hat{n}$ given in eqn 8.26 is obviously of key importance in the number representation. We can verify that the definitions of the creation and annihilation operators given in eqns 8.29 and 8.30 are consistent with this form of $\hat{n}$ by applying the operator $\hat{a}^\dagger\hat{a}$ to $|n\rangle$:

$$\hat{a}^\dagger\hat{a}|n\rangle = n^{1/2}\hat{a}^\dagger|n-1\rangle$$
$$= n^{1/2}\big((n-1)+1\big)^{1/2}|n\rangle$$
$$= n|n\rangle \,. \tag{8.33}$$

This confirms that the number states are eigenfunctions of the number operator $\hat{a}^\dagger\hat{a}$ with eigenvalue n.

The main definitions and results of the number representation are summarized in Table 8.2. Note that the mass of the oscillator does not enter explicitly in any of these formulae, so that the formalism can be transferred directly to massless harmonic oscillators such as photons. In the next section we shall see how this is done for the case that we are interested in here, namely the photon number states of quantum optics.

Table 8.2 Principal definitions of the number state representation.

	Symbol	Definition
Number state	$\|n\rangle$	$\hat{n}\|n\rangle = n\|n\rangle$
		$\hat{H}\|n\rangle = E_n\|n\rangle$, $E_n = (n + \frac{1}{2})\hbar\omega$
Ground state	$\|0\rangle$	$\hat{a}\|0\rangle = 0$
Number operator	$\hat{n}$	$\hat{n} = \hat{a}^\dagger\hat{a}$
Creation operator	$\hat{a}^\dagger$	$\hat{a}^\dagger\|n\rangle = (n + 1)^{1/2}\|n + 1\rangle$
Annihilation operator	$\hat{a}$	$\hat{a}\|n\rangle = (n)^{1/2}\|n - 1\rangle$
Commutator	$[\hat{a}, \hat{a}^\dagger]$	$[\hat{a}, \hat{a}^\dagger] \equiv \hat{a}\hat{a}^\dagger - \hat{a}^\dagger\hat{a} = 1$

8.3   Photon number states

The quantization of fields is often called 'second' quantization to distinguish it from the 'first' quantization that refers to the transition from classical to quantum mechanics for massive particles. The theory of field quantization can be traced back to Dirac's work of 1927. See: P.A.M. Dirac, *Proc. R. Soc. Lond. A* **114**, 243 (1927).

The number state representation can be applied to any physical system that has a Hamiltonian of the form equivalent to a simple harmonic oscillator. We saw in Section 7.1 that a single mode of the electromagnetic field in a cavity of volume V falls into this category. The formalism developed in the previous section for the oscillator of finite mass can therefore be applied directly to the quantized light field by making the same definitions as those summarized in Table 8.2.

In applying the number state representation to light fields, we have to forget about trying to write down a wave function for the photon states in terms of a position coordinate x, as we usually do for finite-mass oscillators such as molecular vibrations. Instead, we just concentrate on the properties of the system as described by the number of energy quanta excited. We therefore describe the excitations of the quantized electromagnetic field in terms of the number of photons excited at angular frequency ω. The **photon number state** $|n\rangle$ then represents a monochromatic quantized field of angular frequency ω containing n photons.

Single-mode photon number states are also sometimes called **Fock states**.

In the photon number representation, the creation and annihilation operators correspond to the creation and annihilation of a photon of angular frequency ω, respectively. The ground state $|0\rangle$ corresponds to the electromagnetic vacuum and is called the **vacuum state**. Equation 8.31 implies that we can think of the state $|n\rangle$ as a state in which, n photons have been excited from the vacuum.

The properties of the electromagnetic vacuum were described briefly in Section 7.4.

The dimensionless quadrature fields X_1 and X_2 are very important for the theory of coherent states and squeezed states developed in Chapter 7. In the photon number representation, we can introduce the equivalent quantum operators by the following definitions:

A comparison of eqns 8.34–8.35 and 8.6–8.7 makes it apparent that the quadrature operators are directly related to the position and momentum operators. See Exercise 8.3.

$$\hat{X}_1 = \frac{1}{2}(\hat{a}^\dagger + \hat{a}), \tag{8.34}$$

$$\hat{X}_2 = \frac{1}{2}\mathrm{i}(\hat{a}^\dagger - \hat{a}). \tag{8.35}$$

These operators can be used to work out the quantum uncertainties of the photon fields, as illustrated by the following example.

Example 8.1   Evaluate the commutator $[\hat{X}_1, \hat{X}_2]$, and hence find the uncertainty product $(\Delta X_1)^2(\Delta X_2)^2$.

Solution

The commutator is evaluated from its definition:

$$[\hat{X}_1, \hat{X}_2] = \hat{X}_1\hat{X}_2 - \hat{X}_2\hat{X}_1.$$

On substituting from eqns 8.34 and 8.35 we find:

$$\hat{X}_1\hat{X}_2 = i(\hat{a}^\dagger + \hat{a})(\hat{a}^\dagger - \hat{a})/4 = i(\hat{a}^\dagger\hat{a}^\dagger + \hat{a}\hat{a}^\dagger - \hat{a}^\dagger\hat{a} - \hat{a}\hat{a})/4,$$

and

$$\hat{X}_2\hat{X}_1 = i(\hat{a}^\dagger - \hat{a})(\hat{a}^\dagger + \hat{a})/4 = i(\hat{a}^\dagger\hat{a}^\dagger - \hat{a}\hat{a}^\dagger + \hat{a}^\dagger\hat{a} - \hat{a}\hat{a})/4,$$

Hence:

$$[\hat{X}_1, \hat{X}_2] = i(\hat{a}\hat{a}^\dagger - \hat{a}^\dagger\hat{a})/2 = i[\hat{a}, \hat{a}^\dagger]/2.$$

Then on recalling the result for the commutator of the creation and annihilation operators given in eqn 8.10, we obtain the final result:

$$[\hat{X}_1, \hat{X}_2] = i/2. \tag{8.36}$$

The uncertainty product can be evaluated from the commutator by the standard result given in eqn 3.37:

$$(\Delta X_1)^2(\Delta X_2)^2 \geq \left|[\hat{X}_1, \hat{X}_2]\right|^2 /4.$$

On substituting from eqn 8.36, we then find:

$$(\Delta X_1)^2(\Delta X_2)^2 \geq 1/16. \tag{8.37}$$

This confirms eqn 7.35 which was derived from the Heisenberg uncertainty principle.

8.4   Coherent states

We can now apply the formalism of the photon number representation to describe the properties of the coherent states introduced in Section 7.5. These states are characterized by the complex number α which specifies the complex field amplitude in photon number units. We thus specify a coherent state in Dirac notation as $|\alpha\rangle$.

In the photon number representation, a coherent state is defined by:

$$|\alpha\rangle = \exp\left(-|\alpha|^2/2\right) \sum_{n=0}^{\infty} \frac{\alpha^n}{(n!)^{1/2}} |n\rangle. \tag{8.38}$$

Coherent states are not eigenstates of the Hamiltonian, and neither are they orthogonal to each other. (See Exercise 8.6.) On the other hand, it is apparent from evaluating $\hat{a}|\alpha\rangle$ that they are right eigenstates of the annihilation operator $\hat{a}$:

$$\hat{a}|\alpha\rangle = \mathrm{e}^{-|\alpha|^2/2} \sum_{n=0}^{\infty} \frac{\alpha^n}{(n!)^{1/2}} \, \hat{a}|n\rangle$$

$$= \mathrm{e}^{-|\alpha|^2/2} \sum_{n=1}^{\infty} \frac{\alpha^n}{(n!)^{1/2}} \, n^{1/2}|n-1\rangle$$

$$= \alpha \, \mathrm{e}^{-|\alpha|^2/2} \sum_{n=1}^{\infty} \frac{\alpha^{n-1}}{(n-1)!^{1/2}} \, |n-1\rangle$$

$$= \alpha \, \mathrm{e}^{-|\alpha|^2/2} \sum_{n=0}^{\infty} \frac{\alpha^n}{(n)!^{1/2}} \, |n\rangle$$

$$= \alpha \, |\alpha\rangle, \tag{8.39}$$

where we made use of eqn 8.30, and the fact that $\hat{a}|0\rangle = 0$ (cf. eqn 8.17). If we take the Hermitian conjugate of eqn 8.39, we find that the coherent states are also left eigenstates of the annihilation operator $\hat{a}^\dagger$:

$$\left(\hat{a}|\alpha\rangle\right)^\dagger = \left(\alpha|\alpha\rangle\right)^\dagger,$$

whence $$\langle\alpha|\hat{a}^\dagger = \langle\alpha|\alpha^*. \tag{8.40}$$

Equations 8.39 and 8.40 allow us to work out the expectation value of the number operator $\hat{n}$ using eqn 8.26:

$$\langle\alpha|\hat{n}|\alpha\rangle = \langle\alpha|\hat{a}^\dagger\hat{a}|\alpha\rangle$$

$$= \langle\alpha|\alpha^*\alpha|\alpha\rangle$$

$$= \alpha^*\alpha. \tag{8.41}$$

On equating $\langle\alpha|\hat{n}|\alpha\rangle$ with the mean photon number $\bar{n}$, we see that this result agrees with the one deduced from consideration of the energy of the coherent state (cf. eqn 7.49).

The variance of the photon number is given by:

$$(\Delta n)^2 = \langle\alpha|(\hat{n} - \bar{n})^2|\alpha\rangle$$

$$= \langle\alpha|\hat{n}^2|\alpha\rangle - 2\bar{n}\langle\alpha|\hat{n}|\alpha\rangle + \bar{n}^2\langle\alpha|\alpha\rangle$$

$$= \langle\alpha|\hat{n}^2|\alpha\rangle - \bar{n}^2, \tag{8.42}$$

where we used $\langle\alpha|\alpha\rangle = 1$ in the second line (see Example 8.2). The expectation value of $\hat{n}^2$ can be evaluated with the help of eqn 8.10:

$$\hat{n}^2 = \hat{a}^\dagger\hat{a}\hat{a}^\dagger\hat{a}$$

$$= \hat{a}^\dagger(\hat{a}\hat{a}^\dagger - \hat{a}^\dagger\hat{a} + \hat{a}^\dagger\hat{a})\hat{a}$$

$$= \hat{a}^\dagger([\hat{a}, \hat{a}^\dagger] + \hat{a}^\dagger\hat{a})\hat{a}$$

$$= \hat{a}^\dagger(1 + \hat{a}^\dagger\hat{a})\hat{a}$$

$$= \hat{a}^\dagger\hat{a} + \hat{a}^\dagger\hat{a}^\dagger\hat{a}\hat{a}. \tag{8.43}$$

We therefore find that:

$$\langle\alpha|\hat{n}^2|\alpha\rangle = \langle\alpha|\hat{a}^\dagger\hat{a}|\alpha\rangle + \langle\alpha|\hat{a}^\dagger\hat{a}^\dagger\hat{a}\hat{a}|\alpha\rangle$$
$$= \alpha^*\alpha + \alpha^*\alpha^*\alpha\alpha$$
$$= \bar{n} + \bar{n}^2, \tag{8.44}$$

which implies from eqn 8.42 that:

$$(\Delta n)^2 = (\bar{n} + \bar{n}^2) - \bar{n}^2 = \bar{n}. \tag{8.45}$$

This confirms the Poissonian result found previously in eqn 7.50.

Finally, we can work out the probability $\mathcal{P}(n)$ that there are n photons in the coherent state. This is done by evaluating $\langle n|\alpha\rangle$:

$$\langle n|\alpha\rangle = e^{-|\alpha|^2/2}\sum_{m=0}^{\infty}\frac{\alpha^m}{(m!)^{1/2}}\langle n|m\rangle$$
$$= e^{-|\alpha|^2/2}\sum_{m=0}^{\infty}\frac{\alpha^m}{(m!)^{1/2}}\delta_{nm}$$
$$= e^{-|\alpha|^2/2}\frac{\alpha^n}{(n!)^{1/2}}, \tag{8.46}$$

where we made use of the orthonormality of number states in the second line (cf. eqn 8.28). On equating $\mathcal{P}(n)$ with $|\langle n|\alpha\rangle|^2$, we find:

$$\mathcal{P}(n) \equiv |\langle n|\alpha\rangle|^2 = e^{-|\alpha|^2}\frac{|\alpha^2|^n}{n!}. \tag{8.47}$$

Then, on finding from eqn 8.41 that $|\alpha|^2 = \bar{n}$, we finally obtain a Poisson distribution:

$$\mathcal{P}(n) = \frac{\bar{n}^n}{n!}e^{-\bar{n}}. \tag{8.48}$$

Poissonian photon statistics cause shot noise in photodetection. (See Section 5.9.)

We therefore conclude that coherent states have Poissonian photon statistics, in agreement with the conclusions of Section 7.6.

Example 8.2 Show that the coherent state $|\alpha\rangle$ defined in eqn 8.38 is correctly normalized.

Solution
We confirm that the coherent state is correctly normalized by evaluating the Dirac bracket $\langle\alpha|\alpha\rangle$:

$$\langle\alpha|\alpha\rangle = e^{-|\alpha|^2}\sum_{n=0}^{\infty}\sum_{n'=0}^{\infty}\frac{(\alpha^*)^{n'}\alpha^n}{(n'!)^{1/2}(n!)^{1/2}}\langle n'|n\rangle.$$

We make use of the orthonormality of number states given in eqn 8.28 to write:

$$\langle\alpha|\alpha\rangle = e^{-|\alpha|^2}\sum_{n=0}^{\infty}\sum_{n'=0}^{\infty}\frac{(\alpha^*)^{n'}\alpha^n}{(n'!n!)^{1/2}}\delta_{nn'},$$

where $\delta_{nn'}$ is the Kronecker delta function. Then, on using the definition of $\delta_{nn'}$ given in eqn 3.22, we find:

$$\langle \alpha | \alpha \rangle = e^{-|\alpha|^2} \sum_{n=0}^{\infty} \frac{(\alpha^* \alpha)^n}{n!}.$$

Finally, we recall the Taylor expansion:

$$e^{+|\alpha|^2} = 1 + |\alpha|^2 + \frac{(|\alpha|^2)^2}{2!} + \frac{(|\alpha|^2)^3}{3!} + \cdots = \sum_{n=0}^{\infty} \frac{(\alpha^* \alpha)^n}{n!},$$

to see that:

$$\langle \alpha | \alpha \rangle = e^{-|\alpha|^2} \times e^{+|\alpha|^2} = 1.$$

This shows that the coherent state defined in eqn 8.38 is correctly normalized.

8.5   Quantum theory of Hanbury Brown–Twiss experiments

The Hanbury Brown–Twiss (HBT) experiment and its relevance to quantum optics was described in Chapter 6. We saw there that the experiment has different interpretations, depending on whether the input fields are treated according to classical or quantum optics. In the classical interpretation, the experiment measures the second-order correlation function $g^{(2)}(\tau)$, which is defined in terms of the intensity fluctuations of the incident light. (See Sections 6.2 and 6.3, especially eqn 6.6.) In the quantum interpretation, by contrast, the value of $g^{(2)}(\tau)$ is defined in terms of coincidences between photon counting events. (See Section 6.4.) The comparison of the two interpretations shows that the value of $g^{(2)}(0)$ is particularly significant, since the result $g^{(2)}(0) < 1$ is possible only for quantum states of light that exhibit photon antibunching. It is therefore useful to reanalyse the HBT experiment with the mathematical tools that we have developed in this chapter, in order to explain how the results for different input states can be calculated.

Figure 8.2 shows a schematic diagram of the HBT experiment. The experiment consists of a $50:50$ beam splitter with two input ports labelled 1 and 2, and two output ports labelled 3 and 4. Single-photon counting detectors D3 and D4 are positioned to detect the photons in the output beams, and the time that elapses between a photon count on D3 and another on D4 is recorded. This is done by using the count pulse from D3 to trigger an electronic timer and the count pulse from D4 to stop it. The experiment therefore counts the number of coincidences that occur when a photon is registered on D3 at time t and another is registered on D4 at time $t + \tau$, where τ is the time interval between the start and stop pulses. If the experiment is repeated many times, a histogram of the number of events for each time interval τ can be produced, as in Fig. 6.5(b). After normalizing by the total number of counts registered

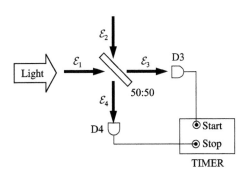

Fig. 8.2 Hanbury Brown–Twiss (HBT) experiment. A 50:50 beam splitter has four ports labelled 1, 2, 3 and 4. The field at the ith port is labelled $\mathcal{E}_i$. Ports 1 and 2 are input ports, while ports 3 and 4 are output ports. The light to be investigated is introduced at port 1. D3 and D4 are single photon counting detectors set to detect the output fields $\mathcal{E}_3$ and $\mathcal{E}_4$, respectively. The count pulses from D3 start an electronic timer that is stopped by a count pulse from D4.

by each detector, the second-order correlation function $g^{(2)}(\tau)$ is then obtained from:

$$g^{(2)}(\tau) = \frac{\langle n_3(t) n_4(t+\tau)\rangle}{\langle n_3(t)\rangle\langle n_4(t+\tau)\rangle}, \tag{8.49}$$

where $n_3(t)$ and $n_4(t)$ are the numbers of photons registered by D3 and D4, respectively, at time t, and the symbol $\langle\cdots\rangle$ indicates the average value found after many repetitions of the experiment.

The definition of $g^{(2)}(\tau)$ given in eqn 8.49 can be cast in a form that is amenable to theoretical analysis by recalling that the photon number operator $\hat{n}$ is given by eqn 8.26. We can therefore rewrite eqn 8.49 as:

$$g^{(2)}(\tau) = \frac{\langle \hat{a}_3^\dagger(t)\hat{a}_4^\dagger(t+\tau)\hat{a}_4(t+\tau)\hat{a}_3(t)\rangle}{\langle \hat{a}_3^\dagger(t)\hat{a}_3(t)\rangle\langle \hat{a}_4^\dagger(t+\tau)\hat{a}_4(t+\tau)\rangle}. \tag{8.50}$$

The ordering with the creation operators to the left and the annihilation operators to the right is called **normal ordering**.

For $\tau = 0$, eqn 8.50 simplifies to:

$$g^{(2)}(0) = \frac{\langle \hat{a}_3^\dagger\hat{a}_4^\dagger\hat{a}_4\hat{a}_3\rangle}{\langle \hat{a}_3^\dagger\hat{a}_3\rangle\langle \hat{a}_4^\dagger\hat{a}_4\rangle}. \tag{8.51}$$

Since the value for $\tau = 0$ is a clear signature of quantum or classical behaviour, we now concentrate on deriving a usable formula for $g^{(2)}(0)$.

In order to evaluate $g^{(2)}(0)$ for an arbitrary input state, we first need to relate the creation and annihilation operators of the output fields of the beam splitter to those of the input fields. This is done by writing down the relationships for the classical fields:

$$\mathcal{E}_3 = (\mathcal{E}_1 - \mathcal{E}_2)/\sqrt{2}, \tag{8.52}$$

$$\mathcal{E}_4 = (\mathcal{E}_1 + \mathcal{E}_2)/\sqrt{2}, \tag{8.53}$$

and then applying the same relationships to the annihilation operators for the output fields:

$$\hat{a}_3 = (\hat{a}_1 - \hat{a}_2)/\sqrt{2},$$

$$\hat{a}_4 = (\hat{a}_1 + \hat{a}_2)/\sqrt{2}. \tag{8.54}$$

The normal ordering in the numerator of eqn 8.50 is a consequence of the photoelectric detection process. See, for example, Loudon (2000, §4.11), or Mandel and Wolf (1995, §12.2.) Note that we preempted this normal ordering in the classical definition of $g^{(2)}(\tau)$ by the ordering of the electric fields in eqn 6.6.

The minus sign in eqn 8.52 originates from the π phase change that occurs for one of the reflections in the beam splitter. Exercise (7.14) explains that this minus sign is a consequence of the need to conserve energy at the beam splitter. The minus sign could equally well have gone into eqn 8.53, and the same result as eqn 8.63 would have been obtained. (See Exercise 8.10.) The factor of $1/\sqrt{2}$ arises from the 50:50 power splitting ratio and the fact that the intensity is proportional to the square of the field.

The corresponding creation operators are found by taking the Hermitian conjugates of these equations.

In the HBT experiment, the light is introduced through just one of the input ports, as shown in Fig. 8.2. This means that the field at port 2 is the vacuum, and that the input states are therefore of the form:

$$|\Psi\rangle = |\psi_1, 0_2\rangle, \tag{8.55}$$

where $|\psi_1\rangle$ is the arbitrary input state at port 1 and $|0_2\rangle$ denotes the vacuum state input to port 2. The denominators in eqn 8.51 can be evaluated by substituting from eqn 8.54 to find:

$$
\begin{aligned}
\langle \hat{a}_3^\dagger \hat{a}_3 \rangle &= \langle \psi_1, 0_2 | (\hat{a}_1^\dagger \hat{a}_1 - \hat{a}_1^\dagger \hat{a}_2 - \hat{a}_2^\dagger \hat{a}_1 + \hat{a}_2^\dagger \hat{a}_2) | \psi_1, 0_2 \rangle / 2, \\
&= \langle \psi_1 | \hat{a}_1^\dagger \hat{a}_1 | \psi_1 \rangle / 2, \\
&= \langle \psi_1 | \hat{n}_1 | \psi_1 \rangle / 2,
\end{aligned}
\tag{8.56}
$$

and likewise:

$$
\begin{aligned}
\langle \hat{a}_4^\dagger \hat{a}_4 \rangle &= \langle \psi_1, 0_2 | (\hat{a}_1^\dagger \hat{a}_1 + \hat{a}_1^\dagger \hat{a}_2 + \hat{a}_2^\dagger \hat{a}_1 + \hat{a}_2^\dagger \hat{a}_2) | \psi_1, 0_2 \rangle / 2 \\
&= \langle \psi_1 | \hat{a}_1^\dagger \hat{a}_1 | \psi_1 \rangle / 2, \\
&= \langle \psi_1 | \hat{n}_1 | \psi_1 \rangle / 2,
\end{aligned}
\tag{8.57}
$$

where we made use of the definition of the vacuum state given in eqn 8.32, together with its Hermitian conjugate, in the second line of each equation, and the definition of the number operator given in eqn 8.26 in the third. For the numerator we need to evaluate:

$$\langle \hat{a}_3^\dagger \hat{a}_4^\dagger \hat{a}_4 \hat{a}_3 \rangle = \langle \Psi | (\hat{a}_1^\dagger - \hat{a}_2^\dagger)(\hat{a}_1^\dagger + \hat{a}_2^\dagger)(\hat{a}_1 + \hat{a}_2)(\hat{a}_1 - \hat{a}_2) | \Psi \rangle / 4. \tag{8.58}$$

This has 16 terms, but most of them are zero when there is a vacuum state at port 2. We first eliminate the terms with $\hat{a}_2$ at the end and those with $\hat{a}_2^\dagger$ at the beginning, giving:

$$\langle \hat{a}_3^\dagger \hat{a}_4^\dagger \hat{a}_4 \hat{a}_3 \rangle = \langle \psi_1, 0_2 | \hat{a}_1^\dagger (\hat{a}_1^\dagger + \hat{a}_2^\dagger)(\hat{a}_1 + \hat{a}_2) \hat{a}_1 | \psi_1, 0_2 \rangle / 4. \tag{8.59}$$

The same reasoning means that we can drop any terms with $\hat{a}_2$ to the right of $\hat{a}_2^\dagger$ or with $\hat{a}_2^\dagger$ to the left of $\hat{a}_2$, leaving just one term:

$$\langle \hat{a}_3^\dagger \hat{a}_4^\dagger \hat{a}_4 \hat{a}_3 \rangle = \langle \psi_1 | \hat{a}_1^\dagger \hat{a}_1^\dagger \hat{a}_1 \hat{a}_1 | \psi_1 \rangle / 4. \tag{8.60}$$

This can be simplified further by using eqn 8.10 to write:

$$
\begin{aligned}
\hat{a}_1^\dagger \hat{a}_1^\dagger \hat{a}_1 \hat{a}_1 &= \hat{a}_1^\dagger (\hat{a}_1 \hat{a}_1^\dagger - 1) \hat{a}_1 \\
&= \hat{a}_1^\dagger \hat{a}_1 \hat{a}_1^\dagger \hat{a}_1 - \hat{a}_1^\dagger \hat{a}_1 \\
&= \hat{n}_1 \hat{n}_1 - \hat{n}_1 \\
&= \hat{n}_1 (\hat{n}_1 - 1),
\end{aligned}
\tag{8.61}
$$

where we again used eqn 8.26. We then combine eqns 8.56, 8.57, and 8.60 to find:

$$g^{(2)}(0) = \frac{\langle\psi_1|\hat{n}_1(\hat{n}_1-1)|\psi_1\rangle/4}{(\langle\psi_1|\hat{n}_1|\psi_1\rangle/2)^2},\qquad(8.62)$$

and hence obtain the final result:

$$g^{(2)}(0) = \frac{\langle\hat{n}(\hat{n}-1)\rangle}{\langle\hat{n}\rangle^2},\qquad(8.63)$$

where the expectation values are evaluated over the input state at port 1.

Equation 8.63 is in a form that can be readily evaluated for arbitrary inputs. For example, if the input is the photon number state $|n\rangle$, we find:

$$g^{(2)}(0) = \frac{n(n-1)}{n^2}.\qquad(8.64)$$

This means that we expect the highly non-classical value of $g^{(2)}(0) = 0$ for a single-photon source that emits photon number states with $n = 1$. Such states have been produced in the laboratory, and values of $g^{(2)}(0)$ close to zero have been observed. (See Section 6.7.)

In real experiments on single-photon sources, the measured values of $g^{(2)}(0)$ are often slightly larger than zero. This is caused by the inherent difficulty in producing ideal single-photon sources, and also by the finite response time of the detectors. (See Section 6.6.)

Further reading

The subject matter of this chapter is developed in much greater depth in Gerry and Knight (2005), Loudon (2000), Mandel and Wolf (1995), Meystre and Sargent (1999), or Walls and Milburn (1994). Most of the subject matter is also covered in the review article by Loudon and Knight (1987a).

Exercises

(8.1) Verify that the wave function given in eqn 8.19 is a solution of eqn 8.18 with energy $(1/2)\hbar\omega$, and that the correct normalization constant C is given by $C = (m\omega/\hbar\pi)^{1/4}$.

(8.2) Use eqn 8.22 to find the functional form of the wave functions of the first two excited states of the quantum harmonic oscillator.

(8.3) By relating the generalized position and momentum coordinates q and p to the real-space position and momentum coordinates x and p_x according to eqns 7.21 and 7.22, confirm that the previous definitions of the quadrature operators given in eqns 7.29 and 7.30 are consistent with those given in eqns 8.34 and 8.35.

(8.4) Evaluate $\langle\hat{X}_1\rangle$, $\langle\hat{X}_2\rangle$, ΔX_1, and ΔX_2 for the vacuum state $|0\rangle$. Relate these results to the phasor diagram of the vacuum state shown in Fig. 7.4.

(8.5) For the coherent state $|\alpha\rangle$ with $\alpha = |\alpha|e^{i\phi}$, show that $\langle\alpha|\hat{X}_1|\alpha\rangle = |\alpha|\cos\phi$, and $\langle\alpha|\hat{X}_2|\alpha\rangle = |\alpha|\sin\phi$. Show further that $\Delta X_1 = \Delta X_2 = 1/2$. Relate these results to the phasor diagram shown in Fig. 7.5.

(8.6) Prove that for two coherent states $|\alpha\rangle$ and $|\beta\rangle$,

$$|\langle\alpha|\beta\rangle|^2 = \exp(-|\alpha-\beta|^2).$$

Briefly discuss the implication of this result.

(8.7) Show that a photon number state $|n\rangle$ is an eigenstate of the operator $\left(\hat{X}_1^2 + \hat{X}_2^2\right)$ with eigenvalue $(n+\frac{1}{2})$. Hence explain why the phasor for a photon number state has a radius of $(n+\frac{1}{2})^{1/2}$.

(8.8) Evaluate the uncertainty product $\Delta X_1 \Delta X_2$ for a number state $|n\rangle$.

(8.9) (a) Write down the commutators $[\hat{a}_1, \hat{a}_1^\dagger]$, $[\hat{a}_1, \hat{a}_2^\dagger]$, $[\hat{a}_2, \hat{a}_1^\dagger]$, and $[\hat{a}_2, \hat{a}_2^\dagger]$ for the input fields to the beam splitter in Fig. 8.2.

(b) Explain why the output field operators of a general beam splitter with amplitude reflection and transmission coefficients of r and t respectively can be written in the form:

$$\hat{a}_3 = t\hat{a}_1 - r\hat{a}_2,$$
$$\hat{a}_4 = r\hat{a}_1 + t\hat{a}_2.$$

(c) Show the assumption that the commutators of the output fields, namely $[\hat{a}_3, \hat{a}_3^\dagger]$, $[\hat{a}_3, \hat{a}_4^\dagger]$, $[\hat{a}_4, \hat{a}_3^\dagger]$, and $[\hat{a}_4, \hat{a}_4^\dagger]$, obey the usual rules for creation and annihilation operators implies that $|r|^2 + |t|^2 = 1$ and $r^*t - rt^* = 0$.

(8.10) The measurable results derived in Section 8.5 should not depend on the details of the phase shifts in the beam splitter, other than requiring a relative phase shift of π between the two reflections to conserve energy. Show that eqn 8.63 can be derived if we assume that

$$\hat{a}_3 = (\hat{a}_1 + \hat{a}_2)/\sqrt{2},$$
$$\hat{a}_4 = (-\hat{a}_1 + \hat{a}_2)/\sqrt{2},$$

rather than the form given in eqn 8.54.

(8.11) Consider a 50 : 50 beam splitter with single-mode input and output fields of $\hat{a}_1$, $\hat{a}_2$, $\hat{a}_3$, and $\hat{a}_4$ as in Section 8.5. Assume that the relationship between the input and output fields is given by eqn 8.54, and that the commutators are the same as in Exercise (8.9). Assume also that the basis states for the input and output may be written in the form $|n_1\rangle_1 |n_2\rangle_2$ and $|n_3\rangle_3 |n_4\rangle_4$, respectively, where n_i represents the number of photons at port i.

(a) What is the output state for an input of $|0\rangle_1 |0\rangle_2$?

(b) Find the output states for inputs of $|1\rangle_1 |0\rangle_2$ and $|0\rangle_1 |1\rangle_2$. Give a physical interpretation of the results. [Hint: write $|1\rangle_1 |0\rangle_2$ as $\hat{a}_1^\dagger |0\rangle_1 |0\rangle_2$ (and likewise for $|0\rangle_1 |1\rangle_2$), and make use of the result from part (a).]

(c) Find the output for an input state of $|1\rangle_1 |1\rangle_2$, and discuss the implications of the result.

(8.12) Show that $g^{(2)}(0) = 1$ for a coherent state.

(8.13) A squeezed vacuum state with squeezing parameter s can be written as a superposition of number states according to:

$$|s\rangle = (\text{sech } s)^{1/2} \sum_{n=0}^{\infty} \frac{[(2n)!]^{1/2}}{n!} \left(-\tfrac{1}{2}\tanh s\right)^n |2n\rangle.$$

(a) By reference to the method of generation of squeezed vacuum states discussed in Section 7.9.1, explain why $|s\rangle$ only contains number states with n even.

(b) Evaluate $\langle s|\hat{X}_1|s\rangle$ and $\langle s|\hat{X}_2|s\rangle$.

(c) Given that $\langle s|\hat{a}^\dagger\hat{a}^\dagger|s\rangle = \langle s|\hat{a}\hat{a}|s\rangle = -\sinh s \cosh s$, and that $\langle s|\hat{a}^\dagger\hat{a}|s\rangle = \sinh^2 s$, evaluate ΔX_1 and ΔX_2.

(d) Relate the results of parts (b) and (c) to the phasor diagram of the squeezed vacuum state given in Fig. 7.8(a).

Part III

Atom–photon interactions

Introduction to Part III

Having studied the quantum nature of light in itself, we now move on to consider the coupling between atoms and photons, which forms the basis for our understanding of light–matter interactions.

Chapter 9 deals with the interaction between a light beam and an atomic transition of the same frequency. At small light intensities the conventional picture of absorption of photons is appropriate, but at high intensities the behaviour changes and Rabi oscillations can occur. The understanding of Rabi oscillations leads to the concept of the Bloch sphere, which will be very useful for our understanding of quantum gates in Chapter 13.

Chapter 10 begins with a review of the theory of spontaneous emission by atoms in free space, and then describes how the process is altered by a resonant cavity. The weak and strong limits of the atom–cavity coupling are considered, with the latter leading into the subject of cavity quantum electrodynamics (cavity QED).

Finally, Chapter 11 considers the forces that are exerted on an atom by a resonant light beam, and how these forces can be exploited in laser cooling experiments. This naturally leads on to a discussion of Bose–Einstein condensation in atomic systems, and the concept of atom lasers.

The subject matter developed in these chapters assumes an understanding of radiative transitions at the level of an introductory atomic or quantum physics text. A summary of the main aspects of the background theory may be found in Chapter 4.

Resonant light–atom interactions

9.1	**Introduction**	**167**
9.2	**Preliminary concepts**	**168**
9.3	**The time-dependent Schrödinger equation**	**172**
9.4	**The weak-field limit: Einstein's B coefficient**	**174**
9.5	**The strong-field limit: Rabi oscillations**	**177**
9.6	**The Bloch sphere**	**187**
	Further reading	**191**
	Exercises	**191**

In this chapter we shall study the processes that occur when an atom is irradiated with a light beam that is resonant with one of its natural frequencies. We shall start by making a few general observations, and then give a detailed analysis of the process based on the time-dependent Schrödinger equation. This will lead us to the concept of Rabi oscillations, from which we shall be able to obtain a deeper understanding of how absorption transitions work. In so doing we shall introduce the Bloch sphere, which will be very useful for understanding quantum gates in Chapter 13.

The subject material developed here assumes that the reader is familiar with the standard treatment of optical transitions in atoms and the Einstein coefficients. A summary of the main points may be found in Chapter 4.

9.1 Introduction

The treatment of the interaction between light and atoms was pivotal in the development of quantum theory in the first half of the twentieth century. In 1913 Bohr postulated that a quantum of light of angular frequency ω is absorbed or emitted whenever an atom jumps between two quantized energy levels E_1 and E_2 that satisfy:

$$E_2 - E_1 = \hbar\omega. \tag{9.1}$$

The theory was developed by Einstein in 1916–7 when he introduced the **Einstein coefficients** to quantify the rate at which the absorption and emission of quanta occur. In the same paper he discovered the process of stimulated emission, which later proved to be the basis of laser operation.

The picture of the interaction process between light and atoms that emerges from Bohr and Einstein's treatment is illustrated in Fig. 9.1. The absorption process is viewed in terms of the destruction of a photon from the light beam with the simultaneous excitation of the atom, while the emission process corresponds to the addition of a photon to the light field and the simultaneous de-excitation of the atom.

Consider the absorption process shown in Fig. 9.1(a). We imagine that the atom is initially in the lower level, and a light beam of angular frequency ω is turned on at time $t = 0$. At some later time the atom makes the jump to the excited state, and a photon is absorbed from the

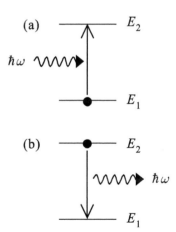

Fig. 9.1 Optical transitions between quantized energy levels: (a) absorption, and (b) spontaneous emission. Stimulated emission processes are not considered here.

light beam. Similarly, in the emission process shown in Fig. 9.1(b), the atom is in the excited state at time $t = 0$. After a time typically equal to the radiative lifetime of the excited state, a photon of angular frequency ω is emitted. Once these processes have been completed, we say that the atom has made a **transition** from level $1 \rightarrow 2$ and vice versa.

The question we wish to address in this chapter is as follows. What happens to the irradiated atom *before* the absorption transition is complete? The Einstein treatment of the process does not address this issue. To do so, we will have to solve the time-dependent Schrödinger equation for the atom with the light included. This will be the subject of Section 9.3, and its consequences will be explored in the following sections. Before we can do this, we must first clarify a few basic concepts and explain the approximations which allow us to set up the problem in a way that can be solved easily. These will be the subject of the next section.

It is, of course, equally interesting to ask about the mechanism of spontaneous emission in more detail. At the level of the Einstein approach, it is viewed as a purely random process governed by decay probabilities. At a deeper level, however, we can consider it as a stimulated emission event triggered by a vacuum photon. The randomness is then attributed to the quantum noise of the zero-point fluctuations of the electromagnetic field. (See Section 7.4.) Further discussion of this topic will be postponed to the next chapter. We shall see in Section 10.3 that the spontaneous emission rate of an atom is not an absolute quantity, but can in fact be controlled by suppressing or enhancing the density of the photon modes by situating the atom within a resonant cavity.

9.2   Preliminary concepts

9.2.1   The two-level atom approximation

The quantum treatment of the interaction between light and atoms is usually developed in terms of the **two-level atom approximation**. This approximation is applicable when the frequency of the light coincides with one of the optical transitions of the atom. The condition is specified in eqn 9.1 and is depicted schematically in Fig. 9.2. The atom will have many quantum levels, and there will be many possible optical transitions between them. However, in the two-level atom approximation we only consider the specific transition that satisfies eqn 9.1 and ignore all the other levels. It is customary to label the lower and upper levels as 1 and 2, respectively.

The physical basis for the two-level approximation is the fact that we are dealing with a **resonance** phenomenon. In the classical picture of light–atom interactions, the light beam induces dipole oscillations in the atom, which then re-radiate at the same frequency. If the light frequency corresponds to the natural frequency of the atom, the magnitude of the dipole oscillations will be large and the interaction between the atom and the light will be strong. (See Exercise 9.1.) On the other hand, if the light frequency is far away from the natural frequency of the atom

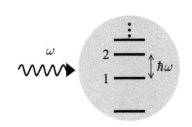

Fig. 9.2 The two-level atom approximation. When the light angular frequency ω coincides with one of the optical transitions of the atom, we have a resonant interaction between that transition and the light field. We can therefore neglect the other levels of the atom, which only weakly interact with the light because they are off-resonance.

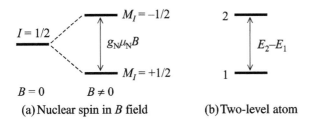

Fig. 9.3 A spin $1/2$ system in a magnetic field of strength B along the z-axis (a) is formally equivalent to a two-level atom (b). For historical reasons, we usually consider nuclear spin systems rather than electronic ones, in which case the field splitting is determined by the nuclear g-factor g_N and the nuclear Bohr magneton μ_N.

(i.e. off-resonance), then the magnitude of the driven oscillations will be small, and the light–atom interaction will be small. In other words, the interaction between the light and the atom is very much stronger for the case of resonant than off-resonant transitions, and so it is a good approximation to ignore the latter.

It is obvious that the assumption that only the resonant levels matter is an approximation. In many cases, the approximation is very good. On the other hand, the presence of the off-resonant levels may become important indirectly. When the atom is in level 2, it could make transitions to other lower levels in addition to level 1. This would cause a loss of atoms from the system considered, and would effectively damp the interaction between the light and the atom. Hence the simplest way to include other levels in the analysis is to include damping terms. We shall see the importance of damping in Section 9.5.

A useful analogy can be made between the properties of two-level atoms and those of spin $1/2$ particles in a magnetic field. Figure 9.3(a) shows how a magnetic field of strength B splits a spin $1/2$ system into a doublet through the Zeeman effect. The Zeeman-split levels are formally equivalent to the two-level atom shown in Fig. 9.3(b). The reason for making the analogy is that the theory of the resonant interaction between microwave radiation and the Zeeman-split nuclear spin states had been developed in the 1940s to account for a whole range of nuclear magnetic resonance (NMR) phenomena. With the advent of the laser in 1960, the same types of phenomena were soon observed in two-level atomic systems at optical frequencies. We shall explore this analogy further in Section 9.6.

9.2.2   Coherent superposition states

The treatment of the resonant interaction between an atom and a light field involves the concept of **coherent superposition states**. Wave function coherence gives rise to many of the phenomena that make quantum systems behave differently from classical ones. For this reason, it is important to remind ourselves of what coherent superposition states are, and how they differ from **statistical mixtures**.

The concept of coherent superposition states lies at the heart of quantum computation. (See Section 13.2.)

Consider a quantum system with two levels, such as a two-level atom or a spin 1/2 nucleus in a magnetic field. (See Fig. 9.3.) We assume that there is some way of measuring whether the system is in the lower or upper level. For example, in the case of the two-level atom we could determine whether the atom is in the upper level by looking for spontaneous emission at angular frequency ω. Similarly, in the case of the nuclear spin we could carry out a Stern–Gerlach experiment to determine I_z. The wave function for the system may in general be written in the form:

$$|\psi\rangle = c_1|1\rangle + c_2|2\rangle, \tag{9.2}$$

where c_1 and c_2 describe the wave function amplitude coefficients for the two states of the atom or nucleus. If we make a measurement we would obtain the result appropriate to level 1 with probability $|c_1|^2$ and that for level 2 with probability $|c_2|^2$.

Now consider a gas of N_0 identical two-level particles (e.g. two-level atoms or spin 1/2 nuclei) with N_1 particles in the lower level and N_2 in the upper level. Such a gas is called a **statistical mixture**. By setting $|c_1|^2 = N_1/N_0$ and $|c_2|^2 = N_2/N_0$ we would obtain the same results as for repeated measurements on a gas of N_0 particles in the superposition state given by eqn 9.2. What, then, is the difference? The answer is that each of the particles in the superposition state is in some sense simultaneously in the $|1\rangle$ and $|2\rangle$ states, and this leads to the possibility of wave function interference. In the statistical mixture, by contrast, a given particle is *either* in level 1 *or* in level 2, and no wave function interference can occur.

A useful analogy can be made here with interference effects in light beams. This analogy also helps to explain why we often include the adjective 'coherent' in the description of the superposition states. Consider two overlapping light beams of the same frequency with phases ϕ_1 and ϕ_2 respectively. The resultant field is given by:

$$
\begin{aligned}
\mathcal{E} &= \mathcal{E}_1 e^{-i(\omega t + \phi_1)} + \mathcal{E}_2 e^{-i(\omega t + \phi_2)} \\
&= \mathcal{E}_1 e^{-i(\omega t + \phi_1)} \left(1 + (\mathcal{E}_2/\mathcal{E}_1) e^{-i(\phi_2 - \phi_1)}\right),
\end{aligned}
\tag{9.3}
$$

where ω is the angular frequency, and $\mathcal{E}_1$ and $\mathcal{E}_2$ are the amplitudes. The beams will interfere if they are *coherent*, that is, when

$$\Delta\phi = \phi_2 - \phi_1 = \text{constant}, \tag{9.4}$$

at a given point in space. In this case, there will be some positions where $\Delta\phi = (\text{even integer} \times \pi)$ and we have a bright fringe, and others where $\Delta\phi = (\text{odd integer} \times \pi)$ and we have a dark fringe. On the other hand, if the beams are *incoherent*, the phases vary randomly with time relative to each other, and there will be no interference. In this case the powers of the beams will just add together, so that the resultant will be given by:

$$|\mathcal{E}|^2 = |\mathcal{E}_1|^2 + |\mathcal{E}_2|^2. \tag{9.5}$$

The analogy with the two-level superposition states can be made apparent by setting $c_1 \rightarrow \mathcal{E}_1 \exp(-i\phi_1)$ and $c_2 \rightarrow \mathcal{E}_2 \exp(-i\phi_2)$. We then see that wave function interference can only occur when there is a definite phase relationship between c_1 and c_2. This occurs in the superposition state, but not in a statistical mixture, where the different particle wave functions all have random phases with respect to each other.

In the treatment of the light–atom interactions given in this chapter, we shall see that a light pulse links the phases of the upper and lower levels of an atom so that wave function coherence is present. This gives rise to new phenomena which are not considered in the Einstein analysis, as it only deals with statistical mixtures. In particular, we shall see that while the light pulse is present, the atom oscillates between the upper and lower levels. This rather striking conclusion appears to be at odds with our simpler picture based on discrete transitions between the upper and lower levels. We shall see how we can reconcile the two approaches when we consider the effects of damping in Section 9.5.2.

9.2.3   The density matrix

More advanced treatments of the phenomena described in this chapter tend to make use of the **density matrix**, ρ. The elements of the density matrix are defined by:

$$\rho_{ij} = \langle c_i c_j^* \rangle, \tag{9.6}$$

where c_i is the wave function amplitude for the ith quantum level, and the subscripts i and j run over all the quantum states of the atom. The symbol $\langle \cdots \rangle$ indicates that we take the average value for the ensemble in a many-particle system.

Let us consider the density matrix of the two-level atoms that we are considering here. In general, the density matrix takes the form:

$$\rho = \begin{pmatrix} \langle |c_1|^2 \rangle & \langle c_1 c_2^* \rangle \\ \langle c_1^* c_2 \rangle & \langle |c_2|^2 \rangle \end{pmatrix}. \tag{9.7}$$

The key difference between statistical mixtures and coherent superpositions is the presence of **off-diagonal terms** in the density matrix, namely ρ_{12} and ρ_{21}. In the case of a statistical mixture, each individual atom will either have $|c_1| = 1$ and $|c_2| = 0$ or vice versa. The off-diagonal terms are therefore zero, and the density matrix for the ensemble is of the form:

$$\rho = \begin{pmatrix} N_1/N_0 & 0 \\ 0 & N_2/N_0 \end{pmatrix}. \tag{9.8}$$

Atoms in coherent superposition states, by contrast, have wave functions in which both $|c_1|$ and $|c_2|$ are non-zero, giving non-zero off-diagonal elements. In this case, the density matrix takes the form given in eqn 9.7 with all four elements non-zero.

We shall make no further use of the density matrix in this book. In simple systems such as two-level atoms, it suffices to discuss the behaviour explicitly in terms of the wave function amplitudes rather than the density matrix. We mention ρ_{ij} here for the sake of completeness, and also to identify the importance of off-diagonal terms, which are central to the whole concept of superposition states.

9.3   The time-dependent Schrödinger equation

Having covered the preliminary concepts, we can now get on with the main business of the chapter. Our objective is to solve the time-dependent Schrödinger equation for a two-level atom in the presence of the light. In other words, we must solve:

$$\hat{H}\Psi = i\hbar\frac{\partial\Psi}{\partial t}, \tag{9.9}$$

for an atom with two energy levels E_1 and E_2 in the presence of a light wave of angular frequency ω. We shall assume that the light is very close to resonance with the transition, so that

$$\omega = \omega_0 + \delta\omega, \tag{9.10}$$

where

$$\omega_0 = (E_2 - E_1)/\hbar, \tag{9.11}$$

and $\delta\omega \ll \omega_0$. Exact resonance thus corresponds to $\delta\omega = 0$.

We start by splitting the Hamiltonian into a time-independent part $\hat{H}_0$ which describes the atom in the dark, and a perturbation term $\hat{V}(t)$ which accounts for the light–atom interaction:

$$\hat{H} = \hat{H}_0(\boldsymbol{r}) + \hat{V}(t). \tag{9.12}$$

Since we are dealing with a two-level atom, there will be two solutions for the unperturbed system:

$$\hat{H}_0\Psi_i = i\hbar\frac{\partial\Psi_i}{\partial t}, \tag{9.13}$$

with

$$\Psi_i(\boldsymbol{r}, t) = \psi_i(\boldsymbol{r})\exp(-iE_i t/\hbar) \quad \{i = 1, 2\}, \tag{9.14}$$

and

$$\hat{H}_0(\boldsymbol{r})\psi_i(\boldsymbol{r}) = E_i\,\psi_i(\boldsymbol{r}) \quad \{i = 1, 2\}. \tag{9.15}$$

The general solution to the time-dependent Schrödinger equation (eqn 9.9) is:

$$\Psi(\boldsymbol{r}, t) = \sum_i c_i(t)\psi_i(\boldsymbol{r})\exp(-iE_i t/\hbar), \tag{9.16}$$

where the subscript i runs over all the eigenstates of the system. In the case of a two-level atom, this reduces to:

$$\Psi(\boldsymbol{r}, t) = c_1(t)\psi_1(\boldsymbol{r})e^{-iE_1 t/\hbar} + c_2(t)\psi_2(\boldsymbol{r})e^{-iE_2 t/\hbar}. \tag{9.17}$$

On substituting this wave function into eqn 9.9 with $\hat{H}$ given by eqn 9.12, we obtain:

$$(\hat{H}_0 + \hat{V})\left(c_1\psi_1 e^{-iE_1t/\hbar} + c_2\psi_2 e^{-iE_2t/\hbar}\right)$$
$$= i\hbar\left((\dot{c}_1 - iE_1c_1/\hbar)\psi_1 e^{-iE_1t/\hbar} + (\dot{c}_2 - iE_2c_2/\hbar)\psi_2 e^{-iE_2t/\hbar}\right). \quad (9.18)$$

Now eqn 9.15 implies that

$$\hat{H}_0\left(c_1\psi_1 e^{-iE_1t/\hbar} + c_2\psi_2 e^{-iE_2t/\hbar}\right)$$
$$= \left(c_1E_1\psi_1 e^{-iE_1t/\hbar} + c_2E_2\psi_2 e^{-iE_2t/\hbar}\right), \quad (9.19)$$

so that we can cancel several of the terms in eqn 9.18 to obtain:

$$c_1\hat{V}\psi_1 e^{-iE_1t/\hbar} + c_2\hat{V}\psi_2 e^{-iE_2t/\hbar} = i\hbar\dot{c}_1\psi_1 e^{-iE_1t/\hbar} + i\hbar\dot{c}_2\psi_2 e^{-iE_2t/\hbar}. \quad (9.20)$$

On multiplying by ψ_1^*, integrating over space, and making use of the orthonormality of the eigenfunctions, which requires that:

$$\int \psi_i^*\psi_j \, d^3\boldsymbol{r} = \delta_{ij}, \quad (9.21)$$

where δ_{ij} is the Kronecker delta function, we find that:

$$\dot{c}_1(t) = -\frac{i}{\hbar}\left(c_1(t)V_{11} + c_2(t)V_{12}\,e^{-i\omega_0 t}\right), \quad (9.22)$$

where

$$V_{ij}(t) \equiv \langle i|\hat{V}(t)|j\rangle = \int \psi_i^*\hat{V}(t)\psi_j \, d^3\boldsymbol{r}. \quad (9.23)$$

Similarly, on multiplying by ψ_2^* and integrating, we find that:

$$\dot{c}_2(t) = -\frac{i}{\hbar}\left(c_1(t)V_{21}e^{i\omega_0 t} + c_2(t)V_{22}\right). \quad (9.24)$$

To proceed further we must consider the explicit form of the perturbation $\hat{V}$. In the semi-classical approach, the light–atom interaction is given by the energy shift of the atomic dipole in the electric field of the light:

$$\hat{V}(t) = e\boldsymbol{r}\cdot\boldsymbol{\mathcal{E}}(t). \quad (9.25)$$

See eqns 4.14 and 4.15. Note that e is the *magnitude* of the electron charge.

We arbitrarily choose the x-axis as the direction of the polarization so that we can write:

$$\boldsymbol{\mathcal{E}}(t) = (\mathcal{E}_0, 0, 0)\cos\omega t, \quad (9.26)$$

where $\mathcal{E}_0$ is the amplitude of the light wave. The perturbation then simplifies to:

$$\hat{V}(t) = ex\mathcal{E}_0\cos\omega t$$
$$= \frac{ex\mathcal{E}_0}{2}\left(e^{i\omega t} + e^{-i\omega t}\right), \quad (9.27)$$

and the perturbation matrix elements are given by:

$$V_{ij}(t) = \frac{e\mathcal{E}_0}{2} \left(e^{i\omega t} + e^{-i\omega t} \right) \int \psi_i^* x \psi_j \, d^3 \boldsymbol{r}. \tag{9.28}$$

Now the **dipole matrix element** μ_{ij} is given by:

$$\mu_{ij} = -e \int \psi_i^* x \psi_j \, d^3 \boldsymbol{r} \equiv -e\langle i|x|j\rangle, \tag{9.29}$$

so that we can write:

$$V_{ij}(t) = -\frac{\mathcal{E}_0}{2} \left(e^{i\omega t} + e^{-i\omega t} \right) \mu_{ij}. \tag{9.30}$$

Since x is an odd parity operator and atomic states have either even or odd parities (see Section 4.3), it must be the case that $\mu_{11} = \mu_{22} = 0$. Moreover, the dipole matrix element represents a measurable quantity and must be real, which implies that $\mu_{21} = \mu_{12}$, because $\mu_{21} = \mu_{12}^*$. With these simplifications, eqns 9.22 and 9.24 reduce to:

$$\dot{c}_1(t) = i\frac{\mathcal{E}_0 \mu_{12}}{2\hbar} \left(e^{i(\omega - \omega_0)t} + e^{-i(\omega + \omega_0)t} \right) c_2(t),$$

$$\dot{c}_2(t) = i\frac{\mathcal{E}_0 \mu_{12}}{2\hbar} \left(e^{-i(\omega - \omega_0)t} + e^{i(\omega + \omega_0)t} \right) c_1(t). \tag{9.31}$$

The Rabi frequency defined here is an *angular* frequency.

We now introduce the **Rabi frequency** defined by:

$$\Omega_{\mathrm{R}} = |\mu_{12}\mathcal{E}_0/\hbar|. \tag{9.32}$$

The phase factors that might result from using the modulus sign to ensure that the Rabi frequency defined in eqn 9.32 is real and positive have no physical significance and have been suppressed in eqn 9.33.

We then finally obtain:

$$\dot{c}_1(t) = \frac{i}{2}\Omega_{\mathrm{R}} \left(e^{i(\omega - \omega_0)t} + e^{-i(\omega + \omega_0)t} \right) c_2(t),$$

$$\dot{c}_2(t) = \frac{i}{2}\Omega_{\mathrm{R}} \left(e^{-i(\omega - \omega_0)t} + e^{i(\omega + \omega_0)t} \right) c_1(t). \tag{9.33}$$

These are the equations that we must solve to understand the behaviour of the atom in the light field. It turns out that there are two distinct types of solution that can be found, which correspond to the **weak-field limit** and the **strong-field limit** respectively. We consider the weak-field limit first.

9.4   The weak-field limit: Einstein's *B* coefficient

The weak-field limit applies to low-intensity light sources such as black-body lamps. We assume that the atom is initially in the lower level and that the lamp is turned on at $t = 0$. This implies that $c_1(0) = 1$ and $c_2(0) = 0$.

With a low-intensity source, the electric field amplitude will be small and the perturbation weak. The number of transitions expected is therefore small, and it will always be the case that $c_1(t) \gg c_2(t)$. In these conditions we can put $c_1(t) = 1$ for all t, so that eqn 9.33 reduces to:

$$\dot{c}_1(t) = 0,$$

$$\dot{c}_2(t) = \frac{\mathrm{i}}{2}\Omega_{\mathrm{R}}\left(e^{-\mathrm{i}(\omega-\omega_0)t} + e^{\mathrm{i}(\omega+\omega_0)t}\right). \tag{9.34}$$

The solution for $c_2(t)$ with $c_2(0) = 0$ is:

$$c_2(t) = \frac{\mathrm{i}}{2}\Omega_{\mathrm{R}}\left[\frac{e^{-\mathrm{i}\delta\omega t} - 1}{-\mathrm{i}\delta\omega} + \frac{e^{\mathrm{i}(\omega+\omega_0)t} - 1}{\mathrm{i}(\omega + \omega_0)}\right], \tag{9.35}$$

where we made use of eqn 9.10. According to the **rotating wave approximation**, we now neglect the second term in eqn 9.35. This is justified by the fact that since $\delta\omega \ll (\omega + \omega_0)$, the second term is much smaller than the first. After some manipulation we find:

$$|c_2(t)|^2 = \left(\frac{\Omega_{\mathrm{R}}}{2}\right)^2\left(\frac{\sin \delta\omega t/2}{\delta\omega/2}\right)^2. \tag{9.36}$$

When the beam is tuned to exact resonance with the transition, $\delta\omega$ is equal to zero. We thus find:

$$|c_2(t)|^2 = \left(\frac{\Omega_{\mathrm{R}}}{2}\right)^2 t^2, \tag{9.37}$$

leading to the unsatisfactory conclusion that the probability that the atom is in the upper level increases as t^2. This is at odds with the Einstein approach in which the transition probability is time-independent, so that $|c_2(t)|^2$ should increase linearly with time.

The way around this apparent contradiction is to re-examine the assumptions of our analysis. We have assumed throughout that the atomic transition line is perfectly sharp. However, we know in fact that all spectral lines have a finite width $\Delta\omega$. (See Section 4.4.) Furthermore, we are considering the interaction between the atom and a broad-band source such as a black-body lamp. Such a broad-band source can be specified by the spectral energy density $u(\omega)$, which must satisfy:

$$\frac{1}{2}\epsilon_0\mathcal{E}_0^2 = \int u(\omega)\,\mathrm{d}\omega. \tag{9.38}$$

We therefore integrate eqn 9.36 over the spectral line:

$$|c_2(t)|^2 = \frac{\mu_{12}^2}{2\epsilon_0\hbar^2}\int_{\omega_0-\Delta\omega/2}^{\omega_0+\Delta\omega/2} u(\omega)\left(\frac{\sin(\omega-\omega_0)t/2}{(\omega-\omega_0)/2}\right)^2\,\mathrm{d}\omega, \tag{9.39}$$

where we used eqns 9.32 and 9.38 to substitute for Ω_{R} and $\mathcal{E}_0^2$, respectively. We now make the approximation that the spectral line is sharp compared to the broad-band spectrum of the lamp, so that $u(\omega)$ does not vary significantly within the integral. This allows us to replace $u(\omega)$ by a constant value $u(\omega_0)$, and thus to evaluate the integral. The limiting value for $t\Delta\omega \to \infty$ is $u(\omega_0)2\pi t$. Hence we finally obtain:

$$|c_2(t)|^2 = \frac{\pi}{\epsilon_0\hbar^2}\mu_{12}^2\,u(\omega_0)\,t, \tag{9.40}$$

which is a much more satisfactory result because it implies that the probability that the atom is in the upper level increases linearly with time.

We can now relate eqn 9.40 to the Einstein B coefficient defined by:

$$\frac{\mathrm{d}N_2}{\mathrm{d}t} = B_{12}^{\omega} u(\omega_0) N_1 \,, \tag{9.41}$$

Note that the energy density can be defined either in terms of the frequency ν or the angular frequency ω, with the two values differing by a factor 2π. Two different Einstein B coefficients B_{12}^{ν} and B_{12}^{ω} can be defined accordingly, with $B_{12}^{\nu} = B_{12}^{\omega}/2\pi$. See eqn 9.43.

which implies that the transition probability per unit time per atom is $B_{12}^{\omega} u(\omega_0)$. In the analysis leading up to eqn 9.33, we assumed that the atomic dipole moment was aligned parallel to the polarization vector of the light. However, in a gas of atoms, the direction of the atomic dipoles will be random. If the angle between the polarization and a particular dipole is θ, then we need to take the average of $(\mu_{12} \cos \theta)^2$ for all the atoms in the gas. On using $\langle \cos^2 \theta \rangle = 1/3$, we then replace μ_{12}^2 by $\mu_{12}^2/3$ throughout to obtain the transition probability rate W_{12}:

$$W_{12} \equiv B_{12}^{\omega} u(\omega_0) = \frac{|c_2|^2}{t} = \frac{\pi}{3\epsilon_0 \hbar^2} \mu_{12}^2 u(\omega_0), \tag{9.42}$$

which finally gives:

$$B_{12}^{\omega} = \frac{\pi}{3\epsilon_0 \hbar^2} \mu_{12}^2,$$

$$B_{12}^{\nu} = \frac{1}{6\epsilon_0 \hbar^2} \mu_{12}^2. \tag{9.43}$$

This shows that the weak-field limit is equivalent to the Einstein analysis, and allows us to calculate explicit values of the B coefficient from the atomic wave functions.

Example 9.1   Calculate the Einstein B coefficient B_{12}^{ω} for the 1s $\rightarrow$ 2p atomic transition in hydrogen for light polarized along the z-axis.

Solution
For light polarized along the z-axis the selection rules permit $\Delta m = 0$ transitions only. (See Section 4.3.) Hence we are considering the transition between two hydrogenic states with quantum numbers (n, l, m_l) of $(1,0,0)$ and $(2,1,0)$, respectively. The initial and final wave functions are therefore:

$$\psi_1(r, \theta, \phi) = \frac{1}{\sqrt{\pi}} \left(\frac{1}{a_0} \right)^{3/2} \mathrm{e}^{-r/a_0},$$

$$\psi_2(r, \theta, \phi) = \frac{1}{\sqrt{\pi}} \left(\frac{1}{2a_0} \right)^{5/2} r \cos \theta \, \mathrm{e}^{-r/2a_0},$$

where a_0 is the Bohr radius. In analogy to eqn 9.29, the transition dipole moment μ_{12} for z-polarized light is given by:

$$\mu_{12} = -e \int \psi_1^* z \psi_2 \, \mathrm{d}^3 \boldsymbol{r}$$

$$= -e \int_{r=0}^{\infty} \int_{\theta=0}^{\pi} \int_{\phi=0}^{2\pi} \psi_1^* \, r \cos\theta \, \psi_2 \, r^2 \sin\theta \, \mathrm{d}r \, \mathrm{d}\theta \, \mathrm{d}\phi$$

$$= -\frac{128\sqrt{2}}{243} e a_0$$

$$= -6.32 \times 10^{-30} \, \mathrm{C\,m}.$$

We then obtain $B_{12}^{\omega} = 4.25 \times 10^{20} \, \mathrm{m}^3 \, \mathrm{rad\,J}^{-1}\,\mathrm{s}^{-2}$ on substituting into eqn 9.43.

9.5   The strong-field limit: Rabi oscillations

9.5.1   Basic concepts

In the previous section we assumed that the light field was weak so that the population of the excited state was always small and the approximation $c_1(t) \approx 1$ was valid for all t. This allowed us to find a simple solution to eqn 9.33. We now wish to return to the more general case in which the population of the upper level is significant. It is intuitively obvious that this condition applies when the light–atom interaction is strong. In other words, we are dealing with the case of strong electric fields, such as those found in powerful laser beams.

In order to find a solution to eqn 9.33 in the strong-field limit we make two simplifications. First, we apply the rotating wave approximation to neglect the terms that oscillate at $\pm(\omega + \omega_0)$, as in the previous section. Second, we only consider the case of exact resonance with $\delta\omega = 0$. With these simplifications, eqn 9.33 reduces to:

The mathematics of non-exact resonance is more complicated, and is considered in Exercise 9.5.

$$\dot{c}_1(t) = \frac{\mathrm{i}}{2}\Omega_{\mathrm{R}} c_2(t),$$

$$\dot{c}_2(t) = \frac{\mathrm{i}}{2}\Omega_{\mathrm{R}} c_1(t). \tag{9.44}$$

We differentiate the first line and substitute from the second to find:

$$\ddot{c}_1 = \frac{\mathrm{i}}{2}\Omega_{\mathrm{R}} \dot{c}_2 = \left(\frac{\mathrm{i}}{2}\Omega_{\mathrm{R}}\right)^2 c_1. \tag{9.45}$$

We thus obtain

$$\ddot{c}_1 + \left(\frac{\Omega_{\mathrm{R}}}{2}\right)^2 c_1 = 0, \tag{9.46}$$

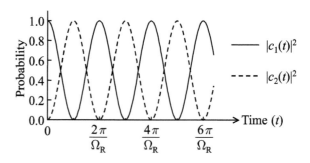

Fig. 9.4 Probability for finding the atom in either the upper or lower level in the strong-field limit in the absence of damping. The electron oscillates back and forth between the two levels at the Rabi angular frequency, Ω_R. This phenomenon is either called Rabi flopping or Rabi oscillation.

which describes oscillatory motion at angular frequency $\Omega_R/2$. If the particle is in the lower level at $t = 0$ so that $c_1(0) = 1$ and $c_2(0) = 0$, the solution is:

$$c_1(t) = \cos(\Omega_R t/2),$$
$$c_2(t) = i\sin(\Omega_R t/2). \tag{9.47}$$

The probabilities for finding the electron in the upper or lower levels are then given by:

$$|c_1(t)|^2 = \cos^2(\Omega_R t/2),$$
$$|c_2(t)|^2 = \sin^2(\Omega_R t/2). \tag{9.48}$$

See I. I. Rabi, *Phys. Rev.* **51**, 652 (1937). Rabi's original derivation applied to oscillating electromagnetic fields tuned to resonance with the Zeeman-split levels of a spin-1/2 nucleus. The RF field tips the spin vector from down to up and then back to down again, a process entirely equivalent to the Rabi flopping considered here. Rabi's work was the precursor to modern NMR techniques, and its importance was recognized by the awarding of the Nobel Prize for Physics in 1944. A brief discussion of the phenomenon of NMR may be found in Appendix E.

The time dependence of these probabilities is shown in Fig. 9.4. At $t = \pi/\Omega_R$ the electron is in the upper level, whereas at $t = 2\pi/\Omega_R$ it is back in the lower level. The process then repeats itself with a period equal to $2\pi/\Omega_R$. The electron thus oscillates back and forth between the lower and upper levels at a frequency equal to $\Omega_R/2\pi$. This oscillatory behaviour in response to the strong-field is called **Rabi oscillation** or **Rabi flopping**.

When the light is not exactly resonant with the transition, it can be shown that the second line of eqn 9.48 is modified to (see Exercise 9.5):

$$|c_2(t)|^2 = \frac{\Omega_R^2}{\Omega^2}\sin^2(\Omega t/2), \tag{9.49}$$

where

$$\Omega^2 = \Omega_R^2 + \delta\omega^2, \tag{9.50}$$

$\delta\omega$ being the detuning. This shows that the frequency of the Rabi oscillations increases but their amplitude decreases as the light is tuned away from resonance.

For transitions in the visible-frequency range, the experimental observation of Rabi flopping requires powerful laser beams. In many cases, these lasers will be pulsed, so that the electric field amplitude $\mathcal{E}_0$ varies with time. Equation 9.32 then tells us that the Rabi frequency $\Omega_R/2\pi$

also varies with time, and so it is useful to define the **pulse area** Θ according to:

$$\Theta = \left| \frac{\mu_{12}}{\hbar} \int_{-\infty}^{+\infty} \mathcal{E}_0(t)\,\mathrm{d}t \right| . \qquad (9.51)$$

The pulse area is a dimensionless parameter which is determined by the pulse energy and serves the same purpose as $\Omega_R t$ in the analysis above. A pulse which has an area equal to π is called a **π-pulse**. An atom in the ground state with $c_1 = 1$ at $t = 0$ will thus be promoted to the excited state with $c_2 = 1$ by a π-pulse, but will end up back in the ground state if it interacts with a 2π-pulse. We shall see in Section 9.6 that the pulse area can be given a geometric interpretation in terms of rotation angles of the Bloch vector.

The startling oscillatory behaviour predicted by eqn 9.48 has been observed in many systems, as will be discussed in Section 9.5.3. The following example illustrates the time-scales involved in Rabi flopping processes, and introduces the importance of damping effects, which are considered next.

Example 9.2   A powerful beam of light is incident on a gas of monatomic hydrogen and is tuned to resonance with the 1s $\rightarrow$ 2p transition at 137 nm.

(a) Calculate the Rabi oscillation period when the optical intensity is $10\ \mathrm{kW\,m^{-2}}$ and the light is polarized in the z-direction.

(b) Calculate the optical intensity required to make the Rabi oscillation period equal to the radiative lifetime of the 2p level, namely 1.6 ns.

Solution

(a) The atomic dipole moment for this transition was calculated in Example 9.1 to be $-0.74ea_0 = -6.32 \times 10^{-30}\ \mathrm{C\,m}$. The intensity of a light beam is related to its electric field amplitude according to (see eqn 2.28):

$$I = \frac{1}{2} c\epsilon_0 n \mathcal{E}_0^2, \qquad (9.52)$$

where n is the refractive index of the medium. In a gas we may take $n \approx 1$, and so we find $\mathcal{E}_0 = 2.7 \times 10^3\ \mathrm{V\,m^{-1}}$. On substituting into eqn 9.32, we find the Rabi frequency:

$$\Omega_R = |-6.32 \times 10^{-30} \times 2.7 \times 10^3/\hbar|$$

$$= 1.6 \times 10^8\ \mathrm{rad\,s^{-1}}.$$

Hence the Rabi flopping period is equal to $2\pi/\Omega_R = 38$ ns.

(b) A Rabi oscillation period of 1.6 ns corresponds to $\Omega_R = 3.9 \times 10^9$ rad s^{-1}. From eqn 9.32 we calculate $\mathcal{E}_0 = 6.5 \times 10^4\ \mathrm{V\,m^{-1}}$, and hence from eqn 9.52 we find $I = 5.7\ \mathrm{MW\,m^{-2}}$.

9.5.2   Damping

Example 9.2 illustrates why it is difficult to observe Rabi oscillations in the laboratory. At low powers, the oscillation period is longer than the radiative lifetime, and we would expect random spontaneous emission events to destroy the coherence of the superposition states, and hence curtail the oscillations. We thus have to work at higher powers to shorten the Rabi flopping period, which can be difficult to achieve in practice.

Spontaneous emission is just one example of a **damping** mechanism for the Rabi oscillations. We shall now see that the consideration of damping is very important for determining the experimental conditions under which Rabi oscillations can be observed. Damping also provides a way to reconcile the rather counter-intuitive phenomenon of Rabi oscillations with the more familiar concept of transition rates which form the basis of the Einstein model.

The damping processes for coherent phenomena such as Rabi flopping are traditionally characterized by two time constants, T_1 and T_2, following Bloch's treatment of nuclear magnetic resonance. (See Appendix E, Section E.3.) As will be explained in Section 9.6, these two types of damping are sometimes called **longitudinal relaxation** and **transverse relaxation**, respectively. In physical terms, T_1 damping is essentially determined by **population decay**, whereas T_2 damping is related to **dephasing** processes.

The T_1 (longitudinal) damping processes are the simplest to understand. If the atom is in the excited state, it will have a spontaneous tendency to decay to lower levels. The decay processes occur stochastically (i.e. according to probabilistic laws) and randomly break the coherence of the electronic wave function. Hence a spontaneous decay would permanently interrupt the Rabi flopping, which relies on the coherence of the superposition states. The rate of these types of damping process is governed by the lifetime τ of the upper level, which itself is determined by both the radiative and non-radiative decay rates:

$$\frac{1}{\tau} = \frac{1}{\tau_{\mathrm{R}}} + \frac{1}{\tau_{\mathrm{NR}}}. \tag{9.53}$$

The upper limit on T_1 is thus set by the radiative lifetime τ_{R} of the excited state, which includes transitions both to the resonant lower level and to other non-resonant levels that have been neglected so far in the two-level atom approximation.

The T_2 (transverse) damping processes are more subtle to understand. It will frequently be the case that an atom in the excited state undergoes an elastic (i.e. energy conserving) or near-elastic collision which breaks the phase of the wave function without altering the population of the excited state. These scattering processes can occur by a number of different mechanisms. In a gas, collisions can occur between the atoms or with the walls of the vessel, whereas in a solid there can be interactions with impurities or lattice vibrations (phonons). By randomizing the phase of the wave function, the collisions destroy any effects such as Rabi flopping which rely on phase coherence.

It is thus apparent that dephasing can occur by two distinct mechanisms: population decay and population-conserving scattering processes. We can therefore write the total dephasing rate, in the presence of both types of decoherence mechanisms, as:

$$\frac{1}{T_2} = \frac{1}{2T_1} + \frac{1}{T_2'}. \tag{9.54}$$

The first term on the right-hand side accounts for dephasing by population decay, while the second accounts for dephasing by population-conserving scattering. The latter process is sometimes called 'pure dephasing' to distinguish it from the dephasing caused by population decay.

It will often be the case, especially in solids at room temperature, that the pure dephasing rate is much faster than the population decay rate (i.e. $T_2' \ll T_1$), and so the decoherence is governed primarily by scattering processes. On the other hand, it can sometimes be the case that $T_2' \gg T_1$, so that the decoherence is then governed primarily by T_1, which itself is determined by the lifetime of the upper level.

Having considered the processes that cause dephasing in quantum systems, we can now study the detailed effects of damping on Rabi oscillations. It can be shown that if the damping rate is γ, the probability that the electron is in the upper level, namely $|c_2(t)|^2$, is given by:

$$|c_2(t)|^2 = \frac{1}{2(1+2\xi^2)} \left[1 - \left(\cos \Omega' t + \frac{3\xi}{(4-\xi^2)^{1/2}} \sin \Omega' t \right) \exp\left(-\frac{3\gamma t}{2} \right) \right], \tag{9.55}$$

where

$$\xi = \gamma/\Omega_R,$$
$$\Omega' = \Omega_R \sqrt{1 - \xi^2/4}. \tag{9.56}$$

It is easily verified that this formula reduces to the undamped case given in eqn 9.48 when $\gamma = 0$.

Figure 9.5 shows graphs of $|c_2(t)|^2$ from eqn 9.55 for three different values of the damping constant. The dotted line shows the undamped case with $\gamma = 0$ considered previously in Section 9.5.1. The two other graphs correspond to light damping ($\gamma/\Omega_R = 0.1$) and strong damping ($\gamma/\Omega_R = 1$), respectively. Let us consider the case of light damping first. The electron performs a few damped oscillations and then approaches the asymptotic limit with $|c_1|^2 = |c_2|^2 = 1/2$. This asymptotic limit is exactly the behaviour we would have expected from the Einstein analysis of a pure two-level system in the strong-field limit. At high optical power levels the spontaneous emission rate is negligible and the rates of stimulated emission and absorption eventually equal out, leading to identical upper and lower level populations. (See Exercise 9.6.)

Now consider the behaviour for strong damping. This is effectively equivalent to the weak-field limit, because we can always make γ/Ω_R

See Allen and Eberly (1975, eqn 3.29), with $\Delta = 0$ and $b \equiv 1/T_2$. The value of T_2' in eqn 9.54 differs from Allen and Eberly's by a factor of two. This change of notation has been made so that, in the limit, $T_2' \ll T_1$, we obtain $T_2 = T_2'$ rather than $T_2 = 2T_2'$.

Pure dephasing processes can usually be suppressed, to a greater or lesser extent, by cooling the system to very low temperatures. Such techniques are important for applications that require long coherence times, for example, quantum computation: see Section 13.4.

See, for example, Loudon (2000, § 2.8).

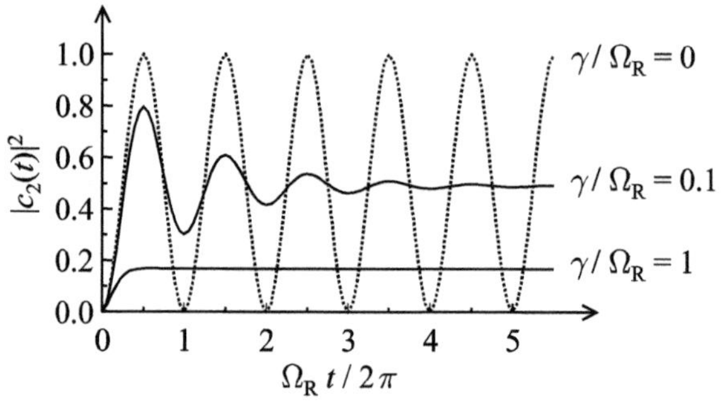

Fig. 9.5 Damped Rabi oscillations for two values of the ratio of the damping rate γ to the Rabi oscillation frequency Ω_R. The dotted curve shows the oscillations when no damping is present.

large by turning down the electric field of the light beam. (See eqn 9.32.) No oscillations are observed, and the asymptotic value of $|c_2|^2$ for very large damping rates (i.e. $\xi \gg 1$) is given by:

$$|c_2|^2 \to \xi^{-2}/4 = \Omega_R^2/4\gamma^2 = \frac{\mu_{12}^2 \mathcal{E}_0^2}{4\hbar^2 \gamma^2}. \tag{9.57}$$

This is consistent with our previous analysis of the weak-field limit, where we found that the transition probability is proportional to the square of both the dipole moment and the electric field (see eqn 9.40, and recall that $u(\omega_0) \propto \mathcal{E}_0^2$.) The time independence of eqn 9.57 compared to eqn 9.40 can be explained by the fact that the former is an asymptotic limit with damping included. In fact, if we solve for the asymptotic populations in the Einstein analysis, we find that N_2 is also independent of time, with $N_2/N_0 \to B_{12}^\omega u_\omega g_\omega(\omega_0)/A_{21}$, where u_ω is the energy density of the radiation and $g_\omega(\omega)$ is the spectral lineshape function. (See Exercise 9.6.) The appearance of A_{21} in the denominator of the asymptotic limit explains one of the factors of γ in the denominator of eqn 9.57, since the damping rate would just be proportional to $\tau_R^{-1} \equiv A_{21}$ for an isolated two-level system. The other factor of γ comes from the fact that the radiative lifetime also causes line broadening.

This simple discussion shows how the inclusion of damping allows us to understand the evolution of the behaviour as the electric field strength is increased. At low fields, we are in the strongly damped regime where there are discrete transitions and the Einstein analysis is valid. As the field is increased, the ratio of the damping rate to the Rabi frequency decreases, and we can eventually reach the case where the oscillations are observable.

9.5.3   Experimental observations of Rabi oscillations

It is apparent from Fig. 9.5 that Rabi oscillations are strongly damped except when

$$\Omega_R \equiv |\mu_{12}\mathcal{E}_0/\hbar| \gg \gamma. \tag{9.58}$$

In gases, the damping rate depends on the collision rate and the radiative lifetime, which gives typical values of γ for optical-frequency transitions in the range 10^7–10^9 s^{-1}. In solids the dephasing times are often shorter due to phonon scattering and scattering by free charge carriers, and γ

can be as high as 10^{12} s^{-1}. These high damping rates make the task of demonstrating Rabi oscillations somewhat difficult, which explains why they are not routinely observed. The observation of the oscillations in the time domain requires a time resolution shorter than $1/\Omega_R$, while the short Rabi oscillation periods demanded by eqn 9.58 require large electric field amplitudes. These conditions are usually satisfied by using high-power pulsed lasers with pulse durations shorter than γ^{-1}.

The first experimental evidence of Rabi oscillations was of an indirect nature and came from the observation of **self-induced transparency** by McCall and Hahn in 1969. They realized that if the pulse area defined in eqn 9.51 is equal to 2π, then the atoms are left in the ground state at the end of the pulse. This implies that there is no net absorption, and so a medium that absorbs strongly at low powers would become transparent to a 2π-pulse: hence the name 'self-induced transparency'. The condition to observe the phenomenon is that the pulse duration should be shorter than the damping time, and that the pulse area should be equal to an integer multiple of 2π. McCall and Hahn performed their experiments on the absorption of a ruby crystal excited resonantly with nanosecond pulses from a ruby laser. The ruby crystal was held at $4.2\,$K in a liquid helium cryostat to suppress damping by phonon scattering. They confirmed that the crystal did indeed become more transparent as the pulse area (determined by the energy of the pulse) approached 2π, although some deviations from the simple theory were observed due to the non-plane-wave nature of the laser beam.

See S. L. McCall and E. L. Hahn, *Phys. Rev.* **183**, 457 (1969).

The first direct evidence of Rabi oscillations came from experiments performed in the 1970s. In 1972–3 Gibbs reported on the fluorescence emitted by Rb atoms excited resonantly by short pulses from a mercury laser. By placing the Rb atoms in a superconducting magnet, one of the hyperfine components of the $M_J = -1/2 \rightarrow +1/2$ line of the $5\,^2S_{1/2} \rightarrow 5\,^2P_{1/2}$ transition with a dipole moment of 1.45×10^{-29} C m could be tuned to resonance with the laser by the Zeeman effect, as shown in Fig. 9.6(a). The upper level could decay either to the $+1/2$ or $-1/2$ levels of the 5s state, with radiative lifetimes of 42 and 84 ns, respectively. This gave a total radiative lifetime of $(1/42 + 1/84)^{-1} = 28$ ns. With a low density of atoms to prevent dephasing by collisions, the right conditions to observe Rabi flopping were present for pulses significantly shorter than 28 ns. The oscillations were then detected by measuring the fluorescence from the upper level as a function of the pulse area Θ.

See H. M. Gibbs, *Phys. Rev. Lett.* **29**, 459 (1972) and *Phys. Rev. A* **8**, 446 (1973).

The fine structure doublet for transitions from the first excited state to the ground state of an alkali atom are often called the D lines. The D_1 and D_2 lines originate from the $^2P_{1/2}$ and $^2P_{3/2}$ levels respectively. See Fig. 3.3.

The results of the experiment are shown in Fig. 9.6(b). The laser operated at 794.466 nm and produced pulses of 7 ns duration. The actual signal recorded was the integrated fluorescence count rate for the time window from 22 to 72 ns after the pulse arrived. This was done to ensure that only incoherent spontaneous emission events occurring after the completion of the pulse were recorded, and was achieved by electronic gating of the photomultiplier used to detect the fluorescence. The fluorescence signal showed a clear oscillatory behaviour as a function of the pulse area. When the pulse area was equal to odd integer multiples of π (i.e. $\Theta = \pi, 3\pi, \ldots$), the atoms ended up in the excited state at the

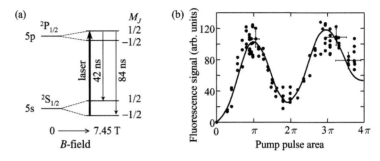

Fig. 9.6 (a) Simplified energy level scheme for the 5s → 5p D_1 transition in Rb at zero field and at $B = 7.45$ T. The wavelength of the transition at $B = 0$ is 794.764 nm, and a field of 7.45 T generated by a superconducting magnet tuned the $(M_J = -1/2 \to +1/2)$ transition to resonance with a mercury laser at 794.466 nm by the Zeeman effect. The upper level can decay spontaneously by the two transitions indicated. (b) Experimental data for the fluorescence intensity from the upper level as a function of pump pulse area, using pulses with a FWHM of 7 ns. The fluorescence signal was integrated from 22 to 72 ns after the pulse arrived at $t = 0$. The solid line is a theoretical fit to the data which includes losses to other levels and also the effects of a weak tail in the laser pulses which persisted to ∼30 ns. (After H. M. Gibbs, *Phys. Rev. Lett.* **29**, 459 (1972) and *Phys. Rev. A* **8**, 446 (1973), © American Physical Society, reproduced with permission.)

completion of the pulse, and then decayed to the ground state by spontaneous emission, thus producing a strong fluorescence signal. On the other hand, when the pulse area was equal to even integer multiples of π (i.e. $\Theta = 2\pi, 4\pi, \ldots$), the atoms were in the ground state at the completion of the pulse, and no fluorescence occurred. These trends were well reproduced in the data. The fluorescence signal did not fall to exactly zero at $\Theta = 2\pi$ and 4π because of loss to the lower $M_J = +1/2$ level during the pump pulse and also due to coherent emission caused by a weak tail in the pump pulses which persisted to ∼30 ns. The solid line in Fig. 9.6(b) shows the results of a theoretical model with the loss mechanism and pulse tail effects included. It is apparent that the fit to the data is excellent, thus confirming the presence of the Rabi oscillations in the Rb atoms.

Another important confirmation of the Rabi flopping process was the observation of **Mollow triplets**. This phenomenon, which was first considered theoretically by B. F. Mollow in 1969, is the frequency-space equivalent of the Rabi oscillations in the time domain. Mollow demonstrated that the coherent oscillatory motion of the electrons in the strong-field limit would beat with the fundamental transition angular frequency ω_0 and produce side bands in the emission spectrum at $\omega_0 \pm \Omega_R$. Hence the fluorescence spectrum would split into a triplet with components at angular frequencies of $(\omega_0 - \Omega_R)$, ω_0, and $(\omega_0 + \Omega_R)$. Several research groups confirmed this behaviour in the 1970s. The lower part of Fig. 9.7(a) shows the fluorescence spectrum measured for one of the hyperfine components of the sodium D_2 ($3\ ^2S_{1/2} \to 3\ ^2P_{3/2}$) line when the atoms are excited resonantly by intense laser light at the transition frequency. The laser light was provided by a continuous wave

See B. R. Mollow, *Phys. Rev.* **188**, 1969 (1969).

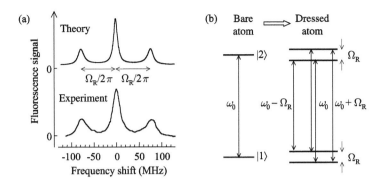

Fig. 9.7 (a) Fluorescence spectrum for one of the hyperfine components of the sodium D_2 line when excited resonantly with intense light from a dye laser. The optical intensity was 6400 Wm^{-2}, and the wavelength was 589.0 nm. The lower part of the figure shows the experimental spectrum, while the upper part shows the theoretical spectrum calculated for a Rabi frequency $\Omega_R = 2\pi \times 78$ MHz. (After R. E. Grove, F. Y. Wu, and S. Ezekiel, *Phys. Rev. A* **15**, 227 (1977), © American Physical Society, reproduced with permission.) (b) Explanation of the Mollow triplet spectrum shown in part (a) using the dressed atom picture. The AC Stark interaction between a two-level atom and an intense resonant light field splits the bare atom states into doublets separated by the Rabi frequency Ω_R. This leads to three emission lines at angular frequencies of ω_0 and $\omega_0 \pm \Omega_R$.

dye laser operating at 589.0 nm and the intensity was 6400 $W\,m^{-2}$. Two peaks on either side of the central peak are clearly observed. The top part of the figure shows the theoretical spectrum calculated for a value of $\Omega_R/2\pi = 78$ MHz. The excellent agreement between theory and experiment is apparent.

Figure 9.7(b) indicates a way to interpret the Rabi oscillation phenomenon in terms of **dressed atoms**. In this picture we consider the states of the coupled resonant light–atom system rather than those of the unperturbed atom. The states of the 'bare' atom are 'dressed' by the intense resonant optical field through the **AC Stark effect** (also called the **dynamic Stark effect**). The Stark effect describes the shift of the levels of an atom in a DC electric field, and the AC Stark effect is the equivalent process for the AC electric field of a light wave. It can be shown that the AC Stark effect splits the bare atom states into doublets separated by the Rabi frequency Ω_R, as shown in Fig. 9.7(b). Hence the emission spectrum of the dressed atom consists of three lines equivalent to the Mollow triplet spectrum shown in Fig. 9.7(a).

There have been many further demonstrations of Rabi-flopping phenomena in the years following these initial experiments. Some of the most interesting recent observations have involved semiconductor quantum dot structures. (See Section D.3 in Appendix D.) Figure 9.8 shows the results for InAs quantum dots embedded within a GaAs photodiode. Figure 9.8(a) shows a schematic diagram of the device used in this experiment. Masks patterned onto the surface allowed individual quantum dots to be excited by short pulses from a mode-locked Ti:sapphire laser. The Rabi oscillations were induced by tuning the laser to the lowest energy

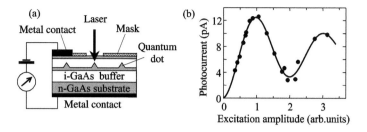

Fig. 9.8 (a) Schematic diagram of the quantum dot photodiode used to demonstrate Rabi oscillations. The device consisted of InAs quantum dots embedded within a GaAs n-i-Schottky diode structure. The device was biassed by an external DC power supply which applied a strong DC electric field to the quantum dots. Free electrons and holes excited in the quantum dots by absorption of laser light tunnel into the GaAs regions under the influence of the DC field and are then swept to the contacts to generate a photocurrent in the external circuit. A mask on the top of the device allowed individual quantum dots to be addressed by the laser beam. (b) Photocurrent measured when exciting the lowest energy transition of the quantum dots resonantly with 1 ps pulses from a Ti:sapphire laser at 1.31 eV. The excitation amplitude has been scaled so that an amplitude of unity corresponds to a π-pulse. (After A. Zrenner *et al.*, *Nature* **418**, 612 (2002), © Nature Publishing Group, reproduced with permission.)

transition of the quantum dot at 1.31 eV. This transition corresponds to the excitation of an electron from the valence band to the conduction band, and has a dipole moment of $\sim 8 \times 10^{-29}$ C m. The experiment consisted in measuring the photocurrent generated in the external circuit as a function of the laser pulse area Θ.

The results of the experiment are shown in Fig. 9.8(b). The photocurrent shows a clear oscillatory behaviour with the pulse excitation amplitude, which has been scaled so that an amplitude of unity corresponds to a pulse area of π. There are peaks for pulse areas of π and 3π, and a minimum for $\Theta = 2\pi$. The results may be understood by considering the state of the system at the completion of the pulse, in an analogous way to the discussion of the data in Fig. 9.6(b). At the end of a π or 3π pulse, the quantum dots are left with one electron in the conduction band and hence one 'hole' in the valence band. These charge carriers can then tunnel out of the quantum dot under the influence of the DC electric field of the photodiode and produce a photocurrent in the external circuit. On the other hand, after a 2π pulse the electron is back in the valence band and there are no free electrons and holes to generate a current. This is exactly what is observed.

The Rabi oscillations shown in Fig. 9.8(b) are partly damped by rapid dephasing processes. The main source of dephasing was the loss of electrons and holes due to tunnelling out of the quantum dots. This process is unavoidable, since it is an integral part of the mechanism to generate the photocurrent used to detect the Rabi oscillations. The estimated tunnelling time was ≈ 10 ps, and it was for this reason that very short pulses with a duration of only 1 ps had to be used in the experiment.

These very short time-scales highlight the difficulty in observing coherent phenomena like Rabi oscillations in the solid state.

9.6   The Bloch sphere

An arbitrary superposition state of a two-level system will have a wave function of the form given by eqn 9.2, namely:

$$|\psi\rangle = c_1|1\rangle + c_2|2\rangle. \tag{9.59}$$

The normalization condition on the wave function requires that:

$$|c_1|^2 + |c_2|^2 = 1, \tag{9.60}$$

which suggests that we can represent the state by a vector of unit length starting at the origin. This geometric interpretation of coherent superposition states is called the **Bloch representation**. The vector that describes the state is called the **Bloch vector**, and the sphere it defines is the **Bloch sphere**.

The Bloch representation was originally developed by Felix Bloch in 1946 to model NMR phenomena, and was first adapted to two-level atoms by Feynman *et al.* in 1957. The equivalence between two-level atoms and Zeeman-split nuclear spin states was noted previously in Section 9.2.1 (see Fig. 9.3) and allows us to share analytic tools such as the Bloch representation between the two subjects. Readers who are unfamiliar with NMR phenomena may therefore find it helpful to refer to Appendix E which gives a summary of the main effects.

The direction of the Bloch vector can be specified either in Cartesian coordinates (x, y, z) or spherical polar coordinates (r, θ, φ), with

$$
\begin{aligned}
x &= r \sin\theta \cos\varphi, \\
y &= r \sin\theta \sin\varphi, \\
z &= r \cos\theta,
\end{aligned} \tag{9.61}
$$

as illustrated in Fig. 9.9. The requirement that the vector has unit length is satisfied when

$$r^2 = (x^2 + y^2 + z^2) = 1. \tag{9.62}$$

We therefore need only two independent variables to define an arbitrary state on the Bloch sphere, for example the angles (θ, φ). This allows us to make a unique mapping between the wave function amplitudes (c_1, c_2) and the direction of the Bloch vector.

The connection between the Bloch vector and the wave function may be made by defining the bottom and top of the sphere to correspond to the $|1\rangle$ and $|2\rangle$ states respectively, as shown in Fig. 9.9. The ground state with $|\psi\rangle = |1\rangle$ thus corresponds to $(0, 0, -1)$ in Cartesian coordinates or $\theta = \pi$ in polar coordinates. Similarly, the pure excited state $|2\rangle$

See R. P. Feynman, *et al. J. Appl. Phys.* **28**, 49 (1957).

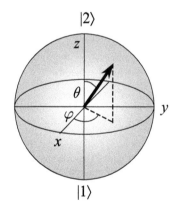

Fig. 9.9 The Bloch sphere. Coherent superposition states lie on the surface of the sphere, with their state defined by the angles (θ, φ) through eqn 9.64.

The choice of assigning the north and south poles of the Bloch sphere, respectively, to the upper and lower level is arbitrary. The analysis works equally well the other way round.

(a)

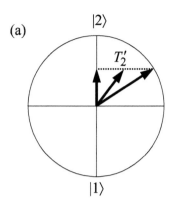

(b)

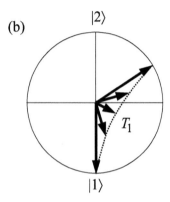

Fig. 9.10 Damping processes in the Bloch representation: (a) transverse (T_2') and (b) longitudinal (T_1) relaxation. T_2' processes conserve z but T_1 processes do not. In part (a) we are assuming that $T_2' \ll T_1$. Slower T_1 processes will eventually cause the excited state population to decay and the Bloch vector to return to the ground state with $\theta = \pi$. Note that the Bloch vector of the relaxed state is only meaningful for the entire ensemble rather than for individual atoms. Note also that longitudinal decay inevitably causes transverse relaxation as well.

corresponds to the point $(0, 0, 1)$ or $\theta = 0$. An arbitrary state is given in Cartesian coordinates as:

$$x = 2\,\mathrm{Re}\langle c_1 c_2 \rangle,$$
$$y = 2\,\mathrm{Im}\langle c_1 c_2 \rangle, \qquad (9.63)$$
$$z = |c_2|^2 - |c_1|^2.$$

In polar coordinates this simplifies to (see Exercise 9.9):

$$c_1 = \sin(\theta/2),$$
$$c_2 = \mathrm{e}^{\mathrm{i}\varphi}\cos(\theta/2). \qquad (9.64)$$

This one-to-one mapping allows us to visualize an arbitrary superposition state of a two-level atom in a geometric way, which is very useful when considering the resonant interaction with an intense optical field.

It is instructive to compare superposition states with statistical mixtures in the Bloch representation. It is apparent that we can apply the Bloch model to individual atoms for the case of superposition states, but not for statistical mixtures. In the latter case the individual atoms are either in level 1 or in level 2, and it is only meaningful to calculate the Bloch vector for the whole ensemble. Furthermore, in a statistical mixture we have $\langle c_1 c_2 \rangle = 0$ for every atom, which implies from eqn 9.63 that $x = y = 0$. Statistical mixtures thus correspond to points *inside* the Bloch sphere on the z-axis. Apart from the ground state with $|c_1|^2 = 1$, statistical mixtures therefore have $r < 1$, in contrast to superposition states, which are always on the surface of the sphere with $r = 1$.

In Section 9.5.2 we discussed the damping processes which destroy coherence and reduce superposition states to statistical mixtures. Since statistical mixtures have $r < 1$, damping processes do not conserve the modulus of the Bloch vector. Figure 9.10 illustrates the two different types of damping that can occur. Figure 9.10(a) illustrates the effect of damping by pure dephasing (T_2') processes. Such processes break the coherence without altering the populations, and they therefore correspond to scattering from the surface of the sphere towards the centre at constant z (cf. eqn 9.63). Figure 9.10(b) illustrates the contrasting effect of damping by population decay (T_1) processes. Since these affect the relative populations, they alter z as well.

The fact that pure dephasing (T_2') processes conserve z whereas population decay (T_1) processes do not explains why they are called transverse and longitudinal relaxation, respectively. Note, however, that Fig. 9.10(b) clearly illustrates the point that longitudinal relaxation simultaneously produces transverse relaxation, even in the absence of pure dephasing processes. This is why the total dephasing rate in eqn 9.54 contains contributions from both pure dephasing and population decay processes.

Up to this point we have been neglecting the explicit time dependence of the wave functions. The two-level atom has an intrinsic angular frequency of ω_0, and in making the transformation to the Bloch representation, we find that the Bloch vector rotates at a constant angular frequency of ω_0 around the z-axis. It is therefore convenient to make a coordinate transformation to a rotating frame so that the Bloch vector is stationary.

Let us now consider the interaction of the Bloch vector with resonant optical pulses at angular frequency ω. As demonstrated in eqn 9.33, the light field produces interaction terms at frequencies of $(\omega - \omega_0)$ and $(\omega + \omega_0)$. In the rotating frame, the term at $(\omega - \omega_0)$ causes a slow precession of the Bloch vector at the difference frequency, and is stationary at exact resonance. By contrast, the term at $(\omega + \omega_0)$ causes a very rapid precession at $\sim 2\omega_0$, and can be neglected because of its highly non-resonant nature. This is why we described the neglect of the terms at $(\omega + \omega_0)$ as the 'rotating wave approximation'.

The application of a short resonant pulse can be considered as a **coherent operation** on the Bloch vector. If the pulse duration is shorter than the damping time, the coherence of the wave function will be retained and the modulus of the Bloch vector conserved. Hence the pulse will only change the direction of the Bloch vector without altering its length. This means that the pulse acts as a **rotation operator**. We showed previously that a π-pulse can convert a system in the ground state $|1\rangle$ to the excited state $|2\rangle$, and vice versa. This corresponds to a change of θ by π radians, and explains the origin of the name 'π-pulse'. In general, it can be shown that the rotation angle is equal to the pulse area defined in eqn 9.51. Hence a $\pi/2$-pulse causes a rotation of $\pi/2$ radians, while a 2π-pulse causes a rotation of 2π radians, leaving the system unchanged. The Bloch vector can thus be manipulated at will by a sequence of resonant pulses of well-defined amplitude and relative phase.

Let us consider the effect of a sequence of exactly resonant pulses at angular frequency ω_0 on a system that is in the ground state at $t = 0$. The azimuthal angles of the Bloch sphere are initially undefined, which means that the choice of the axis of rotation for the first pulse is arbitrary. It is therefore convenient to choose the x and y-axis directions in such a way that the first pulse produces a rotation about, say, the y-axis, leaving the Bloch vector somewhere in x-z plane at the end of the pulse. The axis about which subsequent rotations take place is then determined by the phase of the pulse relative to the first one. For example, a pulse with a phase difference of 90° relative to the first one would rotate the Bloch vector about an axis at 90° to the first one: that is, the $-x$ axis. Combinations of pulses of the appropriate area and phase can then be used to move the Bloch vector to any particular point on the Bloch sphere. Figure 9.11 illustrates how an initial pulse with an area of $3\pi/4$ followed by a $\pi/2$-pulse with a relative phase of $-90°$ moves the system from the ground state to a point within the x-y plane with an azimuthal angle of $\pi/4$.

Coherent operations on the Bloch vector have been used for many years for quantum state preparation in NMR systems. They have also been used extensively in two-level atomic systems at optical frequencies for the description of coherent phenomena such as **photon echoes** and

See, for example, Mandel and Wolf (1995, Chapter 15). The rotating frame transformation is entirely analogous to the one described in Section E.2 for the treatment of Larmor precession in NMR.

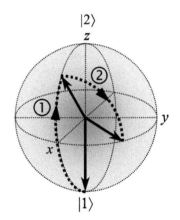

Fig. 9.11 The Bloch vector can be moved to arbitrary positions on the Bloch sphere by appropriate combinations of rotations. This figure illustrates the effect of a rotation by $3\pi/4$ about the y-axis followed by a rotation of $\pi/2$ about the x-axis on a Bloch vector initially in the ground state.

superradiance. In recent years the subject has acquired new importance in the practical implementation of quantum computation, both in NMR systems and also in the atomic systems at optical frequencies. This point will be developed further in Section 13.3.4 of Chapter 13.

Example 9.3   A pulsed laser beam is focussed to a spot of radius 1 μm on a gas of atoms with a dipole moment of 10^{-29} C m at the laser frequency.

(a) Calculate the pulse energy required to rotate the Bloch vector by $\pi/2$ radians for Gaussian pulses with a duration (FWHM) of 1 ps.

(b) If the system is initially in the ground state, find the state of the system at the end of the pulse.

Solution

(a) The time dependence of the electric field and intensity in a Gaussian pulse are given, respectively, by:

$$\mathcal{E}_0(t) = \mathcal{E}_{\text{peak}} \exp(-t^2/\tau^2),$$

$$I(t) = I_0 \exp(-2t^2/\tau^2). \tag{9.65}$$

We can relate τ to the pulse width by finding the time for the intensity to drop to half its peak value:

$$I(t_{1/2})/I_0 = \exp(-2t_{1/2}^2/\tau^2) = 0.5,$$

which implies $t_{1/2} = \sqrt{\ln 2/2}\,\tau$. Hence:

$$\tau_{\text{FWHM}} = 2t_{1/2} = 2\sqrt{\ln 2/2}\,\tau = 1.177\tau.$$

In our example we therefore find $\tau = 0.85$ ps.
The pulse energy E_{pulse} is related to the intensity by:

$$E_{\text{pulse}} = A \int_{-\infty}^{+\infty} I(t)\, dt,$$

where A is the beam area. On making use of eqn 9.52 with Gaussian pulses and refractive index $n = 1$ we find:

$$E_{\text{pulse}} = \frac{1}{2} A c \epsilon_0 \int_{-\infty}^{+\infty} \mathcal{E}_0(t)^2\, dt$$

$$= \frac{1}{2} A c \epsilon_0 \mathcal{E}_{\text{peak}}^2 \int_{-\infty}^{+\infty} \exp(-2t^2/\tau^2)\, dt$$

$$= \sqrt{\frac{\pi}{8}} A c \epsilon_0 \mathcal{E}_{\text{peak}}^2 \tau. \tag{9.66}$$

On the other hand, the pulse area Θ is given from eqn 9.51 by:

$$\Theta = \frac{\mu_{12}}{\hbar} \int_{-\infty}^{+\infty} \mathcal{E}_0(t)\, dt$$

$$= \frac{\mu_{12}}{\hbar} \mathcal{E}_{\text{peak}} \int_{-\infty}^{+\infty} \exp(-t^2/\tau^2)\, dt$$

$$= \sqrt{\pi}\,\mu_{12} \mathcal{E}_{\text{peak}} \tau/\hbar.$$

Thus for $\Theta = \pi/2$, $\mu_{12} = 10^{-29}$ C m and $\tau = 0.85$ ps we find $\mathcal{E}_{\text{peak}} = 11$ MV m^{-1}. On inserting into eqn 9.66 with $A = \pi(10^{-6})^2 = 3.1 \times 10^{-12}$ m^2, we finally find $E_{\text{pulse}} = 0.53$ pJ.

(b) The azimuthal angle for the rotation is arbitrary and so we choose to rotate within the $\varphi = 0$ plane. The state vector initially points downwards with $\theta = \pi$, and after the $\pi/2$-pulse we arrive at the point with $(\theta, \varphi) = (\pi/2, 0)$. We then find from eqn 9.64 that $c_1 = \sin(\pi/4) = 1/\sqrt{2}$ and $c_2 = \cos(\pi/4) = 1/\sqrt{2}$. Hence the final state of the system is:

$$|\psi\rangle = \frac{1}{\sqrt{2}}|1\rangle + \frac{1}{\sqrt{2}}|2\rangle = \frac{1}{\sqrt{2}}(|1\rangle + |2\rangle).$$

Further reading

The classic text on the theory of two-level atoms is Allen and Eberly (1975). The subject material of the chapter is covered in greater depth in the more advanced quantum optics texts such as Loudon (2000) or Mandel and Wolf (1995), while Foot (2005) covers the topics at a similar level from the perspective of atomic physics. Discussions of coherent phenomena such as photon echoes and superradiance may be found in nonlinear optics texts such as Yariv (1989) or Shen (1984).

Exercises

(9.1) The interaction between an atom and a light wave of angular frequency ω may be modelled classically as a driven oscillator system. The displacement of an electron in the atom by a distance x induces a dipole equal to $-ex$. The displacements have their own natural angular frequency ω_0, which is presumed to correspond to the transition frequency. The electric field of the light applies a force to the dipoles, and induces oscillations at its own frequency. The equation of motion for the displacement x of the electron is thus:

$$m_e \frac{\mathrm{d}^2 x}{\mathrm{d}t^2} + m_e \gamma \frac{\mathrm{d}x}{\mathrm{d}t} + m_e \omega_0^2 x = F_0 \cos \omega t,$$

where m_e is the electron mass, γ is a damping constant, and F_0 is the amplitude of the force applied to the electron by the light. With the assumption that $\omega_0 \gg \gamma$, show that the magnitude of the driven oscillations is a maximum when $\omega = \omega_0$. What is the full width at half maximum of the resonance in angular frequency units?

(9.2) Write down the density matrix for the following superposition states:

(a) $|2\rangle$,

(b) $(|1\rangle + |2\rangle)/\sqrt{2}$,

(c) $(1/\sqrt{3})|1\rangle + (\sqrt{2/3})\mathrm{i}|2\rangle$.

(9.3) Write down the density matrix for a gas of two-level atoms at temperature T.

(9.4) The wave functions for hydrogenic states with quantum numbers (n, l, m_l) of $(2,0,0)$ and $(3,1,0)$ are as follows:

$$\psi_1(r, \theta, \varphi) = \frac{1}{4\sqrt{2\pi}a_0^{3/2}}\left(2 - \frac{r}{a_0}\right)e^{-r/2a_0},$$

$$\psi_2(r, \theta, \varphi) = \frac{\sqrt{2}}{81\sqrt{\pi}a_0^{5/2}}\left(6 - \frac{r}{a_0}\right)r\cos\theta\, e^{-r/3a_0}$$

where a_0 is the Bohr radius. Calculate the Einstein B coefficient B_{12}^{ω} for the 2s $\rightarrow$ 3p transition of atomic hydrogen for light polarized along the z-axis. Find also the Einstein A coefficient for the reverse transition.

(9.5) This exercise considers the case of Rabi oscillations when the light is not exactly resonant with the transition frequency. In the rotating wave approximation, eqn 9.33 becomes:

$$\dot{c}_1(t) = \frac{i}{2}\Omega_R e^{i\delta\omega t} c_2(t),$$

$$\dot{c}_2(t) = \frac{i}{2}\Omega_R e^{-i\delta\omega t} c_1(t),$$

where $\delta\omega = \omega - \omega_0$.

(a) Show that:

$$\ddot{c}_2 + i\delta\omega \dot{c}_2 + (\Omega_R^2/4)c_2 = 0.$$

(b) By considering a trial solution of the form $Ce^{-i\zeta t}$, show that the general solution of this differential equation is:

$$c_2(t) = C_+ e^{-i\zeta_+ t} + C_- e^{-i\zeta_- t},$$

where $\zeta_\pm = (\delta\omega \pm \Omega)/2$, $\Omega^2 = \delta\omega^2 + \Omega_R^2$, and C_+ and C_- are constants.

(c) Hence show that the initial conditions of $c_1(0) = 1$ and $c_2(0) = 0$ imply that:

$$|c_2(t)|^2 = \frac{\Omega_R^2}{\Omega^2} \sin^2 (\Omega t/2).$$

(9.6) In this exercise we investigate the behaviour of an ideal two-level atom with Einstein coefficients A_{21} and B_{12}^ω in the presence of a strong resonant field from a narrow bandwidth laser of angular frequency ω. The traditional Einstein analysis discussed in Section 4.1 assumes a broad-band radiation source and a narrow transition line, and therefore has to be modified to account for the situation that we are considering here, namely a narrow bandwidth radiation source.

We assume that the transition probability is proportional to the spectral line shape function $g_\omega(\omega')$, so that the frequency dependence of the absorption and stimulated emission rates are given, respectively, by:

$$W_{12}(\omega')\, d\omega' = N_1 B_{12}^\omega u(\omega') g_\omega(\omega')\, d\omega',$$
$$W_{21}(\omega')\, d\omega' = N_2 B_{21}^\omega u(\omega') g_\omega(\omega')\, d\omega',$$

where N_1 and N_2 are the populations of the lower and upper levels, and $u(\omega')$ is the energy density at frequency ω'.

(a) Consider first the case of a white broad-band source with a slowly varying energy density.

Show that the transition rates written above are consistent with the traditional definitions of the Einstein coefficients.

(b) Now consider the case that we are interested in, namely that the spectral width of the laser is much smaller than the linewidth of the transition. Following Exercise (4.1), we write the spectral energy density as a delta function:

$$u(\omega') = u_\omega\, \delta(\omega' - \omega),$$

where u_ω is the energy density of the laser beam in $J\,m^{-3}$. Show that the absorption and stimulated emission rates are now given, respectively, by:

$$W_{12} = N_1 B_{12}^\omega u_\omega g_\omega(\omega),$$
$$W_{21} = N_2 B_{21}^\omega u_\omega g_\omega(\omega).$$

(c) For simplicity, we now assume that the levels are non-degenerate, so that $B_{12}^\omega = B_{21}^\omega$ (see eqn 4.10). With the initial condition $N_2 = 0$, show that the time dependence of the fractional population of the upper level is given by:

$$\frac{N_2}{N_0} = \frac{B'u_\omega}{2B'u_\omega + A_{21}}$$
$$\times [1 - \exp(-(2B'u_\omega + A_{21})t)],$$

where $B' = B_{12}^\omega g_\omega(\omega)$.

(d) Discuss the asymptotic behaviour for (1) very intense fields, and (2) weak fields.

(9.7) Referring to the data in Fig. 9.6(b), the transition dipole moment was 1.45×10^{-29} C m and the pulse duration (FWHM) was 7 ns. On the assumption that $n \approx 1$, estimate the maximum optical intensity in the pulse when the fluorescence intensity reaches its first maximum for the cases of: (a) a 'top hat' shaped pulse, and (b) a Gaussian pulse.

(9.8) Use the data in Fig. 9.7(a) to estimate the dipole moment of the optical transition in resonance with the laser beam. (Assume that $n \approx 1$.)

(9.9) Verify that eqn 9.64 is consistent with eqn 9.63.

(9.10) Find two-level superposition states that correspond to the following points on the Bloch sphere as defined by their polar angles (θ, φ):

(a) $(90°, 0)$,

(b) $(90°, 90°)$,

(c) $(90°, 180°)$,

(d) $(90°, -90°)$,

(e) $(60°, 45°)$.

(9.11) Find the points on the Bloch sphere corresponding to the following states, quoting your answer in Cartesian coordinates:

(a) $\psi = (\sqrt{1/3})|1\rangle + (\sqrt{2/3})|2\rangle$,

(b) $\psi = (\sqrt{2/3})|1\rangle - (i/\sqrt{3})|2\rangle$,

(c) $\psi = (e^{i\pi/4}|1\rangle + |2\rangle)/\sqrt{2}$.

(9.12) Find the Bloch vector equivalent to an ensemble of two-level atoms with 60% of the atoms in the excited state and 40% in the ground state.

(9.13) A pulsed laser emits Gaussian pulses with a FWHM of 3 ps. The beam is focussed to a spot of radius 2 μm on a quantum dot with a dipole moment of 8×10^{-29} C m at the laser frequency. The refractive index of the crystal containing the quantum dot is 3.5.

(a) Calculate the pulse energy of a π-pulse.

(b) If the system is initially in the state $(1/\sqrt{2})(|1\rangle + |2\rangle)$, find the state of the system at the end of the pulse if its phase is set to rotate about the y-axis of the Bloch sphere.

(9.14) A two-level atom with a transition at angular frequency ω_0 and $\mu_{12} = 2 \times 10^{-29}$ C m is subjected to a sequence of two resonant pulses. The first pulse has an electric field given by $\mathcal{E}(t) = \mathcal{E}_1 \cos(\omega_0 t)$ and a duration of τ_1, while the second has $\mathcal{E}(t) = \mathcal{E}_2 \cos(\omega_0 t + \phi)$ and a duration of τ_2. $\mathcal{E}_1$ and $\mathcal{E}_2$ are both constant during the pulses, and the pulse length is much shorter than the dephasing time T_2. Given that the system starts in the ground state, find the final state when:

(a) $\mathcal{E}_1 = 4139$ V m^{-1}, $\tau_1 = 1$ ns, $\mathcal{E}_2 = 6209$ V m^{-1}, $\tau_2 = 2$ ns, $\phi = 0$;

(b) $\mathcal{E}_1 = 827.8$ V m^{-1}, $\tau_1 = 10$ ns, $\mathcal{E}_2 = 3311$ V m^{-1}, $\tau_2 = 5$ ns, $\phi = 90°$;

(c) $\mathcal{E}_1 = 2.759 \times 10^4$ V m^{-1}, $\tau_1 = 0.3$ ns, $\mathcal{E}_2 = 5.519 \times 10^4$ V m^{-1}, $\tau_2 = 0.3$ ns, $\phi = 45°$.

10

Atoms in cavities

10.1 Optical cavities	194
10.2 Atom–cavity coupling	197
10.3 Weak coupling	200
10.4 Strong coupling	206
10.5 Applications of cavity effects	211
Further reading	213
Exercises	214

In the previous chapter we studied the resonant interaction between photons and an atomic transition of the same frequency. The atoms we considered were in free space and the photons originated from an external source such as a lamp or laser beam. We now wish to re-explore this process in more detail for the special case in which the interaction between the photons and the atom is enhanced by placing the atom inside a resonant cavity. This will naturally lead us into the subject of cavity quantum electrodynamics (cavity QED). We begin our discussion by considering the key parameters that determine the properties of the cavity and the magnitude of the atom–cavity coupling, and then explore the different physical effects that are observed in the limits of weak and strong coupling to the cavity.

10.1 Optical cavities

Before studying the interaction between atoms and cavities, it is helpful to remind ourselves of some of the basic properties of optical cavities. We shall restrict our attention here to the simplest case, namely a planar cavity. This will be sufficient to illustrate the chief points, and the results can then be generalized to other types of cavity. The planar cavity is covered in detail in most classical optics texts, and we give here only a brief summary of the results that are relevant to our discussion.

Consider the planar cavity shown in Fig. 10.1. The cavity consists of two plane mirrors M1 and M2, with reflectivities of R_1 and R_2, respectively, separated by an adjustable length L_{cav}. The space between the mirrors is filled with a medium of refractive index n and the mirrors are aligned parallel to each other so that light inside the cavity bounces backwards and forwards between the mirrors. Planar cavities of the type shown in Fig. 10.1 are frequently used in high-resolution spectroscopy, in which case the instrument is called a **Fabry–Perot interferometer**.

The properties of the planar cavity can be analysed by considering the effect of introducing light of wavelength λ from one side and calculating how much gets transmitted through to the other side. On the assumption that there are no absorption or scattering losses within the cavity, the transmission $\mathcal{T}$ is given by:

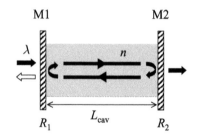

Fig. 10.1 A planar cavity of length L_{cav} with two parallel end mirrors M1 and M2 of reflectivity R_1 and R_2, respectively. The medium inside the cavity has a refractive index n. The cavity acts as a Fabry–Perot interferometer when light of wavelength λ is introduced through one of the end mirrors.

See, for example, Brooker (2003), Hecht (2002), or Yariv (1997).

$$\mathcal{T} = \frac{1}{1 + (4\mathcal{F}^2/\pi^2)\sin^2(\phi/2)} \qquad (10.1)$$

where

$$\phi = \frac{4\pi n L_{\text{cav}}}{\lambda} \qquad (10.2)$$

is the round-trip phase shift, and

$$\mathcal{F} = \frac{\pi (R_1 R_2)^{1/4}}{1 - \sqrt{R_1 R_2}} \qquad (10.3)$$

is the **finesse** of the cavity. It is easy to see from eqn 10.1 that the transmission is equal to unity whenever $\phi = 2\pi m$, where m is an integer. In this situation the cavity is said to be **on-resonance**. From eqn 10.2 we see that the resonance condition occurs when the cavity length L_{cav} is equal to an integer number m of intracavity half wavelengths:

$$L_{\text{cav}} = m\lambda/2n. \qquad (10.4)$$

The resonance condition thus occurs when the light bouncing around the cavity is in phase during each round trip.

Figure 10.2 shows the transmission of a lossless planar cavity with $R_1 = R_2 = 0.9$, giving $\mathcal{F} = 30$. The transmission is a sharply peaked function of the round-trip phase shift ϕ, with maxima at the resonance values of $\phi = 2\pi m$. The width of the peaks can be calculated by finding the condition for $\mathcal{T} = 50\%$. In the limit of large $\mathcal{F}$, this is easily calculated from eqn 10.1 and gives

$$\phi = 2\pi m \pm \pi/\mathcal{F}. \qquad (10.5)$$

The full width at half maximum (FWHM) is therefore equal to

$$\Delta\phi_{\text{FWHM}} = 2\pi/\mathcal{F}, \qquad (10.6)$$

which implies:

$$\mathcal{F} = \frac{2\pi}{\Delta\phi_{\text{FWHM}}}. \qquad (10.7)$$

The finesse of the cavity determines the resolving power when using the instrument for high-resolution spectroscopy.

The cavity resonance condition naturally leads to the concept of **resonant modes**. These are modes of the light field that satisfy the resonance condition and are preferentially selected by the cavity. Since the light fields bouncing around the cavity are all in phase, the waves interfere constructively and have much larger amplitudes than at non-resonant frequencies. The resonant modes have intensities inside the cavity enhanced by a factor $4/(1 - R)$ compared to an incoming wave, while the out of resonance frequencies have their intensity suppressed by a factor $(1 - R)$. (See Exercise 10.2.) The properties of the resonant modes play an essential part in determining the emission spectra of lasers, and will also be very important for the discussion of the emission properties of atoms in cavities.

The angular frequencies of the resonant modes are easily worked out from eqn 10.4 and are given by:

$$\omega_m = m \frac{\pi c}{n L_{\text{cav}}}. \qquad (10.8)$$

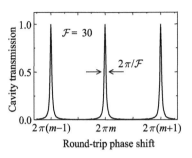

Fig. 10.2 Transmission of a lossless planar cavity with mirror reflectivities of 90%, giving a finesse $\mathcal{F}$ of 30. Resonance occurs whenever the round-trip phase equals $2\pi m$, where m is an integer.

The cavity finesse is usually defined as the ratio of the separation of adjacent maxima to the half width, as in eqn 10.7.

The cavity length is typically tuned by moving one of the mirrors with a piezo-electric transducer. The method used for tuning the refractive index depends on whether the cavity is filled with a gas or a solid. In the former case, the refractive index can be controlled through the gas pressure, and in the latter, by heating the cavity and using the temperature dependence of n.

The mode frequencies can be tuned either by changing L_{cav} or n. Since the mode frequency is proportional to the phase, we can use eqn 10.7 to relate the spectral width $\Delta\omega$ of the resonant modes to the properties of the cavity:

$$\frac{\Delta\omega}{\omega_m - \omega_{m-1}} = \frac{\Delta\phi_{\text{FWHM}}}{2\pi} = \frac{1}{\mathcal{F}}, \tag{10.9}$$

giving:

$$\Delta\omega = \frac{\pi c}{n\mathcal{F}L_{\text{cav}}}. \tag{10.10}$$

This shows that cavities with high finesse values have sharp resonant modes.

The final quantity that we need to consider is the **photon lifetime** τ_{cav} inside the cavity. Consider a light source at the centre of a symmetric, high-finesse cavity with $R_1 = R_2 \equiv R \approx 1$. We suppose that the source emits a short pulse of light containing N photons into the cavity mode at time $t = 0$ as shown in Fig. 10.3. We assume that the fractional photon number change at each reflection is small due to the high reflectivity of the mirrors. After a time $t = nL_{\text{cav}}/c$, the pulse will have gone around half the cavity, and the photon number will be equal to RN. At $t = 2nL_{\text{cav}}/c$, the pulse will have completed a round trip and the photon number will be equal to $R^2 N$. This process continues until all the photons are lost from the cavity. On average, we lose $\Delta N = (1 - R)N$ photons in a time equal to nL_{cav}/c. We can therefore write:

$$\frac{dN}{dt} = -\frac{\Delta N}{nL_{\text{cav}}/c} = -\frac{c(1 - R)}{nL_{\text{cav}}}N, \tag{10.11}$$

which has solution $N = N_0 \exp(-t/\tau_{\text{cav}})$, where the photon lifetime is given by:

$$\tau_{\text{cav}} = \frac{nL_{\text{cav}}}{c(1 - R)}. \tag{10.12}$$

It is helpful to define the **photon decay rate** (κ) as:

$$\kappa = \frac{1}{\tau_{\text{cav}}}. \tag{10.13}$$

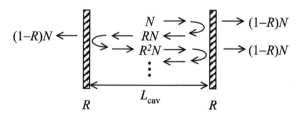

Fig. 10.3 Decay of the cavity photon number following emission from a pulsed source at the centre of the cavity at $t = 0$. The mirror reflectivities are assumed to be high, so that the fractional loss per round trip is small.

We can then combine eqns 10.3, 10.10, and 10.12 with $R \approx 1$ to find:

$$\Delta\omega = (\tau_{\text{cav}})^{-1} \equiv \kappa. \qquad (10.14)$$

This shows that the width of the resonant modes is controlled by the photon loss rate in the cavity, in exactly the same way that the natural width of an atomic emission line is controlled by the spontaneous emission rate (cf. eqn 4.30).

The analysis of the linear cavity shows that there are basically two key parameters that determine the main properties, namely the resonant mode frequency ω_m and the cavity finesse $\mathcal{F}$. The latter parameter controls both the mode width $\Delta\omega$ and the cavity loss rate κ. In dealing with other types of cavity it is helpful to introduce the **quality factor** (Q) of the cavity, defined by:

$$Q = \frac{\omega}{\Delta\omega}. \qquad (10.15)$$

This serves the equivalent purpose for a general cavity as the finesse does for the planar cavity. It is thus convenient to specify the properties of a cavity either by the frequency and finesse, or equivalently by the frequency and quality factor.

10.2 Atom–cavity coupling

Having reminded ourselves of the relevant properties of optical cavities, we can now start to discuss the interaction between the light inside a cavity and an atom, as shown schematically in Fig. 10.4. We assume that the atom is inserted in such a way that it can absorb photons from the cavity modes and also emit photons into the cavity by radiative emission. We are particularly interested in the case where the transition frequency of the atom coincides with one of the resonant modes of the cavity. In these circumstances, we can expect that the interaction between the atom and the light field will be strongly affected, since the atom and cavity can exchange photons in a resonant way.

The transition frequencies of the atom are determined by its internal structure and are taken as fixed in this analysis. The resonance condition is then achieved by tuning the cavity so that the frequency of one of the cavity modes coincides with that of the transition. At resonance we find that the relative strength of the atom–cavity interaction is determined by three parameters:

- the photon decay rate of the cavity κ,
- the non-resonant decay rate γ,
- the atom–photon coupling parameter g_0.

These three parameters each define a characteristic time-scale for the dynamics of the atom–photon system. The interaction is said to be in the **strong coupling** limit when $g_0 \gg (\kappa, \gamma)$, where (κ, γ) represents the larger of κ and γ. Conversely, we have **weak coupling** when $g_0 \ll (\kappa, \gamma)$.

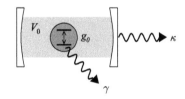

Fig. 10.4 A two-level atom in a resonant cavity with modal volume V_0. The cavity is described by three parameters: g_0, κ, and γ which, respectively quantify the atom–cavity coupling, the photon decay rate from the cavity, and the non-resonant decay rate. Note that the cavity in Fig. 10.4 is drawn with concave mirrors rather than plane ones. If the cavity only had plane mirrors, then off-axis photons emitted by the atom would never re-interact with it. The use of concave mirrors reduces this problem.

In the strong coupling limit, the atom–photon interaction is faster than the irreversible processes due to loss of photons out of the cavity mode. This makes the emission of the photon a *reversible* process in which the photon is re-absorbed by the atom before it is lost from the cavity. In the weak coupling limit, by contrast, the emission of the photon by the atom is an *irreversible* process, as in normal free-space spontaneous emission, but the emission rate is affected by the cavity. To proceed further we therefore need to consider the relative magnitudes of κ, γ, and g_0.

We start with the cavity photon decay rate κ. The photon decay rate was defined in eqn 10.13, and is governed by the properties of the cavity that determine its quality factor Q. This can be seen from eqns 10.14 and 10.15, which show that κ is related to Q by:

$$\kappa = \omega/Q. \tag{10.16}$$

Hence high Q values mean relatively small photon decay rates. In practice, very high Q factors are required before any of the interesting effects described in this chapter are observed.

The non-resonant decay rate γ is determined by several factors. The atom could emit a photon of the resonant frequency in a direction that does not coincide with the cavity mode, for example, sideways, as suggested by Fig. 10.4. Alternatively, the atom could decay to other levels, emitting a photon of a different frequency that is not in resonance with the cavity. Yet again, the atom in the excited state could be scattered to other states and perhaps decay without emission of a photon at all. The first of these processes is a property of the cavity. The second is determined by the internal dynamics of the atom, and represents a breakdown of the two-level atom approximation. The final process is connected with the same sort of scattering events that cause dephasing. (See Section 9.5.2.)

For the case of radiative decay to non-resonant photon modes, we can set γ equal to the transverse dephasing rate:

$$\gamma \equiv 1/T_2 = \gamma_{\parallel}/2, \tag{10.17}$$

where $\gamma_{\parallel}$ is the longitudinal decay rate given by:

$$\gamma_{\parallel} = A_{21}(1 - \Delta\Omega/4\pi), \tag{10.18}$$

A_{21} being the Einstein A coefficient for spontaneous emission into free space, and $\Delta\Omega$ the solid angle subtended by the cavity mode.

This leaves us with the third parameter, namely the atom–cavity coupling rate g_0. In Chapter 9 we studied how two-level atoms interact with resonant light fields originating from external sources such as a lamp or laser. The situation we are considering here is slightly more complicated, because there is no external source to determine the field strength. We therefore have to consider the interaction between the atom and the vacuum field that exists in the cavity due to the zero-point fluctuations of the electromagnetic field. (See Section 7.4.)

See Sections 9.5.2 and 9.6 for explanations of the terms 'transverse' and 'longitudinal' decay rates. The factor of two difference between the rates in eqn 10.17 comes from eqn 9.54 with the pure dephasing rate equal to zero, as appropriate for an isolated atom. The factor of $(1 - \Delta\Omega/4\pi)$ accounts for the fraction of the photons generated by spontaneous emission that are emitted at angles so that they are lost from the cavity.

The interaction energy ΔE between the atom and the cavity vacuum field is set by the electric dipole interaction (cf. eqn 9.25):

$$\Delta E = |\mu_{12}\mathcal{E}_{\text{vac}}|, \tag{10.19}$$

where $\mu_{12} \equiv -e\langle 1|x|2\rangle$ is the electric dipole matrix element of the transition, and $\mathcal{E}_{\text{vac}}$ is the magnitude of the vacuum field as given by eqn 7.37. On setting ΔE equal to $\hbar g_0$ we then find:

$$g_0 = \left(\frac{\mu_{12}^2\omega}{2\epsilon_0\hbar V_0}\right)^{1/2}. \tag{10.20}$$

It is therefore apparent that the atom–photon coupling rate is determined by the dipole moment μ_{12}, the angular frequency ω, and the modal volume V_0.

Equation 10.20 allows us to compare the atom–photon coupling rate directly with the dissipative loss rate, and hence determine whether we are in the strong or weak coupling regime, respectively. If we assume that the cavity loss rate κ is the dominant loss mechanism, we can use eqn 10.16 to see that strong coupling occurs when:

$$g_0 \gg \omega/Q. \tag{10.21}$$

We then find from eqn 10.20 that the condition for strong coupling is:

$$Q \gg \left(\frac{2\epsilon_0\hbar\omega V_0}{\mu_{12}^2}\right)^{1/2}. \tag{10.22}$$

We shall see in Example 10.1 below that this condition is very strict, and requires cavities with very high Q values. In most cases, single atom systems will therefore be in the weak coupling regime, especially when the loss rate to non-resonant modes is significant. The situation improves, however, if we have N atoms in the cavity. The criterion for strong coupling is then given by (cf. eqn 10.49 below):

$$\sqrt{N}g_0 \gg (\kappa, \gamma). \tag{10.23}$$

The factor of $\sqrt{N}$ makes it easier to observe strong coupling.

Example 10.1   An air-spaced symmetric planar cavity of length 60 μm and modal volume 5×10^{-14} m^3 is locked to resonance with a cesium transition at 852 nm which has $|\mu_{12}| = 3 \times 10^{-29}$ C m.

(a) Estimate the smallest values of the cavity Q, the cavity finesse, and the mirror reflectivity required for strong coupling for a single atom.

(b) The radiative lifetime of the transition is equal to 32 ns. Confirm that the atom–cavity coupling is larger than the non-resonant radiative loss rate.

The parameters in this example are based on the experiments described by W. Lange *et al.* in *Microcavities and Photonic Bandgaps*, Ed. J. Rarity and C. Weisbuch, Kluwer Academic Publishers, Dordrecht, 1996, pp. 443–56.

Solution

(a) For strong coupling we require $g_0 \gg \kappa$. We can substitute the values of V_0 and μ_{12} into eqn 10.20 with $\omega = 2\pi c/\lambda = 2.2 \times 10^{15}$ rad s^{-1} to find $g_0 = 1.5 \times 10^8$ rad s^{-1} for this cavity. On using eqn 10.21, we then find $Q \gg 1.5 \times 10^7$. This value of Q implies from eqn 10.15 that the modal angular linewidth must be less than $g_0 = 1.5 \times 10^8$ rad s^{-1}. On substituting this value into eqn 10.10, we then find $\mathcal{F} \gg 1.1 \times 10^5$. Finally, with $\kappa \ll 1.5 \times 10^8$ s^{-1}, we find from eqns 10.12 and 10.13 that $(1-R) \ll 2.9 \times 10^{-5}$. Hence we require mirrors with reflectivities greater than 99.997%.

Although mirrors of such high reflectivities are not readily available, they can nonetheless be obtained from specialist optical coating companies.

(b) We first calculate the longitudinal decay rate from eqn 10.18. Since the length of the cavity is very much larger than the wavelength, we can assume that the solid angle subtended by the cavity mode is small. Hence we can put

$$\gamma_{\parallel} = A_{21} = 1/\tau_{\mathrm{R}} = 3.1 \times 10^7 \text{ s}^{-1}.$$

We then find the non-resonant decay rate from eqn 10.17, giving $\gamma = 1.6 \times 10^7$ s^{-1}. This is an order of magnitude smaller than g_0.

10.3 Weak coupling

10.3.1 Preliminary considerations

In the previous section we saw that the coupling strength between the atom and the cavity can be classified as either strong or weak. In this section we shall investigate the weak coupling limit, leaving strong coupling until Section 10.4.

Weak coupling occurs when the atom–cavity coupling constant g_0 is smaller than the loss rate due to either leakage of photons from the cavity (κ) or decay to non-resonant modes (γ). This means that photons are lost from the atom–cavity system faster than the characteristic interaction time between the atom and the cavity. The emission of light by the atom in the cavity is therefore *irreversible*, just as for emission into free space.

It is important to realize that many of the conclusions of Section 10.3 can be reached by the classical theory of electromagnetism. By treating the atom as an oscillating electric dipole, classical electrodynamics can derive a formula similar to eqn 10.28 for the emission rate into free space and can also explain why the presence of a cavity alters that rate. (See Further Reading.) Moreover, the idea of controlling a radiative transition rate by the environment occurs in other branches of physics, for example, extended X-ray absorption fine structure (EXAFS). On the other hand, most of the results that are presented in Section 10.4 are completely inexplicable in the classical picture, since they depend on the presence of the vacuum field.

Since the effect of the cavity is relatively small in the weak coupling limit, it is appropriate to treat the atom–cavity interaction by perturbation theory. In Section 10.3.2 we shall first use Fermi's golden rule to calculate the emission rate for the atom in free space, and then in Section 10.3.3 we shall calculate the revised rate when the atom is coupled resonantly to a single mode of a high-Q cavity. We shall see that the main effect of the cavity is to enhance or suppress the photon density of states compared to the free-space value, depending on whether the cavity mode is resonant with the atomic transition or not. This then either enhances or suppresses the radiative emission rate through the density of states factor that appears in Fermi's golden rule (see eqn 10.24 below). The spontaneous emission rate from an excited state of an atom is therefore not an absolute number, but can in fact be controlled by suppressing or enhancing the photon density of states by means of a resonant cavity.

10.3.2   Free-space spontaneous emission

Before considering the spontaneous emission of an atom to a single reso- nant cavity mode, it is helpful to remind ourselves of the theory of dipole emission in free space. To do this, it is helpful to consider the properties of an emissive atom in a large cavity of volume V_0. This cavity is consid- ered to be large enough that it has a negligible effect on the properties of the atom, and is merely incorporated to simplify the calculation.

The transition rate for spontaneous emission is given by Fermi's golden rule:

$$W = \frac{2\pi}{\hbar^2}|M_{12}|^2 g(\omega), \tag{10.24}$$

where M_{12} is the transition matrix element and $g(\omega)$ is the density of states. For the density of states we use the standard result for photon modes in free space (see eqn C.11 in Appendix C):

$$g(\omega) = \frac{\omega^2 V_0}{\pi^2 c^3}, \tag{10.25}$$

while for the matrix element we use the electric dipole interaction:

$$M_{12} = \langle \boldsymbol{p} \cdot \boldsymbol{\mathcal{E}} \rangle. \tag{10.26}$$

Since there is no external field source within the cavity, we must use the vacuum field for $\boldsymbol{\mathcal{E}}$. On substituting from eqn 7.37 and averaging over all possible orientations of the atomic dipole with respect to the field direction, we then obtain:

$$M_{12}^2 = \frac{1}{3}\mu_{12}^2 \mathcal{E}_{\text{vac}}^2 = \frac{\mu_{12}^2 \hbar\omega}{6\epsilon_0 V_0}. \tag{10.27}$$

Hence from eqn 10.24 we find the final result:

$$W \equiv \frac{1}{\tau_{\text{R}}} = \frac{\mu_{12}^2 \omega^3}{3\pi\epsilon_0 \hbar c^3}, \tag{10.28}$$

τ_{R} being the radiative lifetime. We thus conclude that the emission rate is proportional to the cube of the frequency and the square of the transition moment.

The result for the radiative emission rate can be related to the treatment based on the Einstein coefficients. (See Section 4.1.) The spontaneous emission rate is given by the Einstein A coefficient:

$$W = A_{21} = \frac{\hbar\omega^3}{\pi^2 c^3}B_{21}^\omega, \tag{10.29}$$

where B_{21}^ω is the Einstein B coefficient derived in Section 9.4. On substituting for B_{21}^ω from eqn 9.43, we obtain the same result as eqn 10.28.

The standard treatment of spontaneous emission based on Fermi's golden rule and the Einstein A coefficient is dis- cussed in more detail in Section 4.2.

10.3.3   Spontaneous emission in a single-mode cavity: the Purcell effect

See E. M. Purcell, *Phys. Rev.* **69**, 681 (1946).

We now come to the main task of this section: to calculate the spontaneous emission rate for a two-level atom coupled to a single-mode resonant cavity in the weak coupling limit. This problem was first considered by E. M. Purcell in 1946, and the resulting change to the emission properties of the atom is now frequently called the **Purcell effect**.

Consider an atom in a single-mode cavity of volume V_0 as shown in Fig. 10.5(a). By 'single-mode' we mean that there is only one resonant mode of the cavity that is close to the emission frequency of the atom. There will of course be other modes in the cavity, but we neglect them in this analysis because they are assumed to be far from resonance. In the weak coupling limit it is possible to use a perturbative approach similar to that for the atom in free space. The emission rate is then given by Fermi's Golden rule as in eqn 10.24.

We assume that the cavity mode has an angular frequency of ω_c with a half width $\Delta\omega_c$ determined by the quality factor Q. The density of states function $g(\omega)$ for the cavity will then take the form shown in Fig. 10.5(b). Since there is only one resonant mode, we must have:

$$\int_0^\infty g(\omega)\,d\omega = 1, \tag{10.30}$$

which is satisfied if we use a normalized Lorentzian function for $g(\omega)$ (cf. eqn 4.29):

$$g(\omega) = \frac{2}{\pi\Delta\omega_c} \frac{\Delta\omega_c^2}{4(\omega-\omega_c)^2 + \Delta\omega_c^2}. \tag{10.31}$$

If the frequency of the atomic transition is ω_0, then we must evaluate eqn 10.31 at ω_0 to obtain:

$$g(\omega_0) = \frac{2}{\pi\Delta\omega_c} \frac{\Delta\omega_c^2}{4(\omega_0-\omega_c)^2 + \Delta\omega_c^2}. \tag{10.32}$$

At exact resonance between the atom and the cavity (i.e. $\omega_0 = \omega_c$), this reduces to:

$$g(\omega_0) = \frac{2}{\pi\Delta\omega_c} = \frac{2Q}{\pi\omega_0}, \tag{10.33}$$

where we have made use of eqn 10.15.

As with the free atom, we use the electric dipole matrix element given in eqn 10.26 and write in analogy to eqn 10.27:

$$M_{12}^2 = \xi^2\mu_{12}^2\mathcal{E}_{vac}^2 = \xi^2\frac{\mu_{12}^2\hbar\omega}{2\epsilon_0 V_0}. \tag{10.34}$$

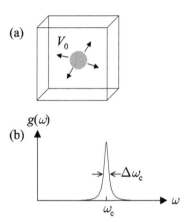

Fig. 10.5 (a) A two-level atom in a single-mode cavity with volume V_0. (b) Density of states function $g(\omega)$ for the cavity. The angular frequency of the cavity mode is ω_c, and $\Delta\omega_c$ is its linewidth.

The factor ξ is the normalized dipole orientation factor defined by:

$$\xi = \frac{|\boldsymbol{p} \cdot \boldsymbol{\mathcal{E}}|}{|\boldsymbol{p}||\boldsymbol{\mathcal{E}}|}. \tag{10.35}$$

We can recall that ξ^2 averaged to $1/3$ for the case of the randomly orientated dipole in free space.

On substituting eqns 10.32 and 10.34 into Fermi's golden rule (eqn 10.24), we obtain:

$$W^{\text{cav}} = \frac{2Q\mu_{12}^2}{\hbar\epsilon_0 V_0}\, \xi^2 \, \frac{\Delta\omega_c^2}{4(\omega_0 - \omega_c)^2 + \Delta\omega_c^2}, \tag{10.36}$$

where we have again made use of eqn 10.15. This rate can be compared to the free-space value given in eqn 10.28. We now introduce the **Purcell factor** F_P defined by:

$$F_P = \frac{W^{\text{cav}}}{W^{\text{free}}} \equiv \frac{\tau_R^{\text{free}}}{\tau_R^{\text{cav}}}. \tag{10.37}$$

On substituting from eqns 10.28 and 10.36 we then find:

$$F_P = \frac{3Q(\lambda/n)^3}{4\pi^2 V_0}\, \xi^2 \, \frac{\Delta\omega_c^2}{4(\omega_0 - \omega_c)^2 + \Delta\omega_c^2}, \tag{10.38}$$

where we have replaced c/ω by $(\lambda/n)/2\pi$, λ being the free-space wavelength of the light and n the refractive index of the medium inside the cavity. At exact resonance and with the dipoles orientated along the field direction, eqn 10.38 reduces to:

$$F_P = \frac{3Q(\lambda/n)^3}{4\pi^2 V_0}. \tag{10.39}$$

This is the main result of the analysis.

The Purcell factor is a convenient parameter that characterizes the effects of the cavity. Purcell factors greater than unity imply that the spontaneous emission rate is enhanced by the cavity, while $F_P < 1$ implies that the cavity inhibits the emission. Equation 10.39 shows that large Purcell factors require high Q cavities with small modal volumes. Furthermore, eqn 10.38 indicates that we need to have close matching between the cavity mode and the atomic transition, and we also need to ensure that the dipole is orientated as near to parallel with the mode field as possible. The enhancement of the emission rate on resonance is related to the relatively large density of states function at the cavity mode frequency. By contrast, the inhibition of emission when the atom is off-resonance is caused by the absence of photon modes into which the atom can emit.

Another useful parameter to describe the effects of the cavity is the **spontaneous emission coupling factor** β. This is the fraction of the number of photons emitted into the cavity mode to the total number of photons emitted. In an ideal cavity, the β-factor would be equal to unity. However, in a realistic cavity, there will still be emission into non-resonant modes and the β-factor will be less than unity.

Consider, for example, the scenario depicted in Fig. 10.6, which shows an atom resonantly coupled to a planar cavity. The atom will emit into the cavity mode at a rate W^{cav} given by eqn 10.36. However, since the direction of spontaneous emission is inherently random, it can also emit into free-space modes. The cavity only affects the density of states for modes in the direction along its axis, and it is therefore reasonable to assume that the density of free-space modes due to emission in all the other directions is largely unaffected by the cavity. If we write this emission rate into free-space modes as W^{free}, then the total emission rate will be equal to $(W^{\text{free}} + W^{\text{cav}})$. The β value is thus given by:

$$\beta = \frac{W^{\text{cav}}}{W^{\text{free}} + W^{\text{cav}}} = \frac{F_{\text{P}}}{1 + F_{\text{P}}}, \tag{10.40}$$

where we have made use of eqn 10.37. We thus conclude that the β-factor only approaches unity for large values of the Purcell factor.

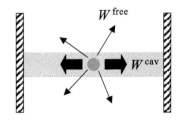

Fig. 10.6 An atom resonantly coupled to a planar cavity can emit both into the cavity mode and also to free-space modes.

This example is based on the experiments reported by Gérard *et al. Phys. Rev. Lett.* **81**, 1110 (1998) and discussed in Section 10.3.4 below.

Example 10.2　A semiconductor quantum dot emits at 900 nm and has a radiative lifetime of 1.3 ns. The dot is placed inside a GaAs microcavity of refractive index 3.5 with modal volume 1×10^{-19} m^{-3} and $Q = 2000$. Calculate the radiative lifetime in the cavity, on the assumption that the dipole moment is parallel to the mode field and that the cavity is exactly on resonance.

Solution
With the dipole parallel to the mode field, we have $\xi = 1$. Furthermore, at exact resonance we have $\omega_0 = \omega_c$. We can thus use eqn 10.39 to calculate the Purcell factor:

$$F_{\text{P}} = \frac{3 \times 2000 \times (9 \times 10^{-7}/3.5)^3}{4\pi^2 \times 10^{-19}} = 26.$$

We can then calculate the lifetime in the cavity using eqn 10.37:

$$\tau_{\text{R}}^{\text{cav}} = \frac{\tau_{\text{R}}^{\text{free}}}{F_{\text{P}}} = \frac{1.3 \text{ ns}}{26} = 0.05 \text{ ns}.$$

The first observations of the Purcell effect at optical frequencies were made in the late 1980s and early 1990s. See D. J. Heinzen *et al.*, *Phys. Rev. Lett.* **58**, 1320 (1987), F. De Martini *et al. Phys. Rev. Lett.* **59**, 2955 (1987), and A. M. Vredenberg *et al.*, *Phys. Rev. Lett.* **71**, 517 (1993). Observations at lower frequencies came earlier. See, for example, P. Goy *et al.*, *Phys. Rev. Lett.* **50**, 1903 (1983).

10.3.4　Experimental demonstrations of the Purcell effect

The observation of lifetime shortening by the Purcell effect at optical frequencies has posed a considerable challenge, since it ideally requires a very high Q cavity with a small modal volume. Some of the clearest observations in recent years have been made with semiconductor **microcavity** structures. These structures are primarily made to provide the wafers for vertical-cavity surface-emitting lasers (VCSELs) for use in fibre optic systems. However, a spin-off of this technology has been the development of monolithic semiconductor structures with extremely small modal volumes ($V_0 \sim (\lambda/n)^3$) which demonstrate strong quantum optical effects. In this way it has been possible to use semiconductor microcavities to demonstrate both weak-coupling effects as discussed here, and also strong coupling: see Section 10.4.2 below.

Figure 10.7 shows a schematic diagram of a generic semiconductor microcavity. The entire structure is grown by techniques of semiconductor epitaxy. The active region usually contains quantum wells or quantum dots (see Appendix D), and the cavity is formed by two distributed Bragg reflector (DBR) mirrors. These mirrors are made by growing alternating layers of semiconductors with different refractive indices to form a highly reflecting quarter-wave stack. The active region is embedded within a spacer layer of thickness λ/n, λ being the vacuum emission wavelength of the active material and n the refractive index of the spacer material. In these conditions the planar cavity between the DBR mirrors supports a resonant mode with field antinodes at the mirrors and at the centre of the cavity. The materials chosen for the DBR mirrors and also the spacer region have larger band gaps than the active material, so that they are transparent at the wavelength of interest.

Figure 10.8 shows results for the Purcell effect observed in InAs quantum dot micropillar structures at 8 K. Figure 10.8(a) gives a schematic diagram of the micropillar samples used for the experiment. The cylindrical micropillars were fabricated by reactive ion etching of a microcavity wafer grown by molecular beam epitaxy. The spacer region was made from GaAs, and the DBR mirrors were made from alternating layers of GaAs and AlAs. Both of these materials are transparent at the emission wavelength of the InAs quantum dots, namely 918 nm. The quantum dots were excited by 1.5 ps pulses at 838 nm from a mode-locked Ti:sapphire laser, and the photoluminescence was detected with a very fast detector called a streak camera.

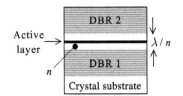

Fig. 10.7 A semiconductor microcavity. The structure consists of an active layer embedded within an inert spacer region of refractive index n, which is itself sandwiched between two DBR mirrors. The thickness of the spacer region is usually chosen to be equal to λ/n, where λ is the emission wavelength of the active material.

Note that the boundary conditions here are different from the usual case with field nodes at the mirrors. This occurs because of the similarity of the spacer region and the mirror materials.

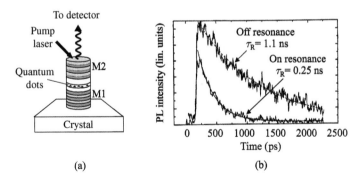

Fig. 10.8 (a) Schematic diagram of a quantum dot micropillar structure. (b) Photoluminescence (PL) decay curves measured for InAs quantum dot micropillar structures with a diameter of 1 μm at 8 K. The two curves compare the decays measured for on-resonance and off-resonance conditions. (After J. M. Gérard *et al.*, *Phys. Rev. Lett.*, **81**, 1110 (1998), ©American Physical Society, reproduced with permission.)

The results presented in Fig. 10.8(b) are for a micropillar with a diameter of 1 μm. The effective modal volume for this structure was estimated to be 1×10^{-19} m^{-3}, and the Q-factor was measured to be 2250, giving a Purcell factor from eqn 10.39 of 32. Two photoluminescence decay curves are shown corresponding to quantum dots of different frequencies that were either in resonance or far out of resonance with the cavity. The decay for the on-resonance dots is more than four times faster than that of the off-resonance dots, which clearly demonstrates the enhancing effect of the cavity. The difference between the measured lifetime reduction and the calculated Purcell factor is caused by inevitable variations in the crystal growth at the microscopic level. This produces an inhomogeneous distribution in the precise position and frequency of the quantum dots, making it impractical to detect the light coming exclusively from optimally coupled dots located exactly at the antinode of the cavity field.

10.4   Strong coupling

10.4.1   Cavity quantum electrodynamics

The conditions for strong coupling were described in Section 10.2. We require that the atom–cavity coupling rate g_0 shall be larger than the cavity decay rate set by the cavity lifetime and also the non-resonant atomic decay rate. In these conditions the interaction between the photons in the cavity mode and the atom is *reversible*. The atom emits a photon into the resonant mode, which then bounces between the mirrors and is re-absorbed by the atom faster than it is lost from the mode. The reversible interaction between the atom and the cavity field is thus faster than the irreversible processes due to loss of photons. This regime of reversible light–atom interactions is called **cavity quantum electrodynamics** (cavity QED).

The interaction between a resonant cavity mode and the atom was first analysed in detail by Jaynes and Cummings in 1963. The **Jaynes–Cummings model** describes the interaction of a two-level atom with a single quantized mode of the radiation field. The workings of the model are beyond the scope of this book and the reader is referred to the bibliography for further details. We summarize here the main conclusions.

We consider first the 'bare' states of an uncoupled resonant system with just a single atom of angular frequency ω in the cavity. The states are labelled by the state of the atom ψ and the number of photons n:

$$\Psi = |\psi; n\rangle. \tag{10.41}$$

The ground state corresponds to the state with the atom in the ground state and no photons in the cavity:

$$\Psi_0 = |g; 0\rangle. \tag{10.42}$$

See E. T. Jaynes and F. W. Cummings, *Proc. IEEE* **51**, 89 (1963).

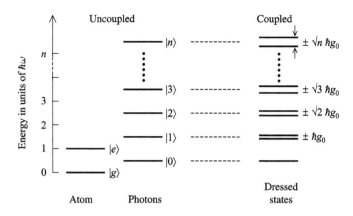

Fig. 10.9 The Jaynes–Cummings ladder. The ladder describes the states of a coupled atom–photon system with a coupling constant of g_0. The states of the uncoupled system are labelled by whether the atom is in the ground state $|g\rangle$ or the excited state $|e\rangle$, and by the number of photons n in the mode.

This has an energy of $(1/2)\hbar\omega$ due to the zero-point energy of the vacuum field in the cavity. The excited states are doubly degenerate. The first excited state is at energy $(3/2)\hbar\omega$, and corresponds to the states with either the atom in the excited state with no photons in the cavity $|e;0\rangle$ or the atom in the ground state with one photon in the cavity $|g;1\rangle$. Similarly, the nth excited state has energy of $(n+1/2)\hbar\omega$ and is derived from the states $|e;n-1\rangle$ and $|g;n\rangle$.

The effect of turning on the atom–cavity interaction is depicted in Fig. 10.9. The electric-dipole interaction between the atom and the photon mixes the degenerate states and lifts the degeneracy. The first excited state now consists of a doublet with energies given by:

$$E_1^{\pm} = (3/2)\hbar\omega \pm \hbar g_0, \tag{10.43}$$

and with corresponding wave functions:

$$\Psi_1^{\pm} = \frac{1}{\sqrt{2}} \left(|g;1\rangle \mp |e;0\rangle \right). \tag{10.44}$$

The nth level consists of a doublet with energies given by

$$E_n^{\pm} = (n+1/2)\hbar\omega \pm \sqrt{n}\hbar g_0, \tag{10.45}$$

and wave functions:

$$\Psi_n^{\pm} = \frac{1}{\sqrt{2}} \left(|g;n\rangle \mp |e;n-1\rangle \right). \tag{10.46}$$

Each level thus consists of a doublet with splitting:

$$\Delta E_n = 2\sqrt{n}\hbar g_0. \tag{10.47}$$

The mixed atom–photon states are called **dressed states**, and the ladder of doublets is called the **Jaynes–Cummings ladder**.

It is informative to compare the dressed states of the Jaynes–Cummings ladder to the dressed states discussed previously when considering Rabi flopping in Section 9.5. The Rabi model considers the resonant interaction between an atom and a *classical* field of high intensity, while the Jaynes–Cummings model considers the same phenomenon for quantized lights field with small photon numbers. Figure 10.9 can be reconciled with Fig. 9.7(b) by equating the Rabi frequency in the latter with $2\sqrt{n}g_0$. The factor of two arises from difference between standing waves in a cavity and travelling waves from a laser beam, while the factor of $\sqrt{n}$ arises from the scaling of the energy ($\propto n$ for large n) with the square of the electric field amplitude. The splittings of the upper and lower levels in Fig. 9.7(b) are the same because we are considering adding or subtracting a photon to the field when n is already large and we can ignore the difference between $\sqrt{n}$ and $\sqrt{n\pm1}$. The interesting point of the cavity system is that there is a splitting even for the first rung of the ladder. This splitting is called the **vacuum Rabi splitting**, and its magnitude is equal to (see eqn 10.20):

$$\Delta E^{\text{vac}} \equiv 2\hbar g_0 = \left(\frac{2\mu_{12}^2\hbar\omega}{\epsilon_0 V_0}\right)^{1/2}. \tag{10.48}$$

The vacuum Rabi splitting can be understood as the AC Stark effect induced by the *vacuum* field.

The Jaynes–Cummings model can be adapted to the case of N atoms interacting with the single-mode field of a resonant cavity. In this case the vacuum Rabi splitting scales as the square root of N:

$$\Delta E^{\text{vac}}(N) = \sqrt{N}\Delta E^{\text{vac}} = \sqrt{N}\left(\frac{2\mu_{12}^2\hbar\omega}{\epsilon_0 V_0}\right)^{1/2}. \tag{10.49}$$

If the medium within the cavity has a relative permittivity of ϵ_r, we should replace ϵ_0 with $\epsilon_r\epsilon_0 \equiv n^2\epsilon_0$, where n is the refractive index, to obtain:

$$\Delta E^{\text{vac}}(N) = \sqrt{N}\left(\frac{2\mu_{12}^2\hbar\omega}{n^2\epsilon_0 V_0}\right)^{1/2}. \tag{10.50}$$

These results are important for understanding the experimental results presented below. It is very challenging to observe the vacuum Rabi splitting from a single atom, and many experiments in fact measure the splitting for multi-atom systems.

The splitting of the modes of the atom–cavity system can be given a quasi-classical explanation by considering the properties of two coupled classical oscillators as shown in Fig. 10.10. If ω_1 and ω_2 are the natural angular frequencies of the uncoupled oscillators, (i.e. the cavity and atom,) and Ω is the coupling strength, then it can be shown that the frequencies of the coupled modes are given by

$$\omega_{\pm} = (\omega_1 + \omega_2)/2 \pm \left(\Omega^2 + (\omega_1 - \omega_2)^2\right)^{1/2}. \tag{10.51}$$

At resonance, with $\omega_1 = \omega_2 \equiv \omega$, this reduces to (see Exercise 10.9):

$$\omega_{\pm} = \omega \pm \Omega, \tag{10.52}$$

which can be compared to the Jaynes–Cummings result with $n = 1$ which gives essentially the same result. Of course, the classical model does not explain why there is a photon oscillator in the cavity in the

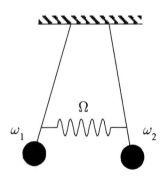

Fig. 10.10 Coupled oscillators. ω_1 and ω_2 are the natural frequencies of the uncoupled oscillators and Ω is the coupling strength.

first place: the vacuum field is a purely quantum effect with no classical analogue. However, if we take the vacuum field for granted, then the vacuum Rabi splitting can be understood as the frequency splitting of the normal modes of the coupled atom–cavity oscillators.

Example 10.3   A cavity of length 3.2 mm and modal volume 1.9×10^{-11} m^3 is tuned to resonance with one of the hyperfine lines of the sodium D$_2$ transition at 589 nm with $\mu_{12} = 2.1 \times 10^{-29}$ C m. Calculate the vacuum Rabi splitting frequency for a cavity containing 200 sodium atoms.

This example is based on the data presented in Fig. 10.11.

Solution
We first calculate the coupling parameter g_0 from eqn 10.20 for angular frequency $\omega_0 = 2\pi c/\lambda = 3.2 \times 10^{15}$ rad s^{-1}. This gives:

$$g_0 = \left(\frac{(2.1 \times 10^{-29})^2 (3.2 \times 10^{15})}{2\hbar\epsilon_0 (1.9 \times 10^{-11})} \right)^{1/2} = 6.3 \times 10^6 \text{ rad s}^{-1}.$$

Then from eqns 10.48 and 10.49 we find:

$$\Delta\Omega^{\text{vac}} = 2\sqrt{N}g_0 = 1.8 \times 10^8 \text{ rad s}^{-1}.$$

In an experiment we measure frequency rather than angular frequency, and we thus expect a vacuum Rabi splitting of 28 MHz.

10.4.2   Experimental observations of strong coupling

The observation of strong coupling requires cavities with small volumes to enhance the coupling constant g_0 and high Q-factors to reduce the photon loss rate. We also require that other dissipative rates due to dephasing and non-resonant emission should be minimized. Finally, we require that the cavity should support only a single mode in resonance with the atom. These requirements are very challenging, and it has taken many years to develop the techniques to achieve the goal of observing the vacuum Rabi splitting predicted by Jaynes–Cummings ladder.

Figure 10.11(a) shows a schematic diagram of an arrangement for observing the vacuum Rabi splitting in atomic physics. The apparatus consisted of a high-finesse resonant cavity through which a beam of sodium atoms was passed. A tunable probe laser was introduced through one of the mirrors and the intensity of the transmitted beam was measured with a detector. The resonant frequency of the cavity was locked to one of the hyperfine lines of the $3S_{1/2} \rightarrow 3P_{3/2}$ transition of the sodium atoms at 589.0 nm and the frequency of the probe laser was scanned through the resonant mode. The intensity of the probe laser was kept small so that the average photon number in the cavity was small. The finesse of the cavity was 26 000 and the cavity length was 3.2 mm. Although the cavity supported many modes, the finesse was sufficiently high that only one of them interacted strongly with the atomic transition.

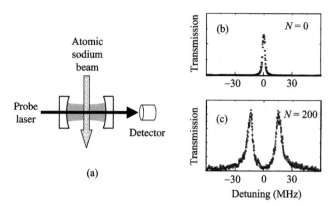

Fig. 10.11 Experimental demonstration of the vacuum Rabi splitting in atomic physics. (a) Experimental arrangement. (b) Transmitted intensity as a function of detuning from the resonant frequency with no atoms in the cavity. (c) Cavity transmission with the atomic beam turned on and the cavity locked to resonance with a transition at 589.0 nm. The average number N of atoms within the resonant mode was 200. (After H. J. Kimble in *Cavity quantum electrodynamics* (ed. P. R. Berman). Academic Press, San Diego, CA (1994), p. 203.)

The observation of the vacuum Rabi splitting for a *single* atom is reported in A. Boca *et al.*, *Phys. Rev. Lett.* **93**, 233603 (2004) and in P. Maunz *et al.*, *Phys. Rev. Lett.* **94**, 033002 (2005).

The results of the experiment are shown in Figs 10.11(b) and (c). Part (b) shows the transmission of the cavity when the atomic beam is switched off. In general, the cavity reflects the probe laser except when its frequency coincides with the resonant mode. We thus observe a single transmission peak at zero detuning. Part (c) shows the transmission with an average number of 200 atoms in the cavity. Two transmission peaks are now observed with their frequency shifted up and down from the empty-cavity value. The magnitude of the splitting was found to scale as the square root of the number of atoms, in agreement with eqn 10.49, and the absolute magnitude was in agreement with the experimental parameters of the cavity and the atoms. (See Example 10.3.) The results therefore clearly demonstrate strong coupling of the atoms to the cavity.

Strong coupling has also been observed in solid-state physics using semiconductor quantum well microcavities. The structures typically contain several semiconductor quantum wells embedded at the centre of the cavity formed between two mirrors as shown in Fig. 10.7. The whole structure is grown as a semiconductor wafer by techniques of advanced epitaxial crystal growth. The length of the cavity is generally chosen to match the wavelength of the strongest exciton line at the band edge of the quantum well. (See Fig. D.2.) Exact resonance is achieved by tuning the frequency of the exciton line to the fixed-length cavity by external parameters such as the temperature or electric field. Since the cavity is only one wavelength long, it supports just a single mode near the exciton line and we do not have to worry about other cavity modes. On the other hand, the presence of thermal excitations in the crystal causes rapid dephasing, and it is usually necessary to work at low temperatures to ensure that the atom–cavity coupling exceeds the broadening due to dephasing.

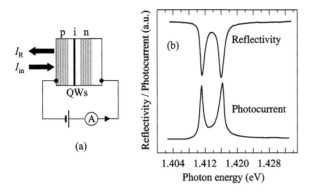

Fig. 10.12 (a) Experimental arrangement for reflectivity and photocurrent measurements on a GaAs/AlGaAs microcavity containing three InGaAs quantum wells (QWs). The top and bottom mirrors were p- and n-type doped respectively to form a p-n junction, with the quantum wells contained in a thin undoped intrinsic (i) region at the centre of the cavity, thus giving a p-i-n structure. I_{in} and I_R represent the incident and reflected light intensities, respectively. (b) Reflectivity and photocurrent spectra at 5 K. (After T. A. Fisher *et al.*, *Solid State Electronics* **40**, 493 (1996), © Elsevier, reproduced with permission.)

Figure 10.12 shows experimental data for a high Q microcavity containing three InGaAs/GaAs quantum wells. Tuning to exact resonance was achieved in this experiment by doping the mirrors to form a p-i-n structure as shown in part (a) of the figure. The quantum wells were undoped and were inserted within the depletion region of the junction, thereby experiencing a strong electric field when reverse bias was applied. Then by varying the bias voltage, the frequency of the exciton lines could be shifted to resonance by the Stark effect. The use of the p-i-n junction also permitted the measurement of the photocurrent generated after absorption of photons in the cavity.

Figure 10.12(b) shows the reflectivity and photocurrent spectra measured at resonance with $T = 5$ K. The reflectivity spectrum should show the opposite behaviour to the transmission, with high reflectivity when the cavity is off-resonance and a dip at the resonant frequency. This is observed when the cavity is out of resonance, but at resonance the mode splits into a doublet due to the vacuum Rabi splitting, as demonstrated in Fig. 10.12(b). The photocurrent spectrum gives a measure of the absorption strength within the cavity and shows complementary behaviour to the reflectivity. The fact that both lines generate a photocurrent demonstrates the mixed atomic–photon character of both components of the Rabi doublet.

Strong coupling has also been reported for *single* quantum dots in microresonator cavities at cryogenic temperatures. See J. P. Reithmaier *et al.*, *Nature* **432**, 197 (2004), T. Yoshie *et al.*, *Nature* **432**, 200 (2004), and E. Peter *et al.*, *Phys. Rev. Lett.* **95**, 067401 (2005).

10.5 Applications of cavity effects

The properties of atoms in cavities have found application in both the weak and strong coupling limits. We mention here two examples, one for each type of coupling. Further examples may be found in the collections of papers cited in the bibliography.

The most common application of cavity effects is in the production of low-threshold lasers. The idea is to employ the control of the spontaneous emission rate that can be achieved in a weakly coupled cavity to improve the performance of the laser medium. An interesting recent development of this basic idea is to employ **photonic crystals** to obtain better control of the spontaneous emission. A photonic crystal is a structure in which the dielectric properties are varied periodically on the length-scale of the wavelength of the light. When the light wavelength matches the period of the structure, Bragg reflection occurs and the light is unable to propagate. This creates a **photonic band gap**, which is analogous to the electronic band gaps formed in crystals when the electron wavelength matches the unit cell size.

The distributed Bragg reflector mirrors found in semiconductor microcavities have a periodic variation of the refractive index along the crystal growth direction, and are thus examples of one-dimensional photonic crystals. It is intuitively obvious that the effects of the cavity can be enhanced further by controlling the spontaneous emission in more than one direction. This requires a two- or three-dimensional periodic structure. The challenge of producing these engineered structures at optical wavelengths is the subject of much current research.

The basic idea of the photonic crystal laser is shown schematically in Fig. 10.13. Figure 10.13(a) gives a schematic diagram of a two-dimensional photonic crystal containing a single emissive defect. The hexagonal pattern of black circles might typically represent holes etched into the surface of a semiconductor wafer, while the defect might be a quantum dot at the position where a hole would normally have been. The periodic structure prevents certain frequencies from propagating, and this creates a photonic band gap in the photon density of states as shown schematically in part (b) of the figure. The density of states

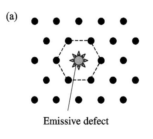
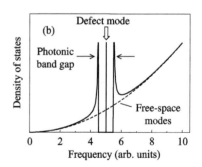

Fig. 10.13 (a) Schematic diagram of a two-dimensional photonic crystal with a hexagonal lattice containing a single emissive defect. The black circles represent air holes etched into the surface of the crystal. (b) Schematic density of states in a photonic crystal (solid) compared to free-space (dashed). The emission frequency of the defect mode is ideally chosen to be at the centre of the photonic band gap.

for free-space modes which follows an ω^2 dependence (cf. eqn 10.25) is shown for comparison.

In a perfect photonic crystal, the density of states drops to zero in the gap, and is enhanced at the edges. The defect creates a new state at the centre of the gap which has a large density of states compared to free space. If this mode coincides with the emission frequency of the defect, we expect enhanced emission properties. Alternatively, we can exploit the enhanced density of states at the edges of the band gap. Many researchers have now demonstrated low threshold lasers by these methods.

Turning now to the strong coupling phenomena, one of the most exciting applications is in the development of single-photon phase gates for use in quantum computation. The idea here is to create an atom–cavity system in which the addition of one atom or one photon produces a measurable alteration to the properties of the system. The key parameters are the **critical atom number** N_0 and **critical photon number** n_0, defined, respectively, by:

$$N_0 = 2\kappa\gamma/g_0^2$$
$$n_0 = 4\gamma^2/3g_0^2. \qquad (10.53)$$

The observation of a single-photon phase gate requires $(n_0, N_0) \ll 1$. In these conditions the addition of a single photon makes a measurable difference to the system.

Figure 10.14 shows a schematic diagram of the experimental arrangement used to demonstrate a single-photon phase gate using cesium atoms. The apparatus comprises of an ultra-high-finesse cavity tuned to resonance with one of the atomic transitions of the cesium atom. The flux of the atomic beam was adjusted so that on average there was only ever one atom in the cavity at a time. The cavity was interrogated with two beams called the control and probe beams, respectively. The frequencies of these two beams were shifted slightly from each other to allow them to be distinguished experimentally. The experiment consisted in measuring the rotation of the polarization of the probe beam induced by a single photon in the control beam. Experimental rotation angles of around 15° were found conditional on the presence of the control beam. The importance of single-photon phase gates is that they can be used as conditional quantum logic gates in a quantum computer. This point is developed further in Chapter 13.

Further reading

The optical properties of planar cavities are considered in many optics texts, for example: Brooker (2003), Hecht (2002), or Yariv (1997).

A comprehensive collection of articles on cavity quantum electrodynamics can be found in Berman (1994). The chapter by Hinds in that

Details of a semiconductor laser operating on a two-dimensional photonic-band-gap defect mode may be found in Painter *et al.*, *Science* **284**, 1819 (1999), while Kopp *et al.* describe a dye-doped liquid crystal laser which exploits the enhanced density of states at the edges of a one-dimensional photonic band gap in *Opt. Lett.* **23**, 1707 (1998).

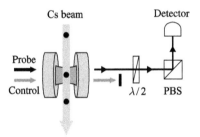

Fig. 10.14 Schematic diagram of a conditional phase gate using a single cesium atom. The polarization state of the transmitted probe beam was measured by using a half wave plate ($\lambda/2$) and a polarizing beam splitter (PBS). This enabled the polarization rotation angle induced by the control beam to be determined. (Adapted from Q. A. Turchette *et al.*, *Phys. Rev. Lett.* **75**, 4710 (1995).)

volume gives a very clear comparison of the classical and quantum-mechanical treatments of radiative emission in free space and cavities, while the chapter by Kimble gives a thorough review on work to observe strong coupling phenomena in atomic physics up to 1994. Rempe (1993) gives a more introductory treatment of the subject, while Kimble (1998) reviews work on single atom strong coupling. Vahala (2003) gives a clear overview of cavity effects in both atomic and solid state physics, while further details about experiments on quantum dots may be found in Michler (2003). A review of the present state-of-the-art for cavity QED experiments in atomic physics may be found in Miller *et al.* (2005).

The Jaynes–Cummings model is described in many theoretical quantum optics texts, for example: Gerry and Knight (2005), Meystre and Sargent (1999), or Yamamoto and Imamoglu (1999). A tutorial review may be found in Shore and Knight (1993).

Collections of review papers on atomic and solid state cavities, and also on photonic structures, may be found in Rarity and Weisbuch (1996) or Benisty *et al.* (1999). A tutorial review on confined electron and photon systems is given by Weisbuch *et al.* (2000), while more specific details about strong coupling in semiconductor microcavities are given by Skolnick *et al.* (1998).

Detailed treatments of photonic crystals made be found, for example, in Joannopoulos *et al.* (1995,1997) or Ozbay *et al.* (2004). Woldeyohannes and John (2003) give a tutorial review of the control of spontaneous emission in photonic materials.

Exercises

(10.1) Show that the quality factor of a planar cavity of finesse $\mathcal{F}$ is equal to $m\mathcal{F}$, where m is the mode number defined in eqn 10.8.

(10.2) By considering the field at an antinode, show that the optical intensity of a resonant mode inside a high-finesse symmetric cavity with mirror reflectivities of R is enhanced by a factor $4/(1-R)$ compared to the incoming intensity. Explain also why the average intensity within the cavity at off-resonant frequencies is diminished by a factor $(1-R)$.

(10.3) Calculate the photon lifetime for an air-spaced symmetric cavity with $R = 99\%$ and $L = 1$ mm.

(10.4) An air-spaced cavity designed for 589 nm has mirrors of reflectivity 99.9% and a length of 1 cm. Calculate the finesse and the Q-factor of the cavity.

(10.5) An air-spaced symmetric planar cavity of length 350 μm has a modal volume of 5×10^{-13} m³.

Estimate the smallest allowed values of (a) the cavity Q, (b) the cavity finesse, and (c) the mirror reflectivity to achieve strong coupling for a transition of a single cesium atom at 852 nm with $\mu_{12} = 3 \times 10^{-29}$ C m.

(10.6) Calculate the maximum mirror separation for a planar cavity that has just a single mode within the emission band of a fluorescent dye with peak emission at 600 nm and emission bandwidth of 40 nm. Assume that the cavity is filled with the dye and that the dye has a refractive index of 1.4.

(10.7) The semiconductor microcavity structures described in this chapter incorporate quarter-wave stack mirrors. (See Fig. 10.7.) Explain why a planar structure consisting of alternating layers of materials with refractive indices of n_1 and n_2 has a high reflectivity for light of air wavelength λ when the material thicknesses are chosen to be $\lambda/4n_1$ and $\lambda/4n_2$, respectively.

(10.8) A quantum dot emitting at 930 nm is placed at the centre of a resonant micropillar containing material of refractive index 3.5. The modal volume is 1.8×10^{-18} m^3 and the spectral width of the resonant mode is 0.18 nm. Calculate the Purcell factor.

(10.9) The equations of motion of two coupled classical oscillators of angular frequency ω may be written:

$$\ddot{x}_1 = -\omega^2 x_1 + 2\omega\Omega x_2,$$
$$\ddot{x}_2 = -\omega^2 x_2 + 2\omega\Omega x_1,$$

where x_1 and x_2 are the displacements of the oscillators, and Ω is the coupling frequency. Find the normal modes and frequencies of the coupled system, on the assumption $\Omega/\omega \ll 1$.

(10.10) Calculate the vacuum Rabi splitting frequency for the cavity in Exercise 10.5 with 100 atoms in the cavity mode.

(10.11) Consider the cavity described in Example 10.3, for which experimental data is presented in Fig. 10.11.

(a) Use the experimental data to estimate the photon lifetime for the cavity.

(b) Estimate the minimum number of atoms that have to be in the cavity to resolve the vacuum Rabi splitting, given that the radiative lifetime of the upper level is 16 ns.

(10.12) (a) The oscillator strength f_{ij} of an atomic transition $i \rightarrow j$ is defined by

$$f_{ij} = \frac{2m_0\omega_{ij}|\mu_{ij}|^2}{e^2\hbar}.$$

Use this definition to show that the vacuum Rabi splitting of a semiconductor microcavity of length L_{cav} containing N_Q quantum well layers at the centre of the cavity is given by:

$$\Delta E^{\text{vac}} = \left(\frac{N_Q e^2 \hbar^2 f_{\text{area}}}{\epsilon_0 m_0 n^2 L_{\text{cav}}} \right)^{1/2},$$

where f_{area} is the oscillator strength per unit area of each quantum well layer and n is the refractive index of the semiconductor.

(b) Evaluate the vacuum Rabi splitting for a resonant microcavity of length 700 nm and refractive index 3.5 containing five quantum wells each with $f_{\text{area}} = 6 \times 10^{16}$ m^{-2}.

(10.13) A cavity is made to the design of the cavity in Exercise 10.5 with a finesse that gives $\kappa/2\pi = 0.6$ MHz. Calculate the critical atom number and photon number for this cavity, given that the free-space radiative lifetime for the transition is 32 ns.

11 Cold atoms

11.1 Introduction   216

11.2 Laser cooling   218

11.3 Bose–Einstein
     condensation   230

11.4 Atom lasers   236

Further reading   238

Exercises   238

In the previous two chapters we have investigated how resonant light beams interact with atomic transitions, and the way this process can be modified by means of cavities. We now wish to explore a different aspect of the light–matter interaction, namely light-induced forces. As we shall see, these forces have been employed to great effect in the techniques of laser cooling, which are now routinely used by many research teams around the world to generate temperatures in the microkelvin range. A great triumph of this work, together with the additional techniques of atom trapping and evaporative cooling, has been the observation of Bose–Einstein condensation in an atomic gas in 1995.

After introducing the basic concepts, our discussion of these subjects will begin with a description of the techniques for laser cooling and atom trapping, and the factors that determine the temperatures that are achieved. We shall then study how the atoms are cooled further to reach the conditions where Bose–Einstein condensation is possible, and conclude with a brief discussion of the subject of atom lasers.

11.1 Introduction

A degree of freedom is defined as a term that is proportional to the square of a coordinate or its first time derivative in the Hamiltonian. Typical examples include the translational motion ($E = m\dot{x}^2/2$), the rotational motion ($E = I_{\mathrm{rot}}\dot{\theta}^2/2$), or the vibrational potential energy ($E = kx^2/2$).

The classical principle of **equipartition of energy** states that the thermal energy per particle per **degree of freedom** at a temperature T is given by:

$$E = \frac{1}{2}k_{\mathrm{B}}T. \tag{11.1}$$

In a gas of non-interacting atoms, the only degrees of freedom that we need to consider are those for the free translational motion. We can therefore set the kinetic energy for each velocity component equal to $k_{\mathrm{B}}T/2$, and then find the r.m.s. thermal velocity from:

$$\frac{1}{2}mv_x^2 = \frac{1}{2}k_{\mathrm{B}}T, \tag{11.2}$$

The mean speeds quoted here are the root-mean-square (r.m.s.) values for a gas. The r.m.s. speed of the atoms in a collimated one-dimensional **atomic beam** is larger than that given in eqn 11.3 by a factor of two. See Example 11.1.

which implies:

$$v_x^{\mathrm{rms}} = \sqrt{\frac{k_{\mathrm{B}}T}{m}}. \tag{11.3}$$

In three dimensions, this generalizes to:

$$v^{\mathrm{rms}} = \sqrt{\frac{3k_{\mathrm{B}}T}{m}}. \tag{11.4}$$

For sodium gas with $m = 23\,m_H$, we find $v_x^{rms} = 330$ m s^{-1} and $v^{rms} = 570$ m s^{-1} at 300 K.

Equations 11.3 and 11.4 indicate that there is a one-to-one relationship between the temperature of a gas and the r.m.s. speeds of the particles that comprise it. This relationship provides a method for determining the temperature from measurements of the average speed. It is also at the basis of **laser cooling**, which uses the mechanical force between a laser beam and the moving atoms in a gas to slow them down and hence to produce very low temperatures. The temperatures that are now routinely achieved by laser cooling are in the microkelvin range, which corresponds to atomic speeds that are about four orders of magnitude smaller than at room temperature.

One of the important points to realize about laser cooling is that it is not the temperature alone that we are interested in: the particle density is also a key parameter. Techniques for achieving very low temperatures have been used for decades by condensed-matter physicists, but the novelty of laser cooling is that it produces of an ultracold *gas*, in contrast to the condensed-matter techniques which all work on liquids or solids. The atoms in the ultracold gas interact only weakly with each other, which makes it possible to observe low-temperature quantum effects in a nearly ideal system.

The most striking effect that has been observed through laser cooling is **Bose–Einstein condensation**, which is a quantum phase transition shown by boson particles at very low temperatures. Einstein discovered the transition in 1924, when he wrote:

From a certain temperature on, the molecules 'condense' without attractive forces, that is, they accumulate at zero velocity. The theory is pretty, but is there some truth to it?

For many years, the subject of Bose–Einstein condensation was mainly the preserve of low-temperature condensed-matter physicists. However, with the development of laser cooling techniques in the 1980s, the possibility of observing the same effect in dilute atomic gases became a realistic goal. It took some years to perfect the techniques, and in fact it transpires that laser cooling alone is unable to produce the effect. The breakthrough eventually came in 1995 when two research groups independently reported the observation of Bose–Einstein condensation in atomic gases at nanokelvin temperatures using the additional technique of evaporative cooling. Since then, the subject has blossomed, and has led to the development of atom lasers that allow the possibility for coherent atom optics.

The importance of laser cooling and Bose–Einstein condensation has been recognized by the awarding of two Nobel Prizes for Physics within four years. Stephen Chu, Claude Cohen-Tannoudji, and William D. Phillips received the Prize in 1997 for their work to develop 'methods to cool and trap atoms with laser light', while Eric A. Cornell, Wolfgang Ketterle, and Carl E. Wieman received the Prize in 2001 for 'the achievement of Bose–Einstein condensation in dilute gases of

The original proposal for laser cooling of neutral atoms was given by T. W. Hänsch and A. L. Schawlow, *Opt. Commun.* **13**, 68 (1975).

Dilution refrigerators routinely achieve temperatures in the millikelvin range, and nuclear spin temperatures in the microkelvin range were first achieved in the 1950s by using adiabatic demagnetization.

The quotation is taken from Einstein's letter to P. Ehrenfest of 29 November, 1924. An historical discussion of how Einstein built on the previous work of Satyendra Bose may be found in Pais (1982). The first successful application of Einstein's theory came in 1938, when Fritz London interpreted the superfluid transition in liquid helium as a Bose–Einstein condensation phenomenon.

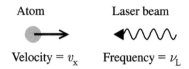

Atom   Laser beam

Velocity = v_x Frequency = ν_L

Fig. 11.1 In Doppler cooling, the laser frequency is tuned below the atomic resonance by δ. The frequency seen by an atom moving towards the laser is Doppler-shifted up by $\nu_0(v_x/c)$.

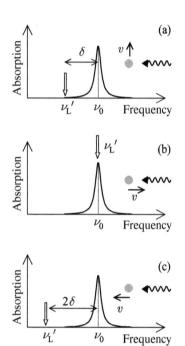

Fig. 11.2 Doppler-shifted laser frequency ν'_L in the rest frame of an atom moving with speed v. When the laser frequency is tuned below ν_0 by v/λ, the Doppler effect shifts the laser into resonance with the atoms if they are moving towards the laser (b), but not if they are moving sideways (a), or away (c) from the laser.

alkali atoms, and for early fundamental studies of the properties of the condensates'.

11.2   Laser cooling

The idea of using a laser to *cool* a gas of atoms is, at first sight, rather surprising: we would normally expect a powerful laser to produce a heating rather than a cooling effect. In fact, the technique only works in a very restricted range of conditions with the laser frequency close to resonance with an atomic transition. In the subsections that follow, we shall study the basic principles of laser cooling, the factors that determine the temperatures that are achieved, and the way in which the experiments are done.

11.2.1   Basic principles of Doppler cooling

The basic principles of laser cooling can be understood with fairly simple arguments that give the correct order of magnitude for the important parameters of the process. The more detailed analysis given in the next subsection reproduces the same basic results but with the numerical factors correctly evaluated.

Let us consider an atom moving in the $+x$-direction with velocity v_x as shown in Fig. 11.1. We assume that the atom interacts with a counter-propagating laser beam with its frequency $\nu_L \equiv c/\lambda$ tuned to near resonance with one of the transitions of the atom. We can then write:

$$\nu_L = \nu_0 + \delta, \tag{11.5}$$

where ν_0 is the atomic transition frequency and $\delta \ll \nu_0$. In the rest frame of the atom, the laser source is moving towards the atom, and its frequency is therefore shifted up by the Doppler effect. The Doppler-shifted frequency is given by:

$$\nu'_L = \nu_L \left(1 + \frac{v_x}{c}\right) = (\nu_0 + \delta)\left(1 + \frac{v_x}{c}\right) \approx \nu_0 + \delta + \frac{v_x}{c}\nu_0, \tag{11.6}$$

where we assumed $v_x \ll c$. It is then apparent that if we choose

$$\delta = -\nu_0\frac{v_x}{c} = -\frac{v_x}{\lambda}, \tag{11.7}$$

we find $\nu'_L = \nu_0$. When this condition is satisfied, the laser will be in resonance with atoms moving in the $+x$-direction, but not with those moving away or obliquely, as depicted schematically in Fig. 11.2.

Now consider what happens after the atom has absorbed a photon from the laser beam. The atom goes into an excited state and then re-emits another photon of the same frequency by spontaneous emission in a random direction. This absorption–emission cycle is illustrated schematically in Fig. 11.3. Each time the cycle is repeated, there is a net change in the momentum of the atom of Δp_x in the x-direction, where:

$$\Delta p_x = -\frac{h}{\lambda}. \tag{11.8}$$

This follows from applying conservation of momentum to both the absorption and emission processes, with the photon momentum equal to h/λ. The momentum change on absorption is always in the $-x$-direction, but the recoil after spontaneous emission averages to zero, because the photons are emitted in random directions.

Equation 11.8 implies that repeated absorption–emission cycles generate a net frictional force in the $-x$-direction. If the laser intensity is large, the probability for absorption will be large, and the time to complete the absorption–emission cycle will be determined by the radiative lifetime τ. The frictional force exerted on the atom is then given by:

$$F_x = \frac{dp_x}{dt} \approx \frac{\Delta p_x}{2\tau} = -\frac{h}{2\lambda\tau}, \qquad (11.9)$$

which corresponds to a deceleration given by

$$\dot{v}_x = \frac{F_x}{m} \approx -\frac{h}{2m\lambda\tau}. \qquad (11.10)$$

The factor of two in the denominator of eqn 11.9 arises from the fact that, at high laser intensities, the population of the upper and lower levels equalize at a value close to $N_0/2$, where N_0 is the total number of atoms. When the atom is in the excited state (step 2 in Fig. 11.3), it can be triggered to undergo stimulated emission by other impinging laser photons. The stimulated photon is emitted in the same direction as the incident photon, and the photon recoil exactly cancels the momentum change due to absorption. This reduces the force in proportion to the number of atoms in the excited state. (See discussion of eqn 11.20.)

The number of cycles required to halt the atoms is given by:

$$N_{stop} = \frac{mu_x}{|\Delta p_x|} = \frac{mu_x\lambda}{h}, \qquad (11.11)$$

where u_x is the initial velocity. (It is, of course, impossible to completely stop the atoms, and we are simply calculating here the conditions required to reduce the velocity to its minimum value, which is assumed to be very much less than u_x.) The minimum time for the laser to slow the atoms is given by:

$$t_{min} \approx N_{stop} \times 2\tau = \frac{2mu_x\lambda\tau}{h}. \qquad (11.12)$$

The distance travelled by the atoms in this time is given by:

$$d_{min} = -\frac{u_x^2}{2\dot{v}_x} \approx \frac{m\lambda\tau u_x^2}{h}. \qquad (11.13)$$

Typical values of the quantities calculated in eqns 11.9–11.13 are given in Example 11.1.

The Doppler cooling process stops working when the detuning required for cooling becomes comparable to the natural width $\Delta\nu$ of the transition line. In these conditions, the thermal energy of the atom will be roughly equal to $h\Delta\nu$, and therefore the minimum temperature will be given by

$$k_B T_{min} \sim h\Delta\nu. \qquad (11.14)$$

The existence of a light-induced mechanical force on an atom was first demonstrated by Frisch in 1933 by measuring the deflection of a sodium beam by light from a sodium lamp. See R. Frisch, *Z. Phys.* **86**, 42 (1933).

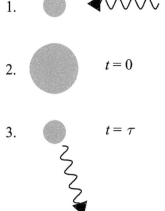

Fig. 11.3 An absorption–emission cycle. (1) A laser photon impinges on the atom. (2) The atom absorbs the photon and goes into an excited state. (3) After an average time equal to the radiative lifetime τ, the atom re-emits a photon in a random direction by spontaneous emission.

On recalling the relationship between the natural line width of the transition and its radiative lifetime (cf. eqn 4.30), we then find:

$$T_{\min} \sim \frac{\hbar}{k_B \tau}. \tag{11.15}$$

This shows that the minimum temperature that can be achieved by the Doppler cooling mechanism is limited by the lifetime of the transition. The rigorous result for the minimum temperature given in eqn 11.37 differs only by a factor of two from eqn. 11.15.

The experimental implementation of the laser cooling process is complicated by the fact that the value of δ required to produce efficient cooling changes as the atoms slow down. In Section 11.2.5 we shall see how a carefully designed magnet can be used to maintain the cooling condition during the slowing process. Alternatively, the frequency of a tunable laser can be scanned in a programmed way to compensate for the deceleration of the atoms. This latter technique is called **chirp cooling** in analogy to the chirping sound made when an audio frequency is rapidly increased, for example, in bird song. Typical tunable lasers used for chirp cooling include dye lasers, Ti : sapphire lasers, and semiconductor diode lasers. In the first two cases the frequency is tuned by scanning an intracavity Fabry–Perot etalon, while the diode lasers can be tuned by varying the temperature of the semiconductor chip.

Example 11.1   A collimated beam of sodium atoms is emitted in the $+x$-direction from an oven at $600\,°C$ and interacts with a counter-propagating laser beam tuned to near resonance with the D_2 line at 589 nm, which has a radiative lifetime of 16 ns. Estimate:

(a) the r.m.s. velocity and most probable velocity of the atoms in the beam as they leave the oven;

(b) the initial detuning required for efficient laser cooling;

(c) the frictional force exerted on the atoms by the laser and the deceleration it produces;

(d) the number of absorption–emission cycles required to bring the atoms to a near halt;

(e) the distance travelled during the laser cooling process.

Solution

(a) The velocity distribution of the atoms within the oven is given by the Maxwell–Boltzmann distribution (eqn 4.33), for which the r.m.s velocity is given by eqn 11.4 and the most probable velocity is given by:

The derivations of eqns 11.4 and 11.16–11.18 may be found, for example, in N. F. Ramsey, *Molecular beams*, Clarendon Press Oxford (1956).

$$v_{\mathrm{mp}} = \sqrt{\frac{2k_B T}{m}}. \tag{11.16}$$

However, the velocity distribution within a collimated atomic beam is different because the atomic flux is proportional to the velocity of the atoms. The r.m.s velocity in the beam is given by:

$$v_{\text{rms}}^{\text{beam}} = \sqrt{\frac{4k_{\text{B}}T}{m}}, \qquad (11.17)$$

while the most probable velocity is given by:

$$v_{\text{mp}}^{\text{beam}} = \sqrt{\frac{3k_{\text{B}}T}{m}}. \qquad (11.18)$$

With $T = 873$ K and $m = 23\, m_{\text{H}}$, we find $v_{\text{rms}}^{\text{beam}} = 1120$ m s^{-1} and $v_{\text{mp}}^{\text{beam}} = 970$ m s^{-1}

(b) The laser detuning required to cool an atom with velocity v_x is given by eqn 11.7. To instigate efficient cooling we need to tune the laser to the appropriate frequency for the most probable velocity in the beam (i.e. 970 m s^{-1}). This gives $\delta = -1.6$ GHz.

(c) The frictional force is given by eqn 11.9 and the deceleration by eqn 11.10. With $\lambda = 589$ nm and $\tau = 16$ ns, we find $F_x \approx -3.5 \times 10^{-20}$ N and $\dot{v}_x \approx -9.1 \times 10^5$ ms^{-2}.

(d) The number of cycles is given by eqn 11.11 with u_x set by the most probable initial velocity within the beam, namely 970 m s^{-1} (cf. part(a)). This gives $N_{\text{stop}} = 3.3 \times 10^4$.

(e) The distance travelled is given by eqn 11.13. On setting $u_x = 970$ m s^{-1}, we find $d_{\text{min}} \approx 51$ cm.

11.2.2 Optical molasses

The results derived in eqns 11.9–11.13 should be considered only as order of magnitude estimations because a number of important processes have been neglected. In this subsection we shall reconsider the cooling process in more detail and derive a value for the limiting temperature that can be achieved.

Let us first consider a laser beam of optical intensity I and detuning $\Delta \equiv 2\pi\delta$ in angular frequency units interacting with an atom of velocity $+v_x$ with respect to the laser. As in eqn 11.9, the frictional force F_x is equal to the momentum change per absorption–emission cycle multiplied by the net rate of such cycles:

$$F_x = -\hbar k \times R(I, \Delta), \qquad (11.19)$$

where $k \equiv 2\pi/\lambda$ is the photon wave vector, and $R(I, \Delta)$ is the *net* absorption rate. $R(I, \Delta)$ is equal to the absorption rate minus the stimulated emission rate, and is given by:

$$R(I, \Delta) = \frac{\gamma}{2} \left(\frac{I/I_s}{1 + I/I_s + [2(\Delta + kv_x)/\gamma]^2} \right), \qquad (11.20)$$

where $\gamma \equiv 1/\tau$ is the natural linewidth in angular frequency units (cf. eqn 4.30), and I_s is the **saturation intensity** of the transition. It is apparent that at very high intensities the net absorption rate limits at $\gamma/2$, which, with $\gamma \equiv 1/\tau$, explains the factor of two in the denominator of eqn 11.9.

The analysis of the cooling process given here roughly follows the paper entitled 'Optical Molasses' by P. D. Lett *et al.* in *J. Opt. Soc. Am. B* **6**, 2084 (1989). The derivation of eqn 11.20 may be found, for example, in Foot (2005) or Shen (1984).

We can understand the general form of eqn 11.20 by first noting that, at low intensities, we can neglect the term in I in the denominator to find that the absorption rate is linearly proportional to the laser intensity as expected. In this low-intensity limit, the frequency dependence is then simply given by a Lorentzian shape (cf. eqn 4.29) with the frequency shift $(\Delta + kv_x)$ equal to the Doppler-shifted laser detuning in the rest frame of the atom. The need for the term in (I/I_s) in the denominator becomes most clearly apparent from considering the behaviour at high intensities at the line centre (i.e. with $(\Delta + kv_x) = 0$). An analysis of the transition rates using the Einstein coefficients quickly establishes the functional form of eqn 11.20. (See Exercise 11.4.)

The arrangement with a single laser beam shown in Fig. 11.1 works well when the laser detuning Δ is much larger than the linewidth. However, as the atoms cool down, it will eventually be the case that the value of Δ required for cooling becomes comparable to the linewidth γ. In these conditions, the atoms moving in the $-x$-direction will experience an accelerating force, and will reheat those moving in the $+x$-direction by collisions. To achieve very low temperatures we therefore need two laser beams as shown in Fig. 11.4. In this situation, the atom experiences separate forces from each laser, giving a net force of:

Fig. 11.4 Two counter-propagating lasers are used to produce the optical molasses cooling effect.

$$F_x = F_+ + F_-, \tag{11.21}$$

where $F_\pm$ refers to the force from the laser beam propagating in the $\pm x$ direction, respectively. When the laser is tuned to the cooling condition given in eqn 11.7, $F_- \gg F_+$ for atoms moving in the $+x$ direction at high temperatures where $k|v_x| \gg \gamma$, and vice versa for those moving in the opposite direction. The two-beam arrangement is therefore able to cool atoms moving in both directions. However, when the atoms get very cold, so that $|v_x|$ is small, we have to analyse the net force more carefully. In the low-temperature limit where $|kv_x| \ll \Delta$, and $|kv_x| \ll \gamma$, the resultant force is given by (see Exercise (11.5)):

$$F_x(I, \Delta) = \frac{8\hbar k^2 \Delta}{\gamma} \left(\frac{I/I_\mathrm{s}}{[1 + I/I_\mathrm{s} + (2\Delta/\gamma)^2]^2} \right) v_x. \tag{11.22}$$

Irrespective of the direction of v_x, the force is of the form:

$$F_x = -\alpha v_x, \tag{11.23}$$

where α is the **damping coefficient**, given by:

'Molasses' is the name given to the thick dark syrup drained from raw sugar during the refining processes. In the United States the word is also used for 'treacle', and it gives a good description of how the Doppler cooling force acts like a viscous medium for the trapped atoms.

$$\alpha = -\frac{8\hbar k^2 \Delta}{\gamma} \left(\frac{I/I_\mathrm{s}}{[1 + I/I_\mathrm{s} + (2\Delta/\gamma)^2]^2} \right). \tag{11.24}$$

When Δ is negative, α is positive, and the motion of the atom is damped in both directions. For this reason, the arrangement with two counter-propagating beams is called the **optical molasses**. At low intensities, the damping force is largest when $\Delta = -\gamma/\sqrt{12}$, but this is not the frequency at which the lowest temperature is achieved, as we shall show below.

The limit to the temperature that is achieved is set by balancing the cooling effect of the damping force with the heating effect associated with the repeated absorption and emission of photons. The cooling rate is given by:

$$\left(\frac{\mathrm{d}E}{\mathrm{d}t}\right)_{\text{cool}} = F_x v_x = -\alpha v_x^2, \tag{11.25}$$

while the heating rate is given by (see eqn 11.34 below):

$$\left(\frac{\mathrm{d}E}{\mathrm{d}t}\right)_{\text{heat}} = \frac{D_p}{m}, \tag{11.26}$$

where D_p is the **momentum diffusion constant** defined in eqn 11.33. On setting the total change of energy equal to zero, we find:

$$-\alpha v_x^2 + \frac{D_p}{m} = 0, \tag{11.27}$$

which implies:

$$v_x^2 = \frac{D_p}{m\alpha}. \tag{11.28}$$

The temperature is then given by eqn 11.2 as:

$$\frac{1}{2}k_{\mathrm{B}}T = \frac{1}{2}mv_x^2 = \frac{D_p}{2\alpha}. \tag{11.29}$$

We therefore obtain:

$$T = \frac{D_p}{\alpha k_{\mathrm{B}}}. \tag{11.30}$$

It thus emerges that the limiting temperature is achieved by minimizing the ratio of D_p to α.

The momentum diffusion introduced into eqn 11.26 is associated with the fact that, even though the damping force reduces the average velocity to zero, the mean squared velocity is not zero. During each absorption–emission cycle, the atom absorbs and emits a photon with momentum $\hbar k$. An atom with zero mean velocity is equally likely to absorb a photon from the positive or negative travelling laser beams, and also to emit in either direction. The atom therefore performs a **random walk** in the x-direction, jolting backwards and forwards each time a photon is absorbed or emitted. If the random walk has N steps, where N is a large number, then the average value of the momentum will be zero, but the average of the square will be given by:

$$\langle p_x^2 \rangle = 2N(\hbar k)^2. \tag{11.31}$$

On counting the interactions with both laser beams, we then have $N = 2Rt$ in time t, so that:

$$\frac{\mathrm{d}\langle p_x^2 \rangle}{\mathrm{d}t} = 4\hbar^2 k^2 R. \tag{11.32}$$

The momentum diffusion due to the random walk is similar to the diffusion of molecules in Brownian motion. The linear increase of $\langle p_x^2 \rangle$ with the number of steps is reminiscent of a Poissonian process: see eqn A.10 in Appendix A. The extra factor of two in eqn 11.31 arises from the one-dimensional nature of the problem.

The momentum diffusion coefficient D_p is defined by:

$$D_p = \frac{1}{2}\frac{\mathrm{d}\langle p_x^2\rangle}{\mathrm{d}t}. \tag{11.33}$$

The heating rate is then given by:

$$\left(\frac{\mathrm{d}E}{\mathrm{d}t}\right)_{\mathrm{heat}} = \frac{1}{2m}\frac{\mathrm{d}\langle p_x^2\rangle}{\mathrm{d}t} = \frac{D_p}{m} = \frac{2\hbar^2 k^2 R}{m}. \tag{11.34}$$

On substituting for R from eqn 11.20 in the limit where $|kv_x| \ll |\Delta|$, we then find:

$$D_p = \hbar^2 k^2 \gamma \left(\frac{I/I_{\mathrm{s}}}{1 + I/I_{\mathrm{s}} + (2\Delta/\gamma)^2}\right). \tag{11.35}$$

We finally substitute eqns 11.24 and 11.35 into eqn 11.30 to obtain:

$$T = -\frac{\hbar\gamma}{8k_{\mathrm{B}}}\frac{(1 + I/I_{\mathrm{s}} + 4\Delta^2/\gamma^2)}{\Delta/\gamma}. \tag{11.36}$$

In the low-intensity limit with $I \ll I_{\mathrm{s}}$, the minimum temperature is given by:

$$T_{\mathrm{min}} = \frac{\hbar\gamma}{2k_{\mathrm{B}}} \equiv \frac{\hbar}{2k_{\mathrm{B}}\tau}, \tag{11.37}$$

at $\Delta = -\gamma/2$. The temperature limit given in eqn 11.37 is called the **Doppler limit**. Through eqn 11.2, it corresponds to a minimum thermal r.m.s. velocity of

$$v_x^{\mathrm{min}} = \sqrt{\hbar/2m\tau}. \tag{11.38}$$

The Doppler temperature in eqn 11.37 puts a fundamental limit to the temperature that can be achieved by the Doppler cooling process in its simplest form.

Example 11.2   Calculate the lowest temperature that can be achieved by the Doppler cooling method using the D_2 line of sodium at 589 nm, which has a radiative lifetime of 16 ns. Calculate also the average velocity of the atoms at this temperature.

Solution
The minimum temperature for Doppler cooling is given by the Doppler limit temperature given in eqn 11.37. With $\tau = 16$ ns, this gives $T_{\mathrm{min}} = 240$ μK. The corresponding minimum thermal velocity from eqn 11.38 with $m = 23m_{\mathrm{H}}$ is 0.29 m s^{-1}.

11.2.3   Sub-Doppler cooling

Equation 11.37 appears to set a fundamental limit to the temperatures that can be achieved by laser cooling. However, careful experiments carried out in the 1980s led to the surprising conclusion that the temperatures that were being achieved could be *lower* than the Doppler limit. It transpires that laser cooling is one of the rare examples of an

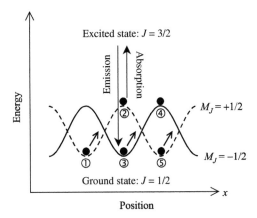

Fig. 11.5 Sisyphus cooling for a $J = 1/2 \to 3/2$ transition in an alkali atom. The atom is moving in the $+x$-direction, and interacts with two counter-propagating laser beams as in Fig. 11.4. The energies of the $M_J = \pm 1/2$ sublevels of the $J = 1/2$ ground state vary sinusoidally with position in the interference pattern of the lasers. The laser frequency is tuned so that the atom can only make a transition to the excited state at the top of one of the potential hills. (Positions 2 and 4.) The atom in the excited state can re-emit to the same sublevel, or to the lower one. (Positions 3 and 5.) In the case of an atom following the path $1 \to 2 \to 3 \to 4 \to 5 \to \cdots$, the difference in the energy of the absorbed and emitted photons is taken from the total energy of the atom, leading to a cooling effect.

experiment that actually works better in the laboratory than the simple theory predicts.

The discrepancy can be explained by realizing that the Doppler cooling mechanism described in Sections 11.2.1 and 11.2.2 is too simplistic. The counter-propagating laser beams in an optical molasses experiment interfere with each other, and this leads to a new type of cooling mechanism called **Sisyphus cooling**.

The detailed mechanism of Sisyphus cooling is too complicated for our level of treatment, but the basic process can be understood with reference to Fig. 11.5. We consider an alkali atom in the $^2S_{1/2}$ ground state moving in the $+x$-direction and making transitions to a $^2P_{3/2}$ excited state under the influence of two counter-propagating resonant laser beams as shown in Fig. 11.4. The interference pattern of the lasers leads to a small periodic modulation of the energies of the ground state levels through the AC Stark effect. The light-induced shifts of the $M_J = \pm 1/2$ magnetic sublevels differ in phase by 180° as shown in Fig. 11.5. As long as the atom stays in the same magnetic sublevel, it moves up and down potential hills, continually converting kinetic to potential energy and back again, but without change of the total energy. (Route $1 \to 2 \to 5 \to \cdots$ in Fig. 11.5.) However, by careful tuning of the laser, we can arrange that some of the atoms follow the route $1 \to 2 \to 3 \to 4 \to 5 \to \cdots$ in Fig. 11.5. In this case, the atoms are constantly losing energy, because they have to climb to the top of the potential hill, and then drop to the valley again, just like Sisyphus.

Sisyphus cooling is named after the character in Greek mythology who was condemned to roll a stone up a hill forever, only for it to roll down again every time he got near the top. The mechanism of Sisyphus cooling is explained in more detail in Foot (2005). See also Cohen–Tannoudji and Phillips (1990). A brief discussion of the AC Stark effect may be found in Section 9.5.3.

The selection rules actually give preferential emission to the lower level, which further improves the efficiency of the cooling process.

The Sisyphus technique works when the laser frequency is tuned so that the atoms can only absorb to the excited state at the top of one of the potential hills. (Positions 2 or 4 in Fig. 11.5.) The selection rules allow the atom to re-emit to either of the magnetic sublevels of the ground state. If the atom goes back to the same level, there is no change of the energy, and no cooling effect. However, if it goes into the lower level (e.g. path 2→, excited state → 3 in Fig. 11.5), the difference in the energies of the absorbed and emitted photons must be taken from the kinetic energy. The atom therefore slows down, thereby producing a cooling effect.

The minimum temperature that can be achieved by Sisyphus cooling is set by the **recoil limit**. The atoms are constantly emitting spontaneous photons of wavelength λ in random directions. The atom recoils each time with momentum h/λ, and so it ends up with a random thermal energy given by:

$$\frac{1}{2}k_\mathrm{B}T_\mathrm{recoil} = \frac{(h/\lambda)^2}{2m} = \frac{h^2}{2m\lambda^2}. \tag{11.39}$$

This gives a minimum temperature of:

Temperatures even lower than the recoil limit have been achieved by a process called **velocity selective coherent trapping** which involves non-absorbing states of the atoms. See, for example, Metcalf and van der Straten (1999) for further details.

$$T_\mathrm{recoil} = \frac{h^2}{mk_\mathrm{B}\lambda^2}. \tag{11.40}$$

Sisyphus cooling experiments on cesium have achieved temperatures as low as 2 μK, which is only an order of magnitude above the recoil limit. (See Example 11.3.)

Example 11.3   Compare the recoil limits for the 3p → 3s transition of sodium at 589 nm and the 6p → 6s transition of cesium at 852 nm.

Solutions
The recoil limit temperature is given in eqn 11.40. For sodium with $m = 23.0\, m_\mathrm{H}$ and $\lambda = 589$ nm we obtain $T_\mathrm{recoil} = 2.4$ μK, while for cesium with $m = 132.9\, m_\mathrm{H}$ and $\lambda = 852$ nm we find $T_\mathrm{recoil} = 0.2$ μK.

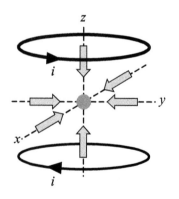

Fig. 11.6 The magneto-optic trap. Two lasers beams travelling in the $\pm x$-directions are used to annul the atom's velocity in both directions along the x-axis. Four other beams do the same for the $\pm y$ and $\pm z$ directions. The magnetic quadrupole field generated by two coils carrying equal currents i flowing in opposite directions traps the atoms with $M_J > 0$ at the intersection point of the beams.

11.2.4   Magneto-optic atom traps

The optical molasses arrangement shown in Fig. 11.4 slows the atoms moving in the $\pm x$-directions to very small velocities, but it has no effect on their motion in the y- and z-directions, and nor does it have the ability to trap them at the same point in space. To stop the atoms for all three velocity components (i.e. the $\pm x$, $\pm y$ and $\pm z$ directions), and to confine them to the same point in space, we need a **magneto-optic trap**.

The most typical type of magneto-optical trap consists of a six-beam arrangement together with a **magnetic quadrupole** field, as shown in Fig. 11.6. The three pairs of orthogonal counter-propagating beams produce an optical molasses effect for all three coordinates, while the quadrupole field creates an attractive potential for atomic states with $M_J > 0$.

The magnetic quadrupole consists of two coils carrying currents flowing in opposite directions. If we define the axis of the coils as the z-direction, with the origin at the centre of the coils as shown in Fig. 11.6, the magnetic field at position (x, y, z) is given by:

$$\boldsymbol{B} = B'(x\hat{\mathbf{i}} + y\hat{\mathbf{j}} - 2z\hat{\mathbf{k}}), \tag{11.41}$$

where B' is the field gradient. The magnitude of the field is accordingly given by:

$$B = B'(x^2 + y^2 + 4z^2)^{1/2}. \tag{11.42}$$

This has a minimum at the centre of the quadrupole, where the fields from the two coils cancel.

The energy of a magnetic sub-level in a magnetic field B is given by the Zeeman energy (cf. eqn 3.82.):

$$E = g_J \mu_B B M_J \tag{11.43}$$

where g_J is the Landé g-factor. States with $M_J > 0$ have their lowest energy when B is smallest, and they are therefore called 'low-field seeking' states. States with $M_J < 0$, by contrast, are high-field seeking. Equation 11.42 shows that a quadrupole trap has a minimum in the field strength at the origin. Quadrupole traps are therefore attractive for states with $M_J > 0$, but repulsive for states with $M_J < 0$. The depth of the potential trap is of magnitude $\sim \mu_B B$, which corresponds to a temperature of only 0.67 K for a trap with a maximum field of 1 T. The trap therefore only works for very cold atoms, which is why it must be combined with laser cooling techniques to work effectively.

Magnetic traps can be used to compress the gas of cold atoms produced by the optical molasses effect, thereby increasing the particle density by several orders of magnitude. This compression process forms an important part of the method used to achieve Bose–Einstein condensation. (See Section 11.3.3.)

Magnetic quadrupole traps are adequate for most applications, but have an important disadvantage for the most demanding requirements. The field at the origin is zero, making all the M_J states degenerate, and allowing easy scattering from positive to negative M_J states. Since the trapping potential is repulsive for negative M_J states, the trap effectively has a hole right at the origin, leading to loss of atoms with time. To compensate for this effect, more sophisticated traps have been designed, for example, the time-averaged orbiting potential trap (TOP trap) and the Ioffe–Pritchard trap. See Foot (2005) or Pethick and Smith (2002) for further details.

11.2.5   Experimental techniques for laser cooling

Figure 11.7 shows a schematic diagram of an experiment to produce a gas of ultracold sodium atoms. The experiment consisted of two parts: the pre-cooling and optical molasses regions. In the pre-cooling region, the sodium atoms were cooled to ~ 2.5 K. Then in the molasses region, a small fraction of these atoms were cooled further to temperatures well below 1 mK.

The laser cooling was performed on the D$_2$ transition (3S$_{1/2}$ → 3P$_{3/2}$) at 589 nm. The source consisted of a sodium oven with a nozzle at 600 °C, together with appropriate apertures to create a collimated beam travelling in the $+x$-direction. The pre-cooling region consisted of the cooling laser and a tapered solenoid. As discussed in Section 11.2.1, the detuning of the laser must be varied as the atoms slow down. (See eqn 11.7.) In the arrangement shown in Fig. 11.7, this was done by

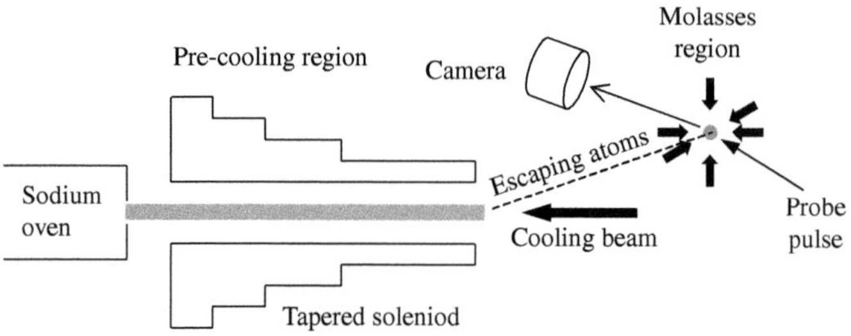

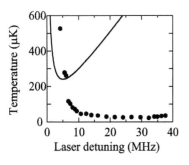

Fig. 11.7 Schematic arrangement of the apparatus used to produce a gas of ultracold atoms using the D$_2$ transition of atomic sodium. (Not to scale.) The first part of the apparatus consisted of a tapered solenoid and a cooling laser to slow the atoms to ~ 30 m s^{-1} ($T \sim 2.5$ K). A small fraction of the slow atoms that escaped traversed the intersection point of the six beams in the molasses region, where they were cooled further to temperatures below 1 mK by the Sisyphus effect. The temperature of the atoms was determined by turning off the molasses beams and imaging the expansion of the gas using a camera and a probe pulse. After P.W. Lett, *et al.*, *J. Opt. Soc. Am. B* **6**, 2084 (1989).

keeping the laser frequency fixed and tuning the transition frequency of the atoms with the solenoid. Let $B(x)$ be the magnetic field at position x along the solenoid. The transition energy is then given by:

$$h\nu(x) = h\nu_0 + \xi\mu_B B(x), \qquad (11.44)$$

where

$$\xi = g_J^{\text{upper}} M_J^{\text{upper}} - g_J^{\text{lower}} M_J^{\text{lower}}. \qquad (11.45)$$

The parameter ξ accounts for the difference of the Zeeman energies of the upper and lower states of the transition. See eqn 3.82.

If the laser is tuned close to ν_0, the cooling condition given in eqn 11.7 is satisfied when $\xi\mu_B B(x) = h\nu_x(x)/\lambda$. Thus by careful design of the solenoid, the reduction of the field strength can be made to compensate for the reduction of v_x during the deceleration process. The average velocity of the atoms that emerged from the solenoid was around 30 m s^{-1}, which corresponds to a temperature of 2.5 K. (See eqn 11.3.)

The optical molasses region was located about 20 cm downstream from the end of the solenoid and about 2.5 cm above its axis. The six beams were generated from a second laser which could be tuned to the optimal frequency for the optical molasses effect. A small fraction of the slow atoms that emerged from the solenoid crossed the centre of the molasses region and were cooled to temperatures as low as 40 µK.

The final temperature was measured by the **time of flight** technique. In this method, the molasses beams were turned off and the gas allowed to expand. (See Fig. 11.14(a).) At a predetermined time later, a probe pulse derived from the same laser was turned on and the fluorescence from the gas was imaged onto a camera. By varying the time between turning off the molasses beams and turning on the probe pulse, the expansion of the gas could be followed and the velocity distribution of the atoms determined. The temperature was then deduced from the velocity distribution.

Figure 11.8 shows the results achieved from the experimental arrangement shown in Fig 11.7. The solid line shows the predictions of the simple

Fig. 11.8 Variation of the temperature with the detuning of the molasses laser for the experiment shown in Fig 11.7. The solid line is the prediction of eqn 11.36 at low intensity (i.e. $I \ll I_s$) for $\gamma/2\pi = 10$ MHz. Note that the detunings are negative with respect to the transition frequency. (After P. W. Lett *et al.*, *J. Opt. Soc. Am. B* **6**, 2084, (1989), © Optical Society of America, reproduced with permission.)

Doppler cooling model of eqn 11.36 in the low-intensity limit (i.e. $I \ll I_s$) for $\gamma/2\pi = 10$ MHz appropriate for the radiative lifetime of 16 ns. The experimental data fall well below the predictions of the Doppler cooling model due to the Sisyphus effect. The minimum temperature achieved was around 40 µK, which is six times lower than the Doppler limit of 240 µK (cf. Example 11.2) and within a factor of 20 of the recoil limit (cf. Example 11.3).

11.2.6   Cooling and trapping of ions

Up to this point, we have only considered the cooling and trapping of neutral atoms. There are, however, a number of important experiments in quantum optics that require techniques to cool and trap charged atoms (i.e. ions). These techniques are of less interest in this present chapter because the repulsive forces between the ions prevent the accumulation of densities sufficient to observe Bose–Einstein condensation. We therefore give here only a very brief discussion of the techniques used with ions, and refer the reader to the bibliography for further details.

The basic principles of Doppler cooling discussed in Section 11.2.1 apply equally well to ions as they do to neutral atoms. An added feature of ion cooling is that the charge on the ions makes it easy to trap them in three dimensions by using electric and magnetic fields. This means that only one laser beam is required to produce efficient cooling. Those ions that are moving towards the laser are cooled, and then these ions are subsequently scattered back into the centre of the trap by the fields, where they exchange energy with the other atoms and hence cool the whole trapped gas. One practical difficulty that occurs with ion cooling experiments is that the transition energies tend to be in the blue or ultraviolet spectral regions, which sometimes makes it harder to find suitable tunable laser sources.

Many of the early experiments on ion cooling were performed with a **Penning trap**, which is illustrated schematically in Fig. 11.9. A static magnetic field in the z-direction confines the motion of the ions in the (x, y) directions, while a static quadrupole electric field provides confinement in the z-direction. Doppler cooling can occur as long as the laser is not directed along any of the principal axes. Using apparatus such as this, Wineland and co-workers first succeeded in cooling a gas of 5×10^4 Mg$^+$ ions with a frequency-doubled dye laser at 280 nm.

Ion cooling experiments can also be carried out in a **Paul trap**, which uses an oscillating electric field to provide the three-dimensional confinement of charged particles. The first experiments of this type were reported in the same year as the Penning trap work, when about 50 Ba$^+$ ions were cooled by using a dye laser operating at 493.4 nm. Subsequent experiments have perfected the techniques to trap and cool *single* ions. These techniques form the basis for the quantum information processing experiments described in Section 13.6.

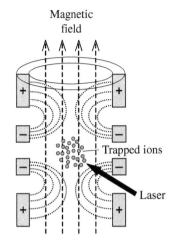

Fig. 11.9 Schematic cross-sectional view of a Penning trap used for ion cooling experiments. The trap incorporates a static magnetic field directed along the z-axis and a static quadrupole electric field. The field lines are shown, respectively, by dashed and dotted lines. The ions are trapped at the centre of the electrodes, and are cooled by an off-axis laser.

The principles of ion cooling were first discussed by David Wineland and Hans Dehmelt in the same year as Hänsch and Schawlow's work on cooling of neutral atoms, namely 1975. (See *Bull. Am. Phys. Soc.* **20**, 637.) Dehmelt later went on to win the Nobel Prize in 1989, together with Wolfgang Paul, for his work on the development of ion traps. The details of the original experiments may be found in D. J. Wineland *et al.*, *Phys. Rev. Lett.* **40**, 1639 (1978), and W. Neuhauser *et al.*, *Phys. Rev. Lett.* **41**, 233 (1978).

11.3   Bose–Einstein condensation

In this section we give a brief description of the phenomenon of Bose–Einstein condensation, and the techniques that are used to produce it. This will pave the way for understanding the basic principles of atom lasers, which are described in the final section of the chapter. A more detailed discussion of the statistical mechanics of Bose–Einstein condensation is provided in Appendix F.

11.3.1   Bose–Einstein condensation as a phase transition

The phenomenon of Bose–Einstein condensation is concerned with observing quantum effects related to the translational kinetic energy of the atoms or molecules in a gas. Let us start by considering a simple example, namely a gas of diatomic molecules. The heat capacity C_V of such a gas is usually one of the first problems studied in statistical mechanics courses, and a discussion of its variation with temperature is instructive for understanding the effects of the quantization of the kinetic energy in which we are interested here.

A diatomic gas has seven degrees of freedom: three for the translational motion, two for the rotations about the axes perpendicular to the bond between the atoms, and two for the kinetic and potential energy of the vibrational oscillations along the bond.

Figure 11.10 shows the generic behaviour of $C_V(T)$ for a typical diatomic gas. At very high temperatures we expect classical behaviour in accordance with the principle of equipartition of energy given in eqn 11.1. Since $C_V = dE/dT$, we therefore expect a contribution of $3k_B/2$ for the translational motion, $2k_B/2$ for the rotational motion, and a further $2k_B/2$ for the vibrations, giving $7k_B/2$ in total.

At lower temperatures, the heat capacity departs from the classical result due to the quantization of the thermal motion. The vibrations of a molecule can be approximated to a simple harmonic oscillator, with quantized energy levels given by (see eqn 3.93):

$$E = (n + \tfrac{1}{2})h\nu_{\text{vib}}, \tag{11.46}$$

Fig. 11.10 Schematic variation of the heat capacity of a gas of diatomic molecules with temperature. The molecule has seven degrees of freedom: three translational, two rotational, and two vibrational. The rotational and vibrational contributions freeze out at characteristic temperatures. The freezing out of the translational motion is, however, never observed in normal circumstances.

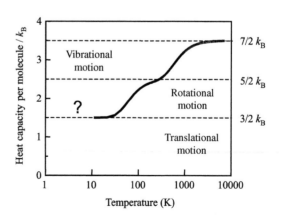

where ν_{vib} is the vibrational frequency. The classical result will only be obtained if the thermal energy is much greater than the vibrational quanta, that is when

$$k_{\mathrm{B}}T \gg h\nu_{\mathrm{vib}}. \tag{11.47}$$

With typical values for ν_{vib} around 10^{13} Hz, the classical behaviour is only observed at temperatures above about 1000 K. At room temperature the vibrational motion is usually 'frozen out', as shown in Fig. 11.10. In the same way we expect the rotational motion to freeze out when the thermal energy is comparable to the quantized rotational energy, that is when

$$k_{\mathrm{B}}T \sim \frac{\hbar^2}{I_{\mathrm{rot}}}, \tag{11.48}$$

where I_{rot} is the moment of inertia about the rotation axis. This typically occurs at temperatures <100 K. Thus the rotational motion is usually classical at room temperature, but freezes out at lower temperatures, as indicated in Fig. 11.10.

We are finally left with the translational motion. The third law of thermodynamics tells us that C_{V} must eventually go to zero at $T = 0$. However, in any normal gas the attractive forces between the molecules cause liquefaction and solidification long before the quantum effects for the translational motion become important. If, however, we could somehow prevent the gas from condensing, we would eventually expect to observe quantum effects related to the translational motion. This is precisely the effect first considered by Einstein in 1924–5. He discovered that even a gas of completely non-interacting particles will undergo a phase transition at a sufficiently low temperature. This phase transition has now come to be known as **Bose–Einstein condensation**.

In statistical mechanics, Bose–Einstein condensation is understood as the accumulation of a macroscopic fraction of the total number of particles in the zero velocity state. (See Appendix F.) The transition temperature T_c at which this effect starts to occur in a gas of free partcicles is given by (cf. eqn F.10):

$$T_c = 0.0839 \frac{h^2}{mk_{\mathrm{B}}} \left(\frac{N}{V}\right)^{2/3}, \tag{11.49}$$

where N is the number of particles in volume V, and m is the particle mass. The fraction of particles in the condensed state is given by (cf. eqn F.12):

$$f(T) = 1 - (T/T_c)^{3/2}. \tag{11.50}$$

This model allow us to complete the discussion of the diatomic gas in Fig. 11.10 in the temperature region indicated by the question mark. Equation 11.50 indicates that the fraction of particles in the zero velocity

Einstein considered the variation of the heat capacity of a crystalline solid in 1906, and showed that $C_{\mathrm{V}}(T)$ goes to zero at $T=0$ in accordance with the third law of thermodynamics. The model was refined a few years later by Peter Debye. The Einstein–Debye model of solids is completely different from the quantization of the translational motion of the free particles in a gas that we are considering here.

Equation 11.49 assumes that the particles in the gas have spin $S=0$. See the derivation in Section F.2 of Appendix F.

state approaches 100% as T goes to zero. The thermal energy of the system, and hence the heat capacity, therefore goes to zero at $T = 0$, finally reaching consistency with the third law of thermodynamics.

The general behaviour predicted by statistical mechanics has been thoroughly established by detailed measurements on well-known Bose–Einstein condensed systems, such as liquid ^{4}He. The difficulty with these conventional examples of Bose–Einstein condensation is that they are not 'non-interacting' systems. The mere fact that helium is a liquid at the Bose–Einstein condensation temperature indicates that there are strong interactions between the atoms over and above any effects due to the quantization of the kinetic energy. In an ideal world we would therefore like to observe the Bose–Einstein condensation in a truly weakly interacting system (i.e. a gas) so that we can study it in isolation. Unfortunately, the variation of T_c with $(N/V)^{2/3}$ indicated by eqn 11.49 implies that low-density systems such as gases have extremely low transition temperatures. This is why it was not possible to observe condensation in gases until the techniques for generating ultracold atoms described in the previous sections were developed.

Example 11.4   Calculate the Bose–Einstein condensation temperature for a free gas of ^{87}Rb atoms with a density of 3×10^{19} m^{-3}.

Solution
The transition temperature is given in eqn 11.49. With $m = 87\ m_H$, we find $T_c = 180$ nK.

11.3.2   Microscopic description of Bose–Einstein condensation

The understanding of Bose–Einstein condensation as an accumulation of particles in the zero velocity state makes it clear that it can only be observed in gases of bosons, which are not subject to the Pauli exclusion principle. The atoms and molecules in a gas are composite particles made up of fermions, namely protons, neutrons, and electrons. The atom or molecule as a whole can therefore be either a fermion or a boson depending on the total spin.

We can find out whether a particular atom is a fermion or boson by working out the total spin according to the rules for the addition of quantum angular momenta: (see discussion of eqn 3.51 in Section 3.1.5)

$$\boldsymbol{S}_{\text{atom}} = \boldsymbol{S}_{\text{electrons}} + \boldsymbol{S}_{\text{nucleus}}. \tag{11.51}$$

Fermions and bosons are particles with half-integer and integer spins, respectively. (See Section F.1 in Appendix F.) Electrons, protons, and neutrons have $S = 1/2$, and are therefore fermions.

It is easy to see that the atom will be a boson if the total number of electrons, protons, and neutrons is an even number, and a fermion if the total number is odd. Consider, for example, the hydrogen atom ^{1}H. This has one proton and one electron. Both the nucleus and electron have spin $1/2$, and so we find $S_{\text{atom}} = 0$ or 1. Hydrogen atoms are therefore bosons. By contrast, deuterium (^{2}H), with one proton, one neutron, and one electron, is a fermion because the total spin is either $1/2$ or $3/2$.

Let us now consider a gas of identical non-interacting atomic bosons of mass m at temperature T. The word 'non-interacting' is very important here. It implies that the particles are completely free, with only kinetic energy. The thermal de Broglie wavelength λ_{deB} is given by :

$$\frac{p^2}{2m} \equiv \frac{1}{2m}\left(\frac{h}{\lambda_{\mathrm{deB}}}\right)^2 = \frac{3}{2}k_{\mathrm{B}}T, \tag{11.52}$$

which implies that:

$$\lambda_{\mathrm{deB}} = \frac{h}{\sqrt{3mk_{\mathrm{B}}T}}. \tag{11.53}$$

The thermal de Broglie wavelength thus increases as T decreases.

The quantum mechanical wave function for the translational motion of a free atom extends over a distance of $\sim \lambda_{\mathrm{deB}}$. As λ_{deB} increases with decreasing T, a temperature will eventually be reached when the wave functions of neighbouring atoms begin to overlap. This situation is depicted in Fig. 11.11. The atoms will interact with each other and coalesce to form an extended state with a common wave function. This is the Bose–Einstein condensed state.

The condition for wave function overlap is that the reciprocal of the effective particle volume determined by the de Broglie wavelength should be equal to the particle density. If we have N particles in volume V, the condition can be written:

$$\frac{N}{V} \sim \frac{1}{\lambda_{\mathrm{deB}}^3}. \tag{11.54}$$

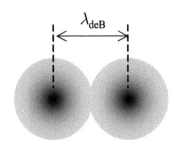

Fig. 11.11 Overlapping wave functions of two atoms separated by λ_{deB}.

By inserting from eqn 11.53 and solving for T, we find:

$$T_{\mathrm{c}} \sim \frac{1}{3}\frac{h^2}{mk_{\mathrm{B}}}\left(\frac{N}{V}\right)^{2/3}. \tag{11.55}$$

This formula is the same as the rigorous one given in eqn 11.49 apart from the numerical factor. The argument used to derive it, by contrast, is far more intuitive, and gives us the insight that the Bose-condensed state consists of an extended state with all the atoms coherent with each other.

11.3.3 Experimental techniques for Bose–Einstein condensation

The conditions required to achieve Bose–Einstein condensation in a gas impose severe technical challenges. The atoms must be kept well apart from each other to prevent complications due to other effects such as liquefaction, but this means that the particle density must be very small, which implies that the transition temperature is very low. (See eqn 11.49.) Most of the successful experiments on gaseous systems have had particle densities in the range 10^{18}–10^{21} m^{-3}, and condensation temperatures below 1 μK.

The laser cooling techniques described in Section 11.2 typically produce a gas with temperatures in the μK range and densities up to around 10^{17} m^{-3}. The condensation temperature at this density is ∼ 10 nK for alkali atoms like sodium or rubidium. It is therefore apparent that laser cooling alone cannot produce Bose–Einstein condensation, and that additional techniques therefore have to be employed.

The general procedure for achieving Bose–Einstein condensation in a gas of atoms usually follows two steps:

1. Cool and trap a gas of atoms to temperatures near the recoil limit by laser-cooling techniques, as discussed in Sections 11.2.4 and 11.2.5.

2. Turn off the cooling laser with the trap still applied, and reduce the trapping potential to cool the gas further by **evaporative cooling** until condensation occurs.

The principle of the evaporative cooling technique is illustrated schematically in Fig. 11.12. When the magnetic trap is turned on, it provides an attractive potential for the atoms with $M_J > 0$, as discussed in Section 11.2.4, and illustrated in Fig. 11.12(a). The magnetic field strength is gradually turned down in order to reduce the depth of the magnetic potential as shown in Fig. 11.12(b). The fastest-moving atoms now have enough kinetic energy to escape, leaving the slower ones behind. This causes an overall reduction in the average kinetic energy, which is equivalent to a reduction in the temperature, as in the cooling of a liquid by evaporation. In the right conditions, the final temperature will be low enough to instigate the Bose–Einstein condensation process.

There are a number of important differences between the physics of Bose–Einstein condensation in a trap potential and in free space. Both the condensation temperature and the variation of the fraction of particles in the condensate are affected. If we assume that the potential is harmonic, with a characteristic angular frequency of ω, then eqn 11.49

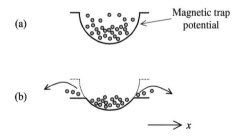

(a)   Magnetic trap potential

(b) $\longrightarrow x$

Fig. 11.12 Evaporative cooling. (a) The laser-cooled atoms are first compressed in a magnetic trap. At this stage the temperature is still above T_c. (b) The laser is then turned off and the trap potential is reduced by decreasing the magnetic field strength. The most energetic atoms escape, and the temperature drops, in analogy to the cooling of a hot liquid by evaporation. In the right conditions, this produces temperatures below T_c.

is modified to:

$$T_c = 0.94 \frac{\hbar\omega}{k_B} N^{1/3}, \qquad (11.56)$$

and 11.50 to:

$$f(T) = 1 - (T/T_c)^3. \qquad (11.57)$$

Note that the condensation temperature now only depends on the total number of particles, rather than the particle density.

The first successful observation of Bose–Einstein condensation by these techniques was made in 1995. ^{87}Rb atoms were used at a density of 2.5×10^{18} m^{-3}, for which the condensation temperature was around 170 nK. The atoms were first cooled to 20 µK on the $5S_{1/2} \to 5P_{3/2}$ transition at 780 nm using diode lasers. The lasers were then turned off and the gas compressed, during which process the temperature rose to around 90 µK. Finally, the gas was evaporatively cooled to temperatures as low as 20 nK.

Figure 11.13 shows typical data obtained during the condensation process. The three images were obtained by the time-of-flight technique illustrated in Fig. 11.14 with an expansion time t_e of 60 ms. In this method, the gas is allowed to expand for a predetermined time, and the shadow images produced under resonant laser excitation allow the velocity distribution to be determined. At 400 nK, the velocity distribution is broad and fits well to a Maxwell–Boltzmann distribution. At 200 nK, the condensate begins to form, and the velocity distribution corresponds to a mixture of condensed atoms with zero velocity and 'normal' atoms with a Maxwell–Boltzmann distribution. Finally, at 50 nK almost all

The derivation of eqn 11.56 may be found, for example, in Pethick and Smith (2002, §2.2.)

See M. H. Anderson *et al.*, *Science* **269**, 198 (1995). Note that the inter-particle distance at the condensation temperature was equivalent to about 1000 atomic radii, so that it is reasonable to assume almost ideal 'non-interacting' conditions. Similar results were obtained for sodium soon afterwards.

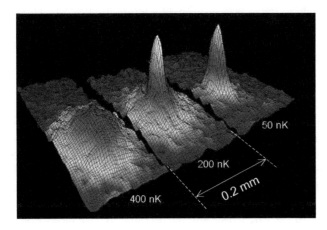

Fig. 11.13 Bose–Einstein condensation in ^{87}Rb atoms. The three figures show the velocity distribution as the gas is cooled through T_c on going from left to right. The velocity distributions were measured by the time of flight technique after a 60 ms free expansion as illustrated in Fig. 11.14. Above T_c, a broad Maxwell–Boltzmann distribution is observed, but as the gas condenses, the fraction of atoms in the zero velocity state at the origin increases dramatically. (Image from http://jilawww.colorado.edu/bec. The experiment is described in M. H. Anderson, *et al.*, *Science* **269**, 198 (1995).)

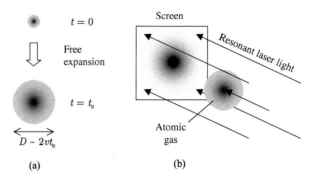

(a)   (b)

Fig. 11.14 Measurement of temperature by the time-of-flight technique. (a) The gas is allowed to expand freely for a controlled time t_e, so that the increase of the cloud diameter D is determined by the velocity v of the atoms in the gas. (b) The expanded gas is illuminated with a resonant laser, which is absorbed by the atoms, thereby creating a shadow on the screen in proportion to the atom density. The velocity distribution is then calculated from the atom distribution deduced from the shadow image. In alternative arrangements, the fluorescence is imaged onto a camera, as in Fig. 11.7.

Table 11.1 Gaseous atomic systems in which Bose–Einstein condensation has been observed as of 2005. Preliminary results on ^{7}Li were first reported in 1995, but it was not until 1997 that conclusive evidence was obtained.

Atom	Isotope	Year of observation
Rubidium	^{87}Rb	1995
Sodium	^{23}Na	1995
Lithium	^{7}Li	1997
Hydrogen	^{1}H	1998
Rubidium	^{85}Rb	2000
Helium	^{4}He	2001
Potassium	^{41}K	2001
Cesium	^{133}Cs	2002
Ytterbium	^{174}Yb	2003
Chromium	^{52}Cr	2005

See Meystre (2001) for further details on the subject of atom optics.

the atoms are in the condensate, as indicated by the sharp peak at the centre of the image.

In the years following the original observation in 1995, there have been many reports of Bose–Einstein condensation in atomic gases. Table 11.1 gives a list of the elements in which Bose–Einstein condensation has been obtained at the time of writing, together with the year of the first observation. The techniques have recently been extended to ^{6}Li$_2$ and ^{40}K$_2$ molecules, and also to atomic ^{40}K. The latter report is very surprising at first sight, because ^{40}K is a fermion. However, careful studies have shown that the ^{40}K atoms can pair up in an analogous way to the electron Cooper pairs in superconductors, creating a collective boson particle that can undergo Bose–Einstein condensation.

11.4   Atom lasers

One of the most remarkable developments of Bose–Einstein condensation has been the demonstration of **atom lasers**. Just as the development of optical lasers in the 1960s revolutionized conventional optics, it is to be expected that the atom lasers that we shall consider here will have a similar impact on the subject of **atom optics**.

Atom optics describes the manipulation of atom waves in a manner analogous to the way lenses and mirrors manipulate light. It is, of course, relatively easy to use magnetic and electric fields to make lenses and mirrors for *charged* particles like electrons, but the development of the equivalent components for *neutral* atoms is far more challenging, and relies on the light–atom force given in eqn 11.19. As we shall discuss briefly below, the development of the atom laser has opened new horizons for the subject by laying the foundations for high intensity coherent atom optics.

The atoms in the Bose–Einstein condensate are trapped by the magnetic potential, and the situation is rather similar to an optical laser with 100% reflectors at either end of the cavity. Although such a laser might oscillate, it has no output, and is of little practical use. The key step in the practical development of the atom laser was therefore the demonstration of the output coupler. The operation of the output coupler relies on the fact that the atoms in the condensate all have their spins parallel to the magnetic field because the trap is only attractive for atoms with $M_J > 0$. (See Section 11.2.4.) Thus by applying a radio frequency (RF) pulse to tip the spins of some of the atoms, the trap suddenly becomes repulsive for those atoms and they are ejected. These ejected atoms then fall downwards under gravity and form a coherent matter pulse. The first successful demonstration of this effect was made in 1997.

Figure 11.15(a) shows an image of the coherent atom pulses produced in this way from a sodium Bose–Einstein condensate. Each pulse contained between 10^5 and 10^6 atoms. The coherence of the matter pulses was established by measuring the interference pattern formed between two such beams. Figure 11.15(b) shows the absorption image obtained when two pulses from the atom laser were overlapped. The interference fringes at the intersection point are clearly visible, and establishes the long range coherence that follows from the coherence of the atomic wave functions in the condensate.

The interference pattern shown in Fig. 11.15 forms the basis of coherent *linear* atom optics with high intensity beams. The next step is to use atom lasers to establish both *nonlinear* and *quantum* atom optics. The subject has advanced very rapidly, and several key proofs of principle have already been made, including the demonstration of four-wave mixing, soliton formation, and atom number squeezing. The reader is referred to the bibliography for further details.

The development of the output coupler for the atom laser is described in M.-O. Mewes, *et al.*, *Phys. Rev. Lett.* **78**, 582 (1997). The existence of gain in the laser medium was demonstrated two years later. (See Further Reading.) A discussion of the principles by which an RF pulse tips the spin through a controlled angle may be found in Section E.2 of Appendix E.

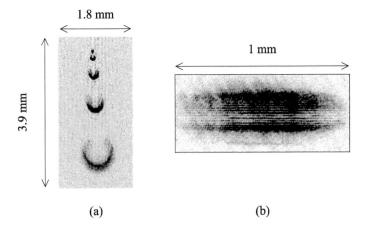

(a) (b)

Fig. 11.15 The atom laser. (a) Coherent matter pulses ejected from a sodium atom laser operating at 200 Hz. (b) Interference fringes with periodicity 15 μm formed at the intersection point of two overlapping matter pulses. Both images were observed by absorption imaging techniques as in Fig. 11.14(b). (After D. S. Durfee and W. Ketterle, *Opt. Express* **2**, 299 (1998), © Optical Society of America, reproduced with permission.)

Further reading

All of the topics covered in this chapter are described in greater depth in Foot (2005). A general overview of the whole subject of cold atoms and their applications may be found in Chu (2002).

An in-depth treatment of the subject of laser cooling may be found in Metcalf and van der Straten (1999). Introductory review articles on the topic may be found in Chu (1992), Cohen-Tannoudji and Phillips (1990), Foot (1991), or Phillips and Metcalf (1987). A more detailed review may be found in Metcalf and van der Straten (2003). A review of the equivalent techniques for the cooling and trapping of ions has been given by Eschner *et al.* (2003). Descriptions of how to build a laser cooling and trapping apparatus for an undergraduate laboratory may be found in Wieman *et al.* (1995) or Mellish and Wilson (2002).

Full-length texts on the subject of Bose–Einstein condensation may be found in Pethick and Smith (2002) or Pitaevskii and Stringari (2003). Overviews of the subject are given in the Nobel Prize lectures of Cornell and Wieman (2002) and Ketterle (2002). Introductory reviews may be found in Burnett (1996), Burnett *et al.* (1999), Collins (2000), Cornell and Wieman (1998), or Ketterle (1999). More advanced reviews are given in Anglin and Ketterle (2002) and Cornish and Cassettari (2003). An introductory review of condensation phenomena in Fermi gases can be found in Chevy and Salomon (2005).

The Nobel Prize lecture in Ketterle (2002) describes the development of the atom laser, and introductory reviews on the topic have been given by Hagley *et al.* (2001) and Helmerson *et al.* (1999). A comprehensive overview of the subject of atom optics may be found in Meystre (2001). The fields of nonlinear and quantum atom optics are reviewed in Anderson and Meystre (2002) and Rolston and Phillips (2002).

Exercises

(11.1) Evaluate the r.m.s. value of the x-component of the velocity in a gas of atoms of mass m with a Maxwell–Boltzmann distribution at temperature T. Hence justify eqn 11.2.

(11.2) A beam of cesium atoms travelling in the $+x$-direction is emitted from an oven with a temperature of 200 °C. A laser beam of wavelength 852 nm propagating in the $-x$-direction is used to cool the atoms. The laser is resonant with the $6P_{3/2} \rightarrow 6S_{1/2}$ transition, which has a lifetime of 32 ns. The relative atomic mass of cesium is 132.9.

(a) What initial frequency detuning of the laser relative to the transition must be used to produce efficient laser cooling?

(b) What is the average momentum change imparted to a cesium atom during an absorption–emission cycle? What is the maximum decelerating force that can be exerted on the atoms by the laser?

(c) Estimate the number of absorption–emission cycles required to cool the atoms to their minimum temperature. Estimate the time taken for the atoms to reach this temperature, and the distance they would travel during the cooling process.

(d) Calculate the final temperature that the atoms reach after this experiment, on the assumption that they are cooled to the Doppler limit.

(11.3) The cold atoms described in the previous question are transferred to a magneto-optical trap where they are cooled by the Sisyphus process to sub-Doppler temperatures. Calculate the lowest temperature that can be achieved by this method and find the r.m.s. velocity of the atoms at this temperature.

(11.4) Consider a non-degenerate two level atom with Einstein coefficients of A_{21}, B_{21}, and B_{12} irradiated with a laser beam of spectral energy density u tuned to resonance with the transition. Show that the difference between the absorption and the stimulated emission rates per atom is given by:

$$R = \frac{A_{21}}{2} \frac{s}{1+s},$$

where $s = 2u/u_s$ is the saturation parameter and $u_s = A_{21}/B_{21}$. Hence explain the functional form of eqn 11.20 when $(\Delta + kv_x) = 0$.

(11.5) Consider an atom moving in the x-direction with velocity v_x in the presence of two laser beams of intensity I and detuning Δ in angular frequency unit as in Fig. 11.4. In the limit where $|kv_x| \ll \Delta$ and $|kv_x| \ll \gamma$, where k is the photon wave vector and γ is the natural line width in angular frequency units, show that the net force on the atom is given by eqn 11.22.

(11.6) Evaluate the possible values of ξ in eqn 11.45 for a $^2S_{1/2} \rightarrow{}^2 P_{3/2}$ transition in a hot alkali atom with σ^+ polarized light.

(11.7) The 4s $^2S_{1/2} \rightarrow$ 4p $^2P_{1/2}$ transition of the Ca$^+$ ion occurs at 397 nm and has an Einstein A coefficient of 1.32×10^8 s^{-1}. A diode laser operating at this wavelength is used to cool a single Ca$^+$ ion held in a Paul trap by the Doppler cooling method. Calculate the lowest temperature that can be achieved by this method, and find the r.m.s. speed of the ions corresponding to this temperature. The relative atomic mass of calcium is 40.1.

(11.8) (a) Explain why neutral atoms with an even number of neutrons in the nucleus are bosons, while those with an odd number are fermions.

(b) Explain why an elemental diatomic molecule is always a boson.

(11.9) Calculate the Bose–Einstein condensation temperature for a gas of free sodium atoms with a density of 10^{21} m^{-3}. Estimate the de Broglie wavelength of the atoms at this temperature, and compare it to the mean particle separation. (The relative atomic mass of sodium is 23.0.)

(11.10) Calculate the fraction of particles in the Bose–Einstein condensate for a gas of free ^{87}Rb atoms with a density of 5×10^{20} m^{-3} at 500 nK.

(11.11) Evaluate the Bose–Einstein condensation temperature for 10 000 ^{87}Rb atoms in a trap of angular frequency 10^3 rad s^{-1}, and find the temperature at which more than half of the atoms are in the condensate.

(11.12) A gas of sodium atoms (relative atomic mass 23.0) is cooled and compressed to a small volume by magneto-optic trapping and evaporative cooling techniques. At time $t=0$ the trap is turned off, and at time $t=6$ ms a shadow image is taken of the expanding gas cloud. The image shows a Gaussian intensity variation from the centre, with a full width at half maximum of 0.5 mm. Calculate the temperature of the atoms.

Part IV

Quantum information processing

Introduction to Part IV

The quantum properties of light have been put to practical use in recent years in various forms of **quantum information processing**. The basic idea here is to use the laws of quantum mechanics to enhance the capabilities of transferring or manipulating data. The subject has three main subbranches:

Quantum cryptography: the use of quantum mechanics to allow the presence of an eavesdropper to be detected when confidential information is being transferred between two parties.

Quantum computing: the use of quantum mechanics to enhance the computational power of a computer.

Quantum teleportation: the use of quantum mechanics to transfer the quantum state of one particle to another.

Quantum optics plays a key role in the practical implementations of all three of these applications. We begin our discussion by considering quantum cryptography in Chapter 12. This is the easiest type of quantum information processing to understand, and the most advanced in terms of progress towards 'real-world' applications. We shall then move on to look at quantum computing in Chapter 13 and quantum teleportation in Chapter 14. Our discussion of quantum teleportation will necessarily lead us to explore the notion of entangled states.

The subject matter in these chapters presumes a basic understanding of the laws of quantum physics. A brief summary of the main ideas may be found in Chapter 3.

Quantum cryptography

12

The fundamental concepts of quantum cryptography were developed in the 1980s and the first experimental proof of principle was given in 1992. Since then, the subject has developed to the point where demonstration systems have been installed that run over long distances down standard telecommunication optical fibre systems. This rapid growth of research activity is partly fuelled by the curiosity of scientists, and partly by the fears of military, government, and financial institutions about data confidentiality and computer security.

We shall begin by first considering the basic principles of classical cryptography, and in particular the concept of public key cryptography. We shall then move on to explain how the laws of quantum physics can be applied to devise a method for transmitting data which is totally safe against eavesdropping attacks. This will lead us to the concept of quantum key distribution, which then forms the basis for secure data transmission. The chapter concludes with a brief description of some of the demonstration quantum cryptography systems that have been implemented, and a discussion of the factors that limit their performance.

12.1	Classical cryptography	243
12.2	Basic principles of quantum cryptography	245
12.3	Quantum key distribution according to the BB84 protocol	249
12.4	System errors and identity verification	253
12.5	Single-photon sources	255
12.6	Practical demonstrations of quantum cryptography	256
Further Reading		261
Exercises		261

12.1 Classical cryptography

Cryptography is the art of encoding a message in such a way that only the person to whom it is addressed can read it. Cryptography is therefore used to send messages that contain secret or confidential information. The techniques of cryptography are widely employed by governments and military organizations, and also in the computer security systems that are used to prevent fraud in financial transactions.

Over the centuries, many ingenious techniques have been devised for encoding secret messages. Let us consider the case of a soldier at the battlefront who wants to send an important radio message to his headquarters without the risk of the enemy learning its contents. He cannot transmit the message in any simple way because the enemy can easily listen in and obtain the message. He has to be more cunning and use codes so that even if the enemy has heard the message they will find it very difficult to decipher. A typical example of how the encoding was done is the ENIGMA code used during the Second World War. This involved the use of a special machine to produce a very sophisticated code. As is now well-known, a secret team working at Bletchley Park in Britain cracked

The team of cryptanalysts who were successful in breaking the ENIGMA code included the mathematician Alan Turing, the pioneer of modern computer science. The work to crack the wartime codes resulted in the development of the first programmable electronic computer.

the code, after making use of vital information from pre-war Polish cryptanalysts. This had a significant effect on the outcome of the war.

The example of the ENIGMA code highlights an inherent weakness of any classical encryption method: there is no way of knowing for certain that unwanted third parties do not have a copy of the code book or encryption machine. Moreover, these third parties might have a team of very clever cryptanalysts who are specially trained in code-breaking skills, with access to very powerful computers that will help them to spot patterns and learn how to decipher the messages.

The only way for the sender to be totally sure that a third party cannot decipher the message is to use a new code for every message. One encryption scheme that follows this method is called the **one-time-pad**, first proposed by Gilbert Vernam during the First World War in 1917. In this cipher the sender and receiver share a common code called the **key**. The key is a random sequence of binary bits (0s and 1s) that is used only once and is at least as long as the message itself. The text of the message is translated into a binary string by some well-known algorithm, and the key is added to produce a new string of bits that comprises the encoded message. The receiver only has to subtract the key from the encoded message to retrieve the text. A simple example may serve to illustrate how this cipher works.

See G. Vernam, *J. Am. Inst. Elect. Eng.* **45** 109 (1926). The one-time-pad is also called the **Vernam cipher**.

Example 12.1   Consider an elementary code in which the letters are represented by five-bit binary numbers from 1 to 26 according to the sequence of the alphabet. Thus A is 00001, B is 00010, and Z is 11010. The encryption process is addition modulo 2, and the random key is '111011000111001'. Decode the message '110111111000001'.

Solution
The message can be deciphered by carrying out subtraction modulo 2:

$$
\begin{array}{r}
110111111000001 \\
\ominus \quad 111011000111001 \\
\hline
001100111111000
\end{array}
$$

The first five bits of the deciphered message are 00110, which we recognize as 'F', the sixth letter of the alphabet. Similarly, the second five bits are 01111 = 15 ('O'), and the last five are 11000 = 24 ('X'). The deciphered message therefore reads 'FOX'.

The one-time-pad cipher is in principle perfectly secure: there are no patterns for the cryptanalysts to recognize because the key is random and unique to each message. However, it is impractical to implement because the sender and receiver must share a common key for each message. This requires a secure and easy method for the sender and receiver to exchange the key without eavesdropping by unwanted third parties. No such method exists for purely classical technology.

There are two basic options for sending a large number of confidential messages quickly. The first option is for the sender and receiver to exchange the secret key in a secure way, for example by a private meeting, and then to use it for all their messages until they next get the chance to exchange a new secret key. This produces insecure messages which are open to deciphering by pattern-spotting through the repeated use of the same key.

The second option is to use **public-key encryption**. Public-key encryption involves two keys: the private key and the public key. The private key is known only to the user, but the public key is known to everyone. The security of the encryption process relies on the fact that certain mathematical functions are very hard to invert. The user generates a private key which is then used to compute the public key. The public key is broadcast openly and used to encrypt the messages that are sent back to the user. Because of the complexity of the encryption process, the encrypted messages can only be deciphered easily with the help of the private key. Since this is only known to the user, only he or she can invert the encryption process with ease.

The RSA encryption scheme used for internet security is a well-known example of public-key encryption. Its security relies on the fact that the time taken to find the prime factors of a large integer increases exponentially with the number of digits. The public key is the product of two large prime numbers which comprise the private key known only to the user. The encryption process cannot be inverted quickly unless these prime numbers are known. However, there is no mathematical proof that an algorithm for finding the prime factors of a large number does not exist. Moreover, if powerful quantum computers should become operational, they would be able to find the prime factors in a manageable time. (See Section 13.5.3.) Thus RSA encryption is only *difficult* to decipher, not *impossible*.

> The RSA encryption scheme is named after its inventors, R. Rivest, A. Shamir, and L. Adleman. It is now known that a similar scheme had secretly been devised some years earlier by British military intelligence researchers. A concise explanation of the principles of RSA encryption may be found in Nielsen and Chuang (2000).

It is in this context that quantum cryptography enters the field. As we shall see in the following sections, quantum cryptography provides a secure method for transmitting private keys across public channels without the risk of undetected eavesdropping by third parties. This is obviously an important issue for military, governmental, and financial organizations, which explains the interest that the subject has aroused in recent years.

12.2   Basic principles of quantum cryptography

We have seen above that present-day cryptographic systems using public-key encoding are not totally secure. For example, the RSA encryption scheme will become obsolete as soon as someone finds an efficient way to factorize large numbers. This inevitably leads us to look for alternative ways to encrypt the data with a higher degree of security.

It is obvious that the whole encryption system would be much safer if the interested parties were to encrypt their message with a secret private key, known only to them, rather than with a public one known

The properties of entangled photon states are described in Chapter 14. For a discussion of their application in quantum cryptography, see Further Reading.

to everyone. The data encrypted with the private key are secure provided that no-one else has the key. The purpose of quantum cryptography is to provide a reliable method for transmitting a secret key and knowing that no-one has intercepted it along the way. The method is founded on the fundamental laws of quantum physics, and the process of sharing a secret key in a secure way is called **quantum key distribution**.

There are two basic schemes that have been devised for carrying out quantum cryptography. The first relies on the basic principles of quantum measurements on single particles, while the second relies on the properties of entangled states. In this chapter we shall only discuss the first type of quantum cryptography, since it is the easiest to understand and is the one which is most commonly implemented in the field.

In discussing quantum cryptography, we invariably encounter three characters: **Alice** (A), **Bob** (B), and **Eve** (E). Alice and Bob are the two people who wish to exchange information. Eve is the eavesdropper who is trying to intercept the message and steal it without disclosing her presence. The task of quantum cryptography is to provide a scheme that enables Eve's activity to be detected.

Quantum cryptography does not protect against eavesdropping attacks, but it does provide a failsafe way for knowing when the message has been intercepted. This allows Alice and Bob to set up a system for transferring private keys with the confidence of knowing that the key really is private. If they detect the presence of an eavesdropper, they can simply discard the bits transferred while Eve was listening in, and start again. Once they have successfully shared the private key, they can use it for encrypting a secret message that can be transmitted across public channels at high data rates. Provided they encrypt with a new key for every message, then they are effectively using a one-time-pad cipher and their message is totally secure against eavesdropping attacks by unwanted third parties.

Let us suppose that Alice wants to send a message to Bob by using a conventional telecommunications system as shown in Fig. 12.1(a). The data signals will be sent as pulses of light along the optical fibre. Strong pulses represent binary '1', while weak pulses, or no pulse at all, represents binary '0'. In this arrangement, there is nothing that Alice and Bob can do to prevent Eve from stealing a copy of the data while it is being transferred down the fibre. All Eve has to do is to intercept the signal, and keep a copy of it without disclosing her presence to Bob. Figure 12.1(b) shows one way in which this might be done. Eve inserts a 50:50 beam splitter (BS) followed by an optical amplifier with a gain of 2 into the fibre. The signal received by Bob is unaffected by Eve's presence, but Eve has obtained a copy which she can then process using her own detection system.

In classical data transmission systems such as the ones shown in Fig. 12.1, there is in principle no way that Alice and Bob can know of Eve's presence. This is because there is no physical law that prevents us from measuring the data signal and making an exact duplicate without affecting it in the process. On the other hand, we know that

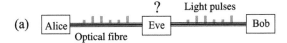

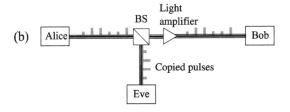

(a)

(b)

Fig. 12.1 (a) In a classical telecommunication system, Alice sends a message to Bob by transmitting high power pulses of light down an optical fibre. Alice and Bob have no way of knowing whether Eve has intercepted the signal along the way or not. (b) Eve's apparatus might consist of a 50 : 50 beam splitter (BS) and an optical amplifier with a gain of 2. This allows her to steal a copy of the data without Bob knowing that she has done so.

quantum mechanics tells us that in general it is not possible to make measurements on single particles without affecting their state in some way or other. For example, we cannot detect a photon, extract all the quantum information from it, and then transmit another photon which is an exact quantum copy of the first one. This is called the **quantum no-cloning theorem**. Now an eavesdropper will have to make some form of measurement on the data stream in order to extract information from it. This means that if we encode the data in a quantum-mechanical way, the eavesdropper will in principle have to reveal her presence by the invasive way in which she makes the measurement. This is the basic principle behind quantum cryptography.

We can illustrate this point by considering the experimental arrangement shown in Fig. 12.2. This arrangement is designed to measure the polarization state of a single photon. As we shall see below, this is in fact one of the methods that are used for the data encoding in practical quantum cryptography systems. The apparatus consists of a polarizing beam splitter (PBS) and two single-photon detectors D1 and D2. The PBS has the property that it transmits vertically polarized light but diverts horizontally polarized light through 90°. This arrangement is conceptually similar to the Stern–Gerlach experiment in which a magnet is used to deflect a particle with a spin quantum number of 1/2. (See Section 3.4.) It is found experimentally that the particle is either deflected up or down depending on the initial state of the incoming particle. The spin up and spin down states of the spin–1/2 particle in the Stern–Gerlach experiment are analogous to the vertical and horizontal polarization states of the photon considered here.

Let us suppose that the incoming photon is linearly polarized with its polarization vector at an unknown angle of θ with respect to the vertical axis. If $\theta = 0°$, we have vertically polarized light and the photon will

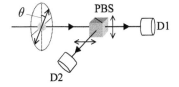

Fig. 12.2 Apparatus to measure the polarization state of a single photon using a polarizing beam splitter (PBS) and two single-photon detectors D1 and D2. The incoming photon is linearly polarized with its polarization vector at an angle of θ with respect to the vertical axis.

be registered by detector D1. Similarly, if $\theta = 90°$, we have horizontally polarized light and the photon will be registered by detector D2. In all other cases we have to resolve the polarization vector into its horizontal and vertical components. Let us represent the quantum state for vertically and horizontally polarized photons by $|\updownarrow\rangle$ and $|\leftrightarrow\rangle$, respectively. We can then write the quantum state $|\theta\rangle$ of a photon with arbitrary polarization angle as a superposition of the two orthogonal polarization states according to:

$$|\theta\rangle = \cos\theta|\updownarrow\rangle + \sin\theta|\leftrightarrow\rangle. \tag{12.1}$$

The probability that the photon is transmitted towards D1 is then given by:

$$\mathcal{P}_{\text{v}} = |\langle\updownarrow|\theta\rangle|^2 = \cos^2\theta. \tag{12.2}$$

Similarly, the probability that the photon is diverted towards D2 is equal to $|\langle\leftrightarrow|\theta\rangle|^2 = \sin^2\theta$.

Now let us suppose that we are trying to determine θ and then transmit another photon with the same polarization angle, as shown in Fig. 12.3. This is exactly what the eavesdropper has to do in the quantum cryptography systems that we shall be discussing below. The measurement could be made by using the arrangement shown in Fig. 12.2. In each measurement the only information Eve receives is whether detector D1 or D2 registers. Detector D1 will register with a probability equal to $\cos^2\theta$ and D2 with probability $\sin^2\theta$. If detector D1 registers then the most sensible thing Eve can do is to send on a vertically polarized photon. Similarly, if D2 registers she will transmit a horizontally polarized photon. However, the state of the second photon is only the same as the first one for the special cases where $\theta = 0°$ or $90°$. For all other values of θ, the act of trying to extract the information about the polarization angle leads Eve to transmit the second photon with a different polarization angle θ' to the first one. This implies that, if measurements are made on the outgoing photon generated by Eve, they can give different results from the ones obtained on the original photon.

The conclusion of this argument is that it is not possible to extract information from a quantum system without altering its state in the

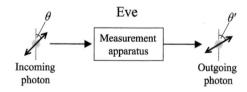

Fig. 12.3 Schematic arrangement for eavesdropping on data encoded as the polarization state of a single photon. In order to extract useful information, Eve must try to determine the unknown polarization angle θ of the incoming photon and send out a second photon with the same value of θ. In general this is not possible according to the laws of quantum mechanics. This means that the polarization angle θ' of the photon sent by Eve will not be equal to θ in most cases.

process. This is a consequence of the invasive nature of quantum measurements. The eavesdroppers must reveal their presence though the disturbance they make through their measurements, which affects the results of subsequent measurements on the photons that are received at the final destination. It could be argued that the eavesdropping scheme we have considered here is very simple and that Eve might devise a more sophisticated way to tap in to the data stream. However, no matter how hard she tries, she will always be subject to the general principles and must give away something in making the measurement. We shall see how this works in practice in the next section.

12.3 Quantum key distribution according to the BB84 protocol

In the previous section we explained the general point that eavesdroppers must reveal their presence through the invasive nature of the measurements they make. We shall now see how this principle is used in practical implementations of quantum cryptography. The idea is to distribute the private key in a secure way so that Alice and Bob can subsequently use it to encrypt secret messages transmitted over public channels. There have been several schemes proposed in the literature and implemented in the laboratory, the two most important of which are:

- the Bennett–Brassard 84 (BB84) protocol,
- the Bennett 92 (B92) protocol.

In what follows we restrict our attention to the BB84 protocol, which will be sufficient to explain the basic principles. The B92 protocol is explored in Exercise 12.3.

In the simplest version of the BB84 protocol, the data are encoded as the polarization states of single photons, with binary '1' and '0' represented by orthogonal polarization states. Thus we could represent 1 by the $\theta = 0$ vertical polarization state and 0 by the $\theta = 90°$ horizontal polarization state, where the polarization angle θ is defined in Fig. 12.2. However, we are not restricted to choosing the axes of the polarization states to be horizontal or vertical. Any orthogonal pair of angles will do. In the BB84 protocol two sets of polarization states called the $\oplus$ and $\otimes$ bases are used:

The $\oplus$ basis: Binary 1 and 0 corresponds to photons with polarization angles of 0° and 90°, respectively.

The $\otimes$ basis: Binary 1 and 0 corresponds to photons with polarization angles of 45° and 135°, respectively.

The two polarization states for the $\oplus$ basis can be represented in Dirac notation by $|\updownarrow\rangle$, $|\leftrightarrow\rangle$, while the two states for the $\otimes$ basis are represented by $|\nearrow\rangle$, and $|\searrow\rangle$ respectively. These assignments are summarized in Table 12.1.

Table 12.1 Data representation values in the BB84 protocol for the two choices of polarization basis. θ is the polarization angle as defined in Fig. 12.2.

Basis	Binary 1	Binary 0
$\oplus$	$\|\updownarrow\rangle$ $\theta = 0°$	$\|\leftrightarrow\rangle$ $\theta = 90°$
$\otimes$	$\|\nearrow\rangle$ $\theta = 45°$	$\|\searrow\rangle$ $\theta = 135°$

See C. H. Bennett and G. Brassard in *Proceedings of IEEE International Conference on Computers, Systems and Signal Processing, Bangalore, India, December 1984*, IEEE, New York (1984), p 175, and C. H. Bennett, *Phys. Rev. Lett.* **68**, 3121 (1992).

The orthogonal polarization states form the foundation for considering the photon as a quantum bit (qubit). See Section 13.2.

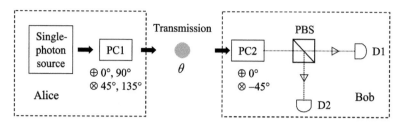

Fig. 12.4 Data encoding scheme according to the BB84 protocol. Alice has a source of vertically polarized photons and a Pockels cell PC1. PC1 rotates the polarization vector by angles of 0°, 45°, 90°, or 135° for each photon at Alice's choice. The photon that has passed through PC1 is then transmitted to Bob who detects it by using a PBS and two single-photon detectors D1 and D2 similar to the arrangement shown in Fig. 12.2. Bob's apparatus includes a second Pockels cell PC2 which can rotate the polarization vector of the incoming photon by an angle of either 0° or −45° at Bob's choice.

A Pockels cell is an electro-optic device which rotates the polarization vector of the light passing through it in proportion to the applied voltage. Many recent implementations of the BB84 protocol do not use Pockels cells any more. See Exercises 12.4 and 12.5.

An experimental scheme for quantum cryptography according to the BB84 protocol is shown in Fig. 12.4. Alice's apparatus consists of a source of vertically polarized photons and a Pockels cell PC1. Alice synchronizes her Pockels cell with the single-photon source and applies the correct voltages to produce polarization rotations of 0°, 45°, 90°, or 135°. In this way she can send a string of binary data which is encoded in either of the two polarization bases at her choice.

The photons emerging from Alice's apparatus are received by Bob who has a polarization measurement arrangement similar to the one shown in Fig. 12.2. Bob's apparatus includes a second Pockels cell PC2 in front of the PBS. Bob applies the correct voltage to this Pockels cell to rotate the polarization vector of the incoming photon by either 0° or −45° at his choice. These two choices are equivalent to detecting in the ⊕ and ⊗ bases, respectively.

Bob does not know the basis that Alice has chosen to encode the individual photons. He therefore has to choose the detection basis at random. If he guesses the right basis, he will register the correct result. This occurs when Alice chooses the ⊕ basis and Bob chooses the 0° detection angle, and also when Alice chooses the ⊗ basis and Bob chooses the −45° rotation angle. If Alice's choice of basis is random, this correct matching of bases will occur 50% of the time. For the remaining 50% of the time Bob will be detecting in the wrong basis and will get random results. Thus, for example, if the incoming photon is polarized at +45° and Bob is detecting in the ⊕ basis (rotation angle = 0°), he will register results on either of his detectors with an equal probability of 50%. (cf. eqn 12.2.)

In the BB84 protocol the following steps are taken.

1. Alice encodes her sequence of data bits according to the scheme in Table 12.1, switching randomly between the ⊕ and ⊗ bases without telling anyone what she is doing. She then transmits the photons to Bob with regular time intervals between them.

2. Bob receives the photons and records the results using a random choice of $\oplus$ and $\otimes$ detection bases as determined by the rotation angle of his Pockels cell.

3. Bob communicates with Alice over a public channel (e.g. a telephone line) and tells her his choice of detection bases, without revealing his results.

4. Alice checks Bob's choices against her own and identifies the subset of bits where both she and Bob have chosen the same basis. She tells Bob over the public channel which of the time intervals have the same choice of basis, and both Alice and Bob discard the other bits. This leaves them both with a set of sifted data bits.

5. Bob transmits to Alice over a public channel a subset of his sifted bits. Alice checks these against her own and performs an error analysis on them.

6. If the error rate is less than 25%, Alice deduces that no eavesdropping has occurred and that the quantum communication has been secure. Alice and Bob are then able to retain the remaining bits as their private key.

Table 12.2 shows an example of how these six steps of the protocol are implemented. The first line shows the original set of the data that Alice wishes to send to Bob. The second line shows the random choice of polarization basis that she makes, which gives rise to the polarization angle encoding of the photons shown in the third line using the criteria given in Table 12.1. The fourth line gives Bob's random choice of detection basis. This will coincide with Alice's for half of the bits on average. In these cases Bob will register the correct result, provided no eavesdropper is present (see below). In the other half of the cases, Bob will only get the right result with a probability of 50%. This does not matter, however, because these data are never used for the key.

The next step involves the comparison of the two bases. Bob publicly announces his choice of bases without revealing his results. Alice

Table 12.2 Representative sequence of data choices according to the BB84 protocol. θ is the polarization angle according to the encoding scheme given in Table 12.1.

A's data	1	0	0	1	1	1	0	0	1	0	0	1
A's basis	$\oplus$	$\otimes$	$\oplus$	$\otimes$	$\otimes$	$\oplus$	$\oplus$	$\otimes$	$\oplus$	$\otimes$	$\otimes$	$\oplus$
θ (°)	0	135	90	45	45	0	90	135	0	135	135	0
B's basis	$\otimes$	$\otimes$	$\oplus$	$\oplus$	$\otimes$	$\oplus$	$\otimes$	$\oplus$	$\oplus$	$\otimes$	$\oplus$	$\otimes$
B's result	1	0	0	0	1	1	0	1	1	0	1	1
Same basis ?	n	y	y	n	y	y	n	n	y	y	n	n
Sifted bits		0	0		1	1			1	0		
Data check ?		y	n		y	n			y	n		
Private key			0			1				0		

checks this against her choices and identifies the cases where the two choices coincide. These are identified with the 'y' label in the sixth row of Table 12.2. Alice tells Bob which bits these are, and they discard the other bits. This now leaves them both with the sifted bits shown in the seventh row of the table. Bob now sends a subset of his sifted bits to Alice, again over a public channel. In the example shown, he sends every other bit. Alice can check these against her own list, and carry out an error analysis.

This is the stage at which the eavesdropper reveals her presence. It is easiest to understand what happens if we assume that Eve has the same apparatus as Alice and Bob. She can then detect the photons sent by Alice using a copy of Bob's apparatus, and transmit new photons to Bob using a copy of Alice's apparatus, as shown schematically in Fig. 12.5. Since she cannot know what choice of basis Alice is making, she must choose her detection basis randomly. Half the time she will guess correctly and accurately determine the polarization state of the photon. She can then send an identically polarized photon on to Bob without anyone knowing about it. For the remaining half of the bits, she will guess incorrectly, and register a result on either of her detectors with an equal probability of 50%. She will then send a photon to Bob which is polarized with her choice of detection basis, rather than Alice's. This means that Eve will alter the polarization basis angle by 45° for 50% of the bits. In the cases where Bob has chosen the same basis as Alice and Eve has guessed incorrectly, Bob will register random results on his detectors with a probability of 50%. He will thus register errors even when he has guessed Alice's basis correctly. The error probability $\mathcal{P}_{\text{error}}$ is given by:

$$\begin{aligned}
\mathcal{P}_{\text{error}} &= \mathcal{P}_{\text{Eve has wrong basis}} \times \mathcal{P}_{\text{Bob gets wrong result}}, \\
&= 50\% \times 50\%, \\
&= 25\%.
\end{aligned} \tag{12.3}$$

This high error rate of 25% will be easily recognizable when Alice carries out her error analysis in the final step of the process. She will thus be able to detect the presence of the eavesdropper, and therefore know whether the private key distribution has been secure.

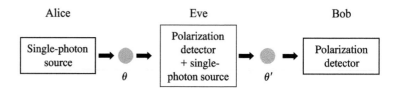

Fig. 12.5 An eavesdropper between Alice and Bob tries to measure the polarization angle θ of the photon sent by Alice and send an identical photon on to Bob. She reveals her presence because the polarization angle θ' of the second photon will be different from θ for 50% of the photons.

12.4   System errors and identity verification

It is apparent from the argument above that the crux of the security of quantum cryptography is the possibility of detecting the errors introduced by Eve's presence. A potential weakness of this line of approach is that there will always be errors even when Eve is not present. In this section we shall deal with the errors introduced by random deletion of photons, birefringence, and detector dark counts. We shall also discuss the problem of identity verification. In the next section we shall deal with the errors related to the fact that the source might emit more than one photon at a time.

12.4.1   Error correction

The easiest type of error to deal with is random deletion of photons between Alice and Bob. It will often be the case that Alice sends a photon and Bob registers no result at all on either of his detectors. This can occur for a number of reasons, including:

- absorption or scattering of the photons as they propagate from Alice to Bob;
- inefficient light collection so that some of the photons miss the detectors;
- detector inefficiency.

These difficulties occur to a greater or lesser extent in all of the experiments carried out so far. We shall discuss them further in the context of the experimental results in Section 12.6 below. At this stage we simply state that random deletion errors do not affect the security of the system. At the time when Bob declares his choice of bases to Alice (step # 4 in the list given in Section 12.3), he must also tell her when he registered a result, without, of course, publicizing what the result was. Alice then checks for the occasions when the bases were the same and Bob registered a result. This subset of the data stream is then used for the error checking analysis. The net result is that Alice ends up throwing away more of her original data set than for the case with no random deletion. This reduces the useful data transfer rate that can be obtained from the quantum cryptography system, but does not affect its security.

The second type of error, namely birefringence, is more serious. If Alice sends a vertically polarized photon, she wants it to stay vertically polarized all the way to Bob. However, if the medium in which the photons travel is birefringent, the polarization angle will change as it propagates. Bob will therefore have a probability of getting the wrong result even when he has chosen the correct detection basis and there is no eavesdropper present. This type of error has to be carefully calibrated out of the system by using classical error-correction algorithms. When Alice carries out her error analysis on the sifted data (step # 4 in the list given

See Section 2.1.4 for a brief discussion of birefringence. A more detailed account may be found, for example, in Hecht (2002).

in Section 12.3) checks have to be made in exactly the same way that is done with classical data transmission systems. This involves grouping the bits together and carrying out a series of systematic parity checks.

See C. E. Shannon, *Bell Syst. Tech. J.*, **27**, 379, 623 (1948).

The number of bits that Alice and Bob have to exchange over the public channel in order to correct for the errors is given by **Shannon's noisy channel coding theorem**. This states that if we have N bits with an error probability of ε, then the number of bits that must be compared to correct for the errors is equal to:

$$N_{\text{correction}} = N \left[-\varepsilon \log_2 \varepsilon - (1 - \varepsilon) \log_2(1 - \varepsilon) \right]. \qquad (12.4)$$

This shows that the larger ε is, the more bits we have to waste in the data correction process. This obviously implies that Alice and Bob have to do everything they can to make ε as small as possible. In practice, we can tolerate error rates up to a certain limit at the cost of a reduced data transmission rate, and with an absolute condition that ε should be significantly smaller than the error rate introduced by the eavesdropper.

The third type of error, namely detector dark counts, has the same effect as the second: Bob can register a wrong result even when he has chosen the correct detection basis. Dark count errors occurs when the photon sent by Alice never reaches Bob and the wrong detector randomly registers due to thermal noise in the photocathode. This type of error again has to be calibrated out by using classical error analysis on a portion of the sifted bits, as described in the previous paragraph.

The combined result of all of these errors is a reduction in the length of the private key that can be shared between Alice and Bob. Bits are lost with the first type of error because only a fraction of Alice's photons are detected. The second and third types of error reduce the key length because we have to waste a portion of the sifted bits in order to carry out error correction algorithms. The end result is effectively the same: the data transmission rate for the private key is reduced. This does not limit the security of the system, but does reduce its efficiency.

12.4.2   Identity verification

Quantum cryptography suffers from another potential weakness. There is nothing Alice and Bob can do to prevent Eve from intercepting the data and then pretending to be Bob. In this way Eve will obtain a copy of the secret key instead of Bob, and will be able to decipher any encrypted confidential messages that are sent afterwards.

This type of 'man-in-the-middle' attack is an old problem and is inherent to all types of cryptography. For example, if Alice uses a trusted courier to send the secret key to Bob, she has to carry out some checks to verify that the key has reached Bob safely and not been intercepted along the way. She might phone up Bob to ask him whether he has received the key or not, after first checking carefully that she really is speaking to Bob and not some impersonator. The technical name for these checks is **identity verification**. Fortunately, there exist well-established classical techniques for this authentication procedure which can be applied to guarantee that the data transfer has been secure. However, it should

See §2.5.2 in Bouwmeester *et al.* (2000) for further details on identity verification.

be pointed out that these authentication procedures do require that Alice and Bob already share a private key. This can only be achieved by a face-to-face communication. This private key can then be used to authenticate the first key-sharing transmission. The new private key obtained by this transmission can then be used to authenticate the next key-sharing transmission, and so on.

12.5 Single-photon sources

In this section we shall explore in more detail the requirements on the light source that Alice uses to generate the photons she sends. We have been supposing all along that Alice sends only one photon at a time to Bob. If she were to send more than one photon, she would risk giving away information to Eve for free. It is easy to see why this is so by considering a simple example.

Let us suppose that Alice sends light pulses containing two photons instead of one, and there is an eavesdropper on the line. If Eve is detecting with the wrong basis, there is a 50% chance that the two photons will be registered by both detectors. If this occurs, Eve knows for sure that she is using the wrong basis. She would then discard this data bit, and it would appear to Bob that a random deletion error has occurred. In practice this reduces the fraction of times that Eve transmits a photon to Bob with the wrong polarization. Hence the error rate introduced by Eve decreases. (See Exercise 12.9.) This problem gets worse with every extra photon that is contained in the light pulse. For example, if there are three photons per pulse, Eve can determine both the basis and bit value for a significant fraction of the data pulses. (See Exercise 12.10.)

The conclusion is that we have to try to make sure that there is only one photon in each light pulse. This is not so easy to achieve in practice. The standard procedure for producing a single-photon source is to take a pulsed laser and attenuate it very strongly so that the mean photon number per pulse $\overline{n}$ is small. We have seen in Section 5.3 that the light from a single frequency laser is expected to have Poissonian photon statistics. When $\overline{n}$ is small, most of the time intervals will contain no photons, a small fraction will contain one photon, and a very small number will contain more than one photon. This fact is illustrated in Example 12.2 below, which shows how the relative probabilities are determined by the Poissonian photon statistics of the attenuated laser pulses.

The value of $\overline{n} = 0.1$ chosen for Example 12.2 is fairly typical of present-day implementations of quantum cryptography. The example shows that the ratio of pulses containing more than one photon to the number with $n = 1$ is alarmingly high at 5%. This is a basic weakness of quantum cryptographic systems using attenuated laser pulses, which have Poissonian photon statistics.

A much better approach is to use a genuine single-photon source. This is a source that emits exactly one photon on demand, as described in Section 6.7. Simple quantum cryptography experiments have been performed with such sources, but so far they have been too inconvenient or slow to be used in the more advanced systems. The development of fast, convenient triggered single-photon sources is therefore a very active area of research at present.

Example 12.2   A laser operating at 800 nm emits pulses at a rate of 4 MHz. The laser is attenuated so that the average power is 0.1 pW. Calculate:

(a) the average number of photons per pulse;

(b) the fraction of the pulses that contain no photons;

(c) the fraction of the pulses that contain one photon;

(d) the fraction of the pulses that contain more than one photon;

(e) the ratio of the number of pulses containing more than one photon to the number with just one.

Solution

(a) The photon energy is 2.5×10^{-19} J, and at a power level of 10^{-13} W the photon flux is 4.0×10^5 s^{-1}. The pulse rate is 4.0×10^6 s^{-1}, and hence $\bar{n}$ is equal to 0.1.

(b) We can calculate the probability from the Poisson distribution given in eqn 5.13 with $\bar{n} = 0.1$. This gives:

$$\mathcal{P}(0) = \frac{0.1^0}{0!} e^{-0.1} = 0.9048.$$

(c) We repeat the procedure for part (b) for $n = 1$ to find:

$$\mathcal{P}(1) = \frac{0.1^1}{1!} e^{-0.1} = 0.0905.$$

(d) The probability that $n \geq 2$ is equal to $1 - [\mathcal{P}(0) + \mathcal{P}(1)]$. Using the results in parts (b) and (c), we then obtain:

$$\mathcal{P}(n \geq 2) = 1 - [0.9048 + 0.0905] = 0.0047.$$

(e) This ratio is given by:

$$\mathcal{P}(n \geq 2)/\mathcal{P}(1) = 0.0047/0.0905 = 5.2\%.$$

12.6   Practical demonstrations of quantum cryptography

The practical demonstrations of quantum cryptography using the BB84 or B92 protocols fall into two broad subcategories:

• free-space quantum cryptography;

• quantum cryptography in optical fibres.

The technical issues for these two types of quantum cryptography are different, and so we shall discuss them separately below, starting with the free-space systems.

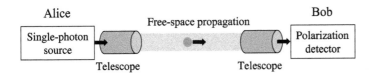

Fig. 12.6 Schematic representation of free-space quantum cryptography. The detailed description of Alice and Bob's apparatus is given in Fig. 12.4.

12.6.1   Free-space quantum cryptography

In free-space quantum cryptography, the photons sent by Alice travel through the air towards Bob's receiver apparatus. The basic arrangement is shown schematically in Fig. 12.6. The data are encoded as the polarization state of the photon, and Alice and Bob both have the same apparatus as shown in Fig. 12.4. Alice fires her photons into a telescope which expands, collimates, and directs the beam towards Bob's receiver. Bob himself has another telescope which allows him to collect the photons efficiently. These telescopes are needed to minimize the effects of beam expansion caused by diffraction when Alice and Bob are separated by long distances. Without the telescopes, the fraction of the photons that would fall upon the detector area would be unacceptably low. (See Exercise 12.11.)

The first practical demonstration of quantum cryptography used free-space propagation and was reported by Bennett, Brassard, and co-workers in 1992. They used strongly attenuated pulses from a light-emitting diode operating at 550 nm and transmitted the photons across an air gap of 0.32 m. Much progress has been made since this first proof-of-principle experiment. Free-space quantum cryptography systems have now been demonstrated across distances of 10 km in both daylight and at night. In another experiment, the quantum key was distributed across 23 km between the summits of two Alpine mountains at night. The long-term aim of these experiments is to develop quantum cryptography systems for communicating with satellites in low earth orbits. Feasibility studies indicate that there are no fundamental obstacles that should prevent this from becoming a reality.

The long-range free-space systems implemented so far have been carried out at wavelengths in the range 600–900 nm. At these wavelengths, the atmospheric losses are small, and low-noise detectors with high quantum efficiencies are readily available. In these conditions there are two main sources of error:

- Air turbulence: this causes random deviations in the direction and timing of the light pulses. The effects of these random deviations are well-known from the twinkling of stars. The effects of air turbulence can be minimized by sending a bright (classical) pulse in front of the encoded photon. This allows Alice and Bob to compensate for the beam wandering and timing jitter.

See C. H. Bennett, *et al.*, *J. Cryptology* **5**, 3 (1992).

The 10 km ground-level system is described in R. J. Hughes *et al.*, *New. J. Phys.* **4**, 43 (2002), while the mountaintop experiment is described in C. Kurtsiefer *et al.*, *Nature* **419**, 450 (2002). The feasibility of ground to satellite quantum cryptography is considered in J. G. Rarity *et al.*, *New. J. Phys.* **4**, 82 (2002).

- Stray light: background light from the sun, moon, or street lamps, causes unwanted detector counts. The stray light signal can be reduced by placing suitable filters in front of the detectors and triggering the detectors so that they are only switched on for the short time interval in which the encoded photon is expected to arrive.

It is important to realize that most of the deleterious effects due to atmospheric turbulence occur in the first few km from the ground. Hence the demonstration of cryptography over similar distances at ground level is a significant step towards the long-term goal of overcoming the problems associated with communicating with satellites.

12.6.2   Quantum cryptography in optical fibres

If quantum cryptography systems are to find widespread applications, it will be necessary to make them compatible with optical fibre telecommunication systems. This has prompted several research groups to set up experimental quantum cryptography systems in which the single photons are transmitted down optical fibres, as indicated schematically in Fig. 12.7. In these arrangements, Alice launches her photons into an optical fibre and Bob receives them after they have propagated to him.

The optical fibre systems are in principle much more convenient than their free-space counterparts because they use standard telecommunication components. Moreover, the beam does not diverge, and therefore the loss of photons due to beam expansion is not an issue. On the other hand, there are two significant disadvantages compared to the free-space systems:

- the fibres introduce losses, which cause the intensity of the optical signals to decay as they propagate;

- the fibres are birefringent, which causes practical problems in implementing the polarization encoding schemes described in Section 12.3.

These two difficulties are discussed in turn below.

Fibre losses are caused both by scattering and absorption. The scattering rate generally varies as $1/\lambda^4$ (Rayleigh scattering), and therefore decreases strongly with wavelength. Absorption losses are high in the ultraviolet and infrared spectral regions below 300 nm and beyond 1600 nm, respectively. There are also high absorption losses at specific wavelengths associated with OH bonds, for example, 1400 nm.

The propagation losses in optical fibres depend strongly on the wavelength. There are three common wavelength bands used in fibre optic systems, namely 850, 1300, and 1550 nm. The 850 nm band has much larger scattering losses, which would seem to suggest that the 1300 and 1550 nm bands would be the optimal ones for quantum cryptography. However, this is not necessarily the case, due to differences in the single-photon detectors that are available for the different wavelengths. Photons of 850 nm can be detected with low noise single-photon avalanche photodiode (SPAD) detectors made from silicon. However, at 1300 and 1550 nm, the photon energy is below the band gap of silicon, and detectors made from materials with smaller band gaps (e.g. germanium or InGaAs) have to be used. These narrow-gap SPADs

Fig. 12.7 Schematic representation of fibre optic quantum cryptography.

have much higher dark count rates, which increases the number of errors. (See Exercise 12.13.) Furthermore, they suffer from an effect called afterpulsing, which severely restricts the bit rate that can be achieved.

The problem of fibre birefringence is not too serious for laboratory-based demonstrations performed over relatively short time-scales. However, for 'real-world' systems operating over long time-scales with fibres buried in the ground, there will inevitably be thermally or mechanically induced changes in the fibre birefringence, which makes it necessary to take a different approach. One solution that is commonly employed is to use **optical phase encoding** instead of polarization encoding. Here, a Mach–Zehnder interferometer is used to encode photons by changing the optical phase in each arm at both Alice and Bob, as shown in Fig. 12.8(a). When the relative phase shift is 0 or π, the photon exits through a definite port of the second fibre coupler, since these phase shifts correspond to the conditions for classical bright or dark fringes. However, for relative phase shifts of $\pi/2$ or $3\pi/2$, the photon can exit at

Optical phase encoding was originally proposed by C. H. Bennett: see *Phys. Rev. Lett.* **68**, 3121 (1992).

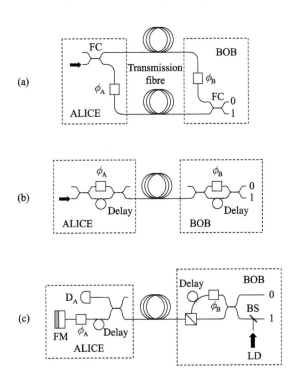

Fig. 12.8 (a) Optical phase encoding cryptography scheme using a single Mach–Zehnder interferometer and two phase shifters, ϕ_A and ϕ_B. (b) Optical phase encoding with two unbalanced interferometers. A time delay is introduced by adding an extra length of fibre in one arm of each interferometer. (c) Plug and play scheme with a single auto-compensating interferometer. The 50 : 50 fibre couplers are labelled FC in part (a), but not in parts (b) and (c) for clarity. In (b), time gating is used to eliminate the photons that take the shortest or longest paths. In (c), a classical pulse is injected at Bob's end from a laser diode (LD) via a weakly reflecting beam splitter (BS). This pulse is then encoded by Alice after reflecting off the Faraday mirror (FM) and being attenuated to the single-photon level. Alice's detector (D_A) is used as a trigger to activate the phase shifter ϕ_A.

Table 12.3 Implementation of the BB84 protocol with phase encoding. When the relative phase difference is equal to $\pi/2$ or $3\pi/2$, Bob's measurement can return the values of 0 or 1 with equal probability.

A's Bit value	ϕ_A	ϕ_B	$|\phi_A - \phi_B|$	B's bit value
0	0	0	0	0
0	0	$\pi/2$	$\pi/2$	0 or 1
1	π	0	π	1
1	π	$\pi/2$	$\pi/2$	0 or 1
0	$\pi/2$	0	$\pi/2$	0 or 1
0	$\pi/2$	$\pi/2$	0	0
1	$3\pi/2$	0	$3\pi/2$	0 or 1
1	$3\pi/2$	$\pi/2$	π	1

either port with 50 : 50 probability, and is thus equivalent to a photon with a polarization angle of 45° entering a polarizing beam splitter. Hence the scheme is equivalent to the polarization encoding version of the BB84 protocol described in Section 12.3 when the phase encoding sequence shown in Table 12.3 is used.

The simple scheme shown in Fig. 12.8(a) is not practical as it requires careful balancing of the arms of an interferometer several kilometres long. For this reason, the scheme using two unbalanced Mach–Zehnder interferometers shown in Fig. 12.8(b) is to be preferred. By disregarding the photons travelling by the shortest and longest routes through the two unbalanced interferometers, it is possible to obtain the phase relationship described in Table 12.3. This approach still relies on the stringent condition of a constant phase relationship between the interferometer arms during the key exchange, but the conditions are significantly relaxed compared to those for a single interferometer.

The scheme shown in Fig. 12.8 (b) was initially demonstrated in 1993 by Townsend *et al.*, *Electron. Lett.* **29**, 634 (1993). Later experiments gave improved results by use of polarization discrimination at Bob's interferometer to identify the paths taken by the photons. See Marand and Townsend, *Opt. Lett.* **20**, 1695 (1995).

The double Mach–Zehnder scheme shown in Fig. 12.8(b) still requires active phase control since it is susceptible to small path length changes in the arms of the interferometer as well as changes in the birefringence of the optical components. For this reason, an auto-compensating technique employing a single interferometer as shown in Fig. 12.8(c) has been developed. In this scheme, a large (multi-photon) pulse is sent from Bob to Alice via a beam splitter (BS) in the path to one of the detectors. The reflectivity of this beam splitter is chosen to be low, so that only a small fraction of the single photons going to the detector at the end of the round trip are lost. After propagating to Alice, the pulse is reflected off a Faraday mirror (FM) consisting of a quarter wave plate and a mirror, before being attenuated to the single-photon level and sent back to Bob along the same path. Provided that the environmentally induced optical changes occur on a much longer time-scale than the transit time, any birefringence in the first transit is exactly compensated during the reflected path. The Mach–Zehnder interferometer at Bob is then used in the same manner as above with phase encoding. This technique has become known as 'plug and play' cryptography, and has been demonstrated on installed fibres at distances of 67 km.

See Stucki *et al.*, *New J. Phys.* **4**, 41 (2002) for technical details of the 67 km experiment. Plug and play cryptography is potentially vulnerable to 'Trojan Horse' eavesdropping attacks. Eve could inject her own bright (i.e. multi-photon) pulse into the quantum channel just before or after Bob and then measure the applied phase shift on the return, thus knowing all of Alice's bit stream. Moreover, the plug and play systems also tend to be relatively slow due to detector saturation by unavoidable scattered light from the high-intensity pulse.

The trade-off between fibre losses and the detector technology means in practice that high-speed quantum cryptography systems tend to operate at 850 nm over modest ranges compatible with local area networks, whereas the long-range systems operate at 1300 or 1550 nm, but at a much slower rate. The importance of the detector technology makes the development of SPADs with low noise, high efficiency, low jitter, and high count rate a very active area of research at present. At the time of writing, the fastest net quantum bit rate (i.e. the quantum bit rate after allowing for error corrections) that has been achieved is 100 kbit s^{-1} at 850 nm over a 4.2 km fibre, while the longest distance over which quantum cryptography has been demonstrated is 122 km at 1550 nm.

See K. J. Gordon *et al.*, *IEEE J. Quantum Electron.* **40**, 900 (2004), and C. Gobby *et al.*, *Appl. Phys. Lett.* **84**, 3762 (2004).

Further reading

The subject of quantum cryptography is explained in greater depth in Bouwmeester *et al.* (2000) and Nielsen and Chuang (2000). A large number of introductory reviews have been published, for example: Bennett *et al.* (1992), Rarity (1994), Hughes *et al.* (1995), Phoenix and Townsend (1995), Tittel *et al.* (1998), and Hughes and Nordholt (1999). A comprehensive review is given in Gisin *et al.* (2002).

The principles of quantum cryptography with entangled states are explained in Bouwmeester *et al.* (2000), and an experimental implementation is described in Jennewein *et al.* (2000). A collection of papers on single-photon sources and their application in quantum cryptography may be found in Grangier *et al.* (2004). Discussions of the single-photon detector technology may be found in Cova *et al.* (1996) and Hiskett *et al.* (2000).

A collection of tutorial articles on quantum cryptography may be found at http://cam.qubit.org. Many other interesting papers on quantum cryptography may be found by searching on the Los Alamos National Laboratory e-print archive at http://xxx.lanl.gov/archive/quant-ph.

Exercises

(12.1) Decode the message:
'1111100001101110011011000100001001101011'
encoded with the key:
'1101001000110011010101100101011101000101'
according to the protocol described in Example 12.1.

(12.2) Suppose that Alice sends the message '001011001011' according to the BB84 protocol with the following sequence of bases: $\oplus \oplus \otimes \oplus \otimes \otimes$ $\otimes \oplus \oplus \otimes \oplus \otimes$. If Bob using the following sequence of detection bases: $\otimes \oplus \oplus \otimes \otimes \oplus \oplus \oplus \otimes \oplus \otimes \otimes$, find the sifted data set.

(12.3) Figure 12.9 gives a schematic representation of a system designed to implement the B92 protocol

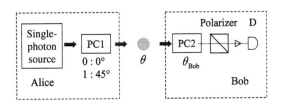

Fig. 12.9 The B92 protocol. Alice has a source of linearly polarized single photons with a polarization angle of 0°. PC1 and PC2 are Pockels cells set to rotate the polarization by the specified angles, and D is a single-photon detector. Bob's polarizer is set so that 100% transmission occurs for input photons with $\theta = 0°$ when PC2 is turned off (i.e. introduces no rotation).

using linearly polarized photons. Alice encodes her data according to the polarization angle θ of the photon, with $0° \equiv 0$ and $45° \equiv 1$. Bob makes measurements with a Pockels cell PC2 randomly set to rotate by an angle θ_{Bob} of either of $45°$ or $90°$. A polarizer set to transmit perfectly for photons with $\theta = 0°$ when $\theta_{\text{Bob}} = 0°$ is placed after PC2, followed by a single-photon detector D.

(a) Describe the possible outcomes for both of Bob's measurement settings. Explain how this arrangement can be used for unambiguous transmission of bits.

(b) In the absence of losses, detector errors, and an eavesdropper, compare the fraction of Alice's bits that Bob receives in the B92 protocol to the fraction in the sifted data set of the BB84 protocol.

(c) How would an eavesdropper be detected in this scheme?

(12.4) Consider a BB84 quantum cryptography system which employs attenuated laser pulses as the source for Alice's photons.

(a) Explain how Alice can produce photons with a particular polarization angle by placing suitable linear optical components after the laser.

(b) Devise a scheme for producing a stream of single photons with their polarization angles switching between angles of $0°$, $45°$, $90°$, or $135°$ at choice by combining four such laser beams. (Assume that Alice can turn the lasers on or off at will.)

(12.5) Consider the detection scheme for the BB84 protocol shown in Fig. 12.10. The apparatus consists of a 50:50 beam splitter which preserves photon polarization (e.g. a half-silvered mirror) and two polarizing beam splitters with single-photon detectors at all output ports. A waveplate which rotates the polarization by $45°$ is inserted in front of one of the polarizing beam splitters. Explain how this arrangement performs the same tasks as Bob's detection apparatus shown in Fig. 12.4.

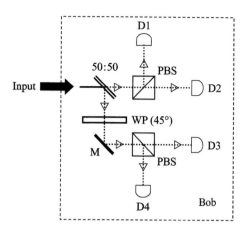

Fig. 12.10 Bob's detection scheme for BB84 quantum cryptography without a Pockels cell. The light is split by a 50:50 non-polarizing beam splitter, and is fed into two polarizing beam splitters (PBS) with single-photon detectors (D) at all output ports. A waveplate (WP) set to rotate the polarization by $45°$ is inserted before one of polarizing beam splitters. The mirror (M) has no significant effect on the results. (After J. G. Rarity *et al.*, *Electron. Lett.* **37**, 512 (2001).)

(12.6) Calculate the fraction of the sifted data that must be used for error correction if the error rate is 1%.

(12.7) Explain why strongly attenuated light pulses always have Poissonian statistics, irrespective of the photon statistics of the original pulse.

(12.8) Consider a Poissonian source with a mean photon number of x. Let $\eta = \mathcal{P}(n \geq 1)/\mathcal{P}(1)$. Find a relationship between η and x when both are $\ll 1$, and hence determine the mean photon number required to obtain $\eta < 1\%$.

(12.9) Consider the case where Alice's pulse contains two photons in the BB84 protocol, and Bob has detectors with 100% quantum efficiency.

(a) Explain why it is sensible for Eve to pretend to be a loss on the line and follow a strategy whereby she sends no photon to Bob when both of her detectors fire.

(b) Calculate the fraction of the pulses transmitted to Bob that have their polarization angle altered by Eve.

(c) Calculate the error rate in the sifted data set caused by Eve's presence.

(12.10) Consider what happens when a pulse containing three photons arrives at the detection system described in Exercise 12.5.

    (a) Calculate the probability that three different detectors fire. (Assume perfect detectors.)

    (b) Show that when this happens, both the basis and the bit value of the incoming photon are revealed.

(12.11) In a free-space quantum cryptography experiment operating at 650 nm over a distance of 20 km, Alice uses a beam collimator with a diameter of 5 cm to send her photons to Bob. On the assumption that other losses are negligible, compare the fraction of the photons that are incident on Bob's detectors when he uses a collecting lens with a diameter of (a) 5 and (b) 25 cm.

(12.12) A ground-to-satellite quantum cryptography system operating at 650 nm is designed with Alice's station on a mountaintop in the desert. In these conditions, a typical value of the angular wander introduced by atmospheric turbulence during the daytime is 10^{-5} radians. At what value of the diameter of the telescope would the divergence of the beam be limited by turbulence rather than diffraction?

(12.13) Single-photon avalanche photodiodes (SPADs) work by multiplying the current produced when a single electron is excited across the band gap of a semiconductor after absorption of a single photon. Explain why a SPAD designed for use at 1300 nm will have a higher dark count rate than one designed for use at 850 nm.

(12.14) In classical fibre-optic communication systems, the signals are amplified at regular intervals by repeaters to compensate for the decay in the intensity due to scattering and absorption losses. Discuss whether it is possible to use repeaters to increase the range of a quantum cryptography system.

13 Quantum computing

13.1	**Introduction**	**264**
13.2	**Quantum bits (qubits)**	**267**
13.3	**Quantum logic gates and circuits**	**270**
13.4	**Decoherence and error correction**	**279**
13.5	**Applications of quantum computers**	**281**
13.6	**Experimental implementations of quantum computation**	**288**
13.7	**Outlook**	**292**
	Further reading	**293**
	Exercises	**294**

In this chapter we shall look at the basic principles of quantum computing and its implementation by optical techniques. Since this is a rapidly developing subject that occupies the attention of many research groups worldwide, we shall concentrate on introducing the fundamental ideas and avoid too many details that will inevitably date very quickly. The reader is referred to the bibliography for more rigorous treatments of the subject and more comprehensive discussions of the present state of the art in the experiments.

13.1 Introduction

Present-day computer technology is based on the silicon microprocessor chip. Silicon technology was first introduced in the 1960s, and has developed at a staggering rate that is familiar to everyone. The rapid development of the technology was noticed as early as 1965, when Gordon Moore, co-founder of the Intel Corporation, enunciated the law which now bears his name. **Moore's law** states that the number of transistors on a chip doubles every 18–24 months. The exponential growth that Moore's law predicts has held true for 30 years. Figure 13.1 shows the exponential progression of chip technology from the Intel 4004 introduced in 1971, which had 2250 transistors, through to the Pentium 4 introduced in 2000, which has 42 000 000.

The optimism that Moore's law engenders seems to hold no bounds. However, a closer look at the underlying principles reveals that the law must eventually break down at some time in the not-too-distant future. The progress in the chip technology has followed developments in the

Fig. 13.1 Evolution of Intel microprocessors from the introduction of the 4004 chip in 1971. The graph shows the number of transistors per microprocessor against year of introduction on a log-linear scale. The straight line fit establishes the exponential growth predicted by Moore's Law.
(*Source*: Intel. See www.intel.com/technology/mooreslaw)

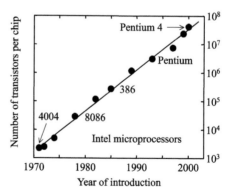

fabrication techniques that make it possible to produce transistors of ever-diminishing size. The transistors used in modern desktop computers are already less than 1 μm in dimension, and to maintain the progress, the size will have to continue to shrink. This makes it more and more difficult to produce the chips, leading to a similar exponential rise in the cost of the fabrication plants, a fact which is sometimes known as **Moore's second law**.

At the more fundamental level, an even more serious problem is going to be encountered soon because quantum effects will begin to become important when the size of the transistors becomes comparable to the de Broglie wavelength of the electrons that carry the signals. On these length-scales the physical laws that govern the circuit design such as Ohm's law no longer hold, and the circuits will no longer operate in the normal way. Even if this problem could be overcome by designing the circuits using quantum transport theory rather than the classical laws, we shall eventually hit another barrier when the size of the transistor becomes comparable to the size of the individual atoms. At this point the progression must stop, because we cannot realistically divide matter into smaller units than its constituent atoms.

Nobody knows for sure when Moore's law will break down. Moore himself has predicted that the end of the road will come around 2020. What is clear is that the law must eventually break down, and this will impose limits on the computational power that can be obtained by improving the existing technologies. For certain types of task, the failure of Moore's law will not lead to particular difficulties. The word-processors of tomorrow will continue to function even though the processing power of the chips will not be improving at the kind of rate that we are used to. However, for number-crunching tasks, the scale of the problems that can be tackled is always ultimately limited by the computer processing power that is available.

Computing tasks are generally classified according to the way in which they scale with the size. If the number of computer operations increases as a polynomial power of the size N, then the problem is said to belong to the **polynomial complexity class**, abbreviated to **P**. If, on the other hand, the number of operations increases faster than a polynomial function, then the problem is said to belong to the **non-polynomial complexity class** (**NP**). This difference is illustrated in Fig. 13.2, which compares the way a polynomial function of N, namely N^4, compares with a non-polynomial function, namely $\exp(N)$. As N gets larger, the non-polynomial functions always win eventually.

Conventional computers are able to handle problems within the **P** class without too much difficulty. If the problem is too hard to solve today, then Moore's law tells us that we should be able to solve it soon, due to the exponential increase in processing power. On the other hand, problems in the **NP** class are always going to prove difficult. We only have to increase the size of the problem by a small amount to need a very large increase in the amount of computing power required. An important example of an **NP** problem is the factorization of large numbers. At

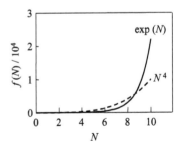

Fig. 13.2 Comparison of the size scaling of a polynomial function (N^4) with a non-polynomial function, namely $\exp(N)$.

It has not been proven mathematically that an algorithm for efficient factorization does not exist. If such an algorithm were to exist, then factorization would reduce to the **P** class.

present, the only way to find the prime factors of a large integer N is to divide N by all odd integers up to $\sqrt{N}$ to see if there is a remainder or not. Since the process of division takes of order N operations, we need an extra N operations each time we increase N by one. In other words, the number of operations required increases exponentially with N. Thus by increasing N, we enforce an exponentially increasing consumption of computer time for finding the factors. This is the basis of the security of the widely used RSA encryption scheme that we encountered in Section 12.1.

The difficulty that computer scientists meet when dealing with problems in the **NP** class stems from the escalating increase in computer time required as N increases, which makes the problem intractable in practice. All these statements presuppose that the computer scientists only have at their disposal a conventional computer which runs according to classical principles. These principles are modelled mathematically according to the operations of **universal Church–Turing machines,** (or **Turing machines** for short). The breakthrough in quantum computation came with the realization that other types of computer might exist that operate on completely different principles to Turing machines. In this case, the Turing machine should only be seen as the limiting case of more general types of computers that operate on the principles of quantum physics rather than classical physics.

See R. P. Feynman, *Int. J. Theor. Phys.* **21**, 467 (1982).

The idea of running a computer according to the laws of quantum mechanics was initially proposed by Richard Feynman in 1982. He pointed out that it gets progressively more difficult to simulate quantum systems with a conventional computer due to the exponential increase in processing power required as the system size increases. He therefore made the radical proposal that we ought to install quantum hardware in the computer, so that the computer's computational power would scale at the same rate as the complexity of the system that was being investigated. Three years later, David Deutsch wrote a theoretical paper which outlined the basic principles of quantum computation. In analogy with the Turing machine, he introduced the notion of a **universal quantum computer,** and showed that it could, in principle, solve problems that are not *efficiently* solvable with a classical computer.

See D. Deutsch, *Proc. R. Soc. London A,* **400**, 97 (1985). The phrase 'information is physical' is usually attributed to Rolf Landauer, and the idea can be traced back to his paper on 'Irreversibility and heat generation in the computing process' in *IBM J. Res. Dev.* **5**, 183 (1961). For a discussion of this concept in the context of quantum computing, see, for example, D. P. DiVincenzo and D. Loss *Superlattices and Microstructures* **23**, 419 (1998).

The revolutionary ideas of quantum computation involve a radical rethink about the way computers work. We have to realize that 'information is physical' in the sense that classical computers encode the bits of information in a variety of physical ways, such as the voltages on a transistor, the magnetization of a ferromagnetic material, or the intensity of a pulse of light. Although the underlying physics of transistors, ferromagnets, and light pulses are governed by quantum mechanics, the way the data is encoded is purely classical. Thus, for example, the voltage on the transistor has a well-defined value that can be uniquely determined according to the laws of classical electromagnetism. Deutsch's idea was to take a leap ahead and encode the information itself as quantum states which have no classical analogue. In doing so, we achieve an exponential increase in the computing power as the system gets larger. This comes about from exploiting the complexity of quantum systems to our advantage.

In the years that immediately followed Deutsch's landmark paper, the subject was mainly restricted to theoretical groups, who worked hard to understand the basic principles and advantages of quantum computing over conventional computational methods. Some groups concentrated on finding specific examples that would establish the general principle that quantum computers can outperform their classical counterparts, at least on paper. Others devoted their attention to designing experiments to prove the principles and establish that the ideas are more than just a theoretical dream.

A key breakthrough was made in 1994 when Peter Shor showed that a quantum computer can factorize a large number in polynomial time rather than exponential time. In this way, he reduced the factorization problem from the **NP** to the **P** complexity class. Since then, more examples have been found where quantum computers have an essential speed advantage over their classical counterparts. Meanwhile, the first generation of experiments has been completed, and several groups have now demonstrated baby quantum computers. Everyone realizes that it will take a long time for these baby quantum computers to grow to maturity and reach the point where they can really outperform their classical counterparts. At the same time, the potential benefits are enormous, and this prompts a forward-looking attitude in which new ideas are explored and developed, both experimentally and theoretically.

See P. W. Shor: Algorithms for quantum computation: Discrete logarithms and factoring, in *Proceedings of the 35th Annual Symposium on Foundations of Computer Science* (ed. S. Goldwasser). IEEE Computer Society Press (1994), Los Alamitos, California, p. 124.

It is interesting to realize that quantum computation is based on the quantum 'weirdness' of superposition states. The superposition principle frequently causes conceptual difficulties when it is first encountered, and could be seen as an obstacle to information processing because it leads to probabilistic outcomes in measurements. In the subject of quantum computation we side-step the conceptual questions and take a pragmatic approach to exploit the parallelism of quantum states in a very practical way. In this way we bypass the philosophical questions and turn quantum mechanics into a practical subject that is used to enhance the possibilities of information science.

In the following sections we shall first introduce the basic concepts of quantum bits (qubits) and quantum logic gates. We shall then look at the problem of decoherence, and discuss some of the potential applications of quantum computers. Finally, we shall give a brief survey of some of the experimental work that has been performed so far, with particular stress on ion-trap systems.

13.2 Quantum bits (qubits)

13.2.1 The concept of qubits

Classical computers store information as binary bits that can take the value of logical 0 or 1. In analogy with their classical counterparts, quantum computers store the information as **quantum bits**, or **qubits** for short. These are quantum-mechanical states of individual particles such as atoms, photons, or nuclei. The key difference between classical and quantum bits is that qubits can not only represent pure 0 and 1 states, but they can also take on superposition states, in which the system is in both the 0 and 1 state at the same time. This is a consequence of the superposition principle of quantum mechanics, and contrasts with classical systems which can only ever be in one of the two possible states at a given moment. (See Section 9.2.2.)

The properties of qubits are governed by their quantum-mechanical wave function ψ. We choose physical systems that have two readily distinguishable quantum states that can be used to represent binary 0 and 1. If we use Dirac notation to label the quantum states corresponding to 0 and 1 as $|0\rangle$ and $|1\rangle$ respectively, then the general state of the qubit can be written in the following form:

$$|\psi\rangle = c_0|0\rangle + c_1|1\rangle, \tag{13.1}$$

where the normalization condition on $|\psi\rangle$ requires that

$$|c_0|^2 + |c_1|^2 = 1. \tag{13.2}$$

Equation 13.1 explicitly expresses the fact that the system is in a superposition of both $|0\rangle$ and $|1\rangle$ states at the same time. The relative proportion of each of the binary states is governed by the amplitude coefficients c_0 and c_1.

In order to clarify what we understand by qubits, it is helpful to discuss the kinds of physical system that might comprise the quantum hardware. Table 13.1 lists some of the most important systems that have been considered in this context. In each case we have an individual quantum system with two clearly distinguishable states. In order for the system to be usable, we require that the chosen property should be easily measurable, and that the two states are orthogonal to each other, such that:

$$\langle 0|1\rangle = 0. \tag{13.3}$$

The examples given in Table 13.1 all satisfy these criteria.

Table 13.1 Some physical realizations of qubits. In the case of the superconducting loop, the direction of the magnetic flux quantum is determined by the direction of the persistent current.

Quantum system	Physical property	$|0\rangle$	$|1\rangle$
Photon	Linear polarization	Horizontal	Vertical
Photon	Circular polarization	Left	Right
Nucleus	Spin	Up	Down
Electron	Spin	Up	Down
Two-level atom	Excitation state	Ground state	Excited state
Josephson junction	Electric charge	N Cooper pairs	$N+1$ Cooper pairs
Superconducting loop	Magnetic flux	Up	Down

Let us suppose that we choose to use the linear polarization of an individual photon as the basis for the qubit states. In this case, we could define the $|0\rangle$ and $|1\rangle$ states to correspond to the horizontal and vertical polarization states, respectively. An arbitrary state of the qubit would then be given by the wave function $|\psi\rangle$ with

$$|\psi\rangle = c_0|0\rangle + c_1|1\rangle$$
$$\equiv c_0|\leftrightarrow\rangle + c_1|\updownarrow\rangle, \qquad (13.4)$$

where we used the same notation for the polarization states as in Table 12.1. The quantum information of the qubit is stored in the amplitude coefficients c_0 and c_1. These coefficients can be calculated precisely, but cannot be measured directly. Thus, for example, measurements using the apparatus shown in Fig. 12.2 give the result 0 with probability $|c_0|^2$ and 1 with probability $|c_1|^2$, so that repeated measurement permit the determination of $|c_i|^2$, but not c_i. It therefore seems that the quantum information is hidden, and that we gain little by moving over to the quantum technology. This is indeed the case if we only have one qubit: the advantages of the quantum technology only emerge when we have several qubits.

A collection of N qubits is called a **quantum register** of size N. Consider a two-qubit register. The wave function for an arbitrary state is specified as a superposition of the four possible combinations of states of the individual qubits:

$$|\psi\rangle = c_{00}|00\rangle + c_{01}|01\rangle + c_{10}|10\rangle + c_{11}|11\rangle, \qquad (13.5)$$

where the notation $|ij\rangle$ implies that qubit 1 is in state i and qubit 2 is in state j. This can be generalized to any number of qubits. Thus a three-qubit register would have a general wave function of the form:

$$|\psi\rangle = c_{000}|000\rangle + c_{001}|001\rangle + c_{010}|010\rangle + c_{011}|011\rangle$$
$$+ c_{100}|100\rangle + c_{101}|101\rangle + c_{110}|110\rangle + c_{111}|111\rangle. \qquad (13.6)$$

It is apparent that an N-qubit register is described by 2^N wave function amplitudes $c_{ijk...}$. The quantum information is stored in these amplitudes, which are complex numbers with a modulus between 0 and 1. The amount of information clearly grows exponentially with the register size, but the information is hidden and a large amount of it is lost when measurements are made. However, provided we only manipulate the qubits and let them interact with each other coherently without making measurements, then all the information is preserved. This is the basis of the huge quantum parallelism that underlies quantum computation. The clever part of the subject is to devise methods to harness the parallelism. We shall give some examples of how this is done in Section 13.5.

13.2.2   Bloch vector representation of single qubits

The normalization condition written in eqn 13.2 suggests that we can represent the state of a single qubit as a vector. This vector is called the

We have already considered photon qubits of this type in the discussion of quantum cryptography in Chapter 12.

The easiest way to make an N-qubit system (at least, conceptually) is to couple together N two-level particles. It is also possible to use different energy levels of a single particle as different qubits. Note that some qubits (e.g. excitons, flux qubits) are collective quantum excitations rather than real particles.

Note that there has been a change of notation here compared to Section 9.6. In the treatment of two-level atoms it is customary to label the lower and upper levels as 1 and 2, respectively, whereas here we are using the labels 0 and 1 instead in order to make the link with binary logic. Note also that some authors put $|1\rangle$ at the South pole and $|0\rangle$ at the North pole, interchanging c_0 and c_1 in eqn 13.7. This choice is purely a matter of convention, and has no physical significance.

Bloch vector and has been discussed previously in Section 9.6 in the context of two-level atoms. The Bloch vector maps out a sphere of unit radius called the **Bloch sphere**, as illustrated in Fig. 13.3. Points on the Bloch sphere are specified by their polar angles (θ, φ). The North pole $(\theta = 0)$ and South pole $(\theta = \pi)$ of the sphere are defined to correspond to the pure $|1\rangle$ and $|0\rangle$ states, respectively. All other values of θ correspond to superposition states of the type given in eqn 13.1.

The correspondence between the amplitude coefficients and polar angles can be made explicit by setting (cf. eqn 9.64)

$$c_0 = \sin(\theta/2),$$
$$c_1 = e^{i\varphi} \cos(\theta/2). \qquad (13.7)$$

We shall see in Section 13.3 below that the Bloch sphere model is very helpful in understanding the effect of quantum operations on qubits.

13.2.3 Column vector representation of qubits

The one- and two- qubit wave functions can also be conveniently represented as the row vectors (c_0, c_1) and $(c_{00}, c_{01}, c_{10}, c_{11})$, respectively.

Another useful way to describe the state of a single qubit with a wave function given by eqn 13.1 is as a column vector of the form:

$$|\psi\rangle = \begin{pmatrix} c_0 \\ c_1 \end{pmatrix}. \qquad (13.8)$$

This column vector notation allows us to use 2×2 matrices to represent the operations that are performed on the qubits, which simplifies the formal treatment. Furthermore, it provides a convenient way to handle multiple qubits. For example, we can represent a two-qubit system with a wave function of the type given in eqn 13.5 as a column vector of the form:

$$|\psi\rangle = \begin{pmatrix} c_{00} \\ c_{01} \\ c_{10} \\ c_{11} \end{pmatrix}, \qquad (13.9)$$

We are then able to use 4×4 matrices to represent the operations that manipulate the two-dimensional qubits, as we shall see in Section 13.3.3.

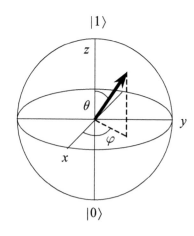

Fig. 13.3 The Bloch sphere representation of qubits. Qubit states correspond to points on the surface of the sphere, with $|0\rangle$ at the South pole, $|1\rangle$ at the North pole, and superposition states everywhere else.

13.3 Quantum logic gates and circuits

13.3.1 Preliminary concepts

A classical computer consists of a memory and a processor. The processor carries out operations on the bits of information stored in the memory according to a program, and outputs the results as a new set of bits. The processing operations are performed by millions of simple **binary logic gates** such as the NOT or NAND gates. These perform operations on either one or two bits at a time. For example, the NOT gate operates on one bit at a time, while the NAND gate operates on

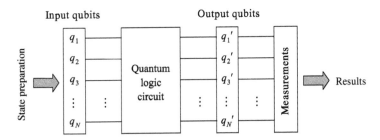

Fig. 13.4 Schematic block diagram of the workings of a quantum computer. The qubits $\{q_1, q_2, q_3, \ldots, q_N\}$ from the input register are set up in the correct initial states and are fed into the quantum logic circuit. The quantum logic circuit performs the processing tasks and outputs a new set of qubits $\{q_1', q_2', q_3', \ldots, q_N'\}$. Measurements are made on the output register and the results are then read out.

two bits. The truth tables for these classical operations are given in Tables 13.2 and 13.3. The program determines how the binary gates are linked together in a logical circuit in order to perform the required task.

The basic idea of a quantum computer is much the same. The information is stored in a register of qubits and the processing tasks are carried out by **quantum logic gates**. These quantum logic gates are connected together in a **quantum circuit** in order to carry out specified processing tasks. Figure 13.4 shows a schematic block diagram of a quantum computer. We start with an N-qubit register $\{q_1, q_2, q_3, \ldots, q_N\}$, in which the qubits have previously been prepared in the required initial states. These input qubits are fed into the quantum logic circuit which then performs the processing task according to the program of the quantum computer. The output of the quantum logic circuit is a new set of qubits $\{q_1', q_2', q_3', \ldots, q_N'\}$. The final results of the computational task are obtained by making measurements on these output qubits, which return a set of N classical bits.

At this stage it appears that we have gained nothing from the quantum calculation. We started with a 'data set' of N qubits and ended up with a result consisting of N classical bits. However, the key point to understand is that with N input qubits we are effectively entering 2^N data points into the computer. If we program the quantum processor intelligently, we can obtain information from the input data set more efficiently than we would with a classical machine. We shall see examples of this in Section 13.5. Thus the benefit of the quantum computer over its classical counterpart comes from the manipulation of the 2^N amplitude coefficients of the N input qubits within the quantum logic circuit before the final measurements are made.

It is clear from the above that the heart of a quantum computer is the quantum logic circuit that performs the information-processing task. The quantum logic circuit consists of a programmed sequence of simple quantum logic gates. Just as with classical computers, it turns out that we only need a very small number of quantum logic gates to perform all the possible computing tasks. We first need a series of **single-qubit**

Table 13.2 Truth table for the classical single-bit NOT gate.

Input bit	Output bit
0	1
1	0

Table 13.3 Truth table for the classical two-bit NAND gate.

Input bits		Output bit
0	0	1
1	0	1
0	1	1
1	1	0

gates which perform operations on one qubit at a time. Then we need one **two-qubit gate** which operates on two qubits at a time. With these basic building blocks we can implement any quantum logic circuit that we may require. Our task therefore is to understand both single- and two-qubit gates, beginning with the single-qubit gates.

See Nielsen and Chuang (2000, §4.5).

13.3.2 Single-qubit gates

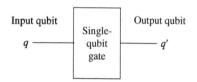

Input qubit — Single-qubit gate — Output qubit
q —————— q'

Fig. 13.5 Schematic diagram of a single-qubit gate. The gate transforms an input qubit q to an output qubit q'.

The operation of a single-qubit gate is shown schematically in Fig. 13.5. A single qubit q is fed into the gate, and the gate outputs another qubit q'. If we write the wave functions of q and q' as $|\psi\rangle$ and $|\psi'\rangle$, respectively, with

$$|\psi\rangle = c_0|0\rangle + c_1|1\rangle, \tag{13.10}$$

and

$$|\psi'\rangle = c_0'|0\rangle + c_1'|1\rangle, \tag{13.11}$$

then we see that the effect of the gate is to change the amplitude coefficients of the qubit in a determined way. By making use of the column vector notation defined in eqn 13.8, we can describe the gate by a 2×2 matrix $\mathbf{M}$ as follows:

$$\begin{pmatrix} c_0' \\ c_1' \end{pmatrix} = \begin{pmatrix} M_{11} & M_{12} \\ M_{21} & M_{22} \end{pmatrix} \begin{pmatrix} c_0 \\ c_1 \end{pmatrix}, \tag{13.12}$$

with

$$c_0' = M_{11}c_0 + M_{12}c_1$$
$$c_1' = M_{21}c_0 + M_{22}c_1. \tag{13.13}$$

It turns out that the only requirement on the gate matrix $\mathbf{M}$ is that it should be **unitary**:

$$\mathbf{MM}^\dagger = \mathbf{I}, \tag{13.14}$$

where $\mathbf{M}^\dagger$ is the **adjoint** matrix of $\mathbf{M}$, and $\mathbf{I}$ is the identity matrix. This condition can be written explicitly as:

$$\begin{pmatrix} M_{11} & M_{12} \\ M_{21} & M_{22} \end{pmatrix} \begin{pmatrix} M_{11}^* & M_{21}^* \\ M_{12}^* & M_{22}^* \end{pmatrix} = \begin{pmatrix} 1 & 0 \\ 0 & 1 \end{pmatrix}. \tag{13.15}$$

The unitarity requirement implies that all quantum gates must be *reversible*. (See Exercise 13.2.)

Three of the most important single-qubit gates are listed in Table 13.4. The **NOT gate**, which is represented by the 'X' symbol, switches the amplitude coefficients around:

$$X \cdot q = \begin{pmatrix} 0 & 1 \\ 1 & 0 \end{pmatrix} \begin{pmatrix} c_0 \\ c_1 \end{pmatrix} = \begin{pmatrix} c_1 \\ c_0 \end{pmatrix}. \tag{13.16}$$

The Z **gate** flips the sign of $|1\rangle$, while leaving $|0\rangle$ unchanged:

$$Z \cdot q = \begin{pmatrix} 1 & 0 \\ 0 & -1 \end{pmatrix} \begin{pmatrix} c_0 \\ c_1 \end{pmatrix} = \begin{pmatrix} c_0 \\ -c_1 \end{pmatrix}. \tag{13.17}$$

Table 13.4 Single-qubit gates. Note that the X and Z gate matrices are identical to their respective Pauli spin matrices, which is one of the reasons why the gates are labelled 'X' and 'Z' in the first place. The other reason relates to their geometric interpretation as rotation operators about the x- and z-axes, respectively.

Quantum gate	Matrix representation
NOT (X)	$\begin{pmatrix} 0 & 1 \\ 1 & 0 \end{pmatrix}$
Z	$\begin{pmatrix} 1 & 0 \\ 0 & -1 \end{pmatrix}$
Hadamard (H)	$\frac{1}{\sqrt{2}}\begin{pmatrix} 1 & 1 \\ 1 & -1 \end{pmatrix}$

Finally, the **Hadamard gate** (H gate) turns basis states into superposition states, and vice versa:

$$H \cdot q = \frac{1}{\sqrt{2}} \begin{pmatrix} 1 & 1 \\ 1 & -1 \end{pmatrix} \begin{pmatrix} c_0 \\ c_1 \end{pmatrix} = \begin{pmatrix} (c_0 + c_1)/\sqrt{2} \\ (c_0 - c_1)/\sqrt{2} \end{pmatrix}. \qquad (13.18)$$

Thus, for example, the H gate maps the $|0\rangle$ and $|1\rangle$ states onto the $(|0\rangle + |1\rangle)/\sqrt{2}$ and $(|0\rangle - |1\rangle)/\sqrt{2}$ superposition states, respectively, whereas it turns superposition states like $(|0\rangle + |1\rangle)/\sqrt{2}$ into basis states:

$$H \cdot \begin{pmatrix} 1/\sqrt{2} \\ 1/\sqrt{2} \end{pmatrix} = \frac{1}{2} \begin{pmatrix} 1 & 1 \\ 1 & -1 \end{pmatrix} \begin{pmatrix} 1 \\ 1 \end{pmatrix} = \begin{pmatrix} 1 \\ 0 \end{pmatrix}. \qquad (13.19)$$

We explained in Section 13.2.2 that the state of a qubit can be mapped onto a point on the surface of the Bloch sphere. By changing the amplitude coefficients of the qubit, a single-qubit gate alters the position of the qubit on the Bloch sphere. The quantum gates can therefore be given a geometrical interpretation. For example, the X gate is equivalent to a rotation of π radians about the x-axis. This can be seen by considering a few examples of the effect of the X gate:

$$|0\rangle \rightarrow |1\rangle,$$
$$|1\rangle \rightarrow |0\rangle,$$
$$(1/\sqrt{2})(|0\rangle + |1\rangle) \rightarrow (1/\sqrt{2})(|0\rangle + |1\rangle),$$
$$(1/\sqrt{2})(|0\rangle + i|1\rangle) \rightarrow (1/\sqrt{2})\, e^{i\pi/2}(|0\rangle - i|1\rangle).$$

In a single-qubit system, phase factors like the one in the output qubit of the fourth example are unmeasurable. Note, however, that *relative* phase shifts are significant in multiple qubit systems.

The equivalent operations in terms of the Bloch vectors are as follows:

$$(0, 0, -1) \rightarrow (0, 0, 1),$$
$$(0, 0, 1) \rightarrow (0, 0, -1),$$
$$(1, 0, 0) \rightarrow (1, 0, 0),$$
$$(0, 1, 0) \rightarrow (0, -1, 0).$$

The first two of these operations are illustrated in Figs 13.6(a) and (b).

In the same way, the Z gate is equivalent to a rotation of π about the z-axis, while the H gate is equivalent to a rotation of π about the z-axis

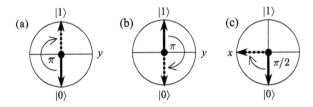

Fig. 13.6 Geometric interpretations of single-qubit operators in the Bloch sphere representation. (a) X operator on the $|0\rangle$ state observed in the y-z plane. (b) X operator on the $|1\rangle$ state observed in the y-z plane. (c) H operator on the $|0\rangle$ state observed in the x-z plane. In (c) the first rotation about the z-axis has no effect in this particular example.

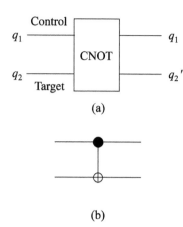

Fig. 13.7 The controlled-NOT (C-NOT) gate. (a) The gate changes the target qubit q_2 depending on the state of the control qubit q_1. (b) Symbol used to represent C-NOT gates in quantum circuits.

Other common examples of two-qubit gates are the controlled rotation gate (C-ROT) and the controlled phase shift gate (C-PHASE).

Table 13.5 Truth table for the controlled-NOT (C-NOT) operation.

Input qubits		Output qubits	
Control	Target	Control	Target
$|0\rangle$	$|0\rangle$	$|0\rangle$	$|0\rangle$
$|0\rangle$	$|1\rangle$	$|0\rangle$	$|1\rangle$
$|1\rangle$	$|0\rangle$	$|1\rangle$	$|1\rangle$
$|1\rangle$	$|1\rangle$	$|1\rangle$	$|0\rangle$

followed by a rotation of $\pi/2$ about the y-axis. Figure 13.6(c) shows the effect of an H gate on the $|0\rangle$ qubit.

Example 13.1   A qubit in the $|0\rangle$ state is input to an H gate followed by a Z gate. What is the output qubit?

Solution
The output qubit is calculated by applying the operation matrices in the correct order to the input qubit:

$$q' = Z \cdot H \cdot q,$$

with $q = (1,0)$. Written explicitly, we have:

$$q' = \begin{pmatrix} 1 & 0 \\ 0 & -1 \end{pmatrix} \frac{1}{\sqrt{2}} \begin{pmatrix} 1 & 1 \\ 1 & -1 \end{pmatrix} \begin{pmatrix} 1 \\ 0 \end{pmatrix} = \begin{pmatrix} 1/\sqrt{2} \\ -1/\sqrt{2} \end{pmatrix}.$$

The output qubit is thus:

$$q' = \frac{1}{\sqrt{2}}|0\rangle - \frac{1}{\sqrt{2}}|1\rangle.$$

13.3.3   Two-qubit gates

We mentioned in Section 13.3.1 that any arbitrary qubit gate can be built up from a sequence of single-qubit gates and one type of two-qubit gate. A particularly useful type of two-qubit gate is the **controlled unitary operator** (C-U) gate.   C-U gates have two input qubits, which are designated as the **control** and **target** qubits, respectively. The gate has no effect on the control qubit, but performs a unitary operation on the target qubit conditionally on the state of the control qubit. Since we only need to develop one type of two-qubit gate to build a quantum computer, we can learn all the basic principles by restricting our discussion to the simplest one, namely the **controlled-NOT gate** (C-NOT gate).

Figure 13.7 shows a schematic diagram of a C-NOT gate, together with the symbol that represents the gate in quantum circuits. In a C-NOT gate, the controlled unitary operation is the NOT gate. The control and target qubits are designated q_1 and q_2, respectively, and the gate carries out the NOT operation on q_2 if $q_1 = |1\rangle$. The truth table for the C-NOT gate is given in Table 13.5.

By making use of the column vector notation for a two-qubit wave function defined by eqn 13.9, we can write down a 4×4 unitary matrix to represent the C-NOT operation (see Exercise 13.6):

$$\hat{U}_{\text{CNOT}} = \begin{pmatrix} 1 & 0 & 0 & 0 \\ 0 & 1 & 0 & 0 \\ 0 & 0 & 0 & 1 \\ 0 & 0 & 1 & 0 \end{pmatrix}. \tag{13.20}$$

The effect of the C-NOT operator on an arbitrary two qubit state can then be found as follows:

$$\hat{U}_{CNOT} \cdot |\psi\rangle = \begin{pmatrix} 1 & 0 & 0 & 0 \\ 0 & 1 & 0 & 0 \\ 0 & 0 & 0 & 1 \\ 0 & 0 & 1 & 0 \end{pmatrix} \begin{pmatrix} c_{00} \\ c_{01} \\ c_{10} \\ c_{11} \end{pmatrix} = \begin{pmatrix} c_{00} \\ c_{01} \\ c_{11} \\ c_{10} \end{pmatrix}. \quad (13.21)$$

It thus becomes apparent that the C-NOT operator has the effect of switching round the amplitude coefficients of the $|10\rangle$ and $|11\rangle$ states.

Example 13.2   What is the output of the quantum circuit shown in Figure 13.8 when both input qubits are in the $|0\rangle$ state?

Solution
The circuit consists of an H gate and a C-NOT gate. We first compute the effect of the H gate on the control bit using eqn 13.18:

$$H \cdot \begin{pmatrix} 1 \\ 0 \end{pmatrix} = \frac{1}{\sqrt{2}} \begin{pmatrix} 1 & 1 \\ 1 & -1 \end{pmatrix} \begin{pmatrix} 1 \\ 0 \end{pmatrix} = \begin{pmatrix} 1/\sqrt{2} \\ 1/\sqrt{2} \end{pmatrix}.$$

We then write the input to the C-NOT gate in the form given in eqn 13.5:

$$|\psi\rangle = \frac{1}{\sqrt{2}} (|0\rangle + |1\rangle) |0\rangle = \frac{1}{\sqrt{2}} |00\rangle + \frac{1}{\sqrt{2}} |10\rangle.$$

Finally we compute the output of the C-NOT gate using the C-NOT operator given in eqn 13.20:

$$|\psi'\rangle = \begin{pmatrix} 1 & 0 & 0 & 0 \\ 0 & 1 & 0 & 0 \\ 0 & 0 & 0 & 1 \\ 0 & 0 & 1 & 0 \end{pmatrix} \cdot \frac{1}{\sqrt{2}} \begin{pmatrix} 1 \\ 0 \\ 1 \\ 0 \end{pmatrix} = \frac{1}{\sqrt{2}} \begin{pmatrix} 1 \\ 0 \\ 0 \\ 1 \end{pmatrix}.$$

The output is therefore:

$$|\psi'\rangle = \frac{1}{\sqrt{2}} (|00\rangle + |11\rangle).$$

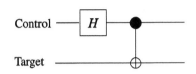

Fig. 13.8 Quantum circuit composed of an H gate and a C-NOT gate.

As we shall see in the next chapter, this output state is called a **Bell state** and is very important in the discussion of entangled states.

13.3.4   Practical implementations of qubit operations

Up to this point our treatment of quantum logic gates has been purely formal. We must now see how these operations can be implemented in the laboratory. As explained above, we can form an arbitrary quantum gate by combining C-NOT gates with single-qubit operations. Our task thus reduces to learning how to implement single-qubit operations and then the C-NOT operation. We begin by considering the single-qubit gates.

In physical terms, a single-qubit gate operates on an input qubit, and returns an output as a new qubit. As explained in Section 13.3.2, the operation of the gate can be given a geometric interpretation in terms of the Bloch vector representing the qubit. In general, it is possible to decompose an arbitrary single-qubit operator $\hat{U}$ into a series of Bloch

See Nielsen and Chuang (2000, §4.2). By symmetry, the operator can equally well be decomposed into a series of rotations about the x- and y-axes.

vector rotations about the y- and z-axes together with multiplication by a phase shift:

$$\hat{U} = e^{i\alpha} R_z(\theta_3) R_y(\theta_2) R_z(\theta_1), \tag{13.22}$$

where α is a real number and $R_i(\theta)$ is the operator representing the rotation through an angle θ about Cartesian axis i. This result can be given an intuitive geometric interpretation in terms of an arbitrary mapping of the Bloch vector angles $(\theta, \varphi) \rightarrow (\theta', \varphi')$. (See Exercise 13.5.)

The absolute phase of a wave function is unobservable, and hence the global phase shift angle α in eqn 13.22 is not significant. The implementation of single-qubit operations thus reduces to carrying out Bloch vector rotations through arbitrary angles about the y- and z-axes. We consider here the specific case where the qubit is based on a two-level atom system. The Bloch vector rotations can then be performed by using the techniques of resonant light–atom interactions developed in Chapter 9.

As explained in Section 9.6, the application of a short electromagnetic pulse at the resonant frequency of the system causes a rotation of the Bloch vector about an axis in the x-y plane. (See Fig. 13.9.) The azimuthal angle φ of the rotation plane is set by the optical phase of the pulse, while the rotation angle Θ is equal to the **pulse area** defined by (cf. eqn 9.51):

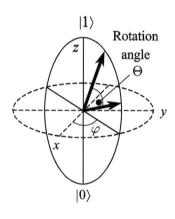

$$\Theta = \left| \frac{\mu_{01}}{\hbar} \int_{-\infty}^{+\infty} \mathcal{E}_0(t) \, dt \right|, \tag{13.23}$$

where μ_{01} is the dipole moment for the $|0\rangle \rightarrow |1\rangle$ transition, and $\mathcal{E}_0(t)$ is the time-dependent electric field amplitude of the pulse. Pulses that produce rotation angles of 180° and 90° are called π- and $\pi/2$-pulses, respectively. These are especially important since they are part of the X and H operators. (See Fig. 13.6.)

Fig. 13.9 The application of a short resonant electromagnetic pulse produces a rotation of the Bloch vector by an angle Θ about an axis within the x-y plane. The azimuthal angle φ of the rotation plane is determined by the phase of the pulse, while the rotation angle is governed by the pulse area given in eqn 13.23.

The fact that the azimuthal angle of the rotation axis is determined by the phase of the pulse means that no explicit pulses are required for the z rotations. Keeping track of the z rotations is in fact effectively a book-keeping exercise. As an example, consider the result of two arbitrary operations $\hat{U}_1$ and $\hat{U}_2$. The combined operation is given from eqn 13.22 as:

$$\hat{U} = \hat{U}_2 \cdot \hat{U}_1$$

$$= e^{i\alpha'} R_z(\theta_3') R_y(\theta_2') R_z(\theta_1') \cdot e^{i\alpha} R_z(\theta_3) R_y(\theta_2) R_z(\theta_1)$$

$$= e^{i(\alpha'+\alpha)} R_z(\theta_3') R_y(\theta_2') R_z(\theta_1' + \theta_3) R_y(\theta_2) R_z(\theta_1). \tag{13.24}$$

Now a rotation about the z-axis of θ_1 followed by a rotation of θ_2 about the y-axis is equivalent to a single rotation by θ_2 about the axis in the x-y plane with an azimuthal angle of $(\pi/2 - \theta_1)$. The first two rotations can thus be performed by a single pulse with a phase of $(\pi/2 - \theta_1)$ and pulse area of θ_2. Similarly the next two rotations can be performed by a pulse of phase $(\pi/2 - \theta_1' - \theta_3)$ and area θ_2'. The final z-axis rotation is simply recorded as a phase shift to be implemented when the next operation is performed. We can therefore perform rotations through arbitrary angles about arbitrary rotation axes by careful choice of the pulse phase and amplitude. This allows us to perform arbitrary single-qubit operations.

The implementation of single-qubit operations therefore presents no fundamental issues. We merely need to irradiate the atoms with short light pulses at the transition frequency with the correct pulse energy and phase to produce the required Bloch vector rotation. The essential physics for these operations has been known and understood for many years now. The key to the practical implementation of quantum computation thus becomes the demonstration of the C-NOT gate, which we now discuss.

C-NOT gates act on two qubits according to the truth table given in Table 13.5. The key point is that we have to flip the target qubit depending on the state of the control qubit. The simplest way to see how this works is to consider the level scheme shown in Fig. 13.10. We have two qubits, q_A and q_B, each with their own resonant angular frequencies ω_A and ω_B. The two qubits interact with each other so that when we put both of them in the $|1\rangle$ state at the same time, the angular frequency of the system is not just equal to $(\omega_A + \omega_B)$, but is shifted to:

$$\omega_{AB} = \omega_A + \omega_B + \Delta \tag{13.25}$$

where $\hbar\Delta$ is the interaction energy. Δ can be either positive or negative depending on whether we have an attractive or repulsive interaction between the qubits. The interaction term has the effect that the resonant frequency of each qubit depends on the state of the other. If $q_A = |0\rangle$, we can perform the NOT operation on q_B by applying a π-pulse at angular frequency ω_B. However, if $q_A = |1\rangle$, the frequency of the π-pulse must be shifted to $\omega'_B = \omega_B + \Delta$. Similarly, q_A can be manipulated with pulses at ω_A if $q_B = |0\rangle$, but the frequency must be shifted to ω'_A when $q_B = |1\rangle$.

Single-qubit operations on spin systems are similarly performed by applying resonant electromagnetic pulses of the required phase, amplitude, and duration. (See Appendix E.)

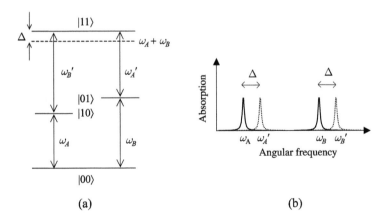

Fig. 13.10 (a) A possible level scheme for the implementation of the C-NOT gate using two qubits q_A and q_B with angular frequencies ω_A and ω_B, respectively. $\hbar\Delta$ is the interaction energy between the two qubits. (b) Absorption spectrum corresponding to the level scheme in part (a). The system only responds at angular frequency ω'_B ($\equiv \omega_B + \Delta$) if $q_A = |1\rangle$. Similarly, the resonant frequency of q_A shifts to ω'_A if $q_B = |1\rangle$.

The argument works equally well the other way round.

This is illustrated by the dotted lines in Fig. 13.10(b), which represent the response of the system when the other qubit is in the $|1\rangle$ state.

Let us suppose that we have a system with the level scheme shown in Fig. 13.10 and we designate q_A as the control and q_B as the target. Starting from the ground state $|00\rangle$, we can demonstrate the four lines of the truth table of the C-NOT operation given in Table 13.5 as follows.

1. $|00\rangle \rightarrow |00\rangle$: this is trivially performed by doing nothing.

2. $|01\rangle \rightarrow |01\rangle$: we apply a π-pulse at ω_B to go from $|00\rangle \rightarrow |01\rangle$ and then follow it with a π-pulse at ω_B'. Nothing happens after the second pulse because the system only responds to frequency ω_B' if $q_A = |1\rangle$.

3. $|10\rangle \rightarrow |11\rangle$: we apply a π-pulse at ω_A to go from $|00\rangle \rightarrow |10\rangle$ and then follow it with a π-pulse at ω_B'. The system can now respond to the second π-pulse and goes into the $|11\rangle$ state.

4. $|11\rangle \rightarrow |10\rangle$: starting with the $|11\rangle$ state prepared in the previous line, we apply a second π-pulse at ω_B'. This causes q_B to flip from $|1\rangle$ to $|0\rangle$, leaving us with the $|10\rangle$ state, as required.

The key point of the demonstration is the response of the system to the π-pulse at ω_B'. This flips the target qubit if $q_A = |1\rangle$, but does nothing if $q_A = |0\rangle$.

It is, of course, apparent that we need to be able to measure the states of both qubits at the end of the operations in order to verify that the required output states have been produced. Hence the fifth item in the DiVincenzo check list. (See Section 13.6.)

Two qubit gates have been demonstrated experimentally in a number of physical systems, and we briefly list here some of the most important ones.

All-optical schemes One- and two-qubit gates can be implemented by encoding the quantum information onto the mode occupied by a single photon, and then manipulating the mode by means of linear optical components such as beam splitters. A single-photon source is required. (See Section 6.7.) This approach differs to the ones described below in that measurements form an intrinsic part of the computational process, instead of being just a method to read out the quantum states at the end of the calculation.

NMR systems The qubits correspond to spin states of specified nuclei within a molecule or crystal, and operations are carried out by RF pulses. The nuclei are in different environments and so have slightly different resonance frequencies. The spins on nearby nuclei interact with each other through the spin–spin interaction. (See Exercise 13.13.)

Ion traps The qubits correspond to the excitation states of a row of single ions held in an ion trap. The ions are all identical and therefore have the same resonance frequency, but can be addressed individually by laser pulses because they are physically separated from each other. The ions interact through the repulsive forces associated with vibrational displacements from the equilibrium positions.

Cavity QED systems The qubits correspond to opposite circular polarization states of two photons interacting with a single atom inside a resonant cavity. The photons interact with each other through their mutual interaction with the atom, which is strongly enhanced by the resonant cavity.

Quantum dots The qubit consists of an exciton confined in a quantum dot. (See Appendix D, especially Section D.3.) The excitons behave like two-level atoms, and the operations are performed by resonant optical pulses. Different types of excitons within an individual dot interact through their Coulomb interaction.

Superconducting systems The quantum information is stored as the charge of a small region of superconducting material called a 'box'. The box is connected to a charge reservoir through a Josephson tunnelling junction. The charge is controlled by the voltage across the junction, and the $|0\rangle$ and $|1\rangle$ states correspond to charges differing by one Cooper pair, with $\Delta q = -2e$. Adjacent boxes are electrostatically coupled via their mutual Coulomb repulsion, and gate operations are performed by sequences of voltage pulses.

Further details of some of these prototype quantum computing systems will be given in Section 13.6. For the other systems, the reader is referred to the bibliography.

13.4 Decoherence and error correction

The operation of a quantum computer relies on the precise manipulation of quantized states of individual quantum systems. We require that the qubits should interact with each other in a controlled way and with nothing else. Unfortunately, this idealistic scenario is impossible to achieve in practice. All quantum systems are fragile because they couple with their environments to a greater or lesser extent. A totally isolated system would in fact be useless for quantum information processing because we would have no means to interact with it and perform the quantum operations that are at the heart of quantum computation.

The 'environment' that we are considering here consists of a very large number of atoms and molecules which obey quantum laws individually, but classical laws collectively. The thermal motion of the particles within the environment acts like a random noise source which can interact with the qubits and introduce uncontrollable random behaviour. For example, the thermal noise could cause a qubit to flip its logical value randomly, and thereby lose its quantum information irretrievably. Since the operation of a quantum computer relies on manipulating *coherent* superposition states, the fragility of qubits with respect to environmental noise is conveniently quantified in terms of **decoherence** rates.

We discussed the various types of process that cause decoherence when we considered the damping of coherent superposition states in Section 9.5.2. The key parameter that quantifies the decoherence is the **dephasing time** T_2. In gases the dephasing time is often limited by collisions between the particles, while in solids or liquids we have to

Superposition states are called 'coherent' because they manifest quantum interference effects analogous to those that can occur between coherent light waves. The coupling of simple quantum systems with the noisy environment is now understood to explain why quantum effects such as the Schrödinger cat paradox are not observed in the macroscopic world. An experiment demonstrating that the coherence of a quantum superposition state is controlled by its coupling to a noisy environment is described in C. J. Myatt, *et al.*, *Nature* **403**, 269 (2000).

Table 13.6 Decoherence times (T_2) and gate operation times (T_{op}) for some of the physical systems considered for quantum computing. N_{op} is the number of gate operations that could be performed before decoherence occurs. It should be emphasized that many of the values quoted in this table represent optimistic upper limits, and only those labelled with an (e) are based on genuine experimental data. Thus, for example, it is known that quantum dot excitons have dephasing times as long as ~ 1 ns, and that T_{op} can be as short as ~ 1 ps, but, as yet, no-one has managed to demonstrate 10^3 gate operations.

System	T_2 (s)	T_{op} (s)	N_{op}	References
Nuclear spin	10^4	10^{-3}	10^7	DiVincenzo, *Phys. Rev A* **50**, 1015 (1995)
Ion trap	10^1 (e)	10^{-6} (e)	10^7	Langer *et al.*, *Phys. Rev. Lett.* **95**, 060502 (2005)
				Steane *et al.*, *Phys. Rev. A* **62**, 042305 (2000)
Exciton (quantum dot)	10^{-9} (e)	10^{-12} (e)	10^3	Langbein *et al.*, *Phys. Rev. B* **70**, 033301 (2004)
				Li *et al.*, *Science* **301**, 809 (2003)
Electron spin (quantum dot)	10^{-7}	10^{-12}	10^5	Pazy *et al.*, *Europhys. Lett.* **62**, 175 (2003)
Superconducting flux qubit	10^{-8} (e)	10^{-10} (e)	10^2 (e)	Chiorescu, I. *et al.*, *Science* **299**, 1869 (2003)

contend with the randomness introduced by interactions with thermally excited vibrations (i.e. phonons).

The number of quantum operations that can be performed before dephasing sets in is given by:

$$N_{op} = \frac{T_2}{T_{op}}, \qquad (13.26)$$

where T_{op} is the time required to perform the operation. Some optimistic values of N_{op} are given in Table 13.6. It is apparent that a certain amount of trade-off takes place. For example, NMR systems have very long dephasing times because nuclear spins only interact very weakly with their environment. At the same time, it also difficult to interact with nuclear spins in a controlled way, and hence the quantum operations tend to be rather slow. Less well-isolated systems decohere faster, but they are easier to interact with and the operations can be performed faster. Thus while it is obvious that we must work as hard as we can to reduce the dephasing rate for any particular system, it does not automatically follow that the systems with the longest dephasing times offer the best possibilities.

Fortunately, the situation is not quite as bad as it might seem at first. In classical data processing, error-checking protocols are used all the time to check and correct for errors. In an analogous way, it is possible to correct for the effects of dephasing on qubits by **quantum error correction** algorithms. The basic principle is essentially the same as for classical error correction, although the details are obviously very different. The idea to use extra qubits to check the fidelity of the data, and then apply quantum algorithms to reconstruct the original states. In this way we can achieve **fault-tolerant quantum computation**: that is, robust quantum computation in the presence of a finite amount of noise from the environment. The relative error rate required is less than about 10^{-5}. The price that is paid is that the processing speed is reduced, since some of the quantum resources are being employed purely for error correction.

The original proposals for quantum error correcting may be found in P. W. Shor, *Phys. Rev. A* **52**, R2493 (1995), and A. M. Steane, *Phys. Rev. Lett.* **77**, 793 (1996). An experimental demonstration of quantum error correction using ion traps is described in J. Chiaverini *et al.*, *Nature* **432**, 602 (2004).

A quick glance at Table 13.6 suggests that the ratio of T_2 to T_{op} is quite promising for some systems. However, it should be stressed that many of the values quoted in Table 13.6 are only theoretical limits. For example, the theoretical limit of T_2 for the 729 nm transition of a Ca$^+$ ion is set by the ~ 1 s radiative lifetime of the upper level, but the coherence time measured experimentally is only ~ 1 ms. Much further work is clearly needed to identify new physical systems and understand the fundamental limits that determine the coherence and gate operation times.

See F. Schmidt-Kaler *et al.*, *J. Phys. B: At. Mol. Opt. Phys.* **36**, 623 (2003).

13.5   Applications of quantum computers

Let us suppose that we had a large quantum computer. What would it be useful for? The general answer to this question has yet to be given. We cannot say for certain whether a quantum computer will always be more powerful than its classical counterpart. On the other hand, there is a growing number of situations where we do know that the quantum computer is more efficient than the classical one, at least in principle.

Before looking at specific examples, it is worth recalling that the fundamental reason why a quantum computer can outperform a classical one is related to the inherent parallelism of quantum systems. (See Section 13.2.1.) A quantum register of size N can hold 2^N numbers simultaneously, whereas a classical register of the same size only contains one number. When we operate on the quantum register, we perform the calculation on many numbers simultaneously, whereas the classical computer only calculates the answer for one given number. Therefore, the quantum computer will eventually beat the classical one provided that we exploit the parallelism effectively.

In the subsections that follow, we shall first illustrate how the benefits of the quantum computer are harnessed in practice for two important quantum algorithms, namely the Deutsch algorithm and the Grover algorithm. We shall then briefly consider a few of the other applications that have been proposed in the literature for quantum computers.

Table 13.7 Possible results for the four possible versions of the one-bit function $f(x)$. The function is described as constant if both outputs are the same, or balanced if the results of 0 and 1 occur with the same frequency.

	$f(0)$	$f(1)$	
f_1	0	0	Constant
f_2	1	0	Balanced
f_3	0	1	Balanced
f_4	1	1	Constant

13.5.1   Deutsch's algorithm

The first algorithm to be proposed that demonstrated that a quantum computer can be more efficient than a classical one is the **Deutsch algorithm**. The algorithm concerns the evaluation of a binary function $f(x)$ that acts on a one-bit binary number. The function has only two possible results: $f(0)$ and $f(1)$, and is defined to be *balanced* or *constant* according to the scheme given in Table 13.7. The task to be performed is to determine whether an unknown function is balanced or constant. A classical computer requires two calls of the function to complete this task, but a quantum computer can do it with just one, as we shall now demonstrate.

The Deutsch algorithm applies specifically to the case of a one-bit function. A more generalized version for an N-bit function is called the **Deutsch–Josza algorithm**. The Deutsch algorithm described here is a slightly improved version of the original one given in Deutsch's paper on the universal quantum computer (*Proc. R. Soc. London A*, **400**, 97 (1985)). See Nielson and Chuang (2000) for further details of the historical development of the algorithm.

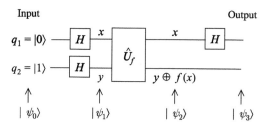

Fig. 13.11 Quantum circuit for Deutsch's algorithm. The circuit has two input qubits q_1 and q_2 which are initialized in the $|0\rangle$ and $|1\rangle$ states, respectively. Both undergo Hadamard operations, before entering the unitary operator $\hat{U}_f$ corresponding to one of the four functions listed in Table 13.7. (The quantum circuits required to perform $\hat{U}_f$ are considered in Exercise 13.9.) q_1 then goes through a second Hadamard gate, and the results are read out by making a measurement on q_1 at the output. The wave functions $|\psi_0\rangle \cdots |\psi_3\rangle$ label the states of the system at the various stages of the circuit.

Deutsch's algorithm can be performed by using the quantum circuit shown in Fig. 13.11. The circuit has two input qubits, q_1 and q_2, and the algorithm ends with a measurement of the state of q_1 at the output. The circuit contains three Hadamard gates together with a unitary operator $\hat{U}_f$ that is determined by the function f. The detailed workings of the algorithm can be understood by following the wave function through the various stages of the circuit.

The input qubits are initialized with $q_1 = |0\rangle$ and $q_2 = |1\rangle$, giving an input wave function of the form:

$$|\psi_0\rangle = |0,1\rangle. \tag{13.27}$$

Both qubits undergo the Hadamard operation, and the outputs are labelled x and y, respectively:

$$x = H \cdot q_1 = \frac{1}{\sqrt{2}}(|0\rangle + |1\rangle),$$

$$y = H \cdot q_2 = \frac{1}{\sqrt{2}}(|0\rangle - |1\rangle). \tag{13.28}$$

The wave function of the qubits at the input of the unitary operator is therefore given by:

$$|\psi_1\rangle = \frac{1}{2}(|0\rangle + |1\rangle)(|0\rangle - |1\rangle) = \frac{1}{2}(|0,0\rangle - |0,1\rangle + |1,0\rangle - |1,1\rangle). \tag{13.29}$$

The unitary operator $\hat{U}_f$ is defined so that it has no effect on the x qubit, but performs the operation $y \oplus f(x)$ on the y qubit, where the $\oplus$ symbol signifies addition modulo two. This unitary operator can be implemented by combinations of single and two qubit gates. (See Exercise 13.9.)

On applying $\hat{U}_f$ to $|\psi_1\rangle$ we obtain:

$$|\psi_2\rangle = \frac{1}{2}(|0, f(0)\rangle - |0, 1 \oplus f(0)\rangle + |1, f(1)\rangle - |1, 1 \oplus f(1)\rangle). \tag{13.30}$$

The need for clarity in eqns 13.30, 13.31, and 13.33 makes it convenient to write the two qubit wave function as $|i,j\rangle$ here instead of $|ij\rangle$.

In the case of a constant function, we have $f(0) = f(1)$, and $|\psi_2\rangle$ is therefore given by:

$$|\psi_2\rangle^{\text{constant}} = \frac{1}{2}\left(|0, f(0)\rangle - |0, 1 \oplus f(0)\rangle + |1, f(0)\rangle - |1, 1 \oplus f(0)\rangle\right)$$

$$= \frac{1}{2}(|0\rangle + |1\rangle)(|f(0)\rangle - |1 \oplus f(0)\rangle). \tag{13.31}$$

On performing the final Hadamard operation, we then obtain:

$$|\psi_3\rangle^{\text{constant}} = |0\rangle \frac{1}{\sqrt{2}}(|f(0)\rangle - |1 \oplus f(0)\rangle). \tag{13.32}$$

On the other hand, if the function is balanced, we have $f(0) \neq f(1)$, and hence $f(1) = 1 \oplus f(0)$. The wave function after $\hat{U}_f$ is therefore:

$$|\psi_2\rangle^{\text{balanced}} = \frac{1}{2}\left(|0, f(0)\rangle - |0, 1 \oplus f(0)\rangle + |1, 1 \oplus f(0)\rangle - |1, f(0)\rangle\right)$$

$$= \frac{1}{2}(|0\rangle - |1\rangle)(|f(0)\rangle - |1 \oplus f(0)\rangle), \tag{13.33}$$

and the output state after the final Hadamard gate is:

$$|\psi_3\rangle^{\text{balanced}} = |1\rangle \frac{1}{\sqrt{2}}(|f(0)\rangle - |1 \oplus f(0)\rangle). \tag{13.34}$$

It is thus apparent that the function is constant when $q_1^{\text{out}} = |0\rangle$ and balanced when $q_1^{\text{out}} = |1\rangle$. A single measurement on q_1 therefore suffices to complete the task. Note that only one call of the function is made throughout the whole algorithm. This is possible because the effect of $\hat{U}_f$ on $|\psi_1\rangle$ is to produce an output wave function $|\psi_2\rangle$ that depends on the value of the function for both possible input bit values.

An example of a situation in which the Deutsch algorithm could be used is in programming a computer to decide whether a coin is fake or genuine. The first step in the procedure would be to check if the coin is different on opposite sides (i.e. balanced) or the same (i.e. constant). A classical computer would have to look at the coin twice to see if the sides are different or not. However, a quantum computer implementing the Deutsch algorithm could perform the task in a single operation, effectively looking at both sides of the coin at the same time. Although this example is rather contrived, it is a simple illustration of how a quantum computer can harness the parallelism of quantum mechanics to perform certain tasks more efficiently than a classical computer.

13.5.2 Grover's algorithm

Grover's algorithm concerns the efficient searching of a database, and is therefore alternatively known as the **quantum search algorithm**. The database is assumed to be unstructured and unsorted. A typical application might be the numbers in a telephone directory which is sorted alphabetically. The London telephone directory might, for example, contain the following entry in row 265,190:

Holmes, Sherlock   221b Baker Street   123 4567

The original reference to Grover's algorithm is given in L. K. Grover, *Proceedings of 28th Annual ACM Symposium on the Theory of Computation*, ACM Press, New York (1996), p 212. See also *Phys. Rev. Lett.* **97**, 325 (1997).

It is easy to find Holmes' telephone number, but it is rather difficult to find the name and address of the person with telephone number 123 4567. Since the telephone number has no connection to the row number, we would probably start at row 1 and laboriously work our way through the directory until we find what we want after 265,190 attempts. The task would be much easier if we had a quantum computer. This is because Grover proved that a quantum computer has the ability to search a database of size N_{data} with $\sim \sqrt{N_{data}}$ operations, in contrast to a classical machine that typically requires $\sim N_{data}/2$ operations. Thus the task of finding a particular number in a telephone directory containing 1 000 000 entries, would take $\sim$1000 operations on a quantum computer, but $\sim$ 500 000 on a classical one.

The detailed workings of the Grover algorithm are quite complicated, and we only present here the gist of the argument, making use of a simple example to illustrate how it works. The quantum circuit required to implement Grover's algorithm is shown in Fig. 13.12. If the data base contains N_{data} entries, the algorithm requires a quantum register comprising N qubits, where N is chosen so that:

$$2^N \geq N_{data}. \tag{13.35}$$

For simplicity, we consider the limiting case where $N_{data} = 2^N$. Each qubit is initialized in the $|0\rangle$ state, and then a Hadamard operation is performed on each one. This produces the state:

$$|\psi_1\rangle = \left(\frac{1}{\sqrt{2}}\right)^N (|0\rangle_1 + |1\rangle_1)(|0\rangle_2 + |1\rangle_2)\cdots(|0\rangle_N + |1\rangle_N)$$

$$= \frac{1}{N_{data}^{1/2}} \sum_{x=0}^{N_{data}-1} |x\rangle, \tag{13.36}$$

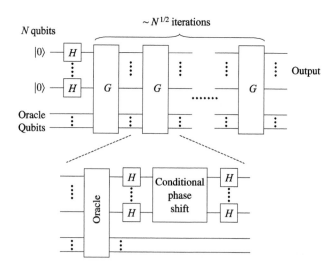

Fig. 13.12 Quantum circuit for Grover's algorithm for a database of size 2^N. The top part of the figure shows the whole circuit which incorporates a series of Grover operators, each labelled G. The bottom part shows the inner workings of one of the operators. The successful completion of the task requires approximately $\sqrt{N}$ Grover operators. (After Nielsen and Chuang 2000, © Cambridge University Press, reproduced with permission.)

where x is a binary number. On writing this in column vector form, we have:

$$|\psi_1\rangle = \frac{1}{N_{\text{data}}^{1/2}}(1,1,1,1,\cdots,1). \qquad (13.37)$$

Note that this state contains all N_{data} numbers with equal phase and with the same probability amplitude of $c_x = 1/N_{\text{data}}^{1/2}$.

The task of the circuit is to change the superposition state of eqn 13.36 into a state that represents the solution. This is done by applying the Grover operator repeatedly until the wave function evolves into a state with a wave function of the form:

$$|\psi_2\rangle = (0,0,1,0,\cdots,0), \qquad (13.38)$$

where the 1 occurs for the binary number of the row corresponding to the solution. The solution is easily detectable by reading out the state of the output qubits. Grover proved that whole operation can be completed by using $\sim \sqrt{N_{\text{data}}}$ calls of the Grover operator.

The Grover operator employs four steps:

(1) apply the **oracle** operator;

(2) apply a Hadamard gate to each qubit in the register;

(3) apply a conditional phase shift;

(4) apply a Hadamard gate to each qubit in the register.

The oracle operator is a unitary operator that employs a number of ancillary qubits called oracle qubits. The oracle can be treated as a black box which has the ability to recognize a solution to the search problem. It is thus the quantum equivalent of looking at the information in the database and checking if it is the desired solution. When the oracle finds a solution it marks it by performing the operation:

$$|x\rangle \to (-1)^{f(x)}|x\rangle, \qquad (13.39)$$

where $f(x)$ is a function defined by:

$$f(x) = 1 \quad \text{if } x \text{ is the solution,}$$

$$f(x) = 0 \quad \text{otherwise.} \qquad (13.40)$$

This shows that the mark that is made is a minus sign. The remaining three steps collectively perform the 'inversion about the mean' operation that we shall discuss below.

We can see how this works by considering the simple example of a database with $N_{\text{data}} = 4$, which requires a quantum register containing $N = 2$ qubits. This problem can be easily solved with just a single call of the Grover operator. The two qubits are initialized in the $|0\rangle$ state, and then undergo the Hadamard operations to give:

$$|\psi_1\rangle = \left(\frac{1}{\sqrt{2}}\right)^2 (|0\rangle_1 + |1\rangle_1)(|0\rangle_2 + |1\rangle_2),$$

$$= \tfrac{1}{2}(|00\rangle + |01\rangle + |10\rangle + |11\rangle),$$

$$= (1/2, 1/2, 1/2, 1/2). \qquad (13.41)$$

In practice, we program the algorithm to stop when the probability that the system is in the solution state is sufficiently high.

It might seem at first that the oracle operator already knows the answer, but this is not the case. In our example of the telephone directory, it merely performs a check to see whether a particular data record has the desired phone number. It does not know where this record lies within the database.

This example was originally presented by Grover in a popular article published in *The Sciences*, July/August 1999, p. 24.

At this stage, the wave function contains all four binary numbers in the database, namely 00, 01, 10, and 11, with equal amplitude and phase. For the sake of argument, let us suppose that the number for which we are searching is 10. We now go through the four steps of the Grover operator. On applying the oracle function, the wave function becomes:

$$|\psi\rangle = (1/2, 1/2, -1/2, 1/2), \tag{13.42}$$

where the solution has been marked by the minus sign. Then, on applying steps 2–4, we invert the probability amplitudes about their mean value. This process is illustrated in Fig. 13.13. The mean amplitude of the wave function before the inversion is $(1/2 + 1/2 - 1/2 + 1/2) \times 1/4 = 1/4$. The numbers with amplitude $+1/2$ are thus mapped to amplitudes of zero, while the one with amplitude $-1/2$ is mapped to probability one. The output wave function is therefore:

$$|\psi_2\rangle = (0, 0, 1, 0). \tag{13.43}$$

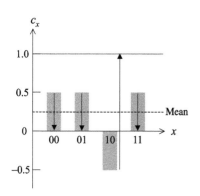

Fig. 13.13 Illustration of the 'invert about the mean' operation for the wave function given by eqn 13.42.

Measurement on the output qubit register would then identify the third item in the database as the solution. Note that the algorithm only uses the oracle operator once, whereas a classical search would usually require 2–3 tests of the database. Note also that in this simple example the output wave function is a genuine pure state, whereas normally the output would contain an admixture of other answers with a low probability set by the tolerance programmed into the algorithm.

13.5.3   Shor's algorithm

Shor's algorithm concerns the Fourier transform operation, which is very widely used in mathematics and computer science. A classical computer requires $\sim N2^N$ operations to take the Fourier transform of 2^N numbers. By contrast, a quantum computer requires only $\sim N^2$ steps, which reduces the difficulty class from exponential (**NP**) to polynomial (**P**). Unfortunately, the result of the quantum Fourier transform is stored as the 2^N amplitudes of the output state, and it is impossible to read out all of these amplitudes directly, since a measurement on the quantum register returns one discrete value for each of the N qubits. This means that a quantum computer *does not* outperform a classical computer for the basic task of taking the Fourier transform of 2^N numbers. On the other hand, the quantum computer *can* find the periodicity of a set of integer numbers very efficiently. In 1994, Peter Shor showed that this efficiency can be exploited to factorize large numbers. The algorithm that he devised is considered a landmark for the subject and is now known as **Shor's algorithm**.

The original reference for Shor's algorithm is given in the *Proceedings of the 35th Annual Symposium on Foundations of Computer Science* (ed. S. Goldwasser). IEEE Computer Society Press, Los Alamitos, California (1994), pp 124–34.

One of the reasons why Shor's algorithm has generated such intense interest is that the security of the RSA encryption method used in classical cryptography is based on the difficulty of factoring large integer numbers. (See Section 12.1.) If anyone possessed a large quantum computer, they would be able to use Shor's algorithm to decode 'secure' data transmissions, which would be a serious cause for concern to military, financial, and governmental organizations. It is important to realize, however, that the size of the quantum computer required to be useful in this context is much larger than anything demonstrated to date, containing perhaps 1000 qubits. At the same time, Shor's algorithm has been tested successfully on a simple NMR-based quantum computer. The results were only very modest, and demonstrated that the factors of 15 were 5 and 3, but it was an important proof-of-principle, and establishes that the security of the RSA encryption method is not assured.

See L. M. K. Vandersypen, *et al.*, *Nature* **414**, 883 (2001).

13.5.4 Simulation of quantum systems

Another class of tasks that is known to benefit from the use of a quantum computer is that of simulating quantum systems. We can recall that this was the problem that prompted Feynman to propose the concept of a quantum computer in the first place. The difficulty in simulating quantum systems on classical computers arises from the exponential increase in computer memory required as the system gets larger. The wave function of N two-level particles contains 2^N amplitudes, and so the full simulation of a relatively small molecule containing 50 atoms requires $2^{50} \sim 10^{15}$ bits of memory. By contrast, a quantum computer would require only N qubits to model the N-particle system, that is: 50 qubits for the 50 atom molecule.

Unfortunately, it is not quite as simple as that. On making the measurements on the output state, we would only obtain N bits of information, and the vast majority of the information would be lost. No one has worked out so far how to harness the full power of the quantum computer in this application. If the problem were to be resolved, then the quantum computer would find very widespread applications in both quantum chemistry and biology.

13.5.5 Quantum repeaters

The final application of quantum computers that we consider here is rather different and concerns the **quantum repeater**. In classical data transmission, repeaters are used all the time to boost the data signals as they become weaker during propagation. For example, in long-distance optical fibre systems, repeaters are used to compensate for the intensity losses due to scattering and absorption as the light pulses propagate down the fibre. The quantum repeater performs an analogous task for the transmission of quantum information.

A typical situation where a quantum repeater would be needed is in long-distance quantum cryptography. Consider the case where Alice and Bob establish a quantum cryptography link over a distance L using lossy optical fibres as shown in Fig. 13.14. With just a single fibre, the probability that the photons reach Bob is given by:

$$\mathcal{P}(L) = \mathrm{e}^{-L/L_0}, \tag{13.44}$$

where L_0 is the distance over which the optical intensity drops by a factor $1/\mathrm{e}$. If no photon arrives, then Bob requests that the transmission

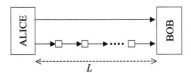

Fig. 13.14 Quantum cryptography using lossy fibres. Alice and Bob can either send the information using one single fibre, or by a compound fibre comprising several segments linked by quantum repeaters.

Quantum repeaters will also be useful for demonstrating quantum teleportation over long distances. See Chapters 12 and 14, respectively, for further details of quantum cryptography and quantum teleportation, and Bouwmeester *et al.* (2000) for a more detailed discussion of quantum repeaters.

be repeated, and thus the number of repetitions required is equal to e^{+L/L_0} on average.

As an alternative strategy, Alice and Bob could use a compound fibre comprising N segments of length L/N, as shown in Fig. 13.14. At the end of each segment is a quantum repeater. The quantum repeater has the ability to detect when a transmission error has occurred, in which case, the transmission across the segment is repeated.

In the compound fibre, the loss per segment is equal to e^{-L/NL_0}, giving an average number of repetitions per segment as e^{+L/NL_0}. Thus the total number of repetitions required is equal to Ne^{+L/NL_0}. This is minimized when $N = L/L_0$, in which case the number of transmissions required is equal to $(L/L_0)e^1$. The use of the repeater is therefore beneficial whenever

$$(L/L_0)e^1 < e^{L/L_0}, \tag{13.45}$$

which applies if $L > L_0$. This corresponds to a few tens of km for 1300 and 1550 nm systems, and even shorter distances for 850 nm systems.

Several schemes have been proposed in the literature for implementing quantum repeaters. One way or another, all of these use quantum error correction-like protocols to compensate for the decoherence of the quantum information during the transmission. Such quantum error correction effectively requires a small quantum computer at each node.

13.6　Experimental implementations of quantum computation

Having studied the basic principles of quantum computation and its applications, we can now give a brief survey of the progress that has been made at the experimental level. David DiVincenzo has given a convenient check list of requirements for the physical hardware:

See, for example, D. P. DiVincenzo *Quantum Information and Computation*, **1**, 1 (2001).

1. The system must possess well-characterized qubits and must be scalable so that it works with large numbers of qubits as well as small ones.

2. It must be possible to prepare the qubits in a simple initial state, such as $|000\ldots\rangle$.

3. The decoherence time must be much longer than the gate operation time.

4. Single- and two-qubit quantum gates must be demonstrated.

5. There must exist a method to measure the state of each individual qubit.

Unfortunately, none of the physical systems that have been investigated so far can satisfy all of these criteria. In the long run, the key issue is scalability: even the optimists admit that it will probably take many years to develop a large quantum computer consisting of hundreds of qubits. The experimentalists working in the field have therefore set about achieving more modest goals such as demonstrating the basic principles on small systems consisting of only a few qubits. The progress at this level has been very rapid, and many encouraging results have been obtained.

In applying the DiVincenzo check list, our first task is to identify suitable quantum systems to act as the qubits. A number of possibilities are listed in Table 13.1. These systems can all be prepared in well-defined initial states (e.g. vertical polarization) and their final states can be measured (e.g. with a polarizing beam splitter and photodetectors). The key task then rests in demonstrating the single qubit and C-NOT operations, and determining how many of these can be performed before dephasing sets in.

One of the most promising systems for quantum computation at optical frequencies is the **ion trap**. The basic principles of ion traps and the techniques used to cool the ions within them were presented in Section 11.2.6. For applications in quantum computation, it is necessary to deal with *single* ions and arrays of them. The qubits can correspond either to electronic states of different ions, or to different sublevels of the electronic ground state of an individual ion.

The C-NOT quantum logic gate was first demonstrated in an ion trap system using a single $^9\text{Be}^+$ ion. Figure 13.15(a) gives a schematic diagram of the experimental arrangement. The ion was excited resonantly with a laser beam and the fluorescence emitted was recorded with photon-counting detectors. The level diagram is shown in part (b) of Fig. 13.15, and the corresponding qubit representation scheme is given in Table 13.8. Part (c) shows the optical transitions used to perform the qubit manipulations and read out their outcomes.

In order to understand how the C-NOT gate works, it is first necessary to consider the vibrational properties of the trapped ion. Voltages

See C. Monroe, *et al.*, *Phys. Rev. Lett.* **75**, 4714 (1995).

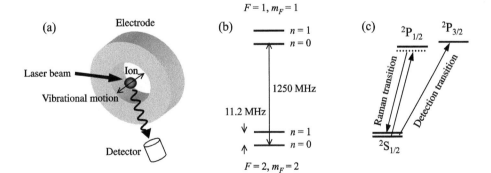

Fig. 13.15 (a) Schematic ion trap for demonstrating the C-NOT quantum gate. The trapped ion is addressed by a laser beam and the fluorescence emitted is recorded with a detector. The full arrangement of electrodes used to produce the three-dimensional trapping potential is not shown for clarity. (b) Level scheme for the C-NOT gate in the $^9\text{Be}^+$ ion trap. The states are all derived from the $^2\text{S}_{1/2}$ electronic ground state and are labelled by the vibrational quantum number n and the hyperfine quantum numbers $|F, m_F\rangle$. The four levels correspond to control and target qubits according to the scheme given in Table 13.8. (c) Optical transitions used to carry out the qubit manipulations and read out their results. All transitions occur at around 313 nm. (Adapted from C. Monroe, *et al.*, *Phys. Rev. Lett.* **75**, 4714 (1995).)

are applied to the electrodes surrounding the ion, leading to an equilibrium configuration with the ion in the centre of the trap, and strong restoring forces for small displacements in all directions. These restoring forces create a simple harmonic oscillator potential with a characteristic frequency determined by the shape of the trapping potential, namely 11.2 MHz for the trap used to demonstrate the C-NOT gate in ^{9}Be$^+$. The thermal agitation of the ion causes it to vibrate about its equilibrium position. At sufficiently low temperatures, the vibrational motion is quantized, and the excitations are governed by the harmonic oscillator quantum number n. (See Section 3.3.)

In the ^{9}Be$^+$ ion trap experiment, the ion was first cooled to temperatures in the μK range by laser cooling techniques. (See Section 11.2.6.) At these very low temperatures, there was a 95% probability that the ion was in the lowest vibrational state of the trap corresponding to the zero-point motion with $n = 0$. In these conditions it is possible to use the quantized vibrational state of the ion as the control qubit, with qubit $|0\rangle$ corresponding to $n = 0$ and qubit $|1\rangle$ to $n = 1$.

The target qubit was formed from two of the hyperfine levels of the ^{9}Be$^+$ ion electronic ground state. The ^{9}Be$^+$ ion has a single valence electron which lies in the 2s atomic shell, giving a $^2S_{1/2}$ ground state term with electron angular momentum $J = 1/2$. The ^{9}Be nucleus has angular momentum $I = 3/2$, and thus we have two possible states for the total angular momentum F, namely $F = 1$ and $F = 2$. This gives eight hyperfine sublevels, with the three sublevels from the $F = 1$ manifold lying above the five $F = 2$ sublevels by 1250 MHz. A weak magnetic field of 0.18 T was applied to split the hyperfine multiplets by the Zeeman effect and hence make the individual m_F sublevels distinguishable. The $(F = 2, m_F = 2)$ and $(F = 1, m_F = 1)$ hyperfine sublevels were then used as the $|0\rangle$ and $|1\rangle$ target qubit states, respectively. The resulting level scheme is given in Fig. 13.15(b), and the physical identification of the qubit states is summarized in Table 13.8.

Table 13.8 Qubit states for the C-NOT quantum logic gate of a single trapped ^{9}Be$^+$ ion. The control qubit corresponded to the first two quantized vibrational levels of the ion, as specified by the harmonic oscillator quantum number n, while the target qubit corresponded to two hyperfine sublevels of the $^2S_{1/2}$ ground-state electronic term of the ^{9}Be$^+$ ion. (After C. Monroe *et al.*, *Phys. Rev. Lett.* **75**, 4714 (1995).)

Control qubit $|n\rangle$	Target qubit	Hyperfine state $|F, m_F\rangle$
$|0\rangle$	$|0\rangle$	$|2, 2\rangle$
$|0\rangle$	$|1\rangle$	$|1, 1\rangle$
$|1\rangle$	$|0\rangle$	$|2, 2\rangle$
$|1\rangle$	$|1\rangle$	$|1, 1\rangle$

The C-NOT gate was demonstrated by a series of steps to establish the truth table given in Table 13.5. The system was first prepared in the $|0,0\rangle$ state, namely $|F, m_F\rangle = |2,2\rangle$ and $n = 0$, by laser cooling. The system was then manipulated between the four states of the two-qubit register by applying electromagnetic pulses of the correct frequency, duration, and phase. Rather than using microwave pulses to excite the transitions directly, two laser beams with their frequency difference tuned to 1250 MHz were employed, as shown in Fig. 13.15(c). With both lasers close to resonance with the $^2S_{1/2} \rightarrow {}^2P_{1/2}$ transition at 313 nm, the qubit manipulations were then driven by the stimulated Raman effect.

The state of the target qubit at the end of the experiment was measured by applying a third laser beam to excite transitions to the $^2P_{3/2}$ atomic term. By using σ^+ excitation and tuning the laser appropriately, transitions from the $(F = 1, m_F = 1)$ sub-level were strongly suppressed. This meant that the fluorescence signal due to spontaneous emission from the $^2P_{3/2}$ level was determined only by the population of the $(F = 2, m_F = 2)$ sublevel of the $^2S_{1/2}$ term. Hence the fluorescence signal registered by a photomultiplier tube gave a measure of the state of the target qubit at the end of the gate operations. The gate operation was completed in 50 µs, some 10 times faster than the decoherence time measured in the experiment. This was sufficient for the proof-of-principle demonstration, but is clearly far from satisfying the third item on the DiVincenzo check list.

Since this first experiment, much progress has been made with ion-trap quantum computing. Simple quantum algorithms have been implemented and quantum coupling between pairs of ions has been demonstrated. The long-term vision is shown schematically in Fig. 13.16. The idea is to have an array of ions held in a linear trap, with each one addressed individually by separate laser beams. The repulsive Coulomb forces lead to an equilibrium configuration with a regular spacing between the ions. The trap is designed so that this spacing is larger than the laser wavelength, which means that it is possible to address each ion separately, and also to resolve the fluorescence from individual

See S. Gulde *et al.*, *Nature* **421**, 48 (2003); F. Schmidt-Kaler *et al.*, *Nature* **422**, 408 (2003); D. Leibfried *et al.*, *Nature* **422**, 412 (2003).

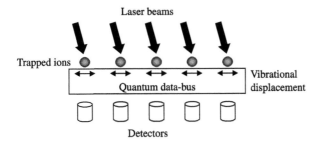

Fig. 13.16 Quantum computation using a linear array of trapped ions. Each individual ion is addressed by a laser beam and the fluorescence emitted is recorded with a detector. The ions are coupled together through their vibrational motion along the trap axis, which serves as a quantum data-bus. The electrodes used to produce the confining forces of the trap are not shown for clarity.

ions. At sufficiently low temperatures, the system forms a multiple qubit quantum computer, with the qubits interacting with each other through the coupling of the vibrations of the individual ions. This occurs because the displacement of one ion affects the others through the repulsive forces between them. Hence the collective vibrational motion of the array acts like a **quantum data-bus**, transferring quantum information in a coherent way between the separate qubits in the register.

It is not clear at present how large a quantum register could be made in this way. Linear arrays with up to ~ 40 trapped ions have been demonstrated, but so far the largest quantum register consists of only a few qubits. In the long run, ion trap systems are always going to be prone to decoherence because the charged nature of the qubits makes them very susceptible to stray electric fields from the noisy environment. One way around this problem is to use neutral atoms instead of ions. Progress in this area has also been very impressive, and the reader is referred to the bibliography for more details.

Many other physical systems are being considered for applications in quantum computation. In Section 10.5, for example, we described a conditional phase gate in which the qubits are photons and the interaction between them is produced by cavity quantum electrodynamics. At the time of writing, some of the most advanced experimental work has been performed by techniques of liquid phase nuclear magnetic resonance (NMR) at microwave frequencies. In this case the qubits are spin 1/2 nuclei in a molecule. The initial states are prepared by applying a strong magnetic field to align the spins along the field direction, and the spins interact with each other through the electrons that form the chemical bonds in the molecule. The quantum operations are performed by sequences of phase-stabilized RF pulses, and their outcomes are read out by standard NMR techniques.

NMR systems suffer from the problem that the energy gap between the spin up and down states is small compared to $k_B T$ at all reasonable working temperatures. The initial states used in the experiments are therefore only 'pseudo-pure'. See the article by J. A. Jones in Bouwmeester *et al.* (2000) for a discussion of the implications of this point.

While impressive results have been achieved with NMR systems, it is known that the present techniques cannot be scaled much further. In the long run, new approaches, perhaps using solid state NMR, will have to be developed if larger NMR–based quantum computers are to be built. Further details may be found in the bibliography.

13.7   Outlook

In the short term, the most likely field for applications of quantum computers is as quantum repeaters, since this only requires a few qubits in the quantum register.

The subject of quantum computation is very young, and it is far too early to make any long-term predictions. Everyone working in the field realizes that the challenges are enormous, the main issue being scalability. It is estimated that the quantum register would have to consist of ~ 50–100 qubits to be useful for quantum simulations, and even larger for many other applications, and no one yet knows how to build a quantum system of this size. In the mean time, the first generation of proof-of-principle experiments has nearly been completed, and the task is now to learn how to scale up the systems to build baby quantum computers with a few tens of qubits in the register.

The fundamental reason why it is going to be difficult to build a very large quantum computer is that we are working at the boundary between classical and quantum physics. It is generally accepted that the reason why quantum effects are only observed in small, isolated systems is that the noisy classical environment causes increasing amounts of decoherence as the system gets larger. The task of building a large quantum computer will thus involve learning how to control the noisy environment on a scale that has never been achieved to date.

One interesting solution to the scalability problem is to learn how to form a network of small quantum computers, thus creating a larger one. This will require the transfer of quantum information from one quantum computer to another, as illustrated schematically in Fig. 13.17. The qubits within the quantum computer are called 'static' qubits, while those carrying the quantum information between the two computers are called 'flying' qubits. Interest in this type of approach has led DiVincenzo to add two further requirements to his check list for the quantum hardware:

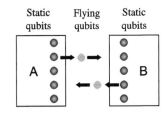

Fig. 13.17 Transfer of quantum information by flying qubits between two quantum computers A and B, each made up of a register of static qubits.

6. The system must have the ability to interconvert stationary and flying qubits.

7. The system must have the ability to transmit flying qubits faithfully between specified locations.

See, for example, D. P. DiVincenzo, *Quantum Information and Computation*, **1**, 1 (2001).

Some preliminary results have already been obtained demonstrating the first steps towards satisfying these requirements, and the results are encouraging. In the long run, only time will tell whether this method will work or not. In the mean time, there is great enthusiasm in the field, and we can expect the very rapid progress to continue for many years to come.

Further reading

Comprehensive treatments of the whole subject of quantum computing may be found in Bouwmeester *et al.* (2000), Nielsen and Chuang (2000), and Stolze and Suter (2004). Introductory overviews written by leading figures in the field may be found in Deutsch and Ekert (1998) or Walmsley and Knight (2002). A more advanced review is given in Bennett and DiVincenzo (2000). Collections of research papers on the subject are to be found in Ekert *et al.* 1998 and in the specialist journal *Quantum information and computation*.

The evolution of information processing technology from classical to quantum computers has been described by Williams (1998), Birnbaum and Williams (2000), and Dowling and Milburn (2003). An introductory discussion of the importance of decoherence in determining the transition from quantum to classical behaviour has been given by Arndt *et al.* (2005), while Zurek (2003) covers the same topic in much greater depth. Introductions to the methods of quantum error correction may be found in DiVincenzo and Terhal (1998) or Preskill (1999).

Introductory reviews on quantum information processing with atoms and ions have been given by Monroe (2002) and Cirac and Zoller (2004). Detailed reviews of ion trap quantum computers may be found in Kielpinski (2003), Leibfried *et al.* (2003), Sasura and Buzek (2002), or Steane (1997). Experimental details for quantum computation by nuclear magnetic resonance are given in Havel *et al.* (2002) or Vandersypen and Chuang (2004). A collection of papers describing the present status of experimental work on quantum information processing is given in Knight *et al.* (2003).

There are many useful internet resources on quantum computing. An index of tutorial articles may be found at http://cam.qubit.org, while many more specialized articles are frequently posted at the Los Alamos National Laboratory e-print archive at http://xxx.lanl.gov/archive/quant-ph.

Exercises

(13.1) Find the coordinates of the Bloch sphere corresponding to the following qubits:

(a) $|\psi\rangle = (1/\sqrt{2})(|0\rangle + |1\rangle)$,

(b) $|\psi\rangle = (1/\sqrt{2})(|0\rangle + \mathrm{i}|1\rangle)$,

(c) $|\psi\rangle = (1/2)|0\rangle + (3/8)^{1/2}(1+\mathrm{i})|1\rangle$.

(13.2) Show that a unitary operation on a qubit is reversible.

(13.3) The matrices to implement single-qubit rotations through an angle ϕ about the x- and z-axes are given, respectively, by:

$$R_x(\phi) = \begin{pmatrix} \cos(\phi/2) & -\mathrm{i}\sin(\phi/2) \\ -\mathrm{i}\sin(\phi/2) & \cos(\phi/2) \end{pmatrix},$$

$$R_z(\phi) = \begin{pmatrix} \mathrm{e}^{-\mathrm{i}\phi/2} & 0 \\ 0 & \mathrm{e}^{\mathrm{i}\phi/2} \end{pmatrix}.$$

Verify that the Hadamard operator can be written in the form:

$$H = \mathrm{e}^{\mathrm{i}\alpha} R_z(\phi_3) \cdot R_x(\phi_2) \cdot R_z(\phi_1),$$

stating the values of ϕ_1, ϕ_2, ϕ_3, and α.

(13.4) Calculate the effect of the following, giving a geometric interpretation of each:

(a) the Z gate on $|1\rangle$,

(b) the X gate on $(1/\sqrt{2})(|0\rangle + \mathrm{i}|1\rangle)$,

(c) the H gate on $(1/\sqrt{2})(|0\rangle + |1\rangle)$.

(13.5) By considering an arbitrary mapping of the qubit Bloch vector polar coordinate angles

$$(\theta, \varphi) \rightarrow (\theta', \varphi'),$$

explain how the sequence of rotations given in eqn 13.22 can perform an arbitrary single qubit operation.

(13.6) Verify that the matrix given in eqn 13.20 correctly reproduces the truth table of the C-NOT operator given in Table 13.5.

(13.7) Calculate the output of the quantum circuit shown in Fig. 13.8 when the input wave function is (a) $|01\rangle$, (b) $|10\rangle$, and (c) $|11\rangle$. Assume that it is the first qubit that undergoes the Hadamard operation.

(13.8) A quantum dot with a transition dipole moment of μ_{01} is irradiated with a resonant laser pulse which has a Gaussian time-dependent electric field of form $\mathcal{E}(t) = \mathcal{E}_0 \exp[-(t/\tau)^2]$.

(a) Show that the Bloch vector rotation Θ caused by this pulse is given by:

$$\Theta = \sqrt{\pi}\mu_{01}\mathcal{E}_0\tau/\hbar.$$

(b) Use eqn 2.28 to relate $\mathcal{E}_0$ to the pulse energy E_p when the laser is focussed to an area of A. (Assume that the pulse is linearly polarized with the field parallel to the dipole of the quantum dot.)

(c) Hence find the energy of a π-pulse for a quantum dot with $\mu_{01} = 1 \times 10^{-28}$ C m and $n = 3.5$, for a pulse with $\tau = 10^{-12}$ s and an area of 10^{-8} m^2.

(13.9) The unitary operator $\hat{U}_f$ for a one-bit function $f(x)$ in the Deutsch algorithm has two input qubits x and y, and the output is equal to x and $y \oplus f(x)$, where $\oplus$ indicates addition modulo two. (See Fig. 13.11.) The four possible versions of $f(x)$, namely f_1, f_2, f_3, and f_4 are defined in Table 13.7. Our task in this exercise is to devise quantum circuits to implement $\hat{U}_f$ for each of these four possible versions of $f(x)$.

(a) Show that the unitary operator for f_1 is the identity operator. Draw the quantum circuit for this operator.

(b) Write down the truth table for $\hat{U}_{f_2}$, and hence show that its matrix is given by:

$$\hat{U}_{f_2} = \begin{pmatrix} \mathbf{X} & \mathbf{0} \\ \mathbf{0} & \mathbf{1} \end{pmatrix},$$

where $\mathbf{X}$, $\mathbf{0}$, and $\mathbf{1}$ are the 2×2 matrices representing the X, zero and identity operators, respectively. Verify that $\hat{U}_{f_2}$ can be performed by the quantum circuit shown in Fig. 13.18(a).

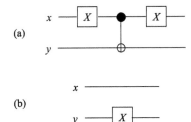

Fig. 13.18 Quantum circuits for (a) $\hat{U}_{f_2}$ and (b) $\hat{U}_{f_4}$ in Deutsch's algorithm.

(c) Find the matrix for $\hat{U}_{f_3}$, and hence show that $\hat{U}_{f_3} = \hat{U}_{\text{CNOT}}$. Draw the quantum circuit for this operator.

(d) Find the matrix for $\hat{U}_{f_4}$. Explain why $\hat{U}_{f_4}$ can be performed by the quantum circuit shown in Fig. 13.18(b).

(13.10) Consider a database containing eight records. Calculate the probability that the system is in the solution state after (a) one iteration, and (b) two iterations, of the Grover operator.

(13.11) Compare the number of transmissions required to transmit one qubit down 100 km of optical fibre with a 1/e loss distance of 20 km when using (a) a single span of fibre, and (b) an optimized compound fibre using quantum repeaters.

(13.12) A Be$^+$ ion trap has a harmonic oscillator potential with a resonant frequency of 11 MHz. The ion is cooled on a transition with a natural line width of 20 MHz. Calculate the probability that the oscillator is in the ground state at the Doppler limit temperature, and find the temperature to which the ion would have to be cooled so that this probability reaches 95%.

(13.13) A two-qubit spin system consists of two spin-1/2 nuclei in a magnetic field. The $|0\rangle$ and $|1\rangle$ states are defined to coincide with the $\sigma_z = -1/2$ and $\sigma_z = +1/2$ states, respectively. The Hamiltonian of the system is of the form:

$$H = \sum_{i=1}^{2} \hbar \omega_i \sigma_z^i + J \sigma_z^1 \sigma_z^2$$

where the superscripts refer to the individual nuclei. $\hbar \omega_i$ is the energy splitting between the up and down spin states in the absence of coupling, and J represents a spin–spin interaction term. Sketch the energy level spectrum of the system, and show that it is of the form shown in Fig. 13.10(a), stating the value of Δ.

14 Entangled states and quantum teleportation

14.1	Entangled states	296
14.2	Generation of entangled photon pairs	298
14.3	Single-photon interference experiments	301
14.4	Bell's theorem	304
14.5	Principles of teleportation	310
14.6	Experimental demonstration of teleportation	313
14.7	Discussion	316
Further reading		**317**
Exercises		**318**

The third branch of quantum information processing is *quantum teleportation*. This is a very new subject, and the aim of researchers working in the field at present is to achieve proof-of-principle demonstrations at the few-particle level. As we shall see, teleportation relies heavily on the properties of *entangled states*. We therefore begin by describing the concept of entangled photon states and explaining how they are generated in the laboratory. This will enable us to describe some recent experiments testing fundamental ideas of interference at the single-photon level. We shall then discuss the Einstein–Podolsky–Rosen (EPR) paradox and Bell's theorem, which will allow us to explain the principles of teleportation, and describe how they have been demonstrated in the laboratory. Finally, we shall briefly discuss a few of the wider issues that arise from the EPR paradox and Bell's theorem.

14.1 Entangled states

Entanglement is one of the most counter-intuitive aspects of the quantum world. The concept is linked to two famous papers in the historical development of quantum theory, and has come to the fore in recent years with the advent of quantum information science. In 1935 Einstein, Podolsky and Rosen published the 'EPR' paper on the properties of an entangled two-particle system formed from the decay of a radioactive source. Soon afterwards, Schrödinger coined the term 'entanglement' in his cat paradox paper that has fuelled the imagination of students and teachers alike for many years.

Let us first consider the EPR paper. We will present the argument in the 'EPRB' form introduced by David Bohm in 1951. The scheme for an optical EPRB experiment is shown in Fig. 14.1. A source S emits a pair of photons arbitrarily labelled 1 and 2, with photon 1 going one way and photon 2 going another. The polarization of each photon is measured with a beam-splitter/detector arrangement similar to the one presented in Fig. 12.2. We designate the polarization states $|\updownarrow\rangle$ and $|\leftrightarrow\rangle$ as $|1\rangle$ and $|0\rangle$, respectively, according to the BB84 scheme in the $\oplus$ basis given in Table 12.1.

The subtlety in the experiment occurs when we use a source that emits **correlated photon pairs**. Correlated photon pairs have the following

See A. Einstein, B. Podolsky, and N. Rosen, *Phys. Rev.* **47**, 777 (1935), and E. Schrödinger, *Die Naturwissenschaften* **23**, 807, 823, 844 (1935). An English translation of the latter is available in *Proc. Am. Philos. Soc.* **124**, 323 (1980). Bohm's variant on the EPR experiment was originally developed in his book *Quantum Theory*, published in 1951 by Prentice-Hall, New Jersey. Bohm actually proposed to make spin measurements on pairs of atoms, but the version we present here is the optical equivalent involving polarization measurements on pairs of photons.

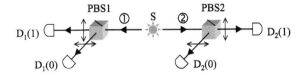

Fig. 14.1 Apparatus for an EPRB experiment. The source S emits two correlated photons arbitrarily labelled 1 and 2 towards polarization detectors involving a polarizing beam splitter (PBS) and single-photon detectors D. The detectors are given a subscript 1 and 2 to identify the photon and the results are designated 0 and 1 according to the scheme presented in Table 12.1 for the ⊕ basis.

properties:

1. The polarization of either photon 1 or photon 2 measured independently of the other is random.

2. The polarization of the pair of photons is perfectly correlated; that is, if $D_1(0)$ fires, then $D_2(0)$ always fires, and if $D_1(1)$ fires, then $D_2(1)$ always fires. Alternatively if $D_1(0)$ fires, then $D_2(1)$ always fires, and vice versa.

The second property follows from internal conservation laws of the source that will be discussed in Section 14.2.

A multi-particle system is described as being in an **entangled state** if its wave function cannot be factorized into a product of the wave functions of the individual particles. The mutual dependence of the results of the polarization measurements on the correlated photon pair means that the wave function has to be written in the form:

$$|\Phi^{\pm}\rangle = \frac{1}{\sqrt{2}}\left(|0_1, 0_2\rangle \pm |1_1, 1_2\rangle\right), \tag{14.1}$$

for the case of perfect positive correlation, and

$$|\Psi^{\pm}\rangle = \frac{1}{\sqrt{2}}\left(|0_1, 1_2\rangle \pm |1_1, 0_2\rangle\right), \tag{14.2}$$

for perfect negative correlation, with the subscripts referring to the individual photons. The wave functions in eqns 14.1 and 14.2 are thus examples of entangled states. They are also called **Bell states** for reasons that will become clear in Section 14.4.

The entangled form of the wave functions in eqns 14.1 and 14.2 implies that a measurement of the polarization of one photon determines the result of a polarization measurement on the other. Thus for the wave function given in eqn 14.1 we will obtain either the result (0,0) or (1,1), each with equal probability. Similarly, eqn 14.2 implies results of (0,1) or (1,0) each with 50% probability. In both cases a measurement on one photon allows us to predict the result of the measurement on the other with 100% certainty.

The **Schrödinger cat paradox** illustrates the concept of entangled states in a graphic way by considering the state of a live cat put into a sealed box containing a radioactive atom as shown Fig. 14.2. The box

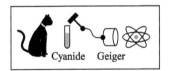

Fig. 14.2 Schrödinger's cat. A live cat is put into a sealed box containing a radioactive atom. The radiation emitted by the decay of the atom is detected by a Geiger counter, which activates a relay on registering a count. The relay is connected to a hammer which smashes a sealed flask of cyanide, and hence kills the cat.

also contains a devious mechanism such that the decay of the atom triggers a device to smash a sealed flask of poison, thereby killing the cat. The state of the cat is therefore entangled with the state of the atom. If we wait for a time such that the probability of the atom decaying is equal to 50%, then we can write the wave function of the system in the form:

$$|\Psi\rangle = \frac{1}{\sqrt{2}} \left(|\text{live}, 1\rangle + |\text{dead}, 2\rangle \right), \qquad (14.3)$$

where $|1\rangle$ and $|2\rangle$ represent the state of the undecayed and decayed atom, respectively. This seems to imply that we have a state inside the box where the cat is both dead and alive at the same time, in clear contrast to our common experience. On opening the box, we would, of course, find the cat dead or alive with probability equal to 50%.

Much to the relief of cat-lovers, there is no need to perform the Schrödinger cat experiment in the laboratory. Paradoxes of this type are not found in the macroscopic world, because large systems consisting of many particles lose their quantum coherence through interactions with the noisy macroscopic environment. (See Section 13.4.) Things are different, however, at the microscopic level of isolated atoms and photons in a well-controlled environment. Entangled photon states of the type required for the EPRB experiment can readily be generated in the laboratory, and photon Schrödinger cat states have been demonstrated.

Quantum entanglement is not restricted to the case of two-particle polarization that we have considered here. Two-particle photon states with time or momentum entanglement can also be generated, and entangled states involving three or more particles have many interesting properties. However, we shall restrict our attention exclusively to two-particle polarization states for simplicity's sake. The reader is referred to the bibliography for details of other types of entangled states.

14.2   Generation of entangled photon pairs

Many of the early optical experiments on entangled states employed atomic cascades in calcium to generate the correlated photon pairs. The experiment consists of a pair of detectors arranged to collect the photons emitted in an atomic cascade from the $4p^2 \ ^1S_0$ excited state of calcium as shown in Fig. 14.3(a). Figure 14.3(b) shows the corresponding level scheme for the transitions involved. The cascade occurs by allowed transitions at 551.3 and 422.7 nm via the $4p4s \ ^1P_1$ intermediate level. Narrow-band interference filters F1 and F2 in front of the photomultiplier tube (PMT) detectors selected these photon wavelengths from others produced by alternative decay routes. In the initial experiment by Kocher and Commins in 1967, the calcium atoms were excited to the $4p^2 \ ^1S_0$ level by absorption of ultraviolet photons from a hydrogen arc lamp. Photons at 227.5 nm from the lamp first excited the atoms from the $4s^2 \ ^1S_0$ ground state to the $3d4p \ ^1P_1$ level, and the atoms then

See C. A. Kocher, and E. G. Commins, *Phys. Rev. Lett.* **18**, 575 (1967).

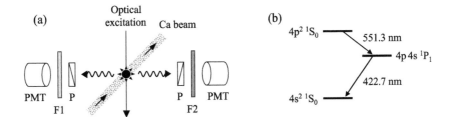

Fig. 14.3 Correlated photon pair generation by atomic cascade in calcium. (a) Experimental arrangement employing two linear polarizers (P) and photomultiplier tube (PMT) detectors. (b) Atomic level scheme. The narrow-band interference filters F1 and F2 used in the experiment were chosen to select the photons at 551.3 and 422.7 nm, respectively. (After C. A. Kocher and E. G. Commins, *Phys. Rev. Lett.* **18**, 575 (1967).)

dropped to the desired $4p^2\ {}^1S_0$ level by spontaneous decay. In the subsequent experiments by Aspect *et al.* described in Section 14.4.3, the atoms were excited directly to the $4p^2\ {}^1S_0$ level by two-photon absorption of photons at 406 and 581 nm from separate laser beams.

The initial and final states for the cascade are both $J = 0$ states with no net angular momentum. This demands that the photon pairs emitted in the cascade carry no net angular momentum. In addition, the rotational invariance of $J = 0$ states, and the fact that the initial and final levels are both of the same even parity, requires that the photon pairs have the polarization correlation properties required for the EPRB experiments. This correlation was confirmed by placing linear polarizers in front of both detectors and checking for coincidences. The experiments clearly demonstrated that the coincidences only occur when the axes of the polarizers are aligned parallel to each other, indicating that Bell states of the type given in eqn 14.1 are being produced.

In the 1980s and 1990s new sources of correlated photon pairs with higher flux rates were developed by techniques of nonlinear optics. (See Section 2.4.) The correlated photon pairs were generated by the **down-conversion** process in which a single photon from a pump laser at angular frequency ω_0 is converted into a pair of signal and idler photons at angular frequencies ω_1 and ω_2, as shown in Fig. 14.4. Conservation of energy and momentum, respectively, require that:

$$\omega_0 = \omega_1 + \omega_2, \qquad (14.4)$$

and

$$\boldsymbol{k}_0 = \boldsymbol{k}_1 + \boldsymbol{k}_2, \qquad (14.5)$$

where $\boldsymbol{k}_i$ is the wave vector of the photon in the crystal. The second of these conditions is equivalent to requiring that the nonlinear waves and the fundamental beam all remain in phase throughout the nonlinear medium. For this reason, the circumstances in which eqns 14.4 and 14.5 are satisfied simultaneously are called **phase-matching** conditions. The down-conversion process is called *degenerate* when $\omega_1 = \omega_2 = \omega_0/2$, and *non-degenerate* otherwise.

At first sight, it might seem that there would be many combinations of frequencies and wave vectors that can be phase-matched. However, this is

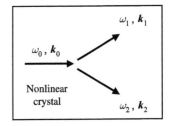

Fig. 14.4 Schematic representation of a down-conversion process within a nonlinear crystal. A single photon of angular frequency ω_0 simultaneously generates a pair of signal and idler photons of angular frequencies ω_1 and ω_2 subject to the phase-matching conditions set out in eqns 14.4 and 14.5.

not the case because of the dispersion in the nonlinear crystal. (See Exercise 14.4.) Dispersion is a general property of all optical materials and refers to the variation of the refractive index with frequency. This means that the refractive indices at the three different frequencies are in general different, making it impossible under normal circumstances to satisfy the phase-matching conditions. Fortunately, the nonlinear crystals are also birefringent, which means that the refractive index depends on the direction of the polarization of the light with respect to the crystal axes. This allows us to balance birefringence against dispersion, and achieve two different types of phase matching. In type-I phase matching the polarizations of the down-converted photons are parallel to each other and orthogonal to the pump photon, while in type-II phase matching the down-converted photons have orthogonal polarizations.

Figure 14.5 illustrates the generation of entangled photon pairs by degenerate down-conversion with type-II phase matching. The principle of the technique in shown in Fig. 14.5(a). Ultraviolet photons from a pump laser are focussed into a β-barium borate (BBO) crystal and are down-converted to two red photons at half the frequency. The phase-matching requirements determine that the down-converted photons emerge in cones of opposite polarization, leading to a double ring pattern with two intersection points, as shown in Fig. 14.5(b). Equation 14.5 demands that if we find a vertically polarized photon at one of the intersection points, then the photon at the other intersection point must be horizontally polarized, and vice versa. However, the photon at each intersection point might have originated from either of the two oppositely polarized rings and can therefore be horizontal or vertical with equal probability. The arrangement therefore produces states of the type:

$$|\Psi\rangle = \frac{1}{\sqrt{2}} \left(|\leftrightarrow_1, \updownarrow_2\rangle + \mathrm{e}^{\mathrm{i}\phi} | \updownarrow_1, \leftrightarrow_2\rangle \right), \qquad (14.6)$$

where ϕ is an optical phase that can be altered with compensator plates. By setting ϕ equal to 0 to π we can then produce either of the Bell

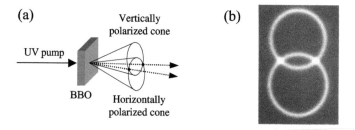

Fig. 14.5 Generation of polarization entangled photon pairs by degenerate down-conversion with type-II phase matching. (a) Experimental arrangement employing an ultraviolet pump laser and a BBO crystal. The phase-matching conditions require that the beams emerge in cones of opposite polarization. (b) Degenerate type II down-conversion as seen through a narrow band filter. The two entangled photons correspond to the intersection points of the rings. (After P. G. Kwiat, *et al., Phys. Rev. Lett.* **75**, 4337 (1995), © American Physical Society, reproduced with permission.)

states given by eqn 14.2. Down-conversion sources of this type have now generally supplanted atomic cascade sources for practically all of the experiments that require polarization-entangled photon states.

Example 14.1 A correlated pair of photons is generated by non-degenerate parametric down conversion using a laser at 502 nm. Given that the wavelength of one of the photons is 820 nm, calculate the wavelength of the other.

Solution
We use eqn 14.4 with $\omega = 2\pi c/\lambda$, which implies:

$$\frac{2\pi c}{\lambda_0} = \frac{2\pi c}{\lambda_1} + \frac{2\pi c}{\lambda_2}.$$

Hence:

$$\frac{1}{\lambda_2} = \frac{1}{\lambda_0} - \frac{1}{\lambda_1} = \left(\frac{1}{502} - \frac{1}{820}\right) \text{ nm}^{-1},$$

giving $\lambda_2 = 1294$ nm.

14.3   Single-photon interference experiments

The main reason for introducing correlated photons pairs in this chapter is to explain how they can be used to test Bell's theorem and to implement quantum teleportation. However, the use of correlated photon pair sources has also enabled the testing of several fundamental ideas about the nature of photon interference, and it is worthwhile to consider some of these briefly here.

Consider first the experimental arrangement shown in Fig. 14.6. Signal and idler beams of the same polarization and frequency are generated by type-I degenerate down conversion in a nonlinear crystal and are made to

The first measurement of a single-photon interference pattern was made as early as 1909, when a Young's slit experiment was performed with only one quantum of energy within the apparatus at a given instant. See G. I. Taylor, *Proc. Camb. Phil. Soc.* **15**, 114 (1909).

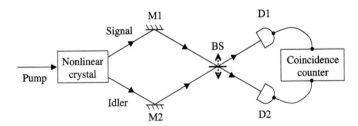

Fig. 14.6 Experimental arrangement for demonstrating single-photon interference effects using correlated photon pairs. M1 and M2 are mirrors, D1 and D2 are single-photon counting detectors, and BS is a 50:50 beam splitter. The path difference between the signal and idler beams can be adjusted by translating BS up and down. (Adapted from C. K Hong, Z. Y. Ou, and L. Mandel, *Phys. Rev. Lett.* **59**, 2044 (1987).)

interfere at a beam splitter BS. The path difference between the beams can be adjusted by translating the beam splitter up and down. With high-intensity classical beams, we would expect to see bright and dark fringes appearing at the output ports as the beam splitter is translated. The total signal on detectors D1 and D2 would be constant, but the magnitude of the signal on the individual detectors would oscillate in anti-phase as BS is translated.

The experiment becomes more interesting when we operate at the single-photon level. The signal and idler beams now contain correlated photon pairs. When the path lengths of the beams are identical, the two photons arrive at the beams splitter at the same time and interfere. When single photons interfere at a 50:50 beam splitter, destructive interference prevents the possibility that the two photons go to different output ports, and both photons therefore emerge at the same output. (See Exercise 8.11.) Hence the only possible results are that both photons go to D1 or both go to D2, leading to no coincidence events on the detectors.

The absence of coincidences when the path lengths are equal was verified experimentally in 1987 by Hong, Ou, and Mandel. For this reason, the arrangement shown in Fig. 14.6 is sometimes called a **Hong–Ou–Mandel interferometer**. An argon ion laser operating at 351.1 nm was used as the pump laser and potassium dihydrogen phosphate (KDP) as the nonlinear crystal. When the difference in the path lengths of the signal and idler beams was larger than the coherence length, no interference occurred. In this situation, each photon randomly exits at either output port, producing coincidences on D1 and D2 for 50% of the events. However, when the path difference was smaller than the coherence length, no coincidences were recorded, confirming the single-photon interference effect.

See C. K. Hong, Z. Y. Ou, and L. Mandel, *Phys. Rev. Lett.* **59**, 2044 (1987).

Consider now the interference experiment shown in Fig. 14.7. The interferometer incorporates two down-converting nonlinear crystals NL1 and NL2, both driven by photons derived from a single pump laser. The overall down-conversion efficiency is rather small, so that it is extremely unlikely that correlated photon pair generation occurs simultaneously in the two nonlinear crystals. The crystals are arranged so that the paths of the two idler beams i_1 and i_2 are coincident. (This is possible because the nonlinear crystal NL2 is transparent at the idler frequency.) A detector D_i registers the combined signal of these two idler beams. At the same time, the signal beams s_1 and s_2 are combined at a 50:50 beam splitter BS2 and the signal at one of the output ports is registered by the detector D_s.

From a classical perspective, the two signal beams s_1 and s_2 should interfere at BS2, and we would therefore expect to observe interference fringes on D_s as the path difference is varied by translating BS2. At the single-photon level, the photon emerges at either of the output ports of BS2 with a probability determined by the classical interference pattern of s_1 and s_2. However, when working with single photons, it is natural to

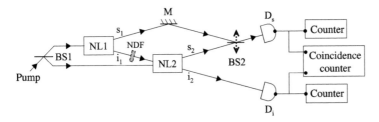

Fig. 14.7 Experimental arrangement for demonstrating that which-path information destroys photon interference. Photons from the pump laser are randomly split by a 50 : 50 beam spitter (BS1) and drive two down-converting nonlinear crystals NL1 and NL2. The crystals are arranged so that the idler photons i_1 and i_2 follow identical paths after NL2, while the signal photons s_1 and s_2 are combined at a second 50 : 50 beam splitter (BS2) by using the mirror (M). Single-photon-counting detectors D_i and D_s are arranged to count photons from the idler path and from one of the output ports of BS2, respectively. A neutral density filter (NDF) can be introduced into the path of i_1 to provide path information about the interference pattern registered by D_s. (After X. Y. Zou, L. J. Wang, and L. Mandel, *Phys. Rev. Lett.*, **67**, 318 (1991).)

ask which of the two possible paths the individual photon took. This is an old issue in quantum mechanics, often called the **which-path** question. Many single-particle interference experiments have been performed, and it is now well-established that all attempts to obtain path information lead to a washing-out of the interference effect.

The arrangement in Fig. 14.7 adds an additional subtlety to the which-path question. The inclusion of the neutral density filter (NDF) on the path of i_1 allows the possibility of determining whether the signal photon that is observed at D_s originated from NL1 or NL2. When the filter is removed, the two paths i_1 and i_2 are coincident, and so a count registered on D_i cannot determine whether the photon was generated in NL1 or NL2. However, when the filter is introduced to block the path i_1, a count on D_i could then only have originated from a photon generated in NL2. In this case, it is known that the photon that was incident on BS2 followed the path s_2, and the interference effect is destroyed. What is more, it is not actually necessary to register a photon on D_i. The mere *possibility* of obtaining which-path information is sufficient to destroy the interference. Moreover, the path information is being obtained from a photon that plays no part in the formation of the interference pattern whatsoever.

Experiments of the type shown in Fig. 14.7 were first performed by Zou, Wang, and Mandel in 1991. They used a pump laser at 351.1 nm and generated signal and idler photons at 788.7 and 632.8 nm, respectively. Colour-sensitive filters in front of the detectors ensured that only signal photons were detected by D_s and idler photons by D_i. The experiment confirmed that the insertion of the neutral density filter washed out the interference effect observed on D_s, thereby elegantly demonstrating that the mere possibility of which-path information is sufficient to destroy the single-photon interference.

The which-path question was originally posed in the explanation of the double-slit interference pattern for individual particles. The appearance of fringes in a double-slit experiment is a trademark feature of wave-like behaviour. When such experiments are performed with individual particles, the probability of detecting the particle at a given position is proportional to the fringe amplitude at that point. Any attempt to determine which slit the particle passed through destroys the effect and a uniform probability pattern is observed.

See X. Y. Zou, L. J. Wang and L. Mandel, *Phys. Rev. Lett.*, **67**, 318, 1991.

14.4 Bell's theorem

14.4.1 Introduction

In Section 14.1 we saw how the EPR paper naturally leads to the concept of entangled states. We now wish to return to the EPR paper and consider why it is considered by some people to constitute a 'paradox'. We shall approach the subject by first considering Einstein's position concerning quantum theory in general, and then move on to the seminal work of John Bell that added a ground-breaking new perspective to the question. In many ways, the EPR paper and Bell's work constitute the foundation of modern quantum information science. At the same time, they have been the inspiration for much philosophical debate, as we shall briefly discuss in Section 14.7.

Einstein's discomfort with the Copenhagen interpretation of quantum theory proposed by Niels Bohr and others is well-documented. The EPR experiment was meant to be a refutation of the Copenhagen approach and a proof of the 'incompleteness' of quantum theory. Quantum mechanics may well be the best theory we have at present, but can we say for sure that it is the last word on the subject? Perhaps there might exist a deeper level of reality, with unknown properties governed by undiscovered laws. The results of present-day experiments would then be determined by **hidden variables** that we do not yet know, and perhaps can never know. The existence of these hidden variables, Einstein believed, was preferable to the probabilistic world implied by the Copenhagen interpretation in which the existence of physical properties such as spin and polarization seems to depend on the measurement process itself.

We now know, in fact, that the crux of the EPR argument is not about hidden variables in general, but rather about *local* hidden variables (LHV). This point became clear with the work of John Bell in 1964 that will be discussed below. Before we can understand Bell's insight, we first have to run through the gist of the EPR argument in favour of LHV.

Let us start by considering the measurement of the polarization of a single photon emitted from an unpolarized source: say, for example, photon 1 in Fig. 14.1. In the Copenhagen interpretation we would say that the polarization of the photon prior to the measurement is undefined. The measurement process then 'collapses' the wave function to produce the particular result of the experiment: that is, 0 or 1, each with 50% probability. In the LHV approach, by contrast, we would argue that the quantum picture is 'incomplete'. We would say that the photon possesses an unknown property governed by hidden variables. The source emits photons with a distribution of these hidden variables that determines that half of them go to detector 0 and the other half to detector 1. We could cite the example of tossing a coin which appears to be a purely chance process, but is in fact governed by well-defined, unknown classical variables such as the initial orientation of the coin, the forces applied to it, etc.

In the case of the single particle, the two approaches lead to the same conclusions and cannot be distinguished. The situation with the dual particle source in the EPR experiment is much more interesting. If we follow Einstein's reasoning, we could argue that trying to apply the Copenhagen approach leads to disconcerting consequences. The measurement of the state of one photon instantly determines the results for the other one. The two sets of detectors can be separated by long distances, and we thus appear to have instantaneous action at a distance in contravention to relativity. We would thus be led to conclude that matter is *non-local* at the microscopic level. This is in fact implicit in the entangled state wave functions given in eqns 14.1–14.2, which are intrinsically non-local in the sense that they depend on the properties of two well-separated particles.

The non-locality implied by the quantum interpretation of the EPR experiment has no counterpart in the classical world. The issue does not immediately arise in the LHV interpretation, (see Exercise 14.1) and this is why Einstein thought that he had proved his point. Bell, however, designed an ingenious variation on the EPR experiment, and succeeded in proving that the LHV and quantum mechanical theories can predict different results in some circumstances. Many experiments have now been performed to test whether the quantum mechanical or LHV approaches give the correct results, and there is almost unanimous evidence that the LHV picture is incorrect. The implication is that microscopic systems exist in nature that are *non-local.* This is why the LHV theories do not predict the correct result. We shall see how this argument works in the following section.

Bell's theorem was originally presented in a paper entitled 'On the Einstein–Podolsky–Rosen paradox', published in *Physics* **1**, 195–200 (1964). This paper is reproduced in Bell (1987), p. 14.

14.4.2   Bell's inequality

Bell's key result was the derivation of an inequality called **Bell's inequality**. Bell's theorem states that the inequality is always obeyed if the LHV picture of the microscopic world is correct. Quantum mechanics, by contrast, predicts violations of Bell's inequality, and we thus have a way to distinguish between the two approaches in the laboratory. The detailed proof of Bell's theorem is quite complicated, and we shall restrict ourselves here to a simple discussion that illustrates the key points of the argument.

Figure 14.8 shows a schematic diagram of the apparatus required to perform a measurement of Bell's inequality on a pair of correlated photons emitted from a source S. The experiment is the basically same as the EPRB arrangement shown in Fig. 14.1 except that there is one additional feature introduced. In the EPRB experiment, the polarization measurements performed on the two photons are identical. This is arranged in the laboratory by ensuring that the axes of the two polarizing beam splitters are parallel to each other. In the Bell experiment, we allow the axes of the two polarizing beam splitter cubes to be *different*. This surprisingly simple variation leads to profound differences in the results.

Let us designate the axes of PBS1 and PBS2 by unit vectors $\vec{a}$ and $\vec{b}$, respectively. We define θ_1 to be the angle between $\vec{a}$ and the vertical, and likewise θ_2 for $\vec{b}$. The EPRB experiment shown in Fig. 14.1 thus corresponds to the case with $\theta_1 = \theta_2$. In the Bell experiment we allow

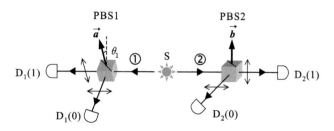

Fig. 14.8 Apparatus for a Bell experiment using correlated photon pairs. Photons 1 and 2 are sent to polarization detectors with their axes defined by unit vectors $\vec{a}$ and $\vec{b}$, respectively. The polarization detectors consist of a polarizing beam splitter (PBS) and two single-photon detectors D(0) and D(1) set to register the orthogonal polarization states. θ_1 and θ_2 define the angles between $\vec{a}$ and $\vec{b}$ and the vertical respectively. The diagram corresponds to the case with $\theta_2 = 0$.

θ_1 and θ_2 to differ. Figure 14.1 illustrates the case where θ_1 is around 30° and $\theta_2 = 0$.

For each setting of the angles θ_1 and θ_2 the Bell experiment has four possible results, which are characterized by their respective probabilities:

Probabilities are defined here as the average outcome of a large number of experiments.

$\mathcal{P}_{11}(\theta_1, \theta_2)$ is the probability that $D_1(1)$ fires and $D_2(1)$ fires,

$\mathcal{P}_{10}(\theta_1, \theta_2)$ is the probability that $D_1(1)$ fires and $D_2(0)$ fires,

$\mathcal{P}_{01}(\theta_1, \theta_2)$ is the probability that $D_1(0)$ fires and $D_2(1)$ fires,

$\mathcal{P}_{00}(\theta_1, \theta_2)$ is the probability that $D_1(0)$ fires and $D_2(0)$ fires.

The probabilities must satisfy two simple check rules. First, the total probability of getting a 1 or 0 result for each photon must be exactly 50%, implying that:

$$\mathcal{P}_{11}(\theta_1, \theta_2) + \mathcal{P}_{10}(\theta_1, \theta_2) = 0.5,$$
$$\mathcal{P}_{01}(\theta_1, \theta_2) + \mathcal{P}_{00}(\theta_1, \theta_2) = 0.5,$$
$$\mathcal{P}_{11}(\theta_1, \theta_2) + \mathcal{P}_{01}(\theta_1, \theta_2) = 0.5,$$
$$\mathcal{P}_{10}(\theta_1, \theta_2) + \mathcal{P}_{00}(\theta_1, \theta_2) = 0.5. \tag{14.7}$$

Second, the perfect correlations for the EPRB experiment must be reproduced when $\theta_1 = \theta_2$, implying for the case of positive correlations that:

$$\mathcal{P}_{11}(\theta, \theta) = 0.5,$$
$$\mathcal{P}_{10}(\theta, \theta) = 0,$$
$$\mathcal{P}_{01}(\theta, \theta) = 0,$$
$$\mathcal{P}_{00}(\theta, \theta) = 0.5, \tag{14.8}$$

and vice versa for negative correlations.

Let us first work out the probabilities according to quantum mechanics. We start by analysing the case with $\theta_2 = 0$ as shown in Fig. 14.8 when the source emits positively correlated Bell states of the type given in eqn 14.1. We choose the horizonatal/vertical measurement basis that coincides with the axes of PBS2. Suppose we obtain the result $D_2(0)$. This means that we are sending a vertically polarized photon to PBS1 and we thus obtain the result $D_1(0)$ with probability $\cos^2 \theta_1$ and $D_1(1)$

with probability $\sin^2\theta_1$. Similarly, if we obtain the result $D_2(1)$, then we have a horizontally polarized photon going to PBS1, meaning that we will obtain the results $D_1(0)$ and $D_1(1)$ with probabilities of $\sin^2\theta_1$ and $\cos^2\theta_1$, respectively. Now the results $D_2(0)$ and $D_2(1)$ both occur with probability 50% and so we have:

$$\mathcal{P}_{11}(\theta_1,0)=\frac{1}{2}\cos^2\theta_1,$$

$$\mathcal{P}_{10}(\theta_1,0)=\frac{1}{2}\sin^2\theta_1,$$

$$\mathcal{P}_{01}(\theta_1,0)=\frac{1}{2}\sin^2\theta_1,$$

$$\mathcal{P}_{00}(\theta_1,0)=\frac{1}{2}\cos^2\theta_1. \tag{14.9}$$

Now suppose that θ_2 is also arbitrary. We are free to choose any pair of orthogonal axes as our measurement basis. We therefore choose axes at angles of θ_2 and $\theta_2+90°$ which coincide with those of PBS2. The argument is then identical, except that the probabilities now depend on $(\theta_1-\theta_2)$ rather than just θ_1, giving:

$$\mathcal{P}_{11}(\theta_1,\theta_2)=\frac{1}{2}\cos^2(\theta_1-\theta_2),$$

$$\mathcal{P}_{10}(\theta_1,\theta_2)=\frac{1}{2}\sin^2(\theta_1-\theta_2),$$

$$\mathcal{P}_{01}(\theta_1,\theta_2)=\frac{1}{2}\sin^2(\theta_1-\theta_2),$$

$$\mathcal{P}_{00}(\theta_1,\theta_2)=\frac{1}{2}\cos^2(\theta_1-\theta_2). \tag{14.10}$$

In the case of negative correlation, the sine and cosine functions are reversed.

Now let us consider the LHV approach. There are, of course, many different LHV models we could propose, but let us choose the simplest, and suppose that the source emits pairs of photons that are either both vertically polarized or both horizontally polarized with equal probability. This will obviously reproduce the results of the EPRB experiment with $\theta_1=\theta_2=0$. For a Bell experiment with both θ_1 and θ_2 arbitrary, the equivalent probabilities are:

This argument is adapted from the one presented by J. G. Rarity, *Science* **301**, 604 (2003).

$$\mathcal{P}_{11}(\theta_1,\theta_2)=\frac{1}{2}(\sin^2\theta_1\sin^2\theta_2+\cos^2\theta_1\cos^2\theta_2),$$

$$\mathcal{P}_{10}(\theta_1,\theta_2)=\frac{1}{2}(\sin^2\theta_1\cos^2\theta_2+\cos^2\theta_1\sin^2\theta_2),$$

$$\mathcal{P}_{01}(\theta_1,\theta_2)=\frac{1}{2}(\cos^2\theta_1\sin^2\theta_2+\sin^2\theta_1\cos^2\theta_2),$$

$$\mathcal{P}_{00}(\theta_1,\theta_2)=\frac{1}{2}(\cos^2\theta_1\cos^2\theta_2+\sin^2\theta_1\sin^2\theta_2). \tag{14.11}$$

For the case shown in Fig. 14.8 with $\theta_2=0$, we obtain the same result as the quantum result given in eqn 14.9. However, in other cases, we obtain different results.

One of the clearest ways to see the difference is to do an EPRB experiment with $\theta_1 = \theta_2 = 0$, and then do another one with $\theta_1 = \theta_2 = 45°$. This should not have any physical effect for a rotationally invariant source, and we should therefore obtain perfectly correlated results in both cases, as experiments confirm. The quantum model predicts the correct outcome, because the choice of measurement basis is arbitrary up to the point when the first measurement is made. By contrast, the LHV approach predicts equal probabilities of 25% for all four possibilities in the second experiment. The reason for the discrepancy is that we are assigning a *local* polarization to each photon as it leaves the source. We then obtain random results in the second experiment when these vertically or horizontally polarized photons are incident on the polarizers angled at 45°. The only way to reconcile the model with the experimental results is to send a faster-than-light signal from the first detector that registers to the other one to create the correlation, which effectively implies non-locality.

The argument presented here is rather simplistic and applies only to a very rudimentary LHV model. The beauty of Bell's theorem is that it is completely general and applies to all possible LHV models. There are several different forms of Bell's theorem and the version we quote here was derived by Clauser, Horne, Shimony, and Holt (CHSH) in 1969. They introduced an experimentally determinable parameter S defined by:

See J. F. Clauser, M. A. Horne, A. Shimony, and R .A. Holt, *Phys. Rev. Lett.* **23**, 880 (1969). The notation that we use here is taken from A. Aspect *et al.*, *Phys. Rev. Lett.* **49**, 1804 (1982).

$$S = E(\theta_1, \theta_2) - E(\theta_1, \theta_2') + E(\theta_1', \theta_2) + E(\theta_1', \theta_2'), \qquad (14.12)$$

where

$$E(\theta_1, \theta_2) = \mathcal{P}_{11}(\theta_1, \theta_2) + \mathcal{P}_{00}(\theta_1, \theta_2) - \mathcal{P}_{10}(\theta_1, \theta_2) - \mathcal{P}_{01}(\theta_1, \theta_2), \qquad (14.13)$$

and proved that the following Bell inequality:

$$-2 \leq S \leq 2, \qquad (14.14)$$

holds for all possible LHV theories. On the other hand, it is not hard to find examples where the quantum predictions violate eqn 14.14. For example, if $\theta_1 = 0°$, $\theta_2 = 22.5°$, $\theta_1' = 45°$, and $\theta_2' = 67.5°$, we find from eqn 14.10 that $S = 2\sqrt{2}$, which violates eqn 14.14 by a substantial margin. The search for violations of Bell's inequality thus provides us with a clear way to test for quantum non-locality in the laboratory.

14.4.3   Experimental confirmation of Bell's theorem

The original reports of the three experiments performed by A. Aspect *et al.* may be found in *Phys. Rev. Lett.* **47**, 460 (1981), **49**, 91 (1982), and **49**, 1804 (1982).

The significance of Bell's theorem was immediately recognized and much experimental work has been devoted to testing its validity. The landmark optical experiments in the field are generally considered to be those of Alain Aspect and co-workers, who completed three beautiful experiments to test for violations of Bell's inequality between 1981 and 1982. All three of these experiments used correlated photon pairs generated by atomic cascades in calcium, as described previously in Section 14.2.

The first experiment checked for violations of a generalized version of the Bell inequality. This type of experiment compares the count rates on the detectors $D_1(1)$ and $D_2(1)$ for different settings of the polarizer angles θ_1 and θ_2. The results were found to be in violation of Bell's inequality and in agreement with the quantum mechanical predictions. The second experiment measured the CHSH inequality of eqn 14.14. The results were again found to be in violation of Bell's inequality.

The final experiment tested the timing of the non-local correlations, in order to eliminate the hypothetical possibility that information-carrying signals were passing from one detector to the other. This was done by changing the polarizer angles in a time shorter than L/c, where L is the distance separating the polarizers. In practice, this was done by means of a fast acousto-optical switch (AOS) which deflected the photons between two polarizers with their axes set at different angles. The experimental arrangement is shown schematically in Fig. 14.9. The apparatus consisted of a correlated photon source of the type shown in Fig. 14.3 with the addition of the switch and extra polarizer/detector on each side. The switching time was less than the value of L/c, namely 40 ns, and the results obtained were again in violation of the Bell inequalities. This experiment therefore confirmed that the non-local correlations occur on a time-scale faster than the speed of light.

Following on from the work of Aspect *et al.*, many new experiments have been performed to test for violations of Bell's inequality with ever greater refinement. The use of polarization-entangled photon pairs generated by down-conversion has increased the sensitivity of the experiments and thus led to even more convincing demonstrations. The violation of Bell's inequality has now been confirmed to very high degrees of accuracy, with considerable distances between the detectors. Furthermore, possible loopholes in the experimental method are gradually being closed by more sophisticated tests. The body of results is very persuasive, and the experimental evidence for non-locality is overwhelming.

See, for example, P. G. Kwiat *et al.*, *Phys. Rev. Lett.* **75**, 4337 (1995); W. Tittel *et al.*, *Phys. Rev. Lett.* **81**, 3563 (1998); G. Weihs, *et al.*, *Phys. Rev. Lett.* **81**, 5039 (1998).

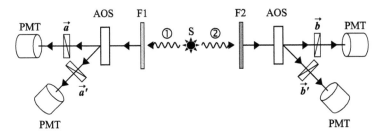

Fig. 14.9 Schematic diagram of the apparatus for the third Aspect experiment. The correlated photon pairs were generated by a calcium cascade source S as described in Fig. 14.3. An acousto-optical switch (AOS) was added on each side of the apparatus to deflect the beam towards different polarizers with axes $\vec{a}$ or $\vec{a}'$ and $\vec{b}$ or $\vec{b}'$ as appropriate. The short switching time of the AOS ensured that the polarization detection angle was being changed faster than any information-carrying signals could pass between the detectors. (Adapted from A. Aspect, *et al.*, *Phys. Rev. Lett.* **49**, 1804 (1982).)

14.5   Principles of teleportation

The starship *Enterprize* in the television series *Star Trek* seemed to possess a teleportation machine capable of 'beaming' (i.e. teleporting) human beings from one place to another. Unfortunately, machines of this complexity are restricted to the realms of science fiction.

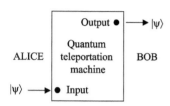

Fig. 14.10 Schematic diagram of the operation of a quantum teleportation machine. A photon in an unknown quantum state $|\psi\rangle$ is fed into the input of the machine and another photon in the same quantum state emerges from the output somewhere else.

See C. H. Bennett *et al.*, *Phys. Rev. Lett.* **70**, 1895 (1993).

The demonstration of quantum non-locality by violation of Bell's inequality lays the foundation for quantum teleportation. The basic idea of teleportation is to transfer the quantum state of one photon to another that is physically separated from it. In principle we can also use other particles such as electrons, atoms, or nuclei, but so far most of the demonstrations have been done with photons, and so we shall restrict our discussion here to the case of photon teleportation.

Figure 14.10 illustrates the basic operation of a quantum teleportation machine. The idea is to send quantum information from one place to another without direct exchange of qubits. As was the case with quantum cryptography, we refer to the sender and recipient of the quantum information as Alice and Bob respectively. The machine has an input in Alice's laboratory and an output in Bob's. A photon is fed into the input in an unknown quantum state $|\psi\rangle$, and Bob produces another photon in the same quantum state $|\psi\rangle$ at the output. One possible long-term application of teleportation is in the transfer of quantum information (i.e. qubits) between the different nodes of a quantum network consisting of quantum computers at different locations. (See Section 13.7.)

Before delving into the details of how such a machine might work, we can first lay down some general principles of its operation.

1. The **quantum no-cloning theorem** says that it is not possible to clone the original photon. The input photon must therefore either be destroyed or lose its initial state in an irretrievable way.

2. The general theory of quantum measurement implies that the fidelity between the output and input wave functions is degraded in proportion to the amount of information gleaned about $|\psi\rangle$ within the teleportation machine. Perfect fidelity can only be achieved when the machine retains no information whatsoever about the unknown quantum state.

3. No *matter* is teleported between the input and output, only *quantum information*.

4. Relativity tells us that we cannot transmit information faster than the speed of light. Therefore, teleportation cannot be used for superluminal information exchange.

With these ideas in mind, let us see how teleportation works in practice. We shall work through a scheme for photon teleportation originally devised by Bennett *et al.* in 1993. Experiments to implement this scheme in the laboratory will then be described in Section 14.6 below.

Figure 14.11 shows the arrangement required for the teleportation of photon polarization. Three photons are required. Photon 1 is the input photon, which is presumed to be in an unknown arbitrary polarization

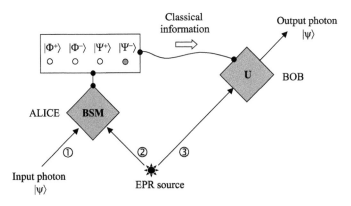

Fig. 14.11 Schematic diagram of an arrangement for photon teleportation. Photon 1 is the input photon whose quantum polarization state $|\psi\rangle$ is to be teleported. Photons 2 and 3 comprise a correlated pair from an EPR source. Alice receives photons 1 and 2 and makes a Bell state measurement (BSM) on them. Bob receives photon 3 and makes a unitary operation (U) on it according to the result of Alice's measurement, which is communicated via a classical channel. Photon 3 then emerges in the same quantum state $|\psi\rangle$ as photon 1.

state given by:

$$|\psi\rangle_1 = C_0|0\rangle_1 + C_1|1\rangle_1, \tag{14.15}$$

where $|C_0|^2 + |C_1|^2 = 1$, and $|0\rangle$ and $|1\rangle$ correspond to the horizontal and vertical polarization states $|\leftrightarrow\rangle$ and $|\updownarrow\rangle$, respectively. Photons 2 and 3 form a correlated photon pair emitted by an EPR source. In general, these two photons could be in any of the four Bell states given by eqns 14.1 and 14.2. We consider here the specific case in which they are in the state:

$$|\Psi^-\rangle_{23} = \frac{1}{\sqrt{2}}\left(|0\rangle_2|1\rangle_3 - |1\rangle_2|0\rangle_3\right). \tag{14.16}$$

This state is readily produced by type II down-conversion (see eqn 14.6 with $\phi = \pi$), and has been employed in the experimental demonstrations of teleportation described in the next section.

   The teleportation protocol proceeds by sending photons 1 and 2 to Alice and photon 3 to Bob. Alice performs a 'Bell-state measurement' (BSM) on her two photons giving one of four possible results. She communicates this result to Bob by a classical channel, and Bob then performs a unitary operation $\hat{U}$ to photon 3 depending on the information he has received from Alice. Then, *hey presto*, the output state of photon 3 becomes

$$|\psi\rangle_3 = C_0|0\rangle_3 + C_1|1\rangle_3, \tag{14.17}$$

which is identical to that of the original photon (cf. eqn 14.15).

To see how this works in detail, we need to consider the full wave function for the three particle system, namely:

$$|\Psi\rangle_{123} = \frac{1}{\sqrt{2}}(C_0|0\rangle_1 + C_1|1\rangle_1)(|0\rangle_2|1\rangle_3 - |1\rangle_2|0\rangle_3)$$

$$= \frac{1}{\sqrt{2}}(C_0|0\rangle_1|0\rangle_2|1\rangle_3 - C_0|0\rangle_1|1\rangle_2|0\rangle_3$$

$$+ C_1|1\rangle_1|0\rangle_2|1\rangle_3 - C_1|1\rangle_1|1\rangle_2|0\rangle_3). \tag{14.18}$$

With the following notation for the four Bell states for particles 1 and 2: (cf. eqns 14.1 and 14.2)

$$|\Phi^+\rangle_{12} = \frac{1}{\sqrt{2}}(|0\rangle_1|0\rangle_2 + |1\rangle_1|1\rangle_2), \tag{14.19}$$

$$|\Phi^-\rangle_{12} = \frac{1}{\sqrt{2}}(|0\rangle_1|0\rangle_2 - |1\rangle_1|1\rangle_2), \tag{14.20}$$

$$|\Psi^+\rangle_{12} = \frac{1}{\sqrt{2}}(|0\rangle_1|1\rangle_2 + |1\rangle_1|0\rangle_2), \tag{14.21}$$

$$|\Psi^-\rangle_{12} = \frac{1}{\sqrt{2}}(|0\rangle_1|1\rangle_2 - |1\rangle_1|0\rangle_2), \tag{14.22}$$

we can rewrite eqn 14.18 as:

$$|\Psi\rangle_{123} = \frac{1}{2}\left(|\Phi^+\rangle_{12}(C_0|1\rangle_3 - C_1|0\rangle_3)\right.$$

$$+ |\Phi^-\rangle_{12}(C_0|1\rangle_3 + C_1|0\rangle_3)$$

$$+ |\Psi^+\rangle_{12}(-C_0|0\rangle_3 + C_1|1\rangle_3)$$

$$\left. - |\Psi^-\rangle_{12}(C_0|0\rangle_3 + C_1|1\rangle_3)\right). \tag{14.23}$$

Alice's BSM device may be considered to be a black box with four lights on it and inputs for photons 1 and 2, as illustrated schematically in Fig. 14.11. When the two input photons are in the Bell state $|\Phi^+\rangle_{12}$, the first bulb lights up. If they are in the state $|\Phi^-\rangle_{12}$, the second one lights up, etc.

The teleportation works by the non-local correlations intrinsic to the entangled state given by eqn 14.23. The measurement by Alice instantly determines the state of photon 3 for Bob. Thus, for example, if Alice's first bulb lights up, then Bob knows that photon 3 must be in the state

$$|\psi\rangle_3 = C_0|1\rangle_3 - C_1|0\rangle_3. \tag{14.24}$$

In the case of teleportation of photon polarization, Bob's unitary operations are mere polarization rotations that can be performed very easily with a half wave plate.

Therefore, if Alice tells Bob that she has measured the state $|\Phi^+\rangle_{12}$, Bob then knows the state of his photon without needing to carry out any measurements on it. He can then produce the desired output state, namely $(C_0|0\rangle_3 + C_1|1\rangle_3)$, by applying a simple unitary operator to photon 3. (See Example 14.2 below.) If Alice obtains other results, all she has to do is tell Bob the result she has obtained, and Bob then knows which unitary operator to use to complete the teleportation process.

Two points are worth emphasizing here. First, the protocol can only work after Alice transmits the result of her measurement to Bob by a classical channel. This is what ensures that no information is transferred faster than the speed of light. Second, photon 1 ends up entangled with photon 2, and neither Alice nor Bob acquire any information about C_0 and C_1. The teleportation process thus clearly adheres to the general principles of quantum measurement and quantum no-cloning.

Example 14.2   Show that a photon in the state $(-C_1|0\rangle + C_0|1\rangle)$ can be transformed to the state $(C_0|0\rangle + C_1|1\rangle)$ by a simple unitary operator.

Solution
We make use of the techniques for manipulating the state of single qubits developed in Section 13.3.2. The input state is written in the form

$$q = \begin{pmatrix} -C_1 \\ C_0 \end{pmatrix},$$

and the output is given by

$$q' = \hat{U} \cdot q.$$

It is apparent that the transformation can be performed if $\hat{U}$ takes the form:

$$\hat{U} = \begin{pmatrix} 0 & 1 \\ -1 & 0 \end{pmatrix}$$

so that:

$$q' = \begin{pmatrix} 0 & 1 \\ -1 & 0 \end{pmatrix} \begin{pmatrix} -C_1 \\ C_0 \end{pmatrix} = \begin{pmatrix} C_0 \\ C_1 \end{pmatrix}.$$

On noting that

$$\begin{pmatrix} 0 & 1 \\ -1 & 0 \end{pmatrix} = \begin{pmatrix} 1 & 0 \\ 0 & -1 \end{pmatrix} \begin{pmatrix} 0 & 1 \\ 1 & 0 \end{pmatrix},$$

and remembering that qubit operators are applied from right to left, we see from Table 13.4 that $\hat{U}$ consists of an X gate followed by a Z gate.

14.6   Experimental demonstration of teleportation

The first two experimental demonstrations of quantum teleportation were completed in 1997–8. In this section we describe one of these, namely that of Bouwmeester *et al.* The reader is referred to the reference for details of the experiment by Boschi *et al.*

Figure 14.12 shows the experimental arrangement, which included two EPR sources producing a total of four photons. Both EPR sources consisted of a nonlinear crystal of the type shown in Fig. 14.5 pumped by

See D. Bouwmeester, *et al.*, *Nature* **390**, 575 (1997) and D. Boschi, *et al.*, *Phys. Rev. Lett.* **80**, 1121 (1998).

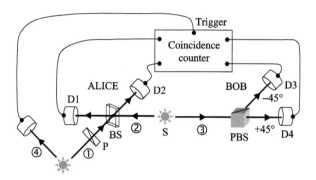

Fig. 14.12 Experimental arrangement for photon teleportation. Photon 1 is prepared in a +45° polarization state by the polarizer P. Photons 2 and 3 form an EPR pair and are generated by degenerate down conversion in a nonlinear crystal pumped by an ultraviolet laser. A fourth photon produced simultaneously with photon 1 is used to trigger the detection electronics. Alice feeds photons 1 and 2 into a non-polarizing 50:50 beam splitter (BS) and looks for coincidences on detectors D1 and D2. Bob sets his PBS to detect the ±45° polarization states and compares the coincidence rates D1D2D3 and D1D2D4. (Adapted from D. Bouwmeester, *et al.*, *Nature* **390**, 575 (1997).)

200 fs pulses from an ultraviolet laser. The first source simultaneously produced photon 1 and a fourth photon labelled 4. Photon 1 was prepared in an arbitrary polarization state by a linear polarizer P, while the detection of photon 4 was used as a trigger to indicate that photon 1 had been sent to Alice. Photons 2 and 3 were produced in the Bell state given by eqn 14.16 by the second EPR source.

A key requirement of the teleportation experiment is for Alice to perform the Bell–state measurement on photons 1 and 2. It transpires that it is only possible to identify two of the four Bell states given in eqns 14.19–14.22 unambiguously, namely $|\Psi^-\rangle_{12}$ and $|\Psi^+\rangle_{12}$. Furthermore, of these two, it is much easier to detect $|\Psi^-\rangle_{12}$. The strategy adopted in the experiment was therefore to look exclusively for the state $|\Psi^-\rangle_{12}$. This was done by bringing both photons onto a 50:50 beam splitter at the same time and looking for signals on detectors D1 and D2 placed at the output ports of the beam splitter. The state $|\Psi^-\rangle_{12}$ is the only one of the four Bell states in which the photons go to separate detectors. For the other three, both photons go to either detector D1 or D2. Thus a simultaneous signal on detectors D1 and D2 unambiguously determined that Alice had detected the $|\Psi^-\rangle_{12}$ state.

Once Alice had detected the state $|\Psi^-\rangle_{12}$, Bob's task was then very easy. It is apparent from eqn 14.23 that, if Alice detects $|\Psi^-\rangle_{12}$, then photon 3 must be in the state $(C_0|0\rangle_3 + C_1|1\rangle_3)$. This means that Bob's unitary operator is the identity. In other words, he has to do nothing. The demonstration of teleportation could therefore be achieved by checking that the polarization of Bob's photon was the same as the polarization of photon 1 set by polarizer P whenever Alice detected the $|\Psi^-\rangle_{12}$ state.

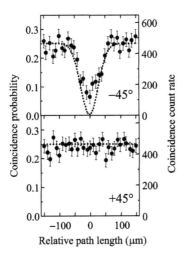

Fig. 14.13 Theoretical coincidence probability ($\cdots$) and experimental coincidence rate ($\bullet$) for teleportation with photon 1 prepared in the $+45°$ polarization state. The results are shown as a function of the relative path lengths for photons 1 and 2 at the beam splitter, with zero path length corresponding to identical arrival times. (After D. Bouwmeester *et al.*, *Nature* **390**, 575 (1997) ©Nature Publishing Group, reproduced with permission.)

The actual test of teleportation was done by setting the polarizer P to $+45°$, and then for Bob to set his polarizing beam splitter at $45°$ so as to detect the $\pm 45°$ polarization states. Teleportation would then be demonstrated by coincidences on detectors D1D2D4 and no coincidences on detectors D1D2D3. The results for a particular set of data are shown in Fig. 14.13. The x-axis of the data graphs corresponds to the relative path length from the source to the beam splitter for photons 1 and 2, with zero corresponding to identical arrival times. When the two photons arrive at different times, there can be no interference at the beam splitter and we would expect to see simultaneous counts on D1 and D2 with probability 50%. In this situation there is no teleportation occurring, and the polarization of photon 3 is random. We thus expect random counts on detectors D3 and D4 with probability 50%. The probability for D1D2D3 and D1D2D4 coincidences is thus 25%. On the other hand, when the two photons arrive at the same time, interference can occur and Alice can make the Bell-state measurement. We would then expect the coincidence rate D1D2D3 to drop to zero, with D1D2D4 remaining at 25%. The experimental data clearly show the basic effect, although the count rate did not drop exactly to zero because of technical difficulties.

The experiment was repeated for other settings of the input polarization, and similar results were obtained. This clearly established that the polarization state of photon 1 had been transferred to photon 3, which originally had a random polarization, and proved that teleportation had occurred. Subsequent experiments have improved on the

See, for example, I. Marcikic, *et al.*, *Nature*, **421**, 509 (2003), or R. Ursin, *et al.*, *Nature* **430**, 849 (2004).

performance, and in recent reports, the distance over which teleportation has demonstrated has increased substantially.

14.7   Discussion

In some older universities, the subject of physics is called 'natural philosophy'. The experiments described in this chapter certainly do raise important questions, and both the EPR paper and Bell's theorem have been very widely discussed in philosophical circles. Given that many great minds have pondered these points at length, it would be rather pretentious to claim to find all the answers in a text such as this. We shall thus briefly review some of the arguments and leave the interested reader to pursue the subject further by referring to the extensive literature that is available on the subject.

The question raised by the EPR paper was whether the quantum mechanical description of physical reality could be considered complete. The EPR argument rests on the fact that a measurement at one place instantly produces an effect at another. This instantaneous cause–effect link appears to be in contravention of relativity. Einstein instinctively rejected such a notion, and even went so far as to describe it as constituting a form of 'spooky' action at a distance. He was thus led to propose that a deeper level of reality must exist that would explain the results without the apparent action at a distance.

See Letter, 3 March 1947, in *The Born–Einstein Letters* (ed. M. Born), Macmillan, London (1971), p. 158.

Bell took the argument a step further. He assumed that the alternative to the 'spooky' effects is to assume that each particle in the entangled pair has well-defined local physical properties before the measurements are made. These properties are quantified by local hidden variables. He then went on to show that such an assumption leads to practical consequences in the form of Bell's inequality. The experimental tests of Bell's inequality now prove that the local hidden variable assumption is incorrect.

Some physicists have sought to combine hidden variables with non-locality by devising *non-local* hidden variable theories. One example is Bohm's **pilot wave theory**, which correctly predicts the violations of Bell's inequality. See D. Bohm, *Phys. Rev.* **85**, 166 (1952).

The obvious question to ask now is: what is wrong with LHVs? Most physicists would assert that the key issue is *locality*, and would conclude that matter is fundamentally *non-local* at the microscopic level. Without necessarily understanding what non-locality might mean, they would accept the concept pragmatically and look for ways to exploit it in the laboratory. Quantum teleportation is a shining example of such an approach.

Even when we accept the discovery of non-locality as a *fait accompli*, we are actually no closer to solving Einstein's basic dilemma concerning the completeness of quantum theory: we still do not know what we have before the measurements are made. It might be argued that because we cannot *know* the values of measurable quantities like photon polarization or electron spin prior to the measurement, then these properties do not actually exist. Bell himself summarized this viewpoint as follows:

See Bell (1987, p. 142).

Making a virtue of necessity, and influenced by positivistic and instrumentalist philosophies, many came to hold not only that it is difficult to find a coherent picture but that it is wrong to look for one – if not actually immoral then certainly unprofessional. Going further still, some asserted that atomic and subatomic particles do not *have* any definite properties in advance of observation. There is nothing, that is to say, in the particles approaching the magnet, to distinguish those subsequently deflecting up from those deflected down. Indeed even the particles are not really there.

Such a position would no doubt be the background to the following anecdote concerning Einstein recounted by Abraham Pais:

See Pais (1982, p. 5).

It must have been around 1950. I was accompanying Einstein on a walk from the Institute of Advanced Study to his home, when he suddenly stopped, turned to me, and asked if I really believed that the moon only exists if I look at it.

The answer to this question is obviously negative. If we apply the same reasoning to the microscopic experiments, we would have to assert that there is definitely some form of objective reality that pre-exists the measurements. At the same time, we are used to the idea that measurements on quantum systems are highly invasive. This means that the quantifiable properties that we assign to a particle such as polarization or spin are inextricably connected to the measurement process that determines them. Moreover, the Bell inequality experiments force us to conclude that these quantifiable properties must possess the feature of non-locality, and that further pursuit of LHVs has become pointless. It seems to the author that this is about as much as we can say on the subject at present.

It is worth closing this discussion by briefly considering whether the non-local correlations have any practical consequences. It might be thought that the apparently instantaneous correlations between separated measurements could provide a mechanism for faster-than-light signalling. However, we have seen that this is impossible. For example, in an EPR experiment the sequence of events registered by either detector is completely random, and the correlations between the two sets of results are only apparent when they are compared by conventional communication channels. At the same time, it is apparent that quantum teleportation, without providing a scheme for super-luminal signalling, is a shining example of quantum non-locality in action. Time will only tell whether teleportation will ever have any commercial applications, but it certainly remains at the present time a *tour de force* of quantum optics at its most fundamental level.

Further reading

A collection of Bell's papers on the EPR paradox and the Bell inequality may be found in Bell (1987). An appreciation of Bell's contribution to science has been given by Whitaker (1998), and a collection of articles on the relevance of his work may be found in Bertlmann and Zeilinger (2002).

A comprehensive treatment of the subject of quantum information processing and teleportation is given in Bouwmeester *et al.* (2000), and an introductory review may be found in Sergienko and Jaeger (2003). A modern perspective on the subject of quantum entanglement may be found in Terhal *et al.* (2003). Haroche (1998) gives an introduction to the relationship between Schrödinger's cat, entanglement and decoherence, and a more detailed account has been given by Raimond *et al.* (2001). An overview of the whole subject of quantum information processing using photons may be found in Zeilinger *et al.* (2005).

Further details of single-photon interference experiments are given in Mandel (1999) or Mandel and Wolf (1995). A discussion of the practical applications of entanglement and teleportation may be found in Zeilinger (1998a) or Walmsley and Knight (2002) at the introductory level, and in Zeilinger (1998b) with more technical detail. Undergraduate experiments on entangled photons and Bell's inequality are described in Dehlinger and Mitchell (2002). Two experiments demonstrating teleportation of atoms are described in Riebe *et al.* (2004) and Barrett *et al.* (2004), while Matsukevich and Kuzmich (2004) present the results of an experiment demonstrating entanglement between matter and light.

Rae (2004) presents an introduction to some of the philosophical issues raised by EPR and Bell. A shorter discussion of the main issues may be found in Mermin (1986) or Hardy (1998). A simplified account of Bohm's pilot wave theory may be found in Albert (1994), and an overview of the status of quantum theory in the light of EPR and Bell has been given by Leggett (1999).

Exercises

(14.1) Explain why a source that emits pairs of photons that are either both horizontally or both vertically polarized each with 50% probability can account for the results obtained in an EPRB experiment.

(14.2) Describe the results that would be obtained in an EPRB experiment using a source which produces entangled photon pairs in the following states:

(a) $\left((2/3)^{1/2}|0_1, 0_2\rangle + (1/3)^{1/2}|1_1, 1_2\rangle\right)$,

(b) $\left((2/5)^{1/2}|0_1, 1_2\rangle - (3/5)^{1/2}|1_1, 0_2\rangle\right)$,

(c) $\left(|0_1, 0_2\rangle + e^{i\pi/4}|1_1, 1_2\rangle\right)/\sqrt{2}$.

(14.3) Explain why the direct promotion of an electron from the $4s^2\,^1S_0$ ground state of calcium to the $4p^2\,^1S_0$ excited state is not possible by absorption of a single photon, but is possible by the simultaneous absorption of two photons.

(14.4) A crystal is said to have *normal* dispersion if the refractive index increases with frequency in the optical spectral range.

(a) Explain why it is not possible to achieve phase-matching for degenerate collinear down-conversion if all three photons have the same polarization.

(b) Explain how the phase-matching condition may be satisfied in a birefringent crystal in which the refractive index depends on the direction of the light polarization with respect to the optic axis of the crystal.

(14.5) The refractive index for the extraordinary ray propagating in a birefringent crystal is given by

$$\frac{1}{n(\theta)^2} = \frac{\sin^2\theta}{n_o^2} + \frac{\cos^2\theta}{n_e^2},$$

where θ is the angle of the polarization vector with respect to the optic axis of the crystal, and n_o and n_e are the ordinary and extraordinary refractive indices, respectively. The orthogonally polarized ordinary ray always has $\theta = 90°$, and thus has a refractive index of n_o.

In type-II down-conversion in BBO crystals, the pump laser at angular frequency ω_0 is an extraordinary ray, while one of the down-converted photons propagates as an extraordinary ray and the other as an ordinary ray.

(a) Prove that phase-matching is achieved for degenerate collinear down-conversion (i.e. all three photons propagating in the same direction) when:

$$2n^{\omega_0}(\theta) = n^{\omega_0/2}(\theta) + n_o^{\omega_0/2},$$

where the superscripts refer to the frequency.

(b) Find the angle of the optic axis with respect to the normal vector from the crystal surface for degenerate collinear type II down-conversion in BBO using a 532 nm pump laser at normal incidence. The values of n_o and n_e are 1.6551 and 1.5425, respectively, at 1064 nm, and 1.6749 and 1.5555, respectively, at 532 nm.

(14.6) A pair of degenerate photons is produced by type II down-conversion using a pump laser at 351.1 nm. The vertically polarized photon is emitted at an angle of $3°$ with respect to the pump laser. Find (a) the wavelength, and (b) the direction of the horizontally polarized photon.

(14.7) In the Hong–Ou–Mandel interferometer experiment shown in Fig. 14.6, the signal and idler photons are generated by degenerate type-I down conversion and therefore have the same polarization and frequency, making the photons indistinguishable. An interesting variation of the experiment can be made by inserting a half wave plate into the idler path so that the polarization of the beams that interfere at the beam splitter BS can be varied.

(a) Discuss what would happen to the interference fringes observed at the output ports of BS as the wave plate is rotated from a classical perspective.

(b) Repeat the explanation from a quantum perspective with single photons by considering the which-path information that can be obtained if polarizers are placed in front of the detectors.

(14.8) Calculate the values of the CHSH parameter S defined in eqn 14.12 predicted by (a) quantum mechanics and (b) the LHV model with probabilities given in eqn 14.11, for the case with $\theta_1 = 0$, $\theta_1' = 45°$, $\theta_2 = 30°$ and $\theta_2' = 60°$.

(14.9) Example 13.2 describes a quantum circuit to produce the $|\Phi^+\rangle$ Bell state starting from two qubits in the $|00\rangle$ state. Find the output of the circuit for inputs of: (a) $|01\rangle$, (b) $|10\rangle$ and (c) $|11\rangle$, and relate the output states to the other three Bell states listed in eqns 14.1 and 14.2. (Assume that the first qubit is the control.)

(14.10) In the Bell experiment depicted in Fig. 14.8, the angles θ_1 and θ_2 are set at $20°$ and $60°$ respectively. 1000 events are recorded. How many events would be expected with the result of (a) 11, (b) 10, (c) 01, (d) 00?

(14.11) Explain how a half wave plate can be used to implement the unitary operation of Example 14.2.

(14.12) Calculate the coincidence probability for the D1D2D4 detector combination in Fig. 14.12 when the polarizer P is set at (a) $+45°$ and (b) $30°$. (Assume that the relative path length of photons 1 and 2 at BS is zero.)

(14.13) In a quantum teleportation experiment of the type shown in Fig. 14.12 using photons of wavelength 788 nm, the bandwidth of photons 1–3 was restricted with a narrow-band filter to 4 nm.

(a) Calculate the coherence time of the photons.

(b) Estimate the precision with which the relative path lengths for photons 1 and 2 must be matched to achieve quantum teleportation.

(c) Compare your answer to the results shown in Fig. 14.13.

Poisson statistics

Poisson statistics apply to random variables where the results can only occur in *positive integer* values. Three well-known examples of physical systems that exhibit Poisson statistics are:

- the number of clicks per second registered by a Geiger counter detecting radioactive emission;
- the number of rain drops falling into a bucket in a specific time interval;
- the count rate per unit time registered by a photomultiplier tube detecting star light.

In each case we can determine the average number of events by performing a large number of measurements. However, the precise result of any individual measurement is unpredictable.

In a Poisson distribution the probability $\mathcal{P}(n)$ for observing n events is given by:

$$\mathcal{P}(n) = \frac{\mu^n}{n!} e^{-\mu}, \tag{A.1}$$

where μ is a constant. We can check that the distribution is correctly normalized by summing over all the possibilities:

$$\sum_{n=0}^{\infty} \mathcal{P}(n) = e^{-\mu} \sum_{n=0}^{\infty} \frac{\mu^n}{n!}$$

$$= e^{-\mu} \left(1 + \mu + \frac{\mu^2}{2!} + \frac{\mu^3}{3!} + \cdots \right)$$

$$= e^{-\mu} \times e^{+\mu}$$

$$= 1. \tag{A.2}$$

The mean value of n is given by:

$$\overline{n} \equiv \langle n \rangle = \sum_{n=0}^{\infty} n \mathcal{P}(n), \tag{A.3}$$

Poisson statistics are the discrete equivalent of the **Gaussian statistics** that generate the **normal distribution**. Gaussian statistics apply to continuous random variables, for example, the length of a piece of string.

which can be evaluated by using eqn A.1:

$$\bar{n} = e^{-\mu} \sum_{n=0}^{\infty} n \frac{\mu^n}{n!}$$

$$= e^{-\mu} \left(0 + \mu + 2\frac{\mu^2}{2!} + 3\frac{\mu^3}{3!} + \cdots \right)$$

$$= \mu e^{-\mu} \left(1 + \mu + \frac{\mu^2}{2!} + \cdots \right)$$

$$= \mu \sum_{n=0}^{\infty} \frac{\mu^n}{n!} e^{-\mu}$$

$$= \mu \sum_{n=0}^{\infty} \mathcal{P}(n)$$

$$= \mu, \tag{A.4}$$

where we made use of eqn A.2 in the last line. We can thus rewrite the distribution in the more familiar form used in quantum optics:

$$\mathcal{P}(n) = \frac{\bar{n}^n}{n!} e^{-\bar{n}}. \tag{A.5}$$

This shows that the probability for obtaining a given result n is a universal function of the mean value $\bar{n}$ alone.

It follows from eqn A.5 that:

$$\mathcal{P}(n) = \frac{\bar{n}}{n} \mathcal{P}(n-1), \tag{A.6}$$

which implies that $\mathcal{P}(n) > \mathcal{P}(n-1)$ if $n < \bar{n}$, and vice versa for $n > \bar{n}$. Poisson distributions therefore peak at the integer value closest to $\bar{n}$ when $\bar{n} > 1$, and decrease monotonically with n when $\bar{n} < 1$.

Examples of Poisson distributions for four different values of $\bar{n}$ are given in Fig. 5.3.

The **variance** of the Poisson distribution is defined by:

$$\mathrm{Var}(n) = \sum_{n=0}^{\infty} (n - \bar{n})^2 \, \mathcal{P}(n)$$

$$= \sum_{n=0}^{\infty} (n^2 - 2n\bar{n} + \bar{n}^2) \, \mathcal{P}(n)$$

$$= \sum_{n=0}^{\infty} n^2 \, \mathcal{P}(n) - 2\bar{n} \sum_{n=0}^{\infty} n \, \mathcal{P}(n) + \bar{n}^2 \sum_{n=0}^{\infty} \mathcal{P}(n)$$

$$= \sum_{n=0}^{\infty} n^2 \, \mathcal{P}(n) - \bar{n}^2, \tag{A.7}$$

where we made use of the definition of $\bar{n}$ given in eqn A.3 in the last line. This can be simplified further by noticing that:

$$\sum_{n=0}^{\infty} n^2 \, \mathcal{P}(n) = \sum_{n=0}^{\infty} (n^2 - n + n) \, \mathcal{P}(n) = \sum_{n=0}^{\infty} n(n-1) \, \mathcal{P}(n) + \bar{n}, \tag{A.8}$$

and that:

$$\sum_{n=0}^{\infty} n(n-1)\,\mathcal{P}(n) = \mathrm{e}^{-\overline{n}} \sum_{n=0}^{\infty} n(n-1)\frac{\overline{n}^n}{n!}$$

$$= \mathrm{e}^{-\overline{n}} \left(0 + 0 + 2\frac{\overline{n}^2}{2!} + 6\frac{\overline{n}^3}{3!} + +12\frac{\overline{n}^4}{4!} + \cdots \right)$$

$$= \overline{n}^2 \mathrm{e}^{-\overline{n}} \left(1 + \overline{n} + \frac{\overline{n}^2}{2!} + \cdots \right)$$

$$= \overline{n}^2 \sum_{n=0}^{\infty} \frac{\overline{n}^n}{n!}\mathrm{e}^{-\overline{n}}$$

$$= \overline{n}^2 \sum_{n=0}^{\infty} \mathcal{P}(n)$$

$$= \overline{n}^2. \tag{A.9}$$

By combining eqns A.7–A.9 we then see that:

$$\mathrm{Var}(n) = (\overline{n}^2 + \overline{n}) - \overline{n}^2 = \overline{n}. \tag{A.10}$$

Hence the variance of a Poisson distribution is equal to the mean value. The standard deviation is defined according to:

$$\sigma^2 = \sum_{n=0}^{\infty}(n - \overline{n})^2 \mathcal{P}(n) \equiv \mathrm{Var(n)}. \tag{A.11}$$

We then see from eqn A.10 that the standard deviation of a Poisson distribution is equal to the square root of the mean value:

$$\sigma = \sqrt{\overline{n}}. \tag{A.12}$$

This result is very widely applied in analysing random processes with integer results.

B

Parametric amplification

B.1 Wave propagation in a nonlinear medium   **324**

B.2 Degenerate parametric amplification   **326**

Further reading   **329**

Parametric amplification is an important process in the generation of quadrature squeezed light, as discussed in Section 7.9.1. In this appendix we give a brief summary of the classical theory of parametric amplification based on Maxwell's equations in the nonlinear optics regime.

B.1 Wave propagation in a nonlinear medium

The propagation of electromagnetic waves through a dielectric medium is governed by the electric displacement $\boldsymbol{D}$ defined by:

$$\boldsymbol{D} = \epsilon_0 \boldsymbol{\mathcal{E}} + \boldsymbol{P}, \tag{B.1}$$

where $\boldsymbol{P}$ is the electric polarization of the medium. In a nonlinear medium we split the polarization into a term that is linear in the electric field and one that is nonlinear according to:

$$\boldsymbol{P} = \epsilon_0 \chi \boldsymbol{\mathcal{E}} + \boldsymbol{P}^{\mathrm{NL}}, \tag{B.2}$$

where χ is the usual linear electric susceptibility. On substituting into eqn B.1 we then find:

$$\boldsymbol{D} = \epsilon_0 \epsilon_{\mathrm{r}} \boldsymbol{\mathcal{E}} + \boldsymbol{P}^{\mathrm{NL}}, \tag{B.3}$$

where $\epsilon_{\mathrm{r}} = (1 + \chi)$ is the relative permittivity.

The propagation of the electromagnetic waves generated by the nonlinear polarization can be described by using the nonlinear displacement of eqn B.3 in the fourth Maxwell equation (eqn 2.12):

We have assumed here that the medium is non-conducting so that the current density $\boldsymbol{j} = 0$. In practice this means that we are assuming that the medium is transparent for the light frequencies of interest.

$$\boldsymbol{\nabla} \times \boldsymbol{H} = \frac{\partial \boldsymbol{D}}{\partial t},$$

$$= \epsilon_0 \epsilon_{\mathrm{r}} \frac{\partial \boldsymbol{\mathcal{E}}}{\partial t} + \frac{\partial \boldsymbol{P}^{\mathrm{NL}}}{\partial t}. \tag{B.4}$$

We take the curl of the third Maxwell equation (eqn 2.11) with $\boldsymbol{B} = \mu_0 \boldsymbol{H}$ and substitute to find:

$$\boldsymbol{\nabla} \times (\boldsymbol{\nabla} \times \boldsymbol{\mathcal{E}}) = -\mu_0 \frac{\partial}{\partial t} \boldsymbol{\nabla} \times \boldsymbol{H},$$

$$= -\mu_0 \epsilon_0 \epsilon_{\mathrm{r}} \frac{\partial^2 \boldsymbol{\mathcal{E}}}{\partial t^2} - \mu_0 \frac{\partial^2 \boldsymbol{P}^{\mathrm{NL}}}{\partial t^2}, \tag{B.5}$$

which, on using the identity:

$$\nabla \times (\nabla \times \boldsymbol{\mathcal{E}}) = \nabla(\nabla \cdot \boldsymbol{\mathcal{E}}) - \nabla^2 \boldsymbol{\mathcal{E}}, \tag{B.6}$$

becomes:

$$\nabla^2 \boldsymbol{\mathcal{E}} = \mu_0 \epsilon_0 \epsilon_{\mathrm{r}} \frac{\partial^2 \boldsymbol{\mathcal{E}}}{\partial t^2} + \mu_0 \frac{\partial^2 \boldsymbol{P}^{\mathrm{NL}}}{\partial t^2} + \nabla(\nabla \cdot \boldsymbol{\mathcal{E}}). \tag{B.7}$$

By substituting the nonlinear displacement from eqn B.3 into the first Maxwell equation (eqn 2.9) with $\varrho = 0$, we find:

$$\nabla \cdot \boldsymbol{D} = \nabla \cdot (\epsilon_0 \epsilon_{\mathrm{r}} \boldsymbol{\mathcal{E}} + \boldsymbol{P}^{\mathrm{NL}}) = 0, \tag{B.8}$$

which implies:

$$\nabla \cdot \boldsymbol{\mathcal{E}} = -\frac{1}{\epsilon_0 \epsilon_{\mathrm{r}}} \nabla \cdot \boldsymbol{P}^{\mathrm{NL}}. \tag{B.9}$$

In a uniform transverse wave, we must have $\nabla \cdot \boldsymbol{P}^{\mathrm{NL}} = 0$, so that eqn B.7 simplifies to:

$$\nabla^2 \boldsymbol{\mathcal{E}} = \mu_0 \epsilon_0 \epsilon_{\mathrm{r}} \frac{\partial^2 \boldsymbol{\mathcal{E}}}{\partial t^2} + \mu_0 \frac{\partial^2 \boldsymbol{P}^{\mathrm{NL}}}{\partial t^2}. \tag{B.10}$$

On defining the direction of propagation as the $+z$-axis, we finally obtain:

$$\frac{\partial^2 \boldsymbol{\mathcal{E}}}{\partial z^2} = \mu_0 \epsilon_0 \epsilon_{\mathrm{r}} \frac{\partial^2 \boldsymbol{\mathcal{E}}}{\partial t^2} + \mu_0 \frac{\partial^2 \boldsymbol{P}^{\mathrm{NL}}}{\partial t^2}. \tag{B.11}$$

This is the nonlinear wave equation that we have to solve.

We restrict our consideration to a second-order nonlinear medium with three waves at angular frequencies ω_1, ω_2, and ω_3. Second-order nonlinear processes mix two fields to generate the third, as discussed in Section 2.4.2. We write the time and spatial dependence of the waves in the form:

> We consider here one of the transverse components of the field and thus drop the vector notation.

$$\mathcal{E}^{\omega_1}(z, t) = \mathcal{E}_1(z) \exp[\mathrm{i}(k_1 z - \omega_1 t)],$$
$$\mathcal{E}^{\omega_2}(z, t) = \mathcal{E}_2(z) \exp[\mathrm{i}(k_2 z - \omega_2 t)],$$
$$\mathcal{E}^{\omega_3}(z, t) = \mathcal{E}_3(z) \exp[\mathrm{i}(k_3 z - \omega_3 t)], \tag{B.12}$$

where $\mathcal{E}_i(z)$ is the amplitude and k_i the wave vector. Let us take the field at ω_3 to be the wave generated by the nonlinear mixing of the other two. With $\mathcal{E}^{\omega_3}(z, t)$ in the form given by eqn B.12, the left-hand side of eqn B.11 becomes:

$$\frac{\partial^2 \mathcal{E}^{\omega_3}}{\partial z^2} = \left(-k_3^2 \mathcal{E}_3 + 2\mathrm{i}k_3 \frac{\mathrm{d}\mathcal{E}_3}{\mathrm{d}z} + \frac{\mathrm{d}^2 \mathcal{E}_3}{\mathrm{d}z^2} \right) \mathrm{e}^{\mathrm{i}(k_3 z - \omega_3 t)}. \tag{B.13}$$

In the **slowly varying envelope approximation** we assume that:

$$\left| k_i \frac{\mathrm{d}\mathcal{E}_i}{\mathrm{d}z} \right| \gg \left| \frac{\mathrm{d}^2 \mathcal{E}_i}{\mathrm{d}z^2} \right|. \tag{B.14}$$

> The slowly varying envelope approximation effectively assumes that the wavelength of the light is much shorter than the length scale over which the electric field amplitude varies.

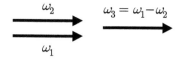

Fig. B.1 The difference-frequency mixing process generates waves at frequency $\omega_1 - \omega_2$ from input wave at frequencies ω_1 and ω_2, where ω_1 is the larger of the two frequencies.

The second-order nonlinear polarization originates from the mixing of two fields according to (cf. eqn 2.58):

$$P^{\text{NL}} = \epsilon_0 \chi^{(2)} \mathcal{E}_1 \mathcal{E}_2.$$

The fields that appear here are real quantities and can be expressed in terms of complex fields as:

$$\mathcal{E}_i = \mathcal{E}^{\omega_i} + \mathcal{E}^{\omega_i *},$$

which implies:

$$\mathcal{E}_1 \mathcal{E}_2 = (\mathcal{E}^{\omega_1} \mathcal{E}^{\omega_2} + \text{c.c.})$$
$$+ (\mathcal{E}^{\omega_1} \mathcal{E}^{\omega_2 *} + \text{c.c.}).$$

The first term gives rise to sum frequency mixing, while the second is the origin of the difference frequency mixing process that we are considering in eqn B.19. With the complex fields varying as $\exp \text{i}(kz - \omega t)$, the complex conjugation of $\mathcal{E}^{\omega_2}$ introduces the minus sign of ω_2 relative to ω_1 and ω_3 in eqn B.18. Note that we are assuming throughout this derivation that the direction of the fields (i.e. their *optical* polarization) has been chosen so that the tensor aspect of the nonlinear susceptibility as given in eqn 2.66 can be ignored.

This allows us to drop the third term in eqn B.13, so that we can rewrite eqn B.11 as:

$$\left(-k_3^2 \mathcal{E}_3 + 2\text{i}k_3 \frac{\text{d}\mathcal{E}_3}{\text{d}z} \right) e^{\text{i}(k_3 z - \omega_3 t)} = -\mu_0 \epsilon_0 \epsilon_{\text{r}} \omega_3^2 \mathcal{E}_3 e^{\text{i}(k_3 z - \omega_3 t)} + \mu_0 \frac{\partial^2 P^{\text{NL}}}{\partial t^2} .$$
(B.15)

Now for electromagnetic waves we have (cf. eqn 2.23 with v given by eqn 2.17):

$$k^2 = \mu_0 \epsilon_0 \epsilon_{\text{r}} \omega^2.$$
(B.16)

We can therefore cancel the first terms on either side and obtain:

$$2\text{i}k_3 \frac{\text{d}\mathcal{E}_3}{\text{d}z} e^{\text{i}(k_3 z - \omega_3 t)} = \mu_0 \frac{\partial^2 P^{\text{NL}}}{\partial t^2} ,$$
(B.17)

allowing us to find $\mathcal{E}_3(z)$ for specific forms of the nonlinear polarization.

We now further restrict our analysis to the **difference frequency mixing** process indicated schematically in Fig. B.1. In this process the field at ω_3 is generated from the fields at ω_1 and ω_2, with:

$$\omega_3 = \omega_1 - \omega_2.$$
(B.18)

We therefore write the nonlinear polarization as:

$$P^{\text{NL}} = \epsilon_0 \chi^{(2)} \mathcal{E}^{\omega_1} \mathcal{E}^{\omega_2 *},$$
(B.19)

where $\chi^{(2)}$ is the nonlinear susceptibility and the complex conjugation of $\mathcal{E}^{\omega_2}$ ensures that the nonlinear polarization has the correct frequency. On substituting this form of P^{NL} into eqn B.17 with the time dependence given by eqn B.12, we find:

$$\frac{\text{d}\mathcal{E}_3}{\text{d}z} = \frac{\text{i}}{2} \frac{\mu_0 \epsilon_0}{k_3} (\omega_1 - \omega_2)^2 \chi^{(2)} \mathcal{E}_1 \mathcal{E}_2^* e^{\text{i}\Delta k z},$$
(B.20)

where

$$\Delta k = k_1 - k_2 - k_3.$$
(B.21)

Finally, on using eqns B.16 and B.18, we obtain:

$$\frac{\text{d}\mathcal{E}_3}{\text{d}z} = \frac{\text{i}}{2} \sqrt{\frac{\mu_0 \epsilon_0}{\epsilon_{\text{r}}}} \omega_3 \chi^{(2)} \mathcal{E}_1 \mathcal{E}_2^* e^{\text{i}\Delta k z}.$$
(B.22)

This shows that the amplitude of the difference-frequency wave grows in proportion to the amplitude of the two waves that generates it.

B.2   Degenerate parametric amplification

In a parametric amplifier we generate two waves called the **signal** and **idler** at angular frequencies ω_{s} and ω_{i}, respectively. A third field called the **pump** has angular frequency ω_p and supplies energy for the nonlinear process. We therefore relabel the fields in the previous section according to the following scheme:

$$\mathcal{E}^{\omega_1} \rightarrow \mathcal{E}^{\omega_p},$$

$$\mathcal{E}^{\omega_2} \rightarrow \mathcal{E}^{\omega_i},$$

$$\mathcal{E}^{\omega_3} \rightarrow \mathcal{E}^{\omega_s}.$$

With this notation we rewrite eqn B.18 as

$$\omega_s = \omega_p - \omega_i, \tag{B.23}$$

and eqn B.21 as

$$\Delta k = k_p - k_i - k_s. \tag{B.24}$$

It is assumed that the amplitude of the pump field is very much larger than the other two. In this regime it is apparent that the idler can mix with the pump to generate the signal, and vice versa, as illustrated schematically in Fig. B.2. In practice, both processes occur simultaneously, and we end up with two coupled nonlinear equations to describe the growth of the signal and idler waves with z.

We restrict our attention here to the case of **degenerate parametric amplification**, in which the signal and idler fields are at the same frequency:

$$\omega_s = \omega_i = \frac{\omega_p}{2} \equiv \omega. \tag{B.25}$$

We assume that the nonlinear crystal has been orientated so that the **phase-matching** condition has been satisfied. (See discussion in Section 2.4.3.) This occurs when:

$$k_p = k_s + k_i, \tag{B.26}$$

so that $\Delta k = 0$. The propagation equation for the signal field is therefore (see eqn B.22):

$$\frac{d\mathcal{E}_s}{dz} = \frac{i}{2}\sqrt{\frac{\mu_0\epsilon_0}{\epsilon_r}}\omega\chi^{(2)}\mathcal{E}_p\mathcal{E}_i^*. \tag{B.27}$$

Since the signal and idler waves are indistinguishable, this further simplifies to:

$$\frac{d\mathcal{E}_\omega}{dz} = ig\mathcal{E}_p\mathcal{E}_\omega^*, \tag{B.28}$$

where $\mathcal{E}_\omega$ is the field at angular frequency ω, and g is the nonlinear coupling given by:

$$g = \frac{1}{2}\sqrt{\frac{\mu_0\epsilon_0}{\epsilon_r}}\omega\chi^{(2)} = \frac{\omega\chi^{(2)}}{2nc}. \tag{B.29}$$

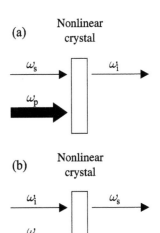

Fig. B.2 In a parametric amplifier, the signal field at angular frequency ω_s mixes with the pump field at angular frequency ω_p to generate the idler field at angular frequency ω_i, as shown in part (a). The reverse process then occurs with the idler generating the signal, as shown in part (b). In practice both processes occur simultaneously.

The degenerate parametric amplification process, in which a pump at angular frequency 2ω generates two photons at angular frequency ω, is the reverse of frequency doubling, where two pump photons at ω generate a single photon at 2ω. The 'parameter' for a degenerate parametric amplifier is the optical phase, as will be demonstrated at the end of this section.

See eqn 2.18 for the relationship between $(\mu_0\epsilon_0)^{1/2}$ and $1/c$, and eqn 2.20 for the relationship between $\epsilon_r^{1/2}$ and the refractive index n.

The complex conjugate of eqn B.28 is:

$$\frac{d\mathcal{E}_\omega^*}{dz} = -ig^*\mathcal{E}_p^*\mathcal{E}_\omega. \tag{B.30}$$

On writing the pump field in the form $\mathcal{E}_p = \mathcal{E}_0 e^{i\phi}$, where $\mathcal{E}_0$ is a real number and ϕ is the phase, we can then put

$$\gamma = ig\mathcal{E}_p = ig\mathcal{E}_0 e^{i\phi}. \tag{B.31}$$

In a non-absorbing medium, g will be real. This means that we can make γ real by setting $\phi = \pm\pi/2$. Equations B.28 and B.30 are now in the form:

$$\frac{d\mathcal{E}_\omega}{dz} = \gamma\mathcal{E}_\omega^*$$

$$\frac{d\mathcal{E}_\omega^*}{dz} = \gamma\mathcal{E}_\omega. \tag{B.32}$$

On adding and subtracting we find:

$$\frac{d}{dz}\left(\mathcal{E}_\omega^* + \mathcal{E}_\omega\right) = \gamma\left(\mathcal{E}_\omega^* + \mathcal{E}_\omega\right)$$

$$\frac{d}{dz}\left(\mathcal{E}_\omega^* - \mathcal{E}_\omega\right) = -\gamma\left(\mathcal{E}_\omega^* - \mathcal{E}_\omega\right). \tag{B.33}$$

By putting $\mathcal{E}^\pm = (\mathcal{E}_\omega^* \pm \mathcal{E}_\omega)$, we finally have:

$$\frac{d\mathcal{E}^+}{dz} = \gamma\mathcal{E}^+$$

$$\frac{d\mathcal{E}^-}{dz} = -\gamma\mathcal{E}^-, \tag{B.34}$$

with solutions:

$$\mathcal{E}^+(z) = \mathcal{E}_0^+ \, e^{\gamma z}$$

$$\mathcal{E}^-(z) = \mathcal{E}_0^- \, e^{-\gamma z}. \tag{B.35}$$

Note that the possibility of de-amplifying one of the field quadratures only occurs for the special case of *degenerate* difference-frequency mixing. In this case, the signal and idler waves are indistinguishable, so that the simplification in eqn B.28 is valid. In *non-degenerate* difference frequency mixing, the two waves both grow exponentially, leading to the possibility of building optical parametric amplifiers and oscillators.

This shows that the field $\mathcal{E}^+$ experiences exponential growth (i.e. amplification) while the field $\mathcal{E}^-$ experiences exponential decay (i.e. deamplification). Since $\mathcal{E}_\omega$ varies with time as $e^{-i\omega t}$, it is apparent that $\mathcal{E}^+$ and $\mathcal{E}^-$ are directly proportional to the field quadratures defined in Section 7.2:

$$\mathcal{E}^- = \mathcal{E}_\omega^* - \mathcal{E}_\omega \equiv 2i\mathcal{E}^{X_1} \sin\omega t$$

$$\mathcal{E}^+ = \mathcal{E}_\omega^* + \mathcal{E}_\omega \equiv 2\mathcal{E}^{X_2} \cos\omega t. \tag{B.36}$$

We therefore see that one of the field quadratures is amplified in the process of degenerate parametric amplification, and the other is de-amplified. This means that the degenerate parametric amplifier acts like a

phase-sensitive amplifier, a fact which is used in the generation of quadrature squeezed light. (See Section 7.9.1.)

Further reading

The detailed principles of parametric amplification are covered in many texts on nonlinear optics, for example: Butcher and Cotter (1990), Shen (1984), or Yariv (1997).

C The density of states

The concept of the density of states arises in many branches of physics. In this appendix we focus on the photon density of states, which is important for the discussion of black-body radiation and for the emission properties of atoms in free space. We also explain how the derivation for photon modes can be adapted to electrons and other massive particles.

We consider the electromagnetic field within a finite volume V of free space as shown in Figure C.1. For simplicity, we assume that the volume comprises a cube of edge length L, so that $V = L^3$. The volume is assumed to be large enough so that its dimensions have no significant effect on the physical result. It then serves just as a computational tool that allows us to find the density of states in an easy way.

The argument can be generalized to volumes of other shapes, without affecting the final result.

The general solution for the electromagnetic field within V can be written as a superposition of travelling waves of the form:

$$\mathcal{E}(\boldsymbol{r}, t) = \sum_{\boldsymbol{k}} \mathcal{E}_{\boldsymbol{k}} e^{i(\boldsymbol{k} \cdot \boldsymbol{r} - \omega t)}, \tag{C.1}$$

with $\omega = c|\boldsymbol{k}|$. The first Maxwell equation (eqn 2.9) in free space reduces to $\boldsymbol{\nabla} \cdot \mathcal{E} = 0$, which is satisfied if:

$$\boldsymbol{k} \cdot \mathcal{E}_{\boldsymbol{k}} = 0, \tag{C.2}$$

and implies that the waves must be transverse: that is, $\mathcal{E}_{\boldsymbol{k}} \perp \boldsymbol{k}$. This transverse condition allows for two independent wave polarizations for each value of $\boldsymbol{k}$.

Equation C.1 gives us a general expression for the field within volume V. The expansion functions are sine waves, and the expression can therefore be thought of as a Fourier series. Since we are dealing with a finite volume, we can write:

$$\mathcal{E}(\boldsymbol{r}, t) = \sum_{k_x, k_y, k_z} \mathcal{E}_{\boldsymbol{k}} e^{ik_x x} e^{ik_y y} e^{ik_z z} e^{-i\omega t}. \tag{C.3}$$

The values of k_x, k_y, and k_z are determined by the dimensions of V, with:

$$
\begin{aligned}
k_x L &= 2\pi n_x, \\
k_y L &= 2\pi n_y, \\
k_z L &= 2\pi n_z,
\end{aligned}
\tag{C.4}
$$

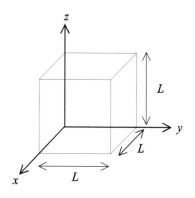

Fig. C.1 Finite volume of free space considered for calculating the electromagnetic density of states.

where n_x, n_y, and n_z are all integers (positive, negative, or zero). The possible values of the wave vector can therefore be written in the form:

$$\boldsymbol{k} \equiv (k_x, k_y, k_z) = \frac{2\pi}{L}(n_x, n_y, n_z). \tag{C.5}$$

Each set of integers (n_x, n_y, n_z) corresponds to two modes of the electromagnetic field: one for each polarization.

Figure C.2 shows a plot of the allowed values of the wave vector in the (x, y) plane of $\boldsymbol{k}$-space. The allowed values form a grid with a spacing of $2\pi/L$ between successive points. Thus each allowed value of the $\boldsymbol{k}$-vector occupies an effective area of $(2\pi/L)^2$ of this two-dimensional slice of $\boldsymbol{k}$-space. We can generalize the argument to the three-dimensional case that we are actually considering to realize that each $\boldsymbol{k}$-state will occupy an effective volume of $(2\pi/L)^3$ of $\boldsymbol{k}$-space.

We now ask the question: how many allowed $\boldsymbol{k}$-states are there with their magnitudes between k and $k + \mathrm{d}k$? We write this number as $g(k)\,\mathrm{d}k$. In the two-dimensional case shown in Fig. C.2, we calculate this number by working out the area of $\boldsymbol{k}$-space enclosed by $\boldsymbol{k}$-vectors with magnitudes between k and $k + \mathrm{d}k$ and then dividing by the effective area per $\boldsymbol{k}$-state:

$$g^{2\mathrm{D}}(k)\,\mathrm{d}k = \frac{2\pi k\,\mathrm{d}k}{(2\pi/L)^2} = L^2 \frac{k}{2\pi}\,\mathrm{d}k. \tag{C.6}$$

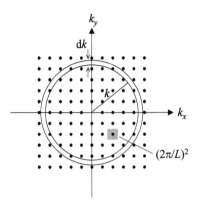

Fig. C.2 Grid of allowed wave vector values in the (x, y) plane of $\boldsymbol{k}$-space. The allowed values of $\boldsymbol{k}$ are given by eqn C.5.

In three dimensions the equivalent result is obtained by dividing the volume of $\boldsymbol{k}$-space enclosed between spherical shells of radius k and $k + \mathrm{d}k$ by the effective volume per $\boldsymbol{k}$-state, namely $(2\pi/L)^3$:

$$g^{3\mathrm{D}}(k)\,\mathrm{d}k = \frac{4\pi k^2 \mathrm{d}k}{(2\pi/L)^3} = L^3 \frac{k^2}{2\pi^2}\,\mathrm{d}k = V\frac{k^2}{2\pi^2}\,\mathrm{d}k. \tag{C.7}$$

We then normalize by V to obtain:

$$g(k) \equiv \frac{g^{3\mathrm{D}}(k)}{V} = \frac{k^2}{2\pi^2}. \tag{C.8}$$

Note that this value does not depend on the volume and confirms that the subdivision of space is merely a computational tool.

Having worked out the state density in $\boldsymbol{k}$-space, we can now work out the number of states per unit volume per unit angular frequency range $g(\omega)$. To do this we map the values of k and $k + \mathrm{d}k$ onto their corresponding angular frequencies, namely ω and $\omega + \mathrm{d}\omega$, and remember that there are two photon polarizations for each $\boldsymbol{k}$-state. We thus write:

$$g(\omega)\,\mathrm{d}\omega = 2 \times g(k)\,\mathrm{d}k, \tag{C.9}$$

implying:

$$g(\omega) = \frac{2g(k)}{\mathrm{d}\omega/\mathrm{d}k}. \tag{C.10}$$

The density of states that appears in Fermi's golden rule (eqn 4.12) is usually defined in terms of energy rather then angular frequency, with:

$$g(E)\,\mathrm{d}E = g(\omega)\,\mathrm{d}\omega.$$

For photons we have $E = \hbar\omega$, and the two quantities are effectively interchangeable:

$$g(E) = g(\omega)/\hbar.$$

With $\omega = ck$ we finally obtain:

$$g(\omega) = \frac{\omega^2}{\pi^2 c^3}. \tag{C.11}$$

This shows that the photon density of states is proportional to the square of the frequency.

When dealing with electrons in crystals, the waves in eqn C.1 are Bloch functions.

The derivation of the density of states for photon modes can be adapted to other branches of physics. In the case of electron waves in crystals, we usually require $g(E)$, the density of states per unit volume per unit energy range. We first work out the density of states in momentum space. The derivation is identical to that given above, with $g(k)$ given by eqn C.8. In analogy with eqn C.10, we then write:

$$g(E) = 2 \times \frac{g(k)}{\mathrm{d}E/\mathrm{d}k}. \tag{C.12}$$

In this case, the factor of two comes from the fact that there are two electron spin states for each available $\boldsymbol{k}$-state, namely spin up and spin down. For free electrons we have:

$$E = \frac{\hbar^2 k^2}{2m_0}, \tag{C.13}$$

For electrons near the bottom of the conduction band in a semiconductor, we can usually apply the effective mass approximation. This allows us to replace the free electron mass m_0 with the electron effective mass m_e^* and measure the energy relative to the bottom of the conduction band. An equivalent approximation can be made for the holes, with the energy measured downwards from the top of the valence band.

which then gives:

$$g(E) = \frac{1}{2\pi^2} \left(\frac{2m_0}{\hbar^2} \right)^{3/2} E^{1/2}. \tag{C.14}$$

For non-interacting particles of spin S and mass m, the factor of 2 in eqn C.12 is replaced by the spin multiplicity $(2S + 1)$, leading to the general result:

$$g(E)\,\mathrm{d}E = \frac{(2S + 1)}{4\pi^2} \left(\frac{2m}{\hbar^2} \right)^{3/2} E^{1/2}\,\mathrm{d}E. \tag{C.15}$$

It is apparent that eqn C.15 reduces to eqn C.14 when $m = m_0$ and $S = 1/2$, as is appropriate for electrons.

Low-dimensional semiconductor structures

D

Low-dimensional semiconductor structures are of considerable importance in the modern electronics and optoelectronics industries. This has led to the development of crystal growth techniques which now routinely make semiconductor layers with atomic precision on the layer thickness. The application of low-dimensional structures in quantum optics comes as a spin-off from this technological progress. In this appendix we briefly explain the general principles of quantum confinement, and then mention some key points on the properties of quantum wells and quantum dots that are relevant to the subject material of this book.

D.1 Quantum confinement   333
D.2 Quantum wells   335
D.3 Quantum dots   337

Further reading   338

D.1 Quantum confinement

Electron waves are characterized by their de Broglie wavelength λ_{deB} defined by:

$$\lambda_{\mathrm{deB}} = \frac{h}{p}, \tag{D.1}$$

where p is the linear momentum. The electrons in the conduction band of a semiconductor are free to move in all three directions, and their de Broglie wavelength is governed by the thermal kinetic energy at temperature T:

$$E_{\mathrm{thermal}} = \frac{p_i^2}{2m_e^*} \sim \tfrac{1}{2}k_{\mathrm{B}}T, \tag{D.2}$$

where m_e^* is the effective mass and the subscript i refers to one of the Cartesian axes x, y, or z. This gives a de Broglie wavelength of order:

$$\lambda_{\mathrm{deB}} \sim \frac{h}{\sqrt{m_e^* k_{\mathrm{B}} T}}. \tag{D.3}$$

In normal circumstances, the de Broglie wavelength is much smaller than the dimensions of the crystal, and the motion is governed by the laws of classical physics. However, when one or more of the dimensions of the crystal is comparable to λ_{deB}, then the motion in that direction will be quantized. This phenomenon is called **quantum confinement**.

Note the difference between Appendix C, where there were no real boundaries, and the volume could be arbitrarily large, and the case considered here, where the boundaries are real.

It is apparent from eqn D.3 that the length scale for the transition from classical to quantum behaviour depends on both the temperature and the effective mass. In a typical semiconductor with $m_e^* \sim 0.1m_0$, we require length scales of about 10 nm or less to observe quantum confinement effects at room temperature.

There are three general classifications of quantum confinement effects. If the motion is confined in one direction (e.g. the z-direction), the structure is called a **quantum well**. The electrons are free to move in the other two directions (i.e. the x- and y-directions) and so we have free motion in two dimensions and quantized motion in the third. If the motion is confined in two directions the structure is called a **quantum wire**. This has free motion in one dimension (e.g. the x-axis) and quantized motion in the other two directions. Finally, if the motion is confined in all three directions the structure is called a **quantum dot**, or alternatively a **quantum box**. The motion of the electrons in a quantum dot is quantized in all three directions. The general scheme of classifying quantum-confined structures is illustrated schematically in Fig. D.1, and summarized in Table D.1.

For the purposes of the discussion in this book, the main effect of the quantum confinement is to modify the energy spectrum and the density of states. The electrons in a bulk semiconductor can have any energy above the band-gap energy E_g and the density of states is proportional to $(E - E_g)^{1/2}$. (See eqn C.14 in Appendix C.) This is a consequence of the free motion in all three dimensions. The effective dimensionality of the system decreases as the electrons are confined in each new direction, which alters the functional form of the density of states and increases the effective band gap.

Let us consider first the properties of a quantum well. This is effectively a two-dimensional system with quantized motion in the z-direction and free motion in other two directions. In the simplest model, we consider the quantum well as a one-dimensional potential well. It is shown in all elementary quantum-mechanics texts that the energy of a particle

Semiconductor quantum wires have not found as many applications as quantum wells and dots because there is no easy way to make them. For this reason, we do not consider them further in this appendix.

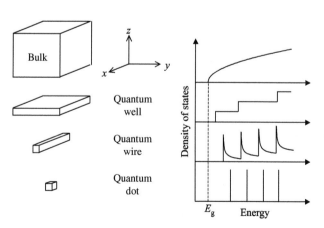

Fig. D.1 Schematic representation of quantum wells, wires, and dots. The generic shape of the density of states function for electrons in the conduction band of a semiconductor with band gap E_g is shown for each type of structure.

of mass m confined in a deep potential well of width L is given by:

$$E = \frac{\hbar^2}{2m}\left(\frac{n\pi}{L}\right)^2,$$ (D.4)

where n is an integer. The energy for the quantized motion in the z-direction of a semiconductor quantum well of thickness L_w will therefore be given approximately by:

$$E = \frac{\hbar^2}{2m_\mathrm{e}^*}\left(\frac{n\pi}{L_\mathrm{w}}\right)^2.$$ (D.5)

In this model the lowest energy state for the electrons in the conduction band is equal to $(E_\mathrm{g} + \hbar^2\pi^2/2m_\mathrm{e}^*L_\mathrm{w}^2)$. This shows that the effective band gap shifts to higher energy as the well width decreases.

The density of states for a quantum well is determined by the two-dimensional free motion in the x- and y-directions. We first derive the density of states per unit area in two-dimensional k-space by a method analogous to that used for three dimensions in Appendix C. This gives:

$$g_\mathrm{2D}(k_{xy}) = k_{xy}/2\pi.$$ (D.6)

We then substitute from eqn C.12 with $E = \hbar^2 k_{xy}^2/2m_\mathrm{e}^*$ to obtain:

$$g_\mathrm{2D}(E) = \frac{m_\mathrm{e}^*}{\pi\hbar^2}.$$ (D.7)

The final result is shown on the right-hand side of Fig. D.1. The band edge is shifted up by the quantum confinement energy, and the density of states is a sequence of steps, with each step adding a constant to the density of states as given by eqn D.7.

These arguments can be repeated for 1-D quantum wire and 0-D quantum dot systems. In the case of quantum wires, the density of states has an $E^{-1/2}$ dependence which leads to peaks at each new quantized state as shown in Fig. D.1. In quantum dots the motion is quantized in all three directions and there are no continuous bands at all. The density of states consists of a series of Dirac-δ functions at each quantized level, as illustrated in Fig. D.1. In this sense, quantum dots behave like 'artificial atoms' in which the electrons have discrete energies rather than continuous bands as is the norm in semiconductor physics.

The holes in the valence band are also confined by the boundaries of the quantum well, and thus have their energies shifted by an analogous quantum confinement energy. This further increases the effective band gap of the quantum well compared to the bulk crystal. See the discussion of the GaAs/AlGaAs quantum well in Section D.2 below.

Table D.1 General scheme of quantum confinement. Quantum dots are alternatively called quantum boxes. Note that the choice of the labelling of the axes for the quantized and free motion is purely conventional.

Structure	Quantum confinement	Free motion
Bulk	None	x, y, z
Quantum well	1-D	x, y
Quantum wire	2-D	x
Quantum dot	3-D	None

D.2   Quantum wells

Quantum wells are now routinely used in optoelectronic devices like light-emitting diodes and laser diodes. They can be grown with great precision by techniques of semiconductor crystal growth called **molecular beam epitaxy** (MBE) or **metalorganic chemical vapour deposition** (MOCVD). These techniques allow the easy production of layered structures containing different semiconductor materials, with precise control of the layer thicknesses down to the atomic level.

Quantum wells are formed by growing a layer of a semiconductor of thickness L_w between layers of another semiconductor with a larger band

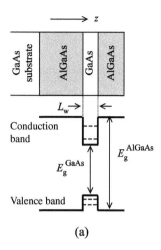

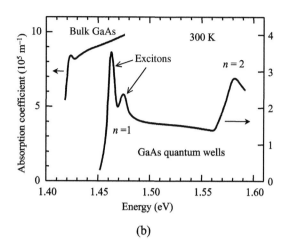

(a) (b)

Fig. D.2 (a) Schematic representation of a GaAs/AlGaAs quantum well of width L_w. The energy band diagram in the growth direction (z) is shown in the lower half of the figure. One-dimensional potential wells are formed in both the conduction and valence bands due to the discontinuity in the band-gap energy E_g at the interfaces between the GaAs and AlGaAs layers. The confined energy levels for electrons in the conduction band and holes in the valence band are indicated by the dashed lines. (b) Comparison of the optical absorption of GaAs/AlGaAs quantum wells and bulk GaAs at room temperature. The quantum well width was 10 nm. (After D.A.B. Miller *et al.*, *Appl. Phys. Lett.* **41**, 679 (1982), © American Institute of Physics, reproduced with permission.)

gap, as illustrated schematically in Fig. D.2(a). A typical combination of materials is the binary III–V semiconductor GaAs for the quantum well layer and the ternary alloy semiconductor AlGaAs as the barrier material. GaAs has a smaller band gap than AlGaAs, and this leads to the formation of finite-depth potential wells for electrons in the conduction band. Furthermore, since the charge carriers in the valence band are positively charged 'holes' with energy decreasing downwards on energy band diagrams, it is apparent that the holes in the valence band of the GaAs are also trapped in a potential well. We thus achieve a situation in which the electrons and holes in the quantum well are both confined in the z-direction. The depth of the conduction and valence band potential wells is typically 0.2 and 0.1 eV, respectively, which leads to strong confinement at room temperature ($k_B T \sim 0.025$ eV) and below.

Figure D.2(b) compares the optical absorption spectrum of GaAs quantum wells with that of bulk GaAs at room temperature. Several features are noteworthy in the data:

- The band edge is shifted to higher energy by the quantum confinement in the z-direction.

- Sharp lines are prominent at the absorption edge due to enhanced excitonic effects.

- The absorption above the exciton energies is approximately constant due to the constant density of states in 2-D materials (cf. eqn D.7.)

Excitons are hydrogen-like systems containing bound electron–hole pairs. See Section 4.6.

The sharp exciton lines which are so prominent in the absorption spectrum are important for the observation of strong-coupling effects in quantum well microcavities. (See Section 10.4.2.) Excitonic effects are enhanced in quantum wells compared to bulk semiconductors because the quantum confinement keeps the electrons and holes closer together, and hence increases their mutual Coulomb attraction.

D.3 Quantum dots

Quantum dots are semiconductor structures in which the motion of the electrons is confined in all three directions. This gives rise to full quantization of the motion, with discrete atom-like states. (See Fig. D.1.) In the context of quantum optics, it is the discrete nature of the energy spectrum that makes quantum dots so interesting. The quantum confinement creates optical states with large dipole moments that can interact very strongly with light. This has led to numerous observations of quantum optical effects, most notably:

- photon antibunching: see Fig. 6.11 in Section 6.6;
- triggered single photon sources: see Fig. 6.13 in Section 6.7;
- Rabi oscillations: see Fig. 9.8 in Section 9.5.3;
- the Purcell effect: see Fig. 10.8 in Section 10.3.4;
- quantum gates: see Section 13.3.4.

Quantum dots are thus important solid state structures for application in quantum optics.

There are two types of quantum dots that are commonly employed in quantum optics experiments. The first type is found in semiconductor doped glasses. These materials have been developed for use in colour-glass filters, and consist of semiconductor microcrystals embedded in a glass matrix. The semiconductor materials used are typically II-VI compounds like ZnS or CdSe, and their alloys. The microcrystals are incorporated into the glass during the melt, and, by adjusting the growth conditions, it is possible to incorporate microcrystals with good size control down to nanometre length-scales. In this way quantum dots are formed within a transparent glass host and their properties can be investigated by techniques of optical spectroscopy.

The second type of dot is the self-organized structures made by epitaxial crystal growth in the Stranski–Krastanow regime. Dots can be formed when a thin layer of a material with a very different unit cell size from that of the main crystal is deposited by MBE or MOCVD. In this situation, the energy required in straining the layer to match the unit cell size of the crystal is so large that the surface breaks up into microscopic clusters with length-scales in the nanometre range. Subsequent growth of further layers on top of the strained layer allows electrical contacts to be applied and cavities to be formed.

Figure D.3 shows transmission electron microscope (TEM) images of InAs quantum dots grown by the Stranski–Krastanow technique. The dots were grown on a GaAs crystal, leading to a 7% difference in the unit cell sizes. Part (a) shows a plan view, while part (b) shows a side

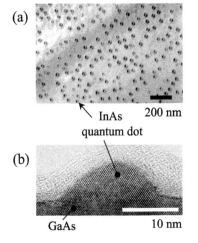

(a)

InAs quantum dot   200 nm

(b)

GaAs   10 nm

Fig. D.3 (a) Plan view of an uncapped layer of InAs quantum dots formed during Stranski–Krastanow growth on a GaAs crystal. (b) Side image of one of the InAs dots looking down the edge of the wafer. The mottled pattern above the dot originates from the adhesive used to hold the sample in position. Both images were taken with a transmission electron microscope. (After P. W. Fry *et al.*, *Phys. Rev. Lett.* **84**, 733 (2000) and M. Hopkinson (unpublished). Part (b) © American Physical Society, reproduced with permission.)

view looking down the wafer edge at higher resolution. The TEM images show that the lateral and vertical dimensions of the dot are both in the nanometre range, leading to strong confinement in all directions. InAs quantum dots grown in this way have been used for the observation of all of the phenomena listed at the start of this section.

Further reading

General introductions to the physics of low-dimensional semiconductor structures may be found in Bastard (1990), Harrison (2005), Mowbray (2005), or Weisbuch and Vinter (1991). The optical properties of quantum wells are described in Fox (2001) and Klingshirn (1995). An introduction to epitaxial quantum dots is given in Petroff (2001), while more detailed information may be found in Bimberg (1999). Woggon (1995) gives a good discussion of quantum dot research prior to the development of epitaxial quantum dots. Review chapters on recent research on single quantum dots, and their application in quantum optics, may be found in Michler (2003).

Nuclear magnetic resonance

The Bloch model of resonant light–atom interactions described in Section 9.6 was adapted from the Bloch model of nuclear magnetic resonance (NMR). It is therefore instructive to summarize the main results of nuclear magnetic resonance phenomena in order to make the analogy with optical systems more apparent.

E.1 Basic principles	339
E.2 The rotating frame transformation	341
E.3 The Bloch equations	344
Further reading	345

E.1 Basic principles

The apparatus used in a typical NMR experiment is shown schematically in Fig. E.1. The technique works for all nuclei with non-zero spins, but we consider here the simplest case of nuclei with spin $I = 1/2$ (e.g. ^{1}H, ^{13}C.) A sample containing the spin $1/2$ nuclei is placed in a strong static magnetic field of strength B_0 pointing in the z direction. The sample is inserted within a coil which produces a much weaker oscillating perpendicular magnetic field of strength B_1 in the x direction. This coil is driven by a pulsed radio-frequency (RF) source, which determines the oscillation frequency ν of B_1. Once the pulse is over, the coil picks up the oscillating magnetic field due to the oscillating magnetization of the sample, and the induced voltage is recorded with sensitive detection electronics.

We first consider the effect of the static field B_0 in the z-direction. The quantized spin states are shifted by the Zeeman effect by an energy equal to:

$$E = -g_N \mu_N B_0 M_I, \tag{E.1}$$

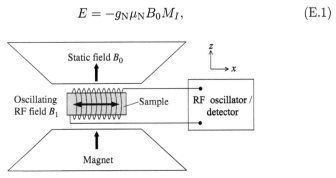

Fig. E.1 NMR apparatus. The sample is placed between the pole pieces of a magnet and is inserted inside a coil, which is driven by a pulsed radio frequency (RF) oscillator. The magnet generates a strong static magnet field of strength B_0 in the z-direction, while the coil generates a pulsed oscillating magnetic field of strength B_1 in the x-direction. A sensitive detector picks up the oscillations of the magnetization of the sample after the RF pulse has ceased.

where g_N is the nuclear g-factor, $\mu_N = e\hbar/2m_p = 5.0508 \times 10^{-27}$ A m^2 is the nuclear magneton, and M_I is the quantum number for the z-component of the nuclear spin. With $M_I = \pm 1/2$, the magnetic sublevels split into a doublet as shown in Fig. E.2. The sublevel with $M_I = +1/2$ corresponds to the spin pointing parallel to B_0 and has the lower energy, while the $M_I = -1/2$ state with spin pointing against the field increases in energy. The energy splitting ΔE between the sub-levels is given by:

$$\Delta E = g_N \mu_N B_0. \tag{E.2}$$

For protons we have $g_N = 5.586$, so that the splitting is 1.76×10^{-7} eV in a field of 1 T.

In the NMR technique we tune the angular frequency ω of the oscillator until the resonance condition with

$$\Delta E = g_N \mu_N B_0 = \hbar\omega \equiv h\nu \tag{E.3}$$

is satisfied. Alternatively, we can keep ω fixed and tune the field strength B_0 until resonance is achieved. These resonant frequencies are typically in the RF spectral range. For example, for the protons in ^{1}H we find $\omega/2\pi = 42.58$ MHz at $B_0 = 1$ T, and $\omega/2\pi = 100.000$ MHz at $B_0 = 2.34866$ T.

The resonance condition given in eqn E.3 can be understood in terms of transitions between the magnetic sublevels as indicated in Fig. E.2. The oscillating magnetic field B_1 generates electromagnetic radiation, and the RF photons have exactly the right energy to induce both absorption and stimulated emission transitions between the levels. These $M_I = -1/2 \leftrightarrow +1/2$ transitions are allowed because they have $\Delta M_I = \pm 1$, which involves a change of one unit of angular momentum (i.e. $\hbar$). In thermal equilibrium there will be more spins pointing along the field than against it, and thus there will be net absorption of the radiation. Hence we expect to observe a net absorption of power from the RF source whenever the resonance condition is achieved.

The resonant frequency can be given another interpretation by means of the classical treatment of the magnetism due to nuclear spin. In this approach the static field exerts a torque $\boldsymbol{\Gamma}$ on the magnetic dipoles within the medium according to:

$$\boldsymbol{\Gamma} = \boldsymbol{\mu} \times \boldsymbol{B}_0, \tag{E.4}$$

where $\boldsymbol{\mu}$ is the magnetic dipole moment of the nucleus, which is directly proportional to its angular momentum $\boldsymbol{I}$ through the **gyromagnetic ratio** γ:

$$\boldsymbol{\mu} = \gamma \boldsymbol{I}. \tag{E.5}$$

In the case of angular momentum due to nuclear spin, the gyromagnetic ratio is given by:

$$\gamma = g_N e/2m_p = g_N \mu_N/\hbar. \tag{E.6}$$

The classical equation of motion is:

$$\frac{d\boldsymbol{I}}{dt} = \boldsymbol{\Gamma}, \tag{E.7}$$

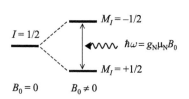

Fig. E.2 Zeeman splitting of the magnetic sublevels of a nucleus with spin $I = 1/2$ in a static field of strength B_0. In the NMR technique the angular frequency ω of the RF oscillator is tuned until the resonance condition with $\omega = g_N \mu_N B_0/\hbar$ is satisfied.

and we can therefore substitute into eqn E.4 using eqn E.5 to obtain:

$$\frac{\mathrm{d}\boldsymbol{\mu}}{\mathrm{d}t} = \gamma\boldsymbol{\mu}\times\boldsymbol{B}_0. \tag{E.8}$$

This equation describes a precession of the magnetic dipole around the field as illustrated in Fig. E.3. The effect is called **Larmor precession** and the precession angular frequency ω_L is given by:

$$\omega_L = -\gamma B_0. \tag{E.9}$$

The $-$ sign in eqn E.9 indicates that particles with positive gyromagnetic ratios (e.g. nuclei) precess in a left-handed sense around the field, whereas those with negative gyromagnetic ratios (e.g. electrons) precess in a right-handed sense. On comparing eqns E.3 and E.9 with γ given by eqn E.6, we find that the resonant frequency of the RF source is exactly equal to the magnitude of the Larmor precession frequency of the spins around the static field.

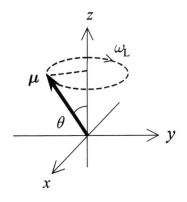

Fig. E.3 Larmor precession of the magnetic dipole about the field direction for the case of a positive gyromagnetic ratio. The polar angle θ is a constant of the motion.

E.2   The rotating frame transformation

The Larmor precession of the spins about the field suggests that we should make a coordinate transformation to a **rotating frame**. We adopt the notation whereby $\boldsymbol{\Omega}$ represents a vector of magnitude Ω pointing along the rotation axis. The transformation from the laboratory to the rotating frame can be made by vector addition of the velocities:

$$\dot{\boldsymbol{r}}_{\text{lab}} = \dot{\boldsymbol{r}}_{\text{rotating}} + \boldsymbol{\Omega}\times\boldsymbol{r}, \tag{E.10}$$

where $\boldsymbol{\Omega}$ is the angular velocity vector of the rotating frame. The first term on the right-hand side of eqn E.10 represents the perceived velocity in the rotating frame, while the second term is the velocity of the rotating frame relative to the laboratory frame. If we let $\boldsymbol{r}$ represent the spin magnetic dipole $\boldsymbol{\mu}$, and make use of eqn E.8 to replace $\dot{\boldsymbol{r}}_{\text{lab}}$ by $\gamma\boldsymbol{\mu}\times\boldsymbol{B}_0$, we then obtain:

$$\dot{\boldsymbol{\mu}}_{\text{lab}} = \gamma\boldsymbol{\mu}\times\boldsymbol{B}_0 = \dot{\boldsymbol{\mu}}_{\text{rotating}} + \boldsymbol{\Omega}\times\boldsymbol{\mu}, \tag{E.11}$$

which can be rearranged to give

$$\left(\frac{\mathrm{d}\boldsymbol{\mu}}{\mathrm{d}t}\right)_{\text{rotating}} = \gamma\boldsymbol{\mu}\times\boldsymbol{B}_{\text{eff}}, \tag{E.12}$$

where

$$\boldsymbol{B}_{\text{eff}} = \boldsymbol{B}_0 + \frac{\boldsymbol{\Omega}}{\gamma}. \tag{E.13}$$

On comparing eqns E.8 and E.12, we see that in the rotating frame the magnetic dipole seems to experience an effective magnetic field equal to $\boldsymbol{B}_{\text{eff}}$.

We now concentrate on the case of special interest where the rotating frame moves at the same rate as the Larmor precession of the spins:

$$\boldsymbol{\Omega} = \omega_L\,\hat{\mathbf{z}} = -\gamma B_0\,\hat{\mathbf{z}}, \tag{E.14}$$

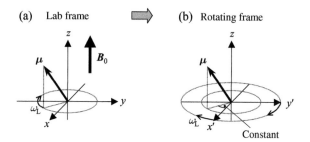

Fig. E.4 Rotating frame transformation. (a) In the laboratory frame, the magnetic dipole precesses about the field $\boldsymbol{B}_0$ at the Larmor frequency ω_L. (b) In a rotating frame in which the x'- and y'-axes rotate about the z-axis at ω_L, the magnetic dipole is stationary and the effective field strength is zero.

where we made use of eqn E.9 in the second equality. On substituting into eqn E.13, we find that $\boldsymbol{B}_{\text{eff}} = 0$, which implies that there is no net torque, and hence that the dipole is static in the resonant rotating frame, as we would of course expect. This transformation is illustrated in figure E.2. The magnetic dipole precesses at angular frequency ω_L around the field direction in the laboratory frame (Fig. E.4(a)), but is static in a frame that rotates at angular frequency ω_L about the z-axis (Fig. E.4(b)).

The transformation to the rotating frame allows us to obtain an intuitive understanding of the effect of the oscillating RF field of frequency ω_L in the x-direction. The field can be written in the form:

$$\boldsymbol{B}_{\text{RF}} = (B_1 \cos \omega_L t)\,\hat{\mathbf{x}}, \tag{E.15}$$

where B_1 is the magnitude of the oscillating field. This linearly polarized electromagnetic wave can be decomposed into left and right circularly polarized waves of equal amplitude $B_1/2$ by writing:

$$\boldsymbol{B}_{\text{RF}} = \frac{B_1}{2}(\cos \omega_L t\,\hat{\mathbf{x}} + \sin \omega_L t\,\hat{\mathbf{y}}) + \frac{B_1}{2}(\cos \omega_L t\,\hat{\mathbf{x}} - \sin \omega_L t\,\hat{\mathbf{y}}). \tag{E.16}$$

In the laboratory frame, the left circular field rotates at exactly the same rate as the precessing dipole, but the right circular field rotates in the opposite sense. Therefore, in the rotating frame the left circular field is static, but the right circular field rotates at frequency $2\omega_L$ and has little effect. We therefore concentrate on the left circular field of magnitude $(B_1/2)\hat{\mathbf{x}}'$ in the rotating frame.

We have seen above that $\boldsymbol{B}_{\text{eff}}$ is zero in the rotating frame in the absence of the RF field. When the RF field is added, the dipole will experience the effect of the left circular field with:

$$\boldsymbol{B}_{\text{eff}} = (B_1/2)\,\hat{\mathbf{x}}'. \tag{E.17}$$

On comparing eqns E.8, E.9, and E.12, we see that in the rotating frame the dipole precesses about $\boldsymbol{B}_{\text{eff}}$ at a rate equal to $-\gamma B_{\text{eff}}$. With $\boldsymbol{B}_{\text{eff}}$ given by eqn E.17, we conclude that in the rotating frame the dipole will precess about the x'-axis at a rate equal to $-\gamma B_1/2$. Therefore, if we apply a pulse of duration T_p, the dipole will rotate about the x'-axis by an angle Θ equal to

$$\Theta = -\gamma B_1 T_p/2. \tag{E.18}$$

This effect is illustrated in Fig. E.5.

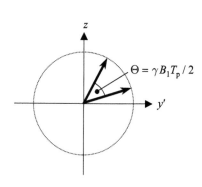

Fig. E.5 Effect of the application of a resonant RF pulse of magnitude B_1 and duration T_p in the rotating frame. The pulse causes a rotation of the magnetic dipole about the x'-axis through an angle Θ given by eqn E.18.

From the analysis above we see that the effect of the resonant RF pulse is to tip the spin vector in the rotating frame. By applying a pulse of duration $\pi/\gamma B_1$, the spin vector will rotate by 90°. Such a pulse is called a $\pi/2$-**pulse** for obvious reasons. Similarly, a π-**pulse** of duration $2\pi/\gamma B_1$ will cause a rotation of 180°. Once the pulses are completed, the spin vector will continue its Larmor precession in the laboratory frame, but with a new polar angle θ with respect to the z-axis.

The model that we have developed here can also be applied to continuous RF fields. Let us assume that the spin vector is initially aligned parallel to the static field and that the RF field is turned on at time $t=0$. We then expect to observe the behaviour shown in Fig. E.6(a). As time progresses, the RF field tips the spin through an ever increasing angle. Three specific examples are shown in Fig. E.6(a), which correspond to tipping angles of $\pi/2$, π, and 2π, respectively. The time dependence of the z-component of the spin vector corresponding to this behaviour is shown in Fig. E.6(b). It is apparent that $I_z/\hbar$ oscillates back and forth between $+1/2$ and $-1/2$ with a period T_{osc} given by

$$T_{\text{osc}} = \frac{2\pi}{\gamma B_1/2} = \frac{4\pi}{\gamma B_1}. \tag{E.19}$$

The tipping of the nuclear spin by a resonant RF field was first demonstrated by I. I. Rabi and co-workers in a pioneering series of molecular beam resonance experiments performed in the late 1930s. For this reason, the oscillations of the spin direction are now called **Rabi oscillations**, and the oscillation frequency implied by eqn E.19, namely $\gamma B_1/4\pi$, is called the **Rabi frequency**. A whole series of experimental techniques have subsequently been developed for manipulating the direction of the magnetic dipole vector using sequences of resonant RF pulses following Rabi's pioneering work. In recent years these techniques have found interesting new applications in quantum computing, as discussed in Section 13.3.4 of Chapter 13.

Note that in this classical model we only know I_z because both I_x and I_y are rapidly changing in the laboratory frame due to the Larmor precession. This is equivalent to the quantum picture in which we cannot know all three components of the spin at the same time.

See I. I. Rabi *et al.*, *Phys. Rev.* **55**, 526 (1939).

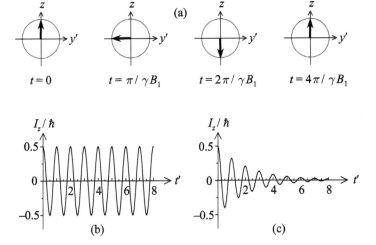

Fig. E.6 Response of a spin vector initially aligned parallel to the static field when a continuous RF field at frequency ω_L of strength B_1 is turned on at time $t=0$. (a) Rotation of the spin vector in the rotating frame with time. (b) Undamped oscillations of I_z in units of reduced time $t' = \gamma B_1 t/4\pi$. (c) Same as (b) but with damping included.

E.3   The Bloch equations

The situation depicted in Figs E.6(a) and (b) in not the end of the story. Up to this point, we have totally neglected the possibility that the response of the spin vector to the RF field might be damped. This is in fact very important, because without damping there would be no net absorption of radiation from the RF field. The spin dipole would just oscillate back and forth between the $+1/2$ and $-1/2$ states, and there would be no change in the time-averaged populations.

See F. Bloch, (1946). *Phys. Rev.* **70**, 460 (1946).

The effects of damping were first considered by Felix Bloch in 1946. He realized that the spin system was subject to two different kinds of damping mechanisms:

- **spin–lattice relaxation**, characterized by a time constant T_1.
- **spin–spin relaxation**, characterized by a time constant T_2.

In this picture, we add together all the nuclear dipoles to obtain the total magnetization vector $\boldsymbol{M}$. In the presence of the static field $\boldsymbol{B_0}$ in the z-direction, there will be an average net magnetization in the z-direction, so that we can write the equilibrium magnetization $\overline{\boldsymbol{M}}$ as:

$$\overline{\boldsymbol{M}} = (0, 0, M_0). \qquad (E.20)$$

We suppose that for some reason the magnetization is suddenly changed from this equilibrium value, for example in response to the RF field in an NMR experiment. The first type of relaxation mechanism describes the time dependence of the z-component of $\mathbf{M}$ as it relaxes back to its equilibrium value:

$$\frac{\mathrm{d}M_z}{\mathrm{d}t} = -\frac{M_z - M_0}{T_1}. \qquad (E.21)$$

Since this first type of relaxation affects the motion along the field direction, it is also called **longitudinal relaxation**. On the other hand, the second type describes the **transverse relaxation** in the direction at right angles to the field:

$$\frac{\mathrm{d}M_x}{\mathrm{d}t} = -\frac{M_x}{T_2} \; ; \; \frac{\mathrm{d}M_y}{\mathrm{d}t} = -\frac{M_y}{T_2}. \qquad (E.22)$$

The time constants that appear in eqns E.21 and E.22 are different because they correspond to physically different processes. The energy of the system depends on the z component of the spin (cf. eqn E.1.), and therefore longitudinal processes change the energy, but transverse processes do not. In the case of longitudinal relaxation, the change in energy accompanying the change in M_z must be taken in or given out to the environment. This energy exchange typically occurs through interactions between the nuclear spins and the crystalline lattice: hence the name 'spin–lattice relaxation'. On the other hand, the energy-conserving changes in the transverse components of $\boldsymbol{M}$, namely M_x or M_y, can occur through interactions within the spin system itself: hence the name

In the language of atomic collisions, we would say that the longitudinal T_1 processes correspond to *inelastic scattering* events, whereas transverse T_2 processes are equivalent to energy-conserving *elastic scattering*.

'spin–spin relaxation'. In general, energy-conserving scattering processes occur more easily, and hence T_2 is usually shorter than T_1.

We can combine the equation of motion in the magnetic field, namely eqn E.8, with the damping effects due to relaxation by writing:

$$\frac{\mathrm{d}M_x}{\mathrm{d}t} = \gamma(\boldsymbol{M}\times\boldsymbol{B}_0)_x - \frac{M_x}{T_2},$$
$$\frac{\mathrm{d}M_y}{\mathrm{d}t} = \gamma(\boldsymbol{M}\times\boldsymbol{B}_0)_y - \frac{M_y}{T_2}, \qquad (E.23)$$
$$\frac{\mathrm{d}M_z}{\mathrm{d}t} = \gamma(\boldsymbol{M}\times\boldsymbol{B}_0)_z - \frac{M_z - M_0}{T_1}.$$

This set of equations is known as the **Bloch equations**. The inclusion of the extra relaxation terms implies that the response of the system to changes of the field will be damped out. This effect is shown schematically in Fig. E.6(c). The oscillatory behaviour at the Rabi frequency is damped out, and the system eventually reverts to the equilibrium situation with $M_z = M_0$ and $M_x = M_y = 0$.

The classical Bloch model of the spins is a very successful prototype for describing the resonant interaction between electromagnetic waves and other two-level systems. For example, it is extensively used in the description of the interaction between resonant laser fields and atomic transitions. (See Chapter 9.) Bloch's seminal contribution to the subject is now recognized by describing the vector that represents the state of the system as the **Bloch vector**, and the sphere that its motion maps out in the rotating frame as the **Bloch sphere**.

Further reading

Nuclear magnetic resonance is covered at an introductory level in many texts on magnetism, for example, Bleaney and Bleaney (1976) or Blundell (2001). The classic treatise on NMR is Abragam (1961), and there are many more recent texts available, for example, Hennel and Klinowski (1993) or Hore (1995).

Bose–Einstein condensation

F.1 Classical and quantum statistics   **346**

F.2 Statistical mechanics of Bose–Einstein condensation   **348**

F.3 Bose–Einstein condensed systems   **350**

Further reading   **351**

The phenomenon of Bose–Einstein condensation was predicted in 1924, and was first successfully applied to explain the superfluid transition in liquid helium in 1938. The observation of Bose–Einstein condensation in a dilute gas of ^{87}Rb atoms in 1995 heralded a new era in the subject and accounts for its inclusion in this book. In this appendix we shall describe the phenomenon from the perspective of thermal physics, and use statistical mechanics to derive general results that are applicable to all Bose–Einstein condensed systems, including those described in Chapter 11.

The appendix begins with a brief review of quantum statistics and the classification of particles by their spin. We then apply the methods of statistical mechanics to derive general formulae for the condensation temperature and the fraction of particles in the condensed state. We finally conclude with a brief overview of systems that are known to exhibit the phenomenon.

F.1 Classical and quantum statistics

The purpose of statistical mechanics is to explain macroscopic phenomena in terms of the distribution of the particles among the microscopic states of the system. A key aspect of this subject is the energy distribution function, which describes the occupancy of the energy levels at temperature T. In dilute systems at high temperatures, the probability for the occupation of any individual quantum state is small. In this regime, the particles obey **Boltzmann statistics**:

The definition of what constitutes a 'high' temperature varies according to the physical system that is considered. See Table F.2.

$$\mathcal{P}(E_i) \propto \exp\left(-\frac{E_i}{k_{\mathrm{B}}T}\right), \tag{F.1}$$

where $\mathcal{P}(E_i)$ is the probability that the particle is in the quantum state with energy E_i. Boltzmann statistics are described as **classical statistics** because the properties do not depend on the quantum spin of the particle. A key assumption of the Boltzmann formula is that the occupation probability $\mathcal{P}(E_i)$ is small for all the energy levels of the system.

If we take a system that obeys Boltzmann statistics and reduce its temperature, the particles will tend to accumulate in the lowest energy levels that are available. It will therefore eventually be the case that the assumption that the occupancy factor is small no longer applies. In this low-temperature regime, the behaviour of a gas of identical particles depends on the spin of the particles that comprise the system. The statistics are then called **quantum statistics**, because they depend on a quantum property of the particles, namely their spin.

Particles with integer spins are called **bosons**, while those with half-integer spins are called **fermions**. Most elementary particles are fermions, but composite particles such as atoms can either be fermions or bosons, depending on the total spin. Table F.1 gives a short list of particles and their spins, together with their classification as either fermions or bosons.

A key aspect of the properties of fermionic particles is that they obey the **Pauli exclusion principle**. This principle states that it is not possible to put more than one particle into a particular quantum state. The application of the Pauli principle to statistical mechanics leads to the **Fermi–Dirac distribution function** for the number n_{FD} of fermions that occupy the quantum level with energy E at temperature T:

$$n_{\mathrm{FD}}(E, T) = \frac{1}{\exp\left(E - \mu\right)/k_{\mathrm{B}}T + 1}. \tag{F.2}$$

The parameter μ that enters here is the **chemical potential**. It is determined by the constraint that the combined occupancy of all the levels of the system must be equal to the total number of particles. Note that the maximum value of the Fermi–Dirac function is unity, in accordance with the Pauli principle. The Pauli exclusion principle precludes the possibility of Bose–Einstein condensation, and we shall therefore not consider fermionic particles further here.

Bosons, with their integer spin, are not subject to the Pauli principle. There is no limit to the number of particles that can be put into any particular level, and their behaviour is therefore totally different to that of fermions at low temperatures. In a gas of non-interacting bosons, the number of particles n_{BE} in the quantum state with energy E at temperature T is given by the **Bose–Einstein distribution function**:

$$n_{\mathrm{BE}}(E, T) = \frac{1}{\exp\left(E - \mu\right)/k_{\mathrm{B}}T - 1}, \tag{F.3}$$

where μ is again the chemical potential. As with fermions, μ is determined by requiring that the combined occupancy of all the levels must be equal to the total number of particles. (See, for example, eqn F.6 below.)

In the following section we shall investigate the properties of boson systems at low temperatures. We shall discover that the simple requirement to conserve the particle number has far-reaching consequences, and leads to the phenomenon of Bose–Einstein condensation that is our interest here.

Table F.1 Classification of common particles as fermions or bosons according to their spin.

Particle	spin	type
Electron	1/2	Fermion
Proton	1/2	Fermion
Neutron	1/2	Fermion
Photon	1	Boson
α	0	Boson
Atom	Integer	Boson
Atom	Half-integer	Fermion

In systems with no constraint on the total number of particles, the chemical potential is zero. This is the case that applies to photons, which are boson particles with zero rest mass, allowing them to be created and destroyed with ease. By setting $E = h\nu$ and $\mu = 0$ in eqn F.3, we obtain the Planck formula given in eqn 5.28. The original derivation by Planck in 1901 was concerned with the thermal properties of black-body radiation, but the formula has general applicability to any boson system that is not subject to conservation of the particle number.

F.2   Statistical mechanics of Bose–Einstein condensation

Let us consider a gas of N non-interacting bosons of mass m in a volume V at temperature T. The non-interacting particle assumption implies that the particles only possess kinetic energy, so that we can define our energy scale with $E = 0$ as the lowest level of the system. The chemical potential is determined by requiring that the combined occupancy of all the energy levels of the system is equal to the total number of particles:

$$\frac{N}{V} = \int_0^\infty n_{\mathrm{BE}}(E)\, g(E)\, \mathrm{d}E, \tag{F.4}$$

Particles are said to be 'non-interacting' when the forces between them are extremely weak. In this limit, the interparticle interactions are sufficient to bring the gas to thermal equilibrium, but so small that the potential energy is negligible. The non-interacting particle approximation generally breaks down when the interparticle separation becomes small.

where $g(E)$ is the **density of states** per unit volume. For non-interacting particles of mass m and spin S, the density of states is given by (see eqn C.15 in Appendix C):

$$g(E)\, \mathrm{d}E = 2\pi(2S+1)\left(\frac{2m}{h^2}\right)^{3/2} E^{1/2}\, \mathrm{d}E . \tag{F.5}$$

Most bosonic systems of interest here have $S = 0$, and so the spin multiplicity $(2S + 1)$ is usually set equal to unity.

Consider now the behaviour of a gas of spin-0 bosons with a fixed particle density N/V as the temperature T is varied. The value of the chemical potential can be calculated by inserting the density of states from eqn F.5 into eqn F.4 and solving the following equation:

$$\frac{N}{V} = 2\pi\left(\frac{2m}{h^2}\right)^{3/2} \int_0^\infty \frac{E^{1/2}}{\exp[(E-\mu)/k_{\mathrm{B}}T] - 1}\, \mathrm{d}E. \tag{F.6}$$

There is no analytic solution to eqn F.6, but the general dependence of μ on T can be understood without recourse to numerical techniques. At sufficiently high temperatures, μ must have a large negative value to compensate for the large value of T. On cooling the gas while keeping N/V constant, μ must increase to compensate for the decrease in $k_{\mathrm{B}}T$. This process continues as we reduce T further, but it cannot continue indefinitely. This is because eqn F.3 shows us that the requirement to keep the occupancy factor positive for all values of $E \geq 0$ implies $\mu < 0$. μ will therefore increase with reducing T until it limits out at a small negative value very close to zero. The critical temperature T_{c} at which μ hits this limit is precisely the Bose–Einstein condensation temperature that we are interested in.

The large negative value of μ in the high T limit can also be deduced by requiring that the Bose–Einstein function reduces to the classical Boltzmann formula in eqn F.1. Equation F.3 shows us that this will be the case if μ has a large negative value, so that $(E - \mu)/k_{\mathrm{B}}T \gg 1$ for all values of E.

When we examine the behaviour of the system in the limit where μ approaches zero, it becomes apparent that the value of the integrand in eqn F.6 is undefined at $E = 0$. We therefore have to treat the zero energy state differently from the others and rewrite eqn F.6 as:

$$\frac{N}{V} = N_0(T) + 2\pi\left(\frac{2m}{h^2}\right)^{3/2} \int_0^\infty \frac{E^{1/2}}{\exp[(E-\mu)/k_{\mathrm{B}}T] - 1}\, \mathrm{d}E, \tag{F.7}$$

where $N_0(T)$ explicitly represents the number density of particles in the zero energy state. On taking the limit of eqn F.3 at $E=0$ for $\mu \to 0$, we find:

$$N_0(T)V = \frac{1}{(1 - (\mu/k_BT) + \cdots) - 1} = -\frac{k_BT}{\mu}. \qquad (F.8)$$

This shows that the limit to how small $|\mu|$ can actually become is determined by the number of particles that occupy the $E=0$ state.

The condensation temperature is defined as that at which it first becomes impossible to accommodate all the particles in the states with $E>0$. The condition for this to happen is given from eqn F.7 as:

$$\frac{N}{V} = 2\pi \left(\frac{2m}{h^2}\right)^{3/2} \int_0^\infty \frac{E^{1/2}}{\exp(E/k_BT_c) - 1} \, dE,$$

$$= 2\pi \left(\frac{2mk_BT_c}{h^2}\right)^{3/2} \int_0^\infty \frac{x^{1/2}}{e^x - 1} \, dx,$$

$$= 2\pi \left(\frac{2mk_BT_c}{h^2}\right)^{3/2} \times 2.315, \qquad (F.9)$$

where $x = E/k_BT$. Equation F.9 can be solved for T_c to obtain:

$$T_c = 0.0839 \frac{h^2}{mk_B} \left(\frac{N}{V}\right)^{2/3}. \qquad (F.10)$$

Equation F.10 shows that the Bose–Einstein condensation temperature is determined by the particle density N/V and mass m.

For temperatures below T_c, a macroscopic fraction of the total number of particles condenses into the state with $E=0$. The remainder of the particles continue to be distributed thermally among the rest of the levels. The number of particles in the states with $E>0$ is still given by eqn F.6, but with the chemical potential fixed at the effective value of zero for all states with $E>0$. Therefore, for $T \leq T_c$, we must write the particle density as:

$$\left(\frac{N}{V}\right) = N_0(T) + 2\pi \left(\frac{2m}{h^2}\right)^{3/2} \int_0^\infty \frac{E^{1/2}}{\exp(E/k_BT) - 1} \, dE$$

$$= N_0(T) + 2\pi \left(\frac{2mk_BT}{h^2}\right)^{3/2} \int_0^\infty \frac{x^{1/2}}{e^x - 1} \, dx$$

$$= N_0(T) + 2\pi \left(\frac{2mk_BT}{h^2}\right)^{3/2} \times 2.315$$

$$= N_0(T) + \left(\frac{N}{V}\right)\left(\frac{T}{T_c}\right)^{3/2}, \qquad (F.11)$$

where $x = E/k_BT$ as before, and we made use of eqn F.9 in the last line. On solving eqn F.11 for $N_0(T)$, we then find:

$$N_0(T) = \left(\frac{N}{V}\right)\left[1 - \left(\frac{T}{T_c}\right)^{3/2}\right]. \qquad (F.12)$$

In eqn F.9 we made use of the fact that:

$$\int_0^\infty \frac{x^{1/2}}{e^x - 1} \, dx = 2.315.$$

We also assumed that the value of the chemical potential is so close to zero that we can take $\mu = 0$ for all states except the one with $E=0$.

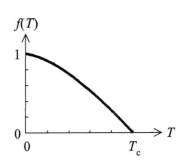

Fig. F.1 Fraction $f(T)$ of the number of particles in the Bose–Einstein condensed state versus temperature. T_c is the condensation temperature given by eqn F.10.

The dependence of the fraction of particles in the condensed state on the temperature according to eqn F.12 is plotted in Fig. F.1. It is apparent that the fraction approaches unity as $T \to 0$.

Work on liquid helium-4 demonstrates that, below T_c, some of the liquid shows superfluid behaviour, while the remainder remains 'normal'. For this reason, it is conventional to describe Bose–Einstein condensed systems according to the **two-fluid model**. In general, the two fluids correspond to the condensed state with $E=0$, and the 'normal' particles with $E>0$, with the fraction of superfluid particles given by eqn F.12 in an ideal system.

The picture which thus emerges from the statistical mechanics of Bose–Einstein condensation is as follows. Above the critical temperature T_c the particles are distributed among the energy states of the system according to the Bose–Einstein distribution function given in eqn F.3. At temperature T_c a phase transition occurs and a substantial fraction of the total number of particles condenses into the state with $E=0$. Since the particles are non-interacting, the only energy they possess is kinetic, and the $E=0$ state corresponds to zero speed. The particles in the zero-speed state are responsible for the spectacular low temperature phenomena associated with Bose–Einstein condensation such as superfluidity. The fraction of particles in the superfluid state is given by eqn F.12, and as the temperature approaches absolute zero, the fraction increases to unity. This picture has been elegantly confirmed by the experiments on ultracold atomic gases described in Section 11.3 of Chapter 11, although with some modifications to account for the fact that the condensation takes place in a trap rather than free space.

Detailed analysis of this phase transition shows that it is second-order with zero latent heat.

F.3   Bose–Einstein condensed systems

Table F.2 gives a partial list of physical systems that exhibit Bose–Einstein condensation phenomena. It is apparent from this list that the definition of 'low' temperature varies dramatically from system to system. The condensed dilute atomic gases described in Section 11.3 are nearly ideal systems in this context. Most of the other systems, by contrast, suffer from the fact that the need to bring the critical temperature up to workable values requires from eqn F.10 that the particle

Table F.2 Physical systems that exhibit (or might exhibit) Bose–Einstein condensation phenomena.

System	Boson	T_c (K)
Liquid helium	^{4}He	2.17 K
Dilute atom gases	See Table 11.1	$\sim 10^{-7}$ K
Superconductors	Electron Cooper pairs	up to ~ 100 K
Semiconductor	Exciton	~ 10 K
Neutron star	Neutron Cooper pairs	$\sim 10^9$ K

density N/V should be relatively large. This means that the particles are relatively close together, and the interparticle forces that have been neglected in the analysis given here become important. The great beauty of the atomic systems is that they are gaseous, so that the interparticle separations are large and it is a genuinely good approximation to treat the particles as 'non-interacting'. The price that has to be paid is the extremely low values of T_c that inevitably follow from the gaseous nature of the system. It was a real triumph of the laser and evaporative cooling methods described in Chapter 11 that the ultra-low temperatures required to observe condensation could actually be achieved in the laboratory.

The best-known example of Bose–Einstein condensation prior to the work on dilute atomic gases was liquid helium. Natural helium contains 99.99% of the ^{4}He isotope, with only 0.01% of ^{3}He, giving it predominantly bosonic properties. Helium is a gas at room temperature and liquefies at 4.2 K. The density of the liquid is $120\,\mathrm{kg\,m^{-3}}$, which implies from eqn F.10 that $T_c = 2.7$ K. Experimental work on liquid helium shows that it undergoes a phase transition to superfluid behaviour at the **lambda point** temperature of 2.17 K. The transition temperature is lower than that calculated from eqn F.10 because the mere fact that the system has condensed to a liquid indicates that there must be strong attractive forces present. It is therefore a gross over-simplification to treat the particles in superfluid helium as non-interacting bosons.

The fermion system ^{3}He shows no equivalent phase transition, which confirms that the lambda-point transition in ^{4}He is a Bose–Einstein condensation effect. Liquid ^{3}He does, however, show a rich variety of other quantum effects, but these occur at very much lower temperatures in the millikelvin range.

The other well-known condensed matter phenomenon that shows two-fluid behaviour is superconductivity. The individual electrons in a metal are fermions, but in a superconductor, two electrons pair up to form a **Cooper pair**. The Cooper pair can either have a spin of 0 of 1, and therefore forms a boson system. The conduction electrons in a superconductor below the transition temperature are either in the 'normal' state or the superconducting state. The fraction in the superconducting state is given by an equation analogous to eqn F.12.

The attractive force that binds the electrons in the Cooper pair is mediated by the electron–lattice interaction.

The discussion of Bose–Einstein condensation in the final two physical systems listed in Table F.2 is much more controversial. There have been a number of alleged reports of excitonic condensation in semiconductors, but most of them are regarded with some scepticism. The discussion of Bose–Einstein condensation in neutron stars is, of course, even more speculative.

Excitons are hydrogen-like atoms formed when a negatively charged electron in the conduction band of a semiconductor binds to a positively charged hole in the valence band. Both the electron and hole have spin 1/2, and so the composite particle is a boson, permitting the possibility of Bose–Einstein condensation.

Further reading

The basic phenomenon of Bose–Einstein condensation is covered in most texts on statistical mechanics, for example: Feynman (1998), Kittel and Kroemer (1980), or Mandl (1988). Following the discovery of Bose–Einstein condensation in dilute atomic gases, a number of new monographs are now available covering the subject in great depth, for example: Pethick and Smith (2002) or Pitaevskii and Stringari (2003).

Solutions and hints to the exercises

Chapter 2

(2.1) Substitute $\mathcal{E}(r, t) = \mathcal{E}_0 \exp i(k \cdot r - \omega t)$ into eqn 2.9 with $D = \epsilon_r \epsilon_0 \mathcal{E}$ and $\varrho = 0$ to find $k \cdot \mathcal{E} = 0$. If $\epsilon_r = 0$, then the waves can be longitudinal. This special condition does occur in plasmas and ionic crystals at certain resonant frequencies (see Fox 2001), but will not be relevant to the topics covered in this book.

(2.2) 2.7×10^4 V m^{-1} and 9.1×10^{-5} T.

(2.3) (a) linear at $+45°$ to x-axis;

   (b) linear at $+30°$ to x-axis;

   (c) left circular;

   (d) right circular;

   (e) left elliptical, major axis along x-axis, major : minor axis length ratio $= \sqrt{3}$;

   (f) left elliptical, major axis at $45°$ to x-axis, major : minor axis length ratio $= (\sqrt{2} + 1)$.

(2.4) 4.8×10^{-11} W. The LISA experiment actually consists of a Michelson interferometer in which the beam fired towards the distant satellite has to be reflected back towards the original one. Without extra tricks, this would give an undetectably small power level at the original satellite of $(4.8 \times 10^{-11})^2 = 2.3 \times 10^{-21}$ W. LISA gets around this problem by using the beam detected at the distant

satellite to seed the phase of a second laser within it. When this second laser is pointed back at the original satellite, the power collected will be the same as that calculated for the single pass of the arm of the interferometer.

(2.5) (a) 4.5×10^{-10} s; (b) 3.2×10^{-11} s.

(2.6) (a) 0.3 m, (b) 50 m.

(2.7) 5×10^{21} W m^{-2}. (This is an extremely large intensity!)

(2.8) 1867 nm.

(2.9) (a) 956 nm; (b) 4.69 μm.

(2.10) $\theta = 41°$.

(2.11) (a) If inversion symmetry applies, the physical properties must be invariant under inversion of the axes. The relationship given in eqn 2.56 will only be maintained when $P \rightarrow -P$ and $\mathcal{E} \rightarrow -\mathcal{E}$ if the terms with even powers of $\mathcal{E}$ are zero. This implies $\chi^{(2)} = 0$.

   (b) Write the intensity dependence of the refractive index as:

$$n = \sqrt{\epsilon_r} = (n_0^2 + \chi^{(3)} \mathcal{E}^2)^{1/2},$$

and expand to find $n = n_0 + \chi^{(3)} \mathcal{E}^2 / 2n_0$. Then relate $\mathcal{E}^2$ to the intensity to find n_2.

Chapter 3

(3.1) The result can be derived from:

$$\int \Psi^* \Psi \, d^3 r = \sum_n \sum_m c_n^* c_m \int \psi_n^* \psi_m \, d^3 r = 1.$$

(3.2) Insert $\hat{O} \varphi_m = O_m \varphi_m$ into

$$\langle \hat{O} \rangle = \sum_n \sum_m c_n^* c_m \int \varphi_n^* \hat{O} \varphi_m \, d^3 r.$$

(3.3) The result follows from:

$$(\Delta O)^2 = \int \psi^*(\hat{O}^2 - 2\langle\hat{O}\rangle\hat{O} + \langle\hat{O}\rangle^2)\psi\, \mathrm{d}^3 r.$$

(3.4) Write $\hat{l}_x = y\hat{p}_z - z\hat{p}_y$, and use the fact that x commutes with $\hat{p}_y$ and $\hat{p}_z$, etc.

(3.5) This is a 1s wave function. $E = -m_0 e^4/8\epsilon_0^2 h^2$.

(3.6) $^1P_1, {}^3P_0, {}^3P_1, {}^3P_2$.

(3.7) (a) Equation 3.78 implies $E_F - E_{F-1} = A(J)\hbar^2 F$.

(b) Equation 3.75 implies $E_J - E_{J-1} = 2C'J$.

(c) (i) The number of levels is equal to the smaller of $2J+1$ or $2I+1$. Since $J = 3/2$, this implies $I \geq 3/2$. (ii) The ratio of the splittings fits best (but not perfectly) to $I = 3/2$, which is the actual value of the nuclear spin in ^{23}Na.

(3.8) 3.4×10^{-5} eV, 0.27 cm^{-1} $(g_J = 7/6)$.

(3.9) $C = (m\omega/\pi\hbar)^{1/4}$.

(3.10) $(\Delta x)^2 = \hbar/2m\omega$, $(\Delta p_x)^2 = \hbar m\omega/2$. Hence result.

(3.11) (a) Use eqn 3.32 with $\langle x\rangle = \langle p_x\rangle = 0$.

(b) Set $\langle p_x^2\rangle/2m = \frac{1}{2}m\omega^2\langle x^2\rangle = (1/2) \times (n + \frac{1}{2})\hbar\omega$ to prove the result.

(3.12) $0.5\cos^2\theta$ and $0.5\sin^2\theta$.

Chapter 4

(4.1) (a) All the energy of the beam is concentrated at ω, and the integral of the spectral energy density over frequency is equal to u_ω.

(b) Set $B_{12}(\omega') = B_{12}^\omega g_\omega(\omega')$, and then substitute for B_{12}^ω.

(c) Use the identity
$$\int_{-\infty}^{+\infty} f(x)\delta(x-a)\,\mathrm{d}x = f(a)$$
together with eqns 4.4, 4.10, and 4.11 to derive the result.

(d) This follows from eqn 4.10 and the definition of the Einstein coefficients.

(4.2) This result follows from the odd parity of the E1 operator.

(4.3) From eqn 4.13 it follows that we must have
$$\int_0^{2\pi} e^{-im'\varphi} H' e^{im\varphi}\,\mathrm{d}\varphi \neq 0$$
for a transition to occur.

(a) $H' \propto z = r\cos\theta$. Hence $m' = m$.

(b) $H' \propto (x + iy) = (r\sin\theta\cos\varphi + ir\sin\theta\sin\varphi) \propto (\cos\theta + i\sin\varphi) = e^{i\varphi}$. Hence $m' = m+1$.

(c) $H' \propto (x - iy) \propto e^{-i\varphi}$. Hence $m' = m-1$.

(d) $H' \propto x$ $(y) \propto (e^{i\varphi} + (-)e^{-i\varphi})$. Hence $m' = m \pm 1$.

(4.4) (a) $I(t) \propto |\mathcal{E}(t)|^2$.

(b) Use $\cos\omega_0 t = (e^{i\omega_0 t} + e^{-i\omega_0 t})/2$ to find:
$$\mathcal{E}(\omega) \propto -\frac{1}{\mathrm{i}(\omega + \omega_0) - 1/2\tau} - \frac{1}{\mathrm{i}(\omega - \omega_0) - 1/2\tau}.$$
Neglect the first term compared to the second, and find $I(\omega)$ from $\mathcal{E}^*\mathcal{E}$.

(c) $\Delta\omega$ is found by working out the FWHM, which gives $\Delta\omega = 1/\tau$. The normalization constant C is found from
$$C\int_0^\infty [(\omega - \omega_0)^2 + (1/2\tau)^2]^{-1}\,\mathrm{d}\omega = 1.$$

(4.5) Parts (a) and (b) are standard results from the kinetic theory of gases. In part (c), put $\tau_{\text{collision}} = L/\bar{c}$ and $(N/V) = P/k_B T$ to derive the result.

(4.6) $\sigma_s = \pi r_{\text{atom}}^2 = 1.3 \times 10^{-19}$ m^2. Hence $\tau_{\text{collision}} \sim 6 \times 10^{-10}$ s.

(4.7) Natural linewidth: 7.8 MHz in both cases. Doppler linewidth: (a) 633 MHz, (b) 770 MHz. Collisional linewidth: (a) $\sim$ 5 kHz, (b) $\sim$ 40 MHz.

(4.8) 0.0015 K. (The neon atoms would have solidified by this temperature.)

(4.9) Normalization requires $g_\omega(\omega_0)\Delta\omega \sim 1$. $C = (2/\pi)$ for a Lorentzian line and $(4\ln 2/\pi)^{1/2}$ for a Gaussian.

(4.10) $E_X = 4.2$ meV.

(4.11) (a) $W_{21}^{net} = W_{21} - W_{12}$.

(b) Intensity = Energy / unit time / unit area = $u_\omega \times c/n$.

(c) Consider a beam increment of unit area. Set the energy added per unit time, namely $W_{21}^{net} \hbar\omega\, dz$, equal to dI.

(d) Substitute the results of parts (a)–(c) into the definition of the gain coefficient, namely $dI = \gamma I dz$.

(4.12) 86.5%.

(4.13) (a) 0.017 m^{-1}; (b) 5.3×10^{14} m^{-3}.

(4.14) (a) 75 MHz; (b) ~ 3 fs; (c) $\sim 6 \times 10^5$.

Chapter 5

(5.1) (a) 9600 counts s^{-1}; (b) 96; (c) 10.

(5.2) $\mathcal{P} = 0.276$ in both cases. With $\bar{n} = 100$, the exact probabilities are hard to compute on a pocket calculator. The Gaussian result for the range 94.5–105.5 gives $\mathcal{P} = 0.418$.

(5.3) The variance is calculated from

$$\text{Var}(n) = \sum_{n=0}^{\infty}(n - \bar{n})^2\, \mathcal{P}_\omega(n)$$
$$= \sum_{n=0}^{\infty} n^2 \mathcal{P}_\omega(n) - \bar{n}^2.$$

$\sum n^2\, \mathcal{P}_\omega(n)$ is worked out by a method similar to eqn 5.27.

(5.4) $\bar{n} = 5.7 \times 10^{-7}$ at 2000 K. $\bar{n} = 1$ at 41 000 K for $\lambda = 500$ nm and at 2100 K for $\lambda = 10$ μm.

(5.5) $N_m/V = 4 \times 10^{16}$. Poissonian statistics unless V very small.

(5.6) $\bar{n} = 4.0 \times 10^7$. $\Delta n = 6.3 \times 10^3$.

(5.7) (a) 0.04; (b) 3.8%; (c) 0.08%.

(5.8) $N = 2000$ in all three cases. ΔN is equal to: (a) 45; (b) 57; (c) 40.

(5.9) 7.3×10^{-20} W Hz^{-1}.

(5.10) 82%.

(5.11) 21 dB.

(5.12) 52 mV.

(5.13) (a) 7.45 mA; (b) 6.8×10^{-3}; (c) 2.15 mA; (d) 0.71; (e) Shot noise = 0.34 fW; photocurrent noise = 0.24 fW.

(5.14) 18 km. Bit error rate $> 10^{-9}$ for $L_{fibre} \geq 18$ km.

Chapter 6

(6.1) The result is derived by equating the angular shift of the fringe pattern from the light at the edge of the source to the angle for the first minimum for light from the centre of the source.

(6.2) (b) $R = 160\ \Omega$.

(6.3) The results all follow directly from the definitions given in the exercise together with the definition of $g^{(2)}(0)$ from eqn 6.10.

(6.4) (b) In a stationary light source, the averages must be the same for all choices of $t = 0$.

(6.5) The result follows from the fact that $\langle \Delta I \rangle = 0$. At $\tau = 0$ we have:

$$g^{(2)}(0) = 1 + \langle(\Delta I(t))^2\rangle/\langle I(t)\rangle^2.$$

The second term must be zero or positive, and therefore $g^{(2)}(0) \geq 1$.

(6.6) 1.04.

(6.7) Gaussian $g^{(2)}(\tau)$ with $\tau_c \sim 0.44$ ns.

(6.8) Lorentzian $g^{(2)}(\tau)$ with $\tau_c \sim 0.16$ ns.

(6.9) (a) $\mathcal{P}(T) = 1 - \exp(-T/\tau_R)$.

(b) We need two photons in time τ_D. This gives $g^{(2)}(0) \sim 1 - \exp(-\tau_D/\tau_R)$.

(6.10) High power: fast excitation time, $g^{(2)}(0) \sim 1 - \exp(-\tau_{\rm D}/\tau_{\rm R})$. Low power: excitation time $\tau_{\rm E}$ significant, $g^{(2)}(0) \sim 1 - \exp(-\tau_{\rm D}/(\tau_{\rm R} + \tau_{\rm E})$, that is, smaller than at high power.

(6.11) $g^{(2)}(0) = 0.5$. This is the probability that the second photon in a particular pulse goes to D2 compared to the probability that either

photon in a different pulses goes to D2. See also eqn 8.64 with $n = 2$.

(6.12) 10^9 photons s^{-1}.

(6.13) (a) The $g^{(2)}(\tau)$ function would consist of a series of regularly spaced peaks of equal height, but with the peak at $\tau = 0$ absent, as in Fig. 6.13(b).

(b) $g^{(2)}(\tau) = 1$ for all τ.

Chapter 7

(7.1) The result follows from eqns 7.12 and 7.15.

(7.2) One way to do this is to use the definition of energy density in eqn 7.11.

(7.3) Substitute for $q(t)$ and $p(t)$ in eqn 7.23 using eqns 7.29 and 7.30.

(7.4) (a) 1.1×10^{-8} m^3; (b) 1.1×10^{-7} m^3.

(7.5) (a) 1.3×10^{-19} N; (b) 1.3×10^{-7} N.

(7.6) (a) 25; (b) 5; (c) 0.1 radians.

(7.7) 8.5×10^{-9} radians.

(7.8) $\sqrt{60} = 7.7$.

(7.9) 2.3×10^{-21}.

(7.10) (a) Reduced noise at the nodes and increased noise at the antinodes; (b) vice versa.

(7.11) The banana state minimizes the radial uncertainty.

(7.12) 1.1×10^8 V m^{-1}.

(7.13) $\Delta X_1 \Delta X_2 \sim n + 1/2$.

(7.14) In (b), the result follows from setting $\mathcal{E}_3 \mathcal{E}_3^* + \mathcal{E}_4 \mathcal{E}_4^* = \mathcal{E}_1^2 + \mathcal{E}_2^2$.

(7.15) 18%.

(7.16) 2.4 dB.

Chapter 8

(8.1) The energy is found by substituting into eqn 8.3 using the Hamiltonian of eqn 8.2.

(8.2) $\psi_1 = C_1 x \exp(-m\omega^2 x^2/2\hbar)$,
$\psi_2 = C_2(2m\omega x^2/\hbar - 1)\exp(-m\omega^2 x^2/2\hbar)$.

(8.3) The results follow from $\hat{X}_1 = (m\omega/2\hbar)^{1/2}\hat{x}$ and $\hat{X}_2 = (1/2m\hbar\omega)^{1/2}\hat{p}_x$.

(8.4) 0, 0, 1/2, 1/2.

(8.5) Phasor of length $|\alpha|$ at angle ϕ with uncertainty circle of diameter 1/2.

(8.6) $\langle \alpha | \beta \rangle = \exp(-|\alpha|^2/2 - |\beta|^2/2 + \alpha^*\beta)$

$= \exp[(\alpha^*\beta - \alpha\beta^*)/2] \exp -|\alpha - \beta|^2/2$.

The first exponential is just a phase factor, and therefore does not appear in the final result. The conclusion is that two coherent states are not orthogonal.

(8.7) The length of a phasor is equal to $(X_1^2 + X_2^2)^{1/2}$.

(8.8) $(n + \frac{1}{2})/2$.

(8.9) (a) 1, 0, 0, 1.

(b) Relative phase shift of π for one of the reflections.

(c) Set $[\hat{a}_i, \hat{a}_j^\dagger] = \delta_{ij}$ to derive the result.

(8.10) $\langle \hat{a}_3^\dagger \hat{a}_3 \rangle = \langle \hat{a}_4^\dagger \hat{a}_4 \rangle = \langle \hat{a}_1^\dagger \hat{a}_1 \rangle/2$, as before, and $\langle \hat{a}_3^\dagger \hat{a}_4^\dagger \hat{a}_4 \hat{a}_3 \rangle = \langle \hat{a}_1^\dagger \hat{a}_1^\dagger \hat{a}_1 \hat{a}_1 \rangle/4$ as before.

(8.11) (a) $|0\rangle_3 |0\rangle_4$.

(b) The output states are given by:

$$|1\rangle_1 |0\rangle_2 \to (|1\rangle_3 |0\rangle_4 + |0\rangle_3 |1\rangle_4)/\sqrt{2},$$

$$|0\rangle_1 |1\rangle_2 \to (-|1\rangle_3 |0\rangle_4 + |0\rangle_3 |1\rangle_4)/\sqrt{2}.$$

The single photon at the input can go to either output port with 50% probability. Hence the output is an entangled state with an equal probability of the two possible outputs, namely $|1\rangle_3 |0\rangle_4$ or $|0\rangle_3 |1\rangle_4$.

(c) $|1\rangle_1 |1\rangle_2 \rightarrow (-|2\rangle_3 |0\rangle_4 + |0\rangle_3 |2\rangle_4)/\sqrt{2}$.

This implies that both photons go to the same output port. The fields add or subtract coherently so that the probability of the photons going to different output ports is zero.

(8.12) $g^{(2)}(0) = \langle\alpha|\hat{n}^2 - \hat{n}|\alpha\rangle/\langle\hat{n}\rangle^2 = \bar{n}^2/\bar{n}^2 = 1$.

Chapter 9

(9.1) γ.

(9.2) (a) $\begin{pmatrix} 0 & 0 \\ 0 & 1 \end{pmatrix}$; (b) $\begin{pmatrix} 1/2 & 1/2 \\ 1/2 & 1/2 \end{pmatrix}$; (c) $\begin{pmatrix} 1/3 & -i\sqrt{2}/3 \\ i\sqrt{2}/3 & 2/3 \end{pmatrix}$.

(9.3) $[1+\exp(-\Delta E/k_{\rm B}T)]^{-1}\begin{pmatrix} 1 & 0 \\ 0 & \exp(-\Delta E/k_{\rm B}T) \end{pmatrix}$, where $\Delta E = E_2 - E_1$.

(9.4) $B_{12}^\omega = 2.39 \times 10^{21}$ m^3 rad J^{-1}s^{-1}; $A_{21} = 2.23 \times 10^7$ s^{-1}.

(9.5) (a) The equation is derived by taking the time derivative of $\dot{c}_2(t)$ and substituting for $\dot{c}_1(t)$.

(b) Substitute to obtain $\zeta^2 - \delta\omega\,\zeta - \Omega_{\rm R}^2/4 = 0$, with $\zeta_\pm$ as the two roots.

(c) The initial conditions lead to $c_2(t) = i(\Omega_{\rm R}/\Omega)\,e^{-i\delta\omega t/2}\sin\Omega t/2$.

(9.6) (a) Integrate over ω' with $u(\omega')=u$ to obtain $W_{12} = N_1 B_{12}^\omega u$, and similarly for W_{21}.

(b) The result is again obtained from $W_{12} = \int W_{12}(\omega')\,d\omega'$.

(c) The result follows from solving:
$$\frac{dN_2}{dt} = W_{12} - W_{21} - A_{21}N_2,$$

(8.13) (a) Parametric conversion produces pair states.

(b) Both 0.

(c) $(1/2)e^{-s}$ and $(1/2)e^{+s}$.

(d) Ellipse centred at the origin of area $(1/4)$. Minor axis length $= e^{-s}/2$, major axis length $= e^{+s}/2$.

subject to the constraint $N_0 = N_1 + N_2 =$ constant.

(d) (1) $N_2/N_0 \rightarrow 1/2$, (2) $N_2/N_0 \rightarrow (B_{12}^\omega g_\omega(\omega)/A_{21})\,u_\omega$. In (1) the populations equalize, while in (2) N_2 is proportional to the light intensity, as expected.

(9.7) (a) 1.4×10^4 W m^{-2}; (b) 6.2×10^3 W m^{-2}.

(9.8) $\mu_{12} = 2.4 \times 10^{-29}$ C m.

(9.9) The result follows from substitution of c_1 and c_2 from eqn 9.64 into eqn 9.63.

(9.10) (a) $(1/\sqrt{2})(|1\rangle + |2\rangle)$.

(b) $(1/\sqrt{2})(|1\rangle + i|2\rangle)$.

(c) $(1/\sqrt{2})(|1\rangle - |2\rangle)$.

(d) $(1/\sqrt{2})(|1\rangle - i|2\rangle)$.

(e) $(1/2)|1\rangle + \sqrt{3/8}(1+i)|2\rangle$.

(9.11) (a) $(\sqrt{8}/3, 0, 1/3)$,

(b) $(0, -\sqrt{8}/3, -1/3)$,

(c) $(1/\sqrt{2}, 1/\sqrt{2}, 0)$.

(9.12) $(0, 0, 0.2)$.

(9.13) (a) 160 fJ; (b) $(1/\sqrt{2})(|1\rangle - |2\rangle)$.

(9.14) (a) $|2\rangle$;

(b) $1/\sqrt{2}(|1\rangle + |2\rangle)$;

(c) $(1/\sqrt{2})(|1\rangle - i|2\rangle)$.

Chapter 10

(10.1) The result follows from combining eqns 10.8, 10.10, and 10.15.

(10.2) When the cavity is on-resonance, all the fields from the waves reflected off the mirrors add up constructively, and the result follows from the summation of these in-phase fields at an antinode. When off resonance, the fields are all out of phase, and so the intensities just add up. With no absorption within the cavity, the average intensity is just equal to the intensity entering the cavity. This is equal to $(1 - R) \times I_{\rm incident}$.

(10.3) 333 ps.

(10.4) $\mathcal{F} = 3140$; $Q = 1.1 \times 10^8$.

(10.5) (a) $Q > 5 \times 10^7$; (b) $\mathcal{F} > 60\,000$; (c) $R > 99.995\%$.

(10.6) 6.4 µm, assuming one mode in the centre of the dye spectrum, and the adjacent modes ±20 nm away.

(10.7) The reflections are all in phase. See Brooker (2003, Chapter 6).

(10.8) 4.1.

(10.9) Mode 1: $(\omega - \Omega)$, $(x_1 + x_2)$; Mode 2: $(\omega + \Omega)$, $(x_1 - x_2)$.

(10.10) 145 MHz.

(10.11) (a) ~ 100 ns, (b) ~ 25.

(10.12) (b) 7 meV.

(10.13) $N_0 = 0.06$; $n_0 = 0.16$.

Chapter 11

(11.1) $\sqrt{k_B T/m}$.

(11.2) (a) $v_{mp}^{beam} = 297$ m s^{-1}, so $\delta = -350$ MHz.

(b) $|\Delta p| = 7.8 \times 10^{-28}$ Ns. $F_{max} = -1.2 \times 10^{-20}$ N.

(c) $N_{stop} = 85\,000$. $t_{min} = 5.4$ ms. $d_{min} = 0.81$ m.

(d) $T_{min} = 1.2 \times 10^{-4}$ K.

(11.3) 2.0×10^{-7} K, 6.1 mm s^{-1}.

(11.4) $A_{21} = \gamma$, $I = cu\delta\nu$, $I_s = cu_s\delta\nu/2$, where $\delta\nu$ is the laser bandwidth.

(11.5) Write

$$F_\pm(\Delta \pm kv_x) = F(\Delta) \pm \frac{dF}{d\Delta}kv_x,$$

with

$$F(\Delta) = -\hbar k \frac{\gamma}{2} \frac{I/I_s}{(1 + I/I_s + 4\Delta^2/\gamma^2)}.$$

Then find F_x from $F_+(\Delta + kv_x) - F_-(\Delta - kv_x)$.

(11.6) 'Hot' means that both M_J levels of the ground state are populated. Hence two values: 5/3 for $-1/2 \rightarrow +1/2$, and 1 for $+1/2 \rightarrow +3/2$.

(11.7) 0.5 mK; 0.56 m s^{-1}.

(11.8) (a) This follows from the fact that the number of protons and electrons in a neutral atom are identical.

(b) The resultant spin of a diatomic molecule is found by combining the spins of the individual atoms. In an elemental molecule, both atoms are identical, so that the resultant must always be an integer.

(11.9) $T_c = 7.0 \times 10^{-6}$ K. $\lambda_{deB} = 2.0 \times 10^{-7}$ m. Particle separation $= 1.0 \times 10^{-7}$ m.

(11.10) $T_c = 1.1$ µK, so $f = 72\%$.

(11.11) $T_c = 155$ nK, 50% in condensate at 123 nK.

(11.12) 3.5 µK.

Chapter 12

(12.1) 'EINSTEIN'

(12.2) '- 0 - - 1 - - 0 - - - 1'

(12.3) (a) When Alice sends a 0, Bob only detects a count when $\theta_{Bob} = 45°$. Similarly, when Alice sends a 1, Bob only receives a count when $\theta_{Bob} = 90°$.

(b) 25% as opposed to 50% in BB84.

(c) If Eve only sends a photon to Bob when she obtains an unambiguous value of Alice's bit, she reduces the number of photons that Bob detects by a factor of 4. If she transmits photons at the same rate as Alice, then she will introduce errors. In both cases, Alice and Bob can detect Eve's presence on checking a subset of data.

(12.4) (a) Use a half waveplate for a polarized laser, and add a polarizer before the waveplate if the laser is unpolarized.

(b) Combine the beams from the four lasers by using beam splitters, and then attenuate strongly. Photons are lost at the beam splitters, but since Alice has to attenuate the beams severely anyway, this makes no difference in practice.

(12.5) The 50 : 50 beam splitter directs the photons randomly to either of the detector pairs with 50% probability. A count on D1 or D2 is like detecting in the 0° basis, whereas a count on D3 or D4 is like detecting in the 45° basis. Thus Bob gets the detection basis correct 50% of the time, just as with the Pockels cell.

(12.6) 8%.

(12.7) See Section 5.7.

(12.8) $\eta \approx x/2$, so $\overline{n} = 0.02$ for $\eta = 1\%$.

(12.9) (a) When both detectors fire, Eve knows that she has chosen the wrong basis. (b) 1/3. (c) 1/6.

(12.10) (a) $\mathcal{P} = 3/16$. (b) The basis and bit value are given by the detector pair where only one count was registered.

(12.11) (a) 0.6%; (b) 15%.

(12.12) 8 cm.

(12.13) Dark counts are caused by the thermal electrons in the detector. In an intrinsic semiconductor of band gap E_g, the density of thermal electrons and holes is proportional to $\exp(-E_g/2k_BT)$. At 850 nm, a detector with a larger band gap, and hence lower dark count, can be used.

(12.14) A classical repeater would not work. Quantum signals require a quantum repeater. See Section 13.5.5.

Chapter 13

(13.1) (a) $(1,0,0)$; (b) $(0,1,0)$; (c) $(\sqrt{3/8}, \sqrt{3/8}, 1/2)$.

(13.2) This can be seen by considering the operation of $\mathbf{M}^{\dagger}$ on the output qubit q':

$$\mathbf{M}^{\dagger}q' = \mathbf{M}^{\dagger}(\mathbf{M}q) = \mathbf{I}q = q,$$

which also shows that $\mathbf{M}^{\dagger}$ acts as the inverse operation for $\mathbf{M}$.

(13.3) $\alpha = \theta_1 = \theta_2 = \theta_3 = \pi/2$.

(13.4) (a) $q' = -|1\rangle$. The operation just produces an unmeasurable phase shift because the $|1\rangle$ state lies along the rotation axis.

(b) $q' = (i/\sqrt{2})(|0\rangle - i|1\rangle)$. $(0,1,0) \rightarrow (0,-1,0)$ on the Bloch sphere due to the π rotation about x-axis.

(c) $q' = |0\rangle$. $(1,0,0) \rightarrow (0,0,-1)$ on the Bloch sphere due to the π rotation about the z-axis and the $\pi/2$ rotation about y-axis.

(13.5) Rotations about the z-axis alter φ without affecting θ, while rotations about the y-axis affect both φ and θ. We can make an arbitrary mapping with the following sequence:

$$(\theta, \varphi) \xrightarrow{R_z(\theta_1)} (\theta, \varphi') \xrightarrow{R_y(\theta_2)} (\theta', \varphi'')$$
$$\xrightarrow{R_z(\theta_3)} (\theta', \varphi'),$$

with $\theta_1 = \varphi' - \varphi$, and $\theta_3 = -(\varphi'' - \varphi')$. θ_2 is chosen to change θ by the desired amount, and the second rotation about z compensates for the change in φ caused by R_y.

(13.6) The results follow by evaluating the output for each of the four basis states: $(1,0,0,0)$, $(0,1,0,0)$, $(0,0,1,0)$, and $(0,0,0,1)$.

(13.7) (a) $(1/\sqrt{2})(|01\rangle + |10\rangle)$;
(b) $(1/\sqrt{2})(|00\rangle - |11\rangle)$;
(c) $(1/\sqrt{2})(|01\rangle - |10\rangle)$.

(13.8) (a) Use eqn 13.23 with the identity:

$$\int_{-\infty}^{+\infty} e^{-x^2}\, dx = \sqrt{\pi}.$$

(b) $E_p = \sqrt{\pi}cn\epsilon_0 A\mathcal{E}_0^2\tau/2\sqrt{2}$.

(c) $E_p = cn\epsilon_0 A\hbar^2\Theta^2/2\sqrt{2\pi}\mu_{01}^2\tau$. This gives $E_p = 0.2$ nJ for $\Theta = \pi$.

(13.9) (a) When $f = f_1$, $f(x) = 0$. Therefore $\hat{U}_1$ maps qubits $x \to x$ and $y \to y$. This is the identity operator.

(b) The truth table is as follows:

Input qubits		Output qubits	
x	y	x	$y \oplus f_2(x)$
$\lvert 0 \rangle$	$\lvert 0 \rangle$	$\lvert 0 \rangle$	$\lvert 1 \rangle$
$\lvert 0 \rangle$	$\lvert 1 \rangle$	$\lvert 0 \rangle$	$\lvert 0 \rangle$
$\lvert 1 \rangle$	$\lvert 0 \rangle$	$\lvert 1 \rangle$	$\lvert 0 \rangle$
$\lvert 1 \rangle$	$\lvert 1 \rangle$	$\lvert 1 \rangle$	$\lvert 1 \rangle$

(c) The matrix is the same as $\hat{U}_{\text{CNOT}}$ given in eqn 13.20.

(d) $\hat{U}_{f_4} = \begin{pmatrix} \mathbf{X} & \mathbf{0} \\ \mathbf{0} & \mathbf{X} \end{pmatrix}$.
When $f = f_4$, $f(x) = 1$, and $y \oplus f(x) = \text{NOT } y$.

(13.10) (a) 78%; (b) 95%.

(13.11) (a) 148; (b) 14.

(13.12) $T^{\text{Doppler}} = 0.48$ mK, so $\mathcal{P}_0 = 0.67$. $\mathcal{P}_0 = 0.95$ at 0.18 mK.

(13.13) $\hbar\Delta = -J$.

Chapter 14

(14.1) The results HH or VV would be obtained with 50% probability.

(14.2) (a) 00 with probability 2/3 and 11 with probability 1/3.

(b) 01 with 40% probability and 10 with probability 60%.

(c) 00 or 11, each with 50% probability.

(14.3) All types of single-photon transitions between $J = 0$ states are forbidden by conservation of angular momentum, because each photon carries away at least $\hbar$ of angular momentum. With two photons, the angular momenta can cancel, making the transition possible.

(14.4) (a) Phase matching requires $n^\omega = n^{\omega/2}$. However, $n^\omega > n^{\omega/2}$ with normal dispersion.

(b) Propagate the beams with different polarizations and use birefringence to cancel the dispersion.

(14.5) (a) Substitute $k = n\omega/c$ into the phase-matching condition $k^{2\omega} = k_1^\omega + k_2^\omega$.

(b) Graphical solution gives $\theta = 57.4°$. This means that we need to cut the crystal with the optic axis at 32.6° to the normal.

(14.6) (a) 702.2 nm; (b) $-3°$.

(14.7) (a) The fringe visibility would vary as $|\cos\theta|$, where θ is the angle between the polarization vectors.

(b) With $\theta = 0°$, a polarizer in front of the detectors cannot distinguish which path the photon followed. However, for $\theta = 90°$, which-path information is possible. For intermediate angles, partial which-path information is obtained, and hence partial interference occurs.

(14.8) (a) 2.73; (b) 1.

(14.9) (a) $\lvert \Psi^+ \rangle$; (b) $\lvert \Phi^- \rangle$; (c) $\lvert \Psi^- \rangle$.

(14.10) (a) 293; (b) 207; (c) 207; (d) 293.

(14.11) Rotate the photon polarization by $\pi/2$ in a clockwise direction.

(14.12) (a) 25%; (b) 23%.

(14.13) (a) $\Delta t = 82$ fs for a Lorentzian lineshape.

(b) The coherence length is 25 μm, and so we expect the interference minimum to have a FWHM of $2L_c \sim 50$ μm.

(c) The experimental half width of ~ 60 μm compares favourably with this estimate.

Bibliography

Abragam, A. (1961). *Principles of nuclear magnetism*. Clarendon Press, Oxford.

Abram, I. and Levenson, J. A. (1994). Quantum noise in parametric amplification. In *Nonlinear spectroscopy of solids—Advances and applications* (ed. B. di Bartolo). NATO ASI Series B, Vol. 339. Plenum Press, New York, p. 251.

Albert, D. Z. (1994). Bohm's alternative to quantum mechanics. *Sci. Am.* **270**(5), 58–67.

Allen, L. and Eberly, J. H. (1975). *Optical resonance and two-level atoms*. Wiley, New York. Reprinted by Dover Publications, New York in 1987.

Anderson, B. P. and Meystre, P. (2002). Nonlinear atom optics. *Optic. & Photon. News* **13**(6), 20–5.

Anglin, J. R. and Ketterle, W. (2002). Bose–Einstein condensation of atomic gases. *Nature* **416**, 211–8.

Aspect, A., Grangier, P., and Roger, G. (1981). Experimental tests of realistic local theories via Bell's theorem. *Phys. Rev. Lett.* **47**, 460–3.

Arndt, M., Hornberger, K., and Zeilinger, A. (2005). Probing the limits of the quantum world. *Phys. World* **18**(3), 35–40.

Bachor, H.-A. and Ralph, T.C (2004). *A guide to experiments in quantum optics* (2nd edn). Wiley-VCH, Weinheim.

Barnett, S. M. and Gilson, C. R. (1988). Manipulating the vacuum: squeezed states of light. *Eur. J. Phys.* **9**, 257–64.

Barrett, M. D., Chiaverini, J., Schaetz, T., Britton, J., Itano, W. M., Jost, J. D., Knill, E., Langer, C., Leibfried, D., Ozeri, R., and Wineland, D. J. (2004). Deterministic quantum teleportation of atomic qubits. *Nature* **429**, 737–9.

Bastard, G. (1990). *Wave mechanics applied to semiconductor heterostructures*. Wiley, New York.

Bell, J. S. (1987). *Speakable and unspeakable in quantum mechanics*. Cambridge University Press, Cambridge.

Benisty, H., Gérard, J.-M., Houdré, R., Rarity, J., and Weisbuch, C. (1999). *Confined photon systems*. Springer-Verlag, Berlin.

Bennett, C. H., Brassard, G., and Ekert, A. K. (1992). Quantum cryptography. *Sci. Am.* **267**(4), 50–7.

Bennett, C. H., and DiVincenzo, D. P. (2000). Quantum information and computation. *Nature* **406**, 247–55.

Berman, Paul R. (1994). *Cavity quantum electrodynamics.* Academic Press, San Diego, CA.

Bertlmann, R. A. and Zeilinger, A. (eds) (2002). *Quantum [un]speakables.* Springer-Verlag, Berlin.

Bimberg, D., Grundmann, M., and Ledentsov, N. N. (1999). *Quantum dot heterostructures.* Wiley, Chichester.

Birnbaum, J. and Wiliams, R. S. (2000). Physics and the information revolution. *Phys. Today* **53**(1), 38–42.

Bleaney, B. I. and Bleaney, B. (1976). *Electricity and magnetism* (3rd edn). Oxford University Press, Oxford. Reissued in two volumes in 1989.

Blundell, S. (2001). *Magnetism in condensed matter.* Oxford University Press, Oxford.

Bouwmeester, Dirk, Ekert, Artur, and Zeilinger, Anton (eds) (2000). *The physics of quantum information.* Springer-Verlag, Berlin.

Brooker, Geoffrey (2003). *Modern classical optics.* Oxford University Press, Oxford.

Burnett, K. (1996). Bose–Einstein condensation with evaporatively cooled atoms. *Contemp. Phys.* **37**, 1–14.

Burnett, K., Edwards, M., and Clark, C. W. (1999). The theory of Bose–Einstein condensation of dilute gases. *Phys. Today* **35**(12), 37–42.

Butcher, P. N. and Cotter, D. (1990). *The elements of nonlinear optics.* Cambridge University Press, Cambridge.

Cirac, J. I. and Zoller, P. (2004). New frontiers in quantum information with atoms and ions. *Phys. Today* **57**(3), 38–44.

Chevy, F. and Salomon, C. (2005). Superfluidity in Fermi gases. *Phys. World* **18**(3), 43–7.

Chu, Stephen (1992). Laser trapping of neutral particles. *Sci. Am.* **266**(2), 70–6.

Chu, S. (2002). Cold atoms and quantum control. *Nature* **416**, 206–10.

Cohen-Tannoudji, C, Diu, B., and Laloë, F. (1987). *Quantum Mechanics,* Volumes I & II. Wiley, New York.

Cohen-Tannoudji, C. N. and Phillips, W. D. (1990). New mechanisms of laser cooling. *Phys. Today* **43**(10), 33–40.

Collins, G. P. (2000). The coolest gas in the universe. *Sci. Am.* **283**(6), 92–9.

Corbitt, T. and Mavalvala, N. (2004). Quantum noise in gravitational-wave interferometers. *J. Opt. B: Quantum Semiclass. Opt.* **6**, S675–83.

Cornell, E. A. and Wieman, C. E. (1998). The Bose–Einstein condensate. *Sci. Am.* **278**(3), 40–5.

Cornell, E. A. and Wieman, C. E. (2002). Nobel lecture: Bose–Einstein condensation in a dilute gas, the first 70 years and some recent experiments. *Rev. Mod. Phys.* **74**, 875–93.

Corney, Alan (1977). *Atomic and laser spectroscopy.* Clarendon Press, Oxford.

Cornish, S. L. and Cassettari, D. (2003). Recent progress in Bose–Einstein condensation experiments. *Phil. Trans. R. Soc. Lond. A* **361**, 2699–713.

Cova, S., Ghioni, M., Lacaita, A., Samori, C., and Zappa, F. (1996). Avalanche photodiodes and quenching circuits for single-photon detction. *Appl. Opt.* **35**, 1956–76.

Dehlinger, D. and Mitchell, M. W. (2002). Entangled photon apparatus for the undergraduate laboratory. *Am. J. Phys.* **70**, 898–902. Entangled photons, nonlocality, and Bell inequalities in the undergraduate laboratory. *Am. J. Phys.* **70**, 902–10.

Deutsch, David and Ekert, Artur (1998). Quantum computation. *Phys. World* **11**(3), 47–52.

DiVincenzo, David and Terhal, Barbara (1998). Decoherence: the obstacle to quantum computation. *Phys. World* **11**(3), 53–7.

Dowling, J. P. and Milburn, G. J. (2003). Quantum technology: the second revolution. *Phil. Trans. R. Soc. Lond. A.* **361**, 1655–74.

Ekert, A., Jozsa, R. Penrose, R. (eds) (1998). Quantum computation: theory and experiment. *Phil. Trans. R. Soc. Lond. A* **356**, 1713–948.

Eschner, J., Morigi, G., Schmidt-Kaler, F., and Blatt, R. (2003). Laser cooling of trapped ions. *J. Opt. Soc. Am. B* **20**, 1003–15.

Feynman, R. P. (1998). *Statistical mechanics.* Perseus Books, Reading, MA.

Foot, C. J. (1991). Laser cooling and trapping of atoms. *Contem. Phys.* **32**, 369–81.

Foot, C. J. (2005). *Atomic physics.* Clarendon Press, Oxford.

Fox, Mark (2001). *Optical properties of solids.* Clarendon Press, Oxford.

Funk, A. C. and Beck, M. (1997). Sub-Poissonian photocurrent statistics: Theory and undergraduate experiment. *Am. J. Phys.* **65**, 492–500.

Gasiorowicz, Stephen (1996). *Quantum physics* (2nd edn). Wiley, New York.

Gerry, C. C. and Knight, P. L. (2005). *Introductory quantum optics.* Cambridge University Press, Cambridge.

Giovannetti, V., Lloyd, S., and Maccone, L. (2004). Quantum-enhanced measurements: beating the standard quantum limit. *Science* **306**, 1330–6.

Gisin, N., Ribordy, G., Tittel, W., and Zbinden, H. (2002). Quantum cryptography. *Rev. Mod. Phys.* **74**, 145–95.

Goodman, Joseph W. (1985). *Statistical optics.* Wiley, New York.

Grangier, P. and Abram, I. (2003). Single photons on demand. *Phys. World* **16**(2), 31–5.

Grangier, P., Sanders, B., and Vuckovic, J. (2004). Focus issue on single photons on demand. *New. J. Phys.* **6**, 85–100, 129, 163.

Hagley, E. W., Deng, L., Phillips, W. D., Burnett, K., and Clark, C. W. (2001). The atom laser. *Opt. Photon. News* **12**(5), 22–6.

Haken, H. and Wolf, H. C. (2000). *The physics of atoms and quanta* (6th edn). Springer-Verlag, Berlin.

Hanbury Brown, R. and Twiss, R. Q. (1956). Correlation between photons in two coherent beams of light. *Nature* **177**, 27–9.

Hanbury Brown, Robert (1974). *The intensity interferometer: its application to astronomy*. Taylor and Francis, London.

Hardy, L. (1998). Spooky action at a distance in quantum mechanics. *Contemp. Phys.* **39**, 419–29.

Haroche, Serge (1998). Entanglement, decoherence and the quantum/classical boundary. *Phys. Today* **51**(7), 36–42.

Harrison, P. (2005). *Quantum wells, wires and dots: theoretical and computational physics* (2nd edn). Wiley, Chichester.

Haus, Hermann A. (2000). *Electromagnetic noise and quantum optical measurements*. Springer-Verlag, Berlin.

Havel, T. F., Cory, D. G., Lloyd, S., Boulant, N., Fortunato, E. M., Pravia, M. A., Tellemariam, G., Weinstein, Y. S., Bhattacharyya, A., and Hou, J. (2002). Quantum information processing by nuclear magnetic resonance spectroscopy. *Am. J. Phys.* **70**, 345–62.

Hecht, Eugene (2002). *Optics* (4th edn). Addison Wesley, Reading, MA.

Helmerson, K., Hutchinson, D., Burnett, K., and Phillips, W. D. (1999). Atom lasers. *Phys. World* **12**(8), 31–5.

Hennel, J. W. and Klinowski, J. (1993). *Fundamentals of nuclear magnetic resonance*. Longman Scientific, Harlow.

Hiskett, P. A., Buller, G. S., Loudon, A. Y., Smith, J. M., Gontijo, I., Walker, A. C., Townsend, P. D., and Robertson, M. J. (2000). Performance and design of InGaAs/InP photodiodes for single-photon counting at $1.55\,\mu$m. *Appl. Opt.* **39**, 6818–29.

Hore, P. J. (1995). *Nuclear magnetic resonance*. Oxford University Press, Oxford.

Hough, J. and Rowan, S. (2005). Laser interferometry for the detection of gravitational waves. *J. Opt. A: Pure Appl. Opt.* **7**, S257–64.

Hughes, R. J., Alda, D. M., Dyer, P., Luther, G. G., Morgan, G. L., and Schauer, M. (1995). Quantum cryptography. *Contemp. Phys.* **36**, 149–63.

Hughes R. and Nordholt J. (1999). Quantum cryptography takes to the air. *Phys. World* **12**(5), 31–5.

Jennewein, T., Simon, C., Weihs, G., Weinfurter, H., and Zeilinger, A. (2000). Quantum cryptography with entangled photons. *Phys. Rev. Lett.* **84**, 4729–32.

Joannopoulos, J. D., Meade, R. D., and Winn, J. N. (1995). *Photonic crystals*. Princeton University Press, Princeton, NJ.

Joannopoulos, J. D., Villeneuve, P. R., and Fan, S. (1997). Photonic crystals: putting a new twist on light. *Nature* **386**, 143–9.

Ketterle, W. (1999). Experimental studies of Bose–Einstein condensation. *Phys. Today* **52**(12), 30–5.

Ketterle, W. (2002). Nobel Lecture: When atoms behave as waves: Bose–Einstein condensation and the atom laser. *Rev. Mod. Phys.* **74**, 1131–51.

Kielpinski, D. (2003). A small trapped-ion quantum register. *J. Opt. B: Quantum Semiclass. Opt.* **5**, R121–35.

Kimble, H. J., Dagenais, M., and Mandel, L. (1977). Photon antibunching in resonance fluorescence. *Phys. Rev. Lett.* **39**, 691–5.

Kimble, H. J. and Walls, D. F. (eds) (1987). Feature issue on squeezed states of the electromagnetic field. *J. Opt. Soc. Am. B* **4**(10), 1449–1741.

Kimble, H. J. (1998). Strong interactions of single atoms and photons in cavity QED. *Physica Scripta* **T76**, 127–37.

Kittel, Charles (1996). *Introduction to solid state physics* (7th edn). Wiley, New York.

Kittel, C. and Kroemer, H. (1980). *Thermal Physics* (2nd edn). W.H. Freeman, New York.

Klingshirn, C. F. (1995). *Semiconductor optics*. Springer-Verlag, Berlin.

Knight, P. L., Hinds, E. A., and Plenio, M. B., Editors, (2003). Practical realizations of quantum information processing. *Phil. Trans. R. Soc. Lond. A* **361**, 1319–1555.

Knill, E., Laflamme, R., and Milburn, G. J. (2001). A scheme for efficient quantum computation with linear optics. *Nature* **409**, 46–52.

Koczyk, P., Wiewiór, P., and Radzewicz, C. (1996). Photon counting statistics—undergraduate experiment. *Am. J. Phys.* **64**, 240–5.

Leggett, Tony (1999). Quantum theory: weird and wonderful. *Phys. World* **12**(12), 73–7.

Leibfried, D., Blatt, R., Monroe, C., and Wineland D. (2003). Quantum dynamics of single trapped ions. *Rev. Mod. Phys.* **75**, 281–324.

Leuchs, G. (1988). Squeezing the quantum fluctuations of light. *Contemp. Phys.* **29**, 299–314.

Lorrain, P., Corson, D. R., and Lorrain, F. (2000). *Fundamentals of electromagnetic phenomena*. W.H. Freeman, Basingstoke.

Loudon, R. and Knight, P. L. (1987a). Squeezed light. *J. Mod. Opt.* **34**, 709–59.

Loudon, R. and Knight, P. L. (eds) (1987b). Special issue on squeezed light. *J. Mod. Opt.* **34**(6/7), 707–1020.

Loudon, Rodney (2000). *The quantum theory of light* (3rd edn). Oxford University Press, Oxford.

Mandel, Leonard and Wolf, Emil (1995). *Optical coherence and quantum optics*. Cambridge University Press, Cambridge.

Mandel, L. (1999). Quantum effects in one-photon and two-photon interference. *Rev. Mod. Phys.* **71**, S274–82.

Mandl, F. (1988). *Statistical Physics* (2nd edn). John Wiley, Chichester.

Matsukevich, D. N. and Kuzmich, A. (2004). Quantum state transfer between matter and light. *Science* **306**, 663–6.

Mellish, A. S. and Wilson, A. C. (2002). A simple laser cooling and trapping apparatus for undergraduate laboratories. *Am. J. Phys.* **70**, 965–71.

Mermin, N. David (1985). Is the moon there when nobody looks? Reality and the quantum theory. *Phys. World* **38**(4), 38–47.

Meschede, Dieter (2004). *Optics, light and lasers*. Wiley-VCH Verlag, Weinheim.

Metcalf, H. J. and van der Straten, P. (1999). *Laser cooling and trapping*. Springer-Verlag, New York.

Metcalf, H. J. and van der Straten, P. (2003). Laser cooling and trapping of atoms. *J. Opt. Soc. Am. B* **20**, 887–908.

Meystre, P. and Sargent III, M. (1999). *Elements of quantum optics* (3rd edn). Springer-Verlag, Berlin.

Meystre, P. (2001). *Atom optics*. Springer-Verlag, New York.

Michler, P. (Ed.) (2003). *Single quantum dots: fundamentals, applications, and new concepts*. Springer-Verlag, Berlin.

Miller, R., Northup, T. E., Birnbaum, K. M., Boca, A., Boozer, A. D., and Kimble, H. J. (2005). Trapped atoms in cavity QED: coupling qunatized light and matter. *J. Phys. B: At. Mol. Opt. Phys.* **38**, S551–65.

Monroe, C. (2002). Quantum information processing with atoms and photons. *Nature* **416**, 238–46.

Mowbray, D. (2005). Inorganic semiconductor nanostructures. In *Nanoscale science and technology* (ed. R. W. Kelsall, M. Geoghegan, and I. W. Hamley). Wiley, Chichester, pp 130–202.

Nielson, Michael A. and Chuang, Issac L. (2000). *Quantum computation and quantum information*. Cambridge University Press, Cambridge.

Ozbay, E., Bulu, I., Aydin, K., Caglayan, H., and Guven, K. (2004). Physics and applications of photonic crystals. *Photon. Nanostruct.* **2**, 87–95.

Pais, Abraham (1982). *Subtle is the Lord*. Oxford University Press, Oxford.

Pegg, D. T. and Barnett, S. M. (1997). Tutorial review: quantum optical phase. *J. Mod. Opt.* **44**, 225–64.

Pethick, C. J. and Smith, H. (2002). *Bose–Einstein condensation in dilute gases*. Cambridge University Press, Cambridge.

Petroff, P. M., Lorke, A., and Imamoglu, A. (2001). Epitaxially self-assembled quantum dots. *Phys. Today* **54**(5), 46–52.

Phillips, W. D. and Metcalf, H. J. (1987). Cooling and trapping of atoms. *Sci. Am.* **256**(3), 50–6.

Pitaevskii, L. P. and Stringari, S. (2003). *Bose–Einstein condensation*. Clarendon Press, Oxford.

Phoenix, S. J. D. and Townsend, P. D. (1995). Quantum cryptography—how to beat the code breakers using quantum mechanics. *Contemp. Phys.* **36**, 165–95.

Preskill, John (1999). Battling decoherence: the fault-tolerant quantum computer. *Phys. Today* **52**(6), 24–30.

Rae, A. I. M. (2004). *Quantum physics: illusion or reality?* (2nd edn). Cambridge University Press, Cambridge.

Raimond, J. M., Brune, M., and Haroche, S. (2001). Manipulating quantum entanglement with atoms and photons in a cavity. *Rev. Mod. Phys.* **73**, 565–82.

Rarity, J. (1994). Dreams of a quiet light. *Phys. World* **7**(6), 46–51.

Rarity, J. and Weisbuch, C. (1996). *Microcavities and photonic bandgaps.* Kluwer, Dordrecht.

Rempe, Gerhard (1993). Atoms in an optical cavity: quantum electrodynamics in confined space. *Contemp. Phys.* **34**, 119–29.

Riebe, M., Häffner, H., Roos, C. F., Hänsel, W., Benhelm, J., Lancaster, G. P. T., Körber, T. W., Becher, C., Schmidt-Kaler, F., James, D. F. V., and Blatt, R. (2004). Deterministic quantum teleportation with atoms. *Nature* **429**, 734–7.

Rolston, S. L. and Phillips, W. D. (2002). Nonlinear and quantum atom optics. *Nature* **416**, 219–24.

Ryan, J. and Fox, M. (1996). Semiconductors put the squeeze on light. *Phys. World* **9**(5), 40–5.

Sakurai, J. J. (1994). *Modern quantum mechanics.* Benjamin Cummings, Menlo Park, California.

Sasura, M. and Buzek, V. (2002). Cold trapped ions as quantum information processors. *J. Mod. Opt.* **49**, 1593–647.

Schiff, L. I. (1969). *Quantum mechanics.* McGraw-Hill, New York.

Sergienko, A. V. and Jaeger, G. S. (2003). Quantum information processing and precise optical measurement with entangled-photon pairs. *Contemp. Phys.* **44**, 341–56.

Shen, Y.-R. (1984). *The principles of nonlinear optics.* Wiley: New York, Chichester.

Shore, B. W. and Knight, P. L. (1993). The Jaynes–Cummings model. *J. Mod. Opt.* **40**, 1195–1238.

Silfvast, W. T. (1996). *Laser fundamentals.* Cambridge University Press, Cambridge.

Singleton, J. (2001). *Band structure and electrical properties of solids.* Clarendon Press, Oxford.

Skolnick, M. S., Fisher, T. A., and Whittaker, D. M. (1998). Strong coupling phenomena in quantum microcavity systems. *Semiconductor Sci. Technol.* **13**, 645–69.

Slusher, R. E. and Yurke, B. (1988). Squeezed light. *Sci. Am.* **258**(5), 32–8.

Slusher, R. E. and Yurke, B. (1990). Squeezed light for coherent communications. *J. Lightwave Technol.* **8**, 466–77.

Smith, F. G. and King, T. A. (2000). *Optics and photonics: an introduction.* John Wiley, Chichester.

Steane, A. (1997). The ion trap quantum information processor. *Appl. Phys. B* **64**, 623–42.

Stolze, J. and Suter, D. (2004). *Quantum computing.* Wiley-VCH, Weinheim.

Suter, Dieter (1997). *The physics of laser-atom interactions*. Cambridge University Press, Cambridge.

Svelto, O. (1998). *Principles of lasers* (4th edn). Plenum Press, New York.

Teich, M. C. and Saleh, B. E. A. (1988). Photon bunching and anti-bunching. *Progress in optics XXVI*, ed. E. Wolf, pp 3–104. North Holland, Amsterdam.

Teich, M. C. and Saleh, B. E. A. (1990). Squeezed and anti-bunched light. *Phys. Today* **43**(6), 26–34.

Terhal, B. M., Wolf, M. M., and Doherty, A. C. (2003). Quantum entanglement: a modern perspective. *Phys. Today* **56**(4), 46–52.

Tittel, W., Ribordy, G., and Gisin, N., (1998). Quantum cryptography. *Phys. World* **11**(3), 41–5.

Thorn, J. J., Neel, M. S., Donato, V. W., Bergreen, G. S., Davies, R. E., and Beck, M. (2004). Observing the quantum behaviour of light in an undergraduate laboratory. *Am. J. Phys.* **72**, 1210–19.

Vahala, K. J. (2003). Optical microcavities. *Nature* **424**, 838–46.

Vandersypen, L. M. K. and Chuang, I. L. (2004). NMR techniques for quantum control and computation. *Rev. Mod. Phys.* **76**, 1037–69.

Walls, D. F. and Milburn, G. J. (1994). *Quantum optics*. Springer-Verlag, Berlin.

Walmsley, Ian and Knight, Peter (2002). Quantum information science. *Opt. Photon. News* **13**(11), 43–9.

Weisbuch, C., Benisty, H., and Houdré, R. (2000). Overview of fundamentals and applications of electrons, excitons and photons in confined structures. *J. Luminescence* **85**, 271–93.

Weisbuch, C. and Vinter, B. (1991). *Quantum semiconductor structures*. Harcourt, San Diego, CA.

Whitaker, Andrew (1998). John Bell and the most profound discovery of science. *Phys. World* **11**(12), 29–34.

Wieman, C., Flowers, G., and Gilbert, S. (1995). Inexpensive laser cooling and trapping experiment for undergraduate laboratories. *Am. J. Phys.* **63**, 317–30.

Williams, R. S. (1998). Computing in the 21st century: nanocircuitry, defect tolerance and quantum logic. *Phil. Trans. R. Soc. Lond. A* **356**, 1783–91.

Woldeyohannes, M. and John, S. (2003). Coherent control of spontaneous emission near a photonic band edge. *J. Opt. B: Quantum Semiclass. Opt.* **5**, R43–R82.

Woggon, U. (1996). *Optical properties of semiconductor quantum dots*. Springer-Verlag, Berlin.

Woodgate, G. K. (1980). *Elementary atomic structure* (2nd edn). Clarendon Press, Oxford.

Yamamoto, Y. and Imamoğlu, A. (1999). *Mesoscopic quantum optics*. Wiley, New York.

Yariv, Amnon (1989). *Quantum electronics* (3rd edn). John Wiley, New York.

Yariv, Amnon (1997). *Optical electronics in modern communications* (5th edn). Oxford University Press, New York.

Zeilinger, Anton (1998a). Fundamentals of quantum information. *Phys. World* **11**(3), 35–40.

Zeilinger, Anton (1998b). Quantum entanglement: a fundamental concept finding its applications. *Physica Scripta* **T76**, 203–9.

Zeilinger, A., Weihs, G., Jennewein, T., and Aspelmeyer, M. (2005). Happy centenary, photon. *Nature* **433**, 230–8.

Zurek, W. H. (2003). Decoherence, einselection, and the quantum origins on the classical. *Rev. Mod. Phys.* **75**, 715–75.

Index

absorption, 48, 49, 60, 167
absorption–emission cycle, 218, 223
AC Stark effect, 185, 208, 225
action at a distance, 316
adiabatic demagnetization, 217
adjoint matrix, 272
AlGaAs LED, 102
algorithm
  Deutsch, 281
  Grover, 283
  quantum, 281, 283, 286
  Shor, 286
Alice, 246
alkali metal, 38, 39
allowed transition, 54
amplifier
  gain, 146
  noise, 146
  noise figure, 147
  noiseless, 148
  nonlinear, 21, 142, 324
  optical, 146
  parametric, 21, 142, 147, 324
  phase-sensitive, 22, 143, 148, 329
amplitude-squeezed light, see squeezed
    state, amplitude
angular momentum, 32
  addition, 34
  coupling schemes, 38
  nuclear, 40, 340
  orbital, 32
  spin, 32
  total, 34, 38, 40
anisotropy, 9, 12
annihilation operator, 155
antibunched light, 87, 105, 115, 117
  experimental observation, 117
  from quantum dot, 119, 122
  from sodium atom, 119
Aspect, Alain, 6, 308
Aspect experiments, 308
atom
  alkali, 38, 39
  artificial, 335
  electronic configuration, 38

  hydrogen, 35
  laser, 236
  multi-electron, 36
  trapping, 226
  two-level, 168
atom–cavity coupling, 197
  strong, 119, 197, 199, 206, 209, 213
  weak, 197, 199, 200, 212
atomic
  beam, 216, 220
  cascade, 120, 298
  fine structure, 39
  gross structure, 35
  hyperfine structure, 39
  level, 39
  shell structure, 38
  states, quantized, 35
  term, 39
atom–photon interaction, 6, 165, 199
atom optics, 236
  coherent, 236
  nonlinear, 237
  quantum, 237
avalanche photodiode, see detector,
    avalanche photodiode

B92 protocol, 249, 256, 261
balanced detection, 97, 139
balanced function, 281
balanced homodyne detector, 139
band gap, 45, 335
  direct, 46, 60
  indirect, 46, 60
  photonic, 212
band theory, 45
basis, complete, 29, 30
basis vector, 29
BB84 protocol, 249, 256
BBO crystal, 23
beam splitter
  50:50, 15, 141, 149, 161, 164
  phase shift, 15
  polarizing, 12
beam spot size, 65
beam velocity, 221

Bell, John, 304, 316
Bell experiment, 305, 308
Bell state, 275, 297, 312
  measurement, 311
Bell's inequality, 6, 305, 308, 317
Bell's theorem, 304
Bennett, C., 249, 259, 310
Bennett–Brassard protocol, 249
Betelgeuse, 107
binary logic, 267, 270
binomial distribution, 79
birefringence, 12, 23, 300
bit, classical, 267
bit, quantum, see qubit
black-body radiation, 4, 5, 50, 83, 347
Bloch
  equations, 344, 345
  representation, 187
  sphere, 187, 270, 273, 345
  vector, 187, 270, 275, 276, 345
Bloch, F., 187, 344
Bob, 246
Bohm, David, 296, 316
Bohr, Niels, 167, 304
Bohr radius, 36
Boltzmann statistics, 50, 83, 346
Bose, Satyendra, 217
Bose–Einstein
  distribution, 85, 347
  statistics, 347
Bose–Einstein condensation, 217, 230,
    346
  atom laser, 236
  atomic, 233
  critical temperature, 349
  examples, 350
  experimental techniques, 233
  fraction, 349
  liquid helium, 350
  microscopic description, 232
  observations of, 235
  phase transition, 230, 350
  quantum statistics, 346
  statistical mechanics of, 348

Bose–Einstein condensation (*contd.*)
  superconductivity, 351
  systems, 350
  temperature, 231, 233, 235, 349
  in a trap, 234
boson, 232, 347
Brassard, G., 249
broadening
  Doppler, 58
  environmental, 59
  homogeneous, 56, 59
  inhomogeneous, 56, 59
  lifetime, 56
  natural, 57
  pressure, 57
  radiative, 57
  in solids, 58
Brownian motion, 223
bunched light, *see* photon bunching
butterfly wing, 212

Casimir force, 133
cavity
  coupling factor, 204
  decay rate, 198
  Fabry–Perot, 194
  finesse, 195
  lifetime, 196
  linewidth, 66
  micro, *see* microcavity
  mode, 64, 195
  optical, 194
  planar, 194
  Purcell effect, 202
  QED, *see* cavity quantum
    electrodynamics
  quality factor (Q), 197
  resonant, 195, 197, 202
  spontaneous emission rate, 202
  tuning, 196
cavity quantum electrodynamics, 119,
    194, 206, 278, 292
CdSe quantum dot, 337
central field approximation, 37
chaotic light, 17, 86, 109, 112, 116
chemical potential, 347
chip, silicon, 264
chirp cooling, 220
CHSH inequality, *see* Clauser, Horne,
    Shimony, Holt inequality
Chu, Stephen, 217
circuit, quantum, 270
circular polarization, 12, 55

classical
  electrodynamics, 200
  light–atom interaction, 168
  optics, 8
  statistics, 346
  theory of light, 3, 8
  theory of radiation, 52
Clauser, Horne, Shimony, Holt
    inequality, 308
Cohen-Tannoudji, Claude, 217
coherence, 16, 109, 170
  damping, 180
  first-order, 17
  length, 16
  longitudinal, 16
  optical, 16
  partial, 17, 18, 86, 109
  perfect, 17, 18, 78, 112, 116
  quantum, 279
  second-order, 111
  spatial, 16, 106
  temporal, 16, 111
  time, 16, 86, 109, 111, 280
  transverse, 16
coherent
  atom optics, 236
  light, 78, 82, 95, 116
  operation, 189
  state, 126, 134, 157
  superposition state, 169
coin tossing, 283
Cold atoms, 216
collision
  broadening, 57
  cross-section, 57
  scattering, 180
  time, 57
commutator, 31
  angular momentum, 32
  creation–annihilation
    operator, 156
  ladder operator, 152
  position–momentum, 31
  quadrature operator, 157
  relationship to uncertainty
    principle, 32
  spin, 33, 44
complexity class
  non-polynomial, 265
  polynomial, 265
Compton effect, 26
computer
  classical, 243, 264, 270
  quantum, *see* quantum computing
condensation, Bose–Einstein, *see*
    Bose–Einstein condensation
conduction band, 45

confinement, quantum, 333
constant function, 281
controlled-NOT gate, 213, 274, 277, 289
controlled unitary operator, 274
cooling
  chirp, 220
  Doppler, 218, 229
  evaporative, 234
  ion, 229
  laser, 218
  Sisyphus, 225, 229
  sub-Doppler, 224
Cooper pair, 236, 279, 351
Copenhagen interpretation, 30, 304
Cornell, Eric A., 217
corpuscular theory of light, 4
correlated intensity fluctuations, 109
correlated photon pair, 296, 298, 301
correlation function
  first-order, 17, 111
  second-order, 17, 111, 114, 160
count rate, 77
coupled oscillator, 208
C-NOT gate, *see* controlled-NOT gate
creation operator, 155
critical atom number, 213
critical photon number, 213
critical temperature, 348
cryptanalysis, 243
cryptography
  classical, 243
  quantum, *see* quantum cryptography
crystal
  birefringent, 12, 23
  nonlinear, 22, 23
  photonic, 212
  symmetry, 23
  uniaxial, 24
cycle, absorption–emission, 218, 223

damping, 180, 188, 344
damping coefficient, 222
database searching, quantum, 283
data-bus, quantum, 292
dBm units, 96
DBR mirror, *see* distributed Bragg
    reflector mirror
dead time, 77
de-amplification, 22, 143, 328
de Broglie wavelength, 233, 333
Debye model, 231
decibel, 96
decoherence, 180, 188, 279, 288, 293
defect mode, 212
degenerate parametric amplifier, *see*
    parametric amplification,
    degenerate

degree of first-order coherence, 17
degree of freedom, 216, 230
degree of second-order coherence, 111
Dehmelt, H., 229
demagnetization, adiabatic, 217
density matrix, 171
density of states, 51, 330, 334, 335, 348
  electron, 52, 332
  energy, 332
  joint, 54
  massive particle, 332
  photon, 51, 201, 202, 212, 332
dephasing, 180, 198, 280
dephasing time
  T_1, 9, 180, 188, 344
  T_2, 180, 188, 198, 279, 344
  flux, 280
  quantum dot, 280
  spin, 280
destruction operator, 155
detailed balance, 50
detector
  avalanche photodiode, 90, 258
  balanced, 97, 139, 142
  dead time, 77
  homodyne, 139
  inefficient, 89, 93, 98
  phase-sensitive, 141
  photodiode, 94
  photomultiplier, 90
  photon-counting, 77
  quantum efficiency, 77, 93, 94, 99, 108, 114
  quantum limit, 96
  response time, 96, 110, 120
  responsivity, 94
  saturation, 102
  single photon, 3, 75, 76, 90, 258
Deutsch, David, 266
Deutsch algorithm, 281
Deutsch–Josza algorithm, 281
dielectric medium, 9, 11, 20
difference frequency mixing, 21, 143, 326
diffraction, 13
  electron, 26
  far-field, 14
  Fraunhofer, 13
  Fresnel, 13
  near-field, 14
  single slit, 14
  spherical aperture, 14
diffusion, atomic, 223
dimensionality, 334
dipole
  matrix element, 53, 174
  moment, 9, 53

orientation factor, 203
Dirac
  bracket, 34
  delta function, 29
  notation, 34
Dirac, Paul, 5, 34, 156
direct band gap, 46, 60
discharge lamp, 17, 118
  low pressure, 57
  sub-Poissonian light generation, 100
dispersion, 23, 300
displacement, electric, 8
distributed Bragg reflector mirror, 205, 212
distribution
  binomial, 79
  Boltzmann, 346
  Bose–Einstein, 85, 347
  Fermi–Dirac, 347
  Maxwell–Boltzmann, 58
  normal, 321
  Poisson, 80, 321
DiVincenzo check list, 288, 293
Doppler
  broadening, 58
  cooling, 218, 229
  effect, 58, 218
  limit, 224
  linewidth, 58
  temperature, 219
down-conversion, 21, 299
  degenerate, 299
  non-degenerate, 299
dressed state, 185, 207
dynamic Stark effect, 185

E1 transition, 52, 54
eavesdropping, 246, 252, 260
effective mass, 46, 332
eigenfunction, 28, 30
eigenvalue, 28, 30
Einstein, Albert, 3, 4, 49, 85, 167, 217, 231, 296, 304, 316
Einstein coefficients, 48, 167, 182, 201
  A, 49, 53, 201
  B, 49, 53, 174, 176
Einstein model, 231
Einstein–Podolsky–Rosen (EPR) paradox, 296
electric
  displacement, 8
  field, 8
  permittivity, 9
  polarization, 9
  quadrupole, 52, 55, 229
  susceptibility, 9

electric dipole
  interaction, 52, 173
  matrix element, 53, 174
  moment, 9, 53
  selection rules, 54
  transition, 52
electrodynamics
  classical, 200
  quantum, 206
electroluminescence, 60
electromagnetic wave, 8, 10
electromagnetism, 4, 8
electronic configuration, 38
elliptical polarization, 12
emission
  radiative, 48, 52
  spontaneous, *see* spontaneous emission
  stimulated, *see* stimulated emission
encryption, 244
  public-key, 245
  RSA, 245, 266, 286
energy
  electromagnetic, 53, 127
  equipartition of, 216, 230
  harmonic oscillator, 42, 129
  kinetic, 231, 348
  quanta, 4
  rotational, 231
  translational, 231
  vibrational, 230
  zero-point, 131, 132, 154
ENIGMA code, 243
ensemble average, 171
entangled states, 296
  generation of, 298
  in quantum cryptography, 246
entanglement, 296
environment, noisy, 279, 293
environmental broadening, 59
EPRB experiment, 296, 305
EPR experiment, 296, 304
EPR paper, 296, 304, 316
equipartition of energy, 216, 230
error checking, 280
error correction, 253
  quantum, 280
evaporative cooling, 234
Eve, 246
EXAFS, 200
exchange interaction, 39
exciton, 61, 279, 336, 351
  Bose–Einstein condensation, 351
expectation value, 31, 35
extended source, 106
extraordinary ray, 12

Fabry–Perot interferometer, 194
factorization, 245, 265, 286
Fano factor, 102
far-field diffraction, 14
faster-than-light signalling, 310, 317
fault-tolerant quantum computation,
    280
feedback, negative, 97
femtosecond laser, 66
Fermi–Dirac statistics, 347
fermion, 232, 347
    wave function, 37
Fermi's golden rule, 51, 201, 331
Feynman, Richard, 21, 187, 266, 287
fibre optics, 146, 258, 287
field
    electric, 8
    electric quadrupole, 229
    electromagnetic, 8
    magnetic, 8
    magnetic quadrupole, 226
    quadrature, 129, 328
    quantization, 26, 156
    vacuum, 132, 133, 141, 198, 200, 201,
        209
filter, colour-glass, 337
finesse, 195
fine structure, 39
    interval rule, 47
first quantization, 26, 156
first-order correlation function, 17
fluctuations
    correlated, 109
    energy, 5, 85
    intensity, 82, 86, 108, 109, 116
    photocurrent, 95
    photon number, 78, 80, 82,
        86, 95, 116
    vacuum, 132
    zero-point, 132
fluorescence, 54
flux, magnetic, 8
flux, photon, 77
Fock state, 156
forbidden transition, 54
force
    Casimir, 133
    interparticle, 348
    laser, 219, 221
    light-induced, 219, 236
Fourier transform,
        quantum, 286
four-level laser, 63
Franck–Hertz tube, 100
Fraunhofer diffraction, 13
free carrier scattering, 182

frequency doubling, *see* second
        harmonic generation
Fresnel diffraction, 13
fringe visibility, 18, 106, 111
Frisch, R., 219
full width at half maximum, 56
FWHM, *see* full width at half
        maximum

g factor, 41, 340
$g^{(1)}$ function, 17
$g^{(2)}$ function, 111, 114, 160, 161
GaAs absorption spectrum, 336
GaAs/AlGaAs quantum well, 336
gain
    bandwidth, 66
    coefficient, 61, 146
    medium, 61
    saturation, 63
GaN absorption and emission spectra,
    61
gate
    binary, 270
    C-NOT, 274, 277, 289
    controlled, 274
    C-PHASE, 274
    C-ROT, 274
    C-U, 274
    Hadamard, 273
    NOT, 272
    quantum, 271, 272, 274, 275
    single qubit, 272
    two qubit, 274
    Z, 272
Gaussian
    beam, 65
    lineshape, 18, 56, 58
    statistics, 321
Geiger counter, 75, 321
generalized coordinates, 128
geometric progression, 84
Gibbs, H.M., 183
glass, semiconductor doped, 337
Glauber, R., 6, 134
golden rule, *see* Fermi's golden rule
gravity wave detector, 25,
        136, 149
gross structure, 35
Grover's algorithm, 283
gyromagnetic ratio, 340

Hänsch, T.W., 217
Hadamard gate, 273
Hahn, S.L., 183

Hamiltonian, 27
    spin–orbit, 39
Hanbury Brown, R., 5, 105, 108
Hanbury Brown–Twiss experiment, 5,
        107, 108, 113, 119, 120, 122, 160
harmonic oscillator
    classical, 126
    electromagnetic, 126
    quantum, 41, 131, 151
heat capacity, 230
Heisenberg uncertainty principle, 32,
        43, 131
helium
    ^{3}He, 351
    ^{4}He, 351
    liquid, 232, 350, 351
Hermite polynomial, 42, 65
heterodyne receiver, 140
hidden variables, 304, 316
high reflector mirror, 61
Hilbert space, 28, 29, 34
history of quantum optics, 4
hole state, 45
homodyne detection, 139
homogeneous broadening, 56, 59
Hong–Ou–Mandel interferometer, 302,
        319
horizontal polarization, 12
hydrogen atom, 35
hyperfine structure, 39–41
    hydrogen, 40
    interval rule, 47

identity verification, 254
idler wave, 21, 143, 326
impedance, wave, 11
impurity scattering, 180
InAs quantum dot, 119, 121, 185,
        205, 337
indirect band gap, 46, 60
inefficient detector, 89, 93, 98
information, quantum, 269
inhomogeneous broadening, 56, 59
Intel Corporation, 264
intensity, optical, 12
intensity interferometer, 105
interband transitions, 59
interference, 13, 15
    optical, 170
    single photon, 301
    wave function, 170
interferometer
    Fabry–Perot, 194
    gravity wave, 25, 136, 149
    Hanbury Brown–Twiss, 107
    Hong–Ou–Mandel, 302

intensity, 105
Michelson, 15, 18
Michelson stellar, 105
sensitivity, 137, 138
interval rule, 47
invasiveness of quantum measurement, 31, 247
inversion about the mean, 285, 286
Ioffe Pritchard trap, 227
ion cooling, 229
ion trap, 229, 278, 289
dephasing time, 280
isotropic medium, 9

Jaynes–Cummings model, 206
Jodrell Bank telescope, 105
Johnson noise, 96, 102, 104
Josephson junction, 279

KDP crystal, 23
ket vector, 34
Ketterle, Wolfgang, 217
key, 244
distribution, quantum, 246, 249
private, 245, 246
public, 245
secret, 246
Kronecker delta function, 29

ladder operator, 152
Lamb shift, 133
lambda point, 351
Landauer, Rolf, 266
Landé *g*-factor, 41
Larmor precession, 341
laser, 6, 61
atom, 236
continuous-wave, 67
femtosecond, 66
force, 221
four-level, 63
mode, 64
mode-locked, 66
modulation, 98
multi-mode, 66
Nd:YAG, 96
noise, 96, 98, 103, 136
oscillation, 61
photonic crystal, 213
properties, 67
pulsed, 67
semiconductor, 102
single-mode, 18, 66
spectrum, 65
sub-Poissonian light generation, 102

three-level, 63
threshold, 63
Ti:sapphire, 96
tunable, 66
types, 68
vertical-cavity surface-emitting, 205
wavelengths, 68
laser cooling, 217, 218, 234
Doppler, 218
basic principles, 218
experimental techniques, 227
ion trap, 229
magneto-optic trapping, 226
optical molasses, 221
Sisyphus effect, 225
sub-Doppler, 224
LED, *see* light-emitting diode
Lewis, Gilbert, 5
lifetime
broadening, 56
collisional, 57
non-radiative, 59
radiative, 49, 52, 57, 201
light
amplitude, 11
antibunched, 6, 87, 105, 115, 117, 160
bunched, 115, 116
chaotic, 17, 86, 109, 112, 116
classical theory, 3, 8
coherence, 16
coherent, 17, 78, 82, 95, 112, 115, 116
corpuscular theory, 4
detection, 76, 89
diffraction, 13
elliptical, 12
as harmonic oscillator, 126, 131
intensity, 12
interference, 15
left circular, 12
non-classical, 6, 82, 87, 99, 105, 115, 117, 138, 142, 144
partially coherent, 18, 86, 109
Poissonian, 78, 79, 82, 116, 136, 159
polarized, 12
quadratures, 129
quantum theory, 3, 131, 151
right circular, 12
semi-classical theory, 3, 76, 90
shift, 185, 225
speed, 10
squeezed, *see* squeezed state
sub-Poissonian, 82, 87, 99, 117, 139, 142, 144
super-Poissonian, 82, 83, 86

thermal, *see* black-body radiation
unpolarized, 12
wave nature, 4, 13
wave–particle duality, 4, 122
light–atom interaction, 165, 167, 173
light-emitting diode, 101
single-photon source, 121
sub-Poissonian light generation, 102
light–atom force, 219, 236
LIGO, 136
limit
Doppler, 220, 224
quantum, 96
recoil, 226
strong coupling, 199, 206
strong-field, 174, 177
weak coupling, 199, 200
weak-field, 174
linear optics, 20
linear polarization, 12
lineshape
function, 56
Gaussian, 18, 56, 58
Lorentzian, 18, 56, 57
spectral, 56
linewidth, 56
cavity, 66, 196
collisional, 57
Doppler, 58
full width at half maximum, 56
homogeneous, 56, 59
inhomogeneous, 56, 59
Lorentzian, 57
natural, 57, 219
radiative, 57
in solids, 58
LISA, 25, 149
local hidden variables, 304, 316
locality, 305, 316
local oscillator, 140
logic
binary, 267, 270
quantum, 270
longitudinal
coherence, 16
mode, 65
relaxation, 180, 188, 198, 344
Lorentzian lineshape, 18, 56, 57
lossy medium, 89
low-dimensional semiconductor structure, 333
lowering operator, 153
LS coupling, 38, 39
luminescence, 60

McCall, S.L., 183
magnetic
    dipole, 52, 55, 340
    field, 8
    flux density, 8
    induction, 8
    permeability, 9
    quadrupole, 226
    quantum number, 33
    susceptibility, 9
magnetization, 9, 344
magneto-optic trap, 226
Mandel, L., 117, 302, 303
matrix element, 35, 51, 52
    electric dipole, 53, 174, 201
matrix representation, 30
Maxwell, James Clerk, 4, 8
Maxwell–Boltzmann distribution, 58
Maxwell's equations, 8, 10
MBE, *see* molecular beam epitaxy
measurement, 304, 316
    Bell state, 311
    invasive, 31, 247, 249
    polarization, 247
    quantum, 30, 43, 170, 247
    spin, 44
mechanical effect of light, 219, 236
mechanics, statistical, 346
metalorganic chemical vapour
        deposition, 335
metastable state, 56
Michelson interferometer, 15, 18
Michelson stellar interferometer, 105
microcavity, 204, 210
microprocessor, silicon, 264
minimum uncertainty state, 133, 134,
        138, 147
mirror
    distributed Bragg reflector, 205, 212
    half-silvered, 15
    quarter-wave, 214
MOCVD, *see* metalorganic chemical
        vapour deposition
modal volume, 199
mode
    cavity, 64
    frequency, 195
    laser, 64
    longitudinal, 65
    resonant, 195
    transverse, 64
    width, 196
mode-locked laser, 66
molasses, *see* optical molasses
molecular beam epitaxy, 335
molecule, diatomic, 230
Mollow triplet, 184

momentum diffusion constant, 223
momentum operator, 27
Moore's law, 264
motion, translational, 231
Mount Wilson telescope, 105
multi-mode laser, 66
multi-mode thermal light, 85
multiplicity, spin, 39

NAND gate, 271
natural
    broadening, 57
    linewidth, 57
    philosophy, 316
    photonic crystal, 212
Nd:YAG laser, 68, 96
near-field diffraction, 14
negative feedback, 97
network, quantum, 293, 310
neutron star, 351
NMR, *see* nuclear magnetic resonance
NMR quantum computer, 278, 280,
        287, 292
no-cloning theorem, *see* quantum
        no-cloning theorem
noise
    amplifier, 146
    classical, 97
    dBm units, 96
    eater, 97
    figure, 147
    Johnson, 96, 102, 104
    laser, 96, 98, 102
    photocurrent, 95
    power, 95
    quantum, 96, 135
    shot, 96, 135, 139, 141, 159
    sub-shot, 101, 139, 145
    thermal, 96, 102, 104, 279
    wave, 86, 87
    white, 96
noiseless amplifier, 148
non-classical light, 6, 82, 87, 99, 105,
        115, 117, 126, 138, 142, 144
non-interacting particles, 348, 351
non-locality, 305, 312, 316
nonlinear
    amplification, 142, 324, 326
    coupling, 327
    medium, 20, 142, 324, 325
    mixing, 21, 326
    optical coefficient tensor, 22
    optics, 19, 20, 68, 142, 145, 299, 324
    polarization, 20, 324
    refractive index, 25
    susceptibility, 19, 326
    wave equation, 325

non-polynomial complexity class, 265
non-radiative transition, 59
normal distribution, 321
normal ordering, 161
normalization, wave function, 28
NOT gate, 270, 272
NP complexity class, 265
nth-order nonlinear susceptibility, 20
n-type doping, 46
nuclear magnetic resonance, 169, 178,
        187, 189, 278, 339
nuclear spin, 40, 169, 178, 339
number crunching, 265
number operator, 154, 155
number–phase uncertainty, 135, 136
number state, 88, 139, 151, 154, 156
    representation, 154

off-diagonal term, 171
one-time-pad, 244, 246
opal photonic crystal, 212
operation
    coherent, 189
    rotation, 189
operator
    angular momentum, 32
    annihilation, 155
    creation, 155
    destruction, 155
    expectation value, 31
    Hamiltonian, 27
    kinetic energy, 27
    ladder, 152
    lowering, 153
    momentum, 27
    number, 154, 155
    position, 27
    potential energy, 27
    quadrature, 156
    quantum mechanical, 27, 30
    raising, 153
    rotation, 30
    spin, 33, 44
    variance, 31
optical
    anisotropy, 12
    cavity, 194
    coherence, 16
    fibre, 258, 287
    intensity, 12
    interference, 15, 170
    loss, 88
    molasses, 221, 222, 225, 226, 228
    parametric amplifier, 324
    phase, 11, 136
    phase encoding, 259

polarization, 12
  signal-to-noise, 97, 98
  transition, 48, 167
optics
  atom, 236
  classical, 8
  linear, 20
  nonlinear, 19, 20, 68, 142,
    145, 299, 324
  quantum, 3
oracle, 285
orbital
  angular momentum, 32
  quantum number, 33
ordinary ray, 12
orientation factor, dipole, 203
orthogonality, 28, 35, 268
orthonormality, 29
oscillator strength, 54, 215
output coupler, 61, 64
  atom laser, 237
overlap integral, 35

p-type doping, 46
parametric amplification, 21, 324
  degenerate, 22, 142, 148,
    326, 327
  non-degenerate, 147, 328
parity selection rule, 54
partial coherence, 17, 86, 109
particle wave, 26
Pauli exclusion principle, 38,
  232, 347
Pauli spin matrices, 33, 272
Paul trap, 229
P complexity class, 265
Penning trap, 229
permeability, magnetic, 9
permittivity, 9
  relative, 9
perturbation, electric dipole, 173
phase
  gate, quantum, 213
  optical, 11
  in quantum optics, 136
  uncertainty, 136
phase matching, 23, 299, 318, 327
  type I, 24, 300
  type II, 24, 300
phase-sensitive amplifier, 22, 143, 148,
  329
phase-squeezed light, 138
phase transition, 231, 350
  Bose–Einstein, 230
  second-order, 350
phasor diagram, 129, 130

Phillips, William D., 217
phonon scattering, 180, 182
phosphorescence, 54
photocurrent, 94
photodetection, 76, 89
  quantum theory, 93
  semi-classical theory, 90, 108
photodiode, 94
photoelectric effect, 3, 4, 90
photoluminescence, 60
photomultiplier, 90
photon, 5
  angular momentum, 55
  antibunching, 6, 87, 105, 115, 117,
    119, 160
  bosonic nature, 101, 116
  bunching, 83, 115, 116, 119
  counting, 75, 76, 80, 89,
    90, 321
  density of states, *see* density of
    states, photon
  echo, 189
  flux, 77
  harmonic oscillator, 131
  interference, 301
  lifetime, 196
  number squeezing, 139, 144
  number state, 88, 139, 151, 156
  number uncertainty, 135
  pair, 296, 298, 301
  polarization, 44, 247, 249
  qubit, 249, 268, 269, 278, 292
  single, 120, 255
  teleportation, 310
  vacuum, 132, 168
photonic band gap, 212
photonic crystal, 212
photon statistics, 75, 76, 78,
  82, 92, 93
  degradation by loss, 88, 93, 101
  Poissonian, 78, 79, 82, 92, 116, 136,
    159, 255, 256
  sub-Poissonian, 82, 87, 92, 99, 117,
    139, 142, 144
  super-Poissonian, 82, 83, 86
pilot wave theory, 316
p-i-n structure, 211
π-pulse, 179, 189, 276, 343
$\pi/2$-pulse, 189, 276, 343
2π-pulse, 179, 189
Planck, Max, 4, 347
Planck formula, 51, 83,
  84, 347
plug and play quantum
  cryptography, 260

Pockels cell, 250
point source, 106
Poissonian distribution, 80,
  92, 321
Poissonian electron statistics, 100
Poissonian photon statistics, 78, 79, 82,
  92, 116, 136, 159
polarization
  circular, 12, 55
  dielectric, 9, 12
  elliptical, 12
  entanglement, 297
  linear, 12
  nonlinear, 20, 324
  optical, 12
  single photon, 44, 247
polarizing beam splitter, 12
polynomial complexity
  class, 265
population decay, 180, 188
population inversion, 62
position operator, 27
potential, chemical, 347
Poynting vector, 12
precession, Larmor, 341
pressure broadening, 57
principal quantum
  number, 36
processor, quantum, 271
public-key encryption, 245
pulse
  2π, 179, 189
  π, 179, 189, 276, 343
  $\pi/2$, 189, 276, 343
  area, 179, 276
  Gaussian, 190
  RF, 343
pump beam, 21, 327
Purcell effect, 202, 204
Purcell factor, 203

QED, *see* quantum electrodynamics
Q factor, *see* quality
  factor
quadrature field, 129, 328
  commutator, 157
  operator, 156
  uncertainty, 132, 133, 157
quadrature-squeezed
  states, 138, 139, 142
quadrupole
  electric, 229
  magnetic, 226
quality factor, 197

quantum
  algorithm, 281, 283, 286, 291
  bit, *see* qubit
  box, *see* semiconductor quantum dot
  circuit, 271, 274, 282, 284
  coherence, 279
  coin tossing, 283
  confinement, 333
  database searching, 283
  data-bus, 292
  dot, *see* semiconductor quantum dot
  efficiency, 77, 93, 94
  electrodynamics, 119, 206
  error correction, 280, 288
  Fourier transform, 286
  gate, 272, 274, 275
  hardware, 268
  harmonic oscillator, 41, 131
  information, 241, 269
  key distribution, 246, 249
  limit, 96
  logic gate, 213, 271
  measurement, 30, 43, 170, 247, 310
  network, 293, 310
  no-cloning theorem, 247, 310
  noise, 96, 135, 146
  non-locality, 308, 309, 316
  parallelism, 269, 281
  phase gate, 213
  processor, 271
  register, 269, 292
  repeater, 287, 292
  simulation, 287
  state preparation, 189
  statistics, 347
  teleportation, 287, 296, 310, 313
  theory of light, 3, 151
  uncertainty, 31, 131, 135
  vacuum, 132, 156, 168
  well, *see* semiconductor quantum well
  wire, 334
quantum computing, 264
  algorithms, 281
  applications, 281
  cavity QED, 213, 278, 292
  circuit, 270
  error correction, 279
  experimental, 288
  ion trap, 278, 289
  Josephson junction, 279
  logic gate, 271
  network, 293
  NMR, 278, 292
  optical, 213, 278
  practical implementations, 275
  sensitivity to decoherence, 279
  using quantum dots, 279

quantum cryptography, 243
  B92 protocol, 249
  BB84 protocol, 249
  birefringence errors, 253
  dark count errors, 254
  demonstrations of, 256
  error correction, 253
  free space, 257
  identity verification, 254
  optical fibres, 258
  optical phase encoding, 259
  plug and play, 260
  principles of, 245
  random deletion errors, 253
  with entangled states, 246
quantum information processing, 241
  quantum computation, 264
  quantum cryptography, 243
  quantum teleportation, 296
quantum mechanics, 26
  interpretation, 30
  matrix representation, 30
  representations, 34
quantum number, 28
  magnetic, 33
  orbital, 33
  principal, 36
quantum optics
  definition, 3
  history, 4
  scope of subject, 4
quarter wave stack mirror, 214
qubit, 267
  atomic, 213, 268, 278
  charge, 268, 279
  control, 274, 290
  dephasing time, 280
  flux, 268
  flying, 293
  gate, 272, 274, 275, 288
  ion trap, 278, 289, 290
  photon, 249, 269, 278, 292
  polarization, 249, 269, 278
  scalability, 288, 292
  spin, 268, 278, 292
  static, 293
  target, 274, 290

Rabi, 178, 343
  flopping, *see* Rabi oscillations
  frequency, 174, 208, 343
  splitting, 208, 209, 211

Rabi oscillations, 177, 178,
    207, 343
  damped, 181, 345
  experimental observation, 182
  quantum dot, 185
  rubidium atoms, 183
radial wave function, 37
radiation, black-body, *see* black-body
    radiation
radiative
  broadening, 57
  emission, 48, 52, 201
  lifetime, 49, 52, 57, 201
  transition, 48
rain drops, 321
raising operator, 153
random process, 76, 80, 100, 116, 321
random sampling, 89
random walk, 223
Rayleigh distance, 14
Rayleigh scattering, 258
recoil limit, 226
recoil temperature, 226
reduced mass, 36
refractive index, 11
  extraordinary, 12
  nonlinear, 25
  ordinary, 12
register, quantum, 269
relative magnetic permeability, 9
relative permittivity, 9
relaxation
  longitudinal, *see* longitudinal
    relaxation
  non-radiative, 59
  spin–lattice, 344
  spin–spin, 344
  transverse, *see* transverse relaxation
repeater, 146, 287
  quantum, 287, 292
representation, number state, 154
residual electrostatic interaction, 37
resolution, angular, 106
resolving power, 195
resonance, 168, 172, 195, 197, 202
  fluorescence, 183
resonant mode, 195, 202
responsivity, 94
reversibility, 272
RF pulse, 343
r.m.s. velocity, *see* root mean square
    velocity
root mean square velocity, 216
rotating frame, 189, 341
rotating wave approximation, 175, 177,
    189

rotation
  molecular, 231
  operator, 189, 276
RSA encryption, 245, 266, 286
Russell–Saunders coupling, 38
Rydberg constant, 36

satellite communications, 257
saturation intensity, 221
scattering, 180, 198
  collisional, 180
  elastic, 344
  free carrier, 182
  impurity, 180
  inelastic, 344
  phonon, 180, 182
Schawlow, A. L., 217
Schrödinger's cat, 279, 297
Schrödinger equation, 26
  harmonic oscillator, 42
  hydrogen atom, 35
  multi-electron atom, 37
  time-independent, 27
Schrödinger, E., 134, 296
second-harmonic generation, 21, 23, 68,
    327
  amplitude squeezing, 144
second-order correlation function, 17,
    111, 160, 161
second-order nonlinear optics, 20
  nonlinear susceptibility, 20, 326
  nonlinear susceptibility tensor, 22
second quantization, 26, 156
selection rules, 54
  J, 55
  L, 55
  S, 55
  $\sigma^{\pm}$, 55
  l, 55
  m, 55
  electric dipole, 54
  nuclear spin, 340
  parity, 54
  spin, 55
self-induced transparency, 183
self-organized quantum dot, 337
semi-classical theory of light, 3, 52, 76,
    90
semiconductor, 45
  low-dimensional structure, 333
  microcavity, 204, 210
  n-type, 46
  optical properties, 59
  p-type, 46
  quantum dot, 119, 121, 185, 205, 211,
      279, 334, 337

quantum well, 205, 210, 334, 335
quantum wire, 334
semiconductor laser, 68, 102
  noise, 103
  single-mode, 103
  sub-Poissonian light generation, 102
  wavelength, 68
shadow image technique, 235
Shannon's theorem, 254
shell structure of atoms, 38
Shor's algorithm, 267, 286
shot noise, 94, 96, 102, 135, 139, 141,
    159
  reduction, 102, 139
signal wave, 21, 22, 142, 326
signal-to-noise ratio, 97, 98, 146
silicon technology, 264
simple harmonic oscillator, 41, 126, 151
single photon
  avalanche photodiode, 90, 122, 258
  detector, 3, 90, 258
  interference, 301
  phase gate, 213
  source, 120, 163, 255, 278
single-mode laser, 18, 66, 103
singlets, 39
Sirius, 108
Sisyphus cooling, 225, 229
Slater determinant, 38
slowly varying envelope approximation,
    325
sodium
  fine structure, 40
  Zeeman effect, 41
solenoid, tapered, 227
solid-state physics, 45, 58, 59
space charge, 100
SPAD, *see* single photon avalanche
    photodiode
spatial coherence, 106
spectral lineshape function, 56
spectroscopic notation, 38, 39
spectrum analyser, 94
spherical harmonic function, 33, 37
spin, 32, 43, 347
  matrices, 272
  nuclear, 169, 170, 339
spin–lattice relaxation, 344
spin–spin relaxation, 344
spin–orbit interaction, 38, 39
spontaneous emission, 48, 51, 54, 60,
    118, 133, 167, 168, 201, 202, 218
  coupling factor, 204
squeezed state, 6, 126, 138
  amplitude, 138, 142, 144
  detection, 139
  generation, 142

phase, 138
photon number, 139
quadrature, 138, 139, 142
vacuum, 138, 142, 164
stabilization, intensity, 97
standard deviation, 80, 323
standard quantum limit, 96
star
  Betelgeuse, 107
  diameter, 106, 108
  light, 106, 321
  neutron, 351
  red giant, 107
  Sirius, 108
Stark effect, 185
Star Trek, 310
state
  banana, 149
  Bell, 275, 297
  coherent, 126, 134, 157
  dressed, 185, 207
  entangled, 296
  Fock, 156
  harmonic oscillator, 154
  Jaynes–Cummings, 206
  metastable, 56
  minimum uncertainty, 133, 134, 138
  number, 88, 139, 154, 156
  photon number, 151, 156
  polarization, 44
  pure, 268
  spin, 44
  squeezed, *see* squeezed state
  superposition, 28, 169, 171, 187, 268,
      279
  vacuum, 132, 142, 156
states, density of, *see* density of states
stationary light source, 124
statistical mechanics, 346, 348
statistical mixture, 169–171, 188
statistics
  Boltzmann, 50, 83, 346
  Bose–Einstein, 347
  classical, 346
  Fermi–Dirac, 347
  Gaussian, 321
  Poisson, 321
  quantum, 347
stellar interferometer, 105
Stern–Gerlach experiment, 43, 247
stimulated emission, 49, 61
Stirling's formula, 79
Stranski–Krastanow crystal growth, 337
strong coupling, 119, 197, 199, 206, 209
strong-field limit, 174, 177, 181
sub-Doppler cooling, 224

sub-Poissonian counting
    statistics, 99
sub-Poissonian electron
    statistics, 101, 102
sub-Poissonian photon statistics, 82,
    87, 92, 99, 117, 139, 142, 144
sub-shot noise, 101, 139, 145
sum frequency mixing, 21, 326
superconductivity, 236, 279, 351
superfluidity, 350, 351
super-Poissonian photon statistics, 82,
    83, 86
superposition
    principle of, 268
    state, 28, 169, 171, 187, 268, 279
superradiance, 190
susceptibility
    electric, 9
    magnetic, 9
    nonlinear, 19
symmetry, crystal, 23

T_1 time, *see* dephasing time, T_1
T_2 time, *see* dephasing time, T_2
Taylor, G.I., 5, 301
telecommunication, 98, 146, 246, 258
teleportation, 287, 296, 310
    experiment, 310, 313
telescope, 105
temperature
    Bose–Einstein, 231, 233, 235, 349
    critical, 231, 348, 350
    Doppler limit, 219, 224
    measurement of, 228, 235
    minimum, 220, 224, 226
    recoil limit, 226
temporal coherence, 111
tensor, nonlinear, 22
term, atomic, 39
thermal light, *see* black-body
    radiation
thermodynamics, third law, 231
third-order nonlinear
    optics, 20, 25
    nonlinear susceptibility, 20
three-level laser, 63
threshold, laser, 63
time of flight technique, 228, 235
time-bandwidth product, 66
Ti:sapphire laser, 96
TOP trap, 227
transition
    absorption, 49, 60, 167
    allowed, 54
    dipole moment, 53
    electric dipole, 52
    electric octupole, 52

electric quadrupole, 52, 55
excitonic, 61
forbidden, 54
interband, 59
magnetic dipole, 52, 55
magnetic quadrupole, 52
non-radiative, 59
probability, 51, 175, 176
radiative, 48
rate, 49–51, 176, 201
selection rules, 54
spontaneous, *see* spontaneous
    emission
stimulated, *see* stimulated
    emission
transverse
    coherence, 16
    mode, 64
    relaxation, 180, 188, 198, 344
trap
    atom, 226
    Ioffe–Pritchard, 227
    ion, 229, 278, 280, 289
    magnetic, 234
    magneto–optic, 226
    Paul, 229
    Penning, 229
    time-averaged potential (TOP), 227
triggered single-photon
    source, 120, 255
triplets, 39
Trojan Horse eavesdropping
    attack, 260
Turing, Alan, 243
Turing machine, 266
Twiss, R.Q., 5, 105, 108
two-fluid model, 350, 351
two-level atom, 168, 206
type I phase matching, 24, 300
type II phase matching, 24, 300

uncertainty principle, 31, 42, 57, 131,
    136, 157
uniaxial crystal, 24
unitarity, 272
unitary operator, 272, 274
universal Church–Turing machine, 266
universal quantum computer, 266
unpolarized light, 12

vacuum
    field, 43, 132, 133, 141, 198, 200, 201,
        209
    fluctuations, 132, 168
    Rabi splitting, 208, 209, 211
    state, 132, 156
    state, squeezed, 138, 142, 164

vacuum tube electronics, 96
valence band, 45
valence electron, 38
variance, 31, 80, 322
VCSEL, *see* vertical-cavity
    surface-emitting laser
velocity
    beam, 221
    component, 216
    distribution, measurement of, 235
    light, 10
    minimum, 224
    most probable, 220
    root mean square, 216, 221, 224
    thermal, 216
velocity selective coherent trapping, 226
Vernam cipher, 244
vertical-cavity surface-emitting laser,
    204, 212
vertical polarization, 12
vibration, molecular, 230
visibility, fringe, 18, 106

wave
    electromagnetic, 8, 10
    function, 26, 28, 34
    impedance, 11
    noise, 86, 87
    particle, 26
    transverse, 11
    vector, 11
wave equation, 10
    electromagnetic, 10
    nonlinear, 325
wavelength, de Broglie, 233, 333
wavenumber, 36
wave–particle duality, 4,
    26, 122
weak coupling, 197, 199, 200, 212, 213
weak-field limit, 174, 181
which-path information, 303
Wieman, Carl E., 217
Wineland, D.J., 229

Young's slit experiment, 4, 5, 15, 301

Zeeman effect, 41, 339
    hyperfine, 41
zero-point energy, 43, 131, 132, 154
Z gate, 272
ZnS quantum dot, 337